Machine Design Elements and Assemblies

Michael B. Spektor

INDUSTRIAL PRESS, INC.

Industrial Press, Inc.
32 Haviland Street, Suite 3
South Norwalk, Connecticut 06854
Phone: 203-956-5593
Toll-Free in USA: 888-528-7852
Fax: 203-354-9391
Email: info@industrialpress.com

Author: Michael Spektor
Machine Design Elements and Assemblies
Library of Congress Control Number: 2018941405

ISBN (print): 978-0-8311-3619-2
ISBN (ePUB): 978-0-8311-9453-6
ISBN (eMOBI): 978-0-8311-9454-3
ISBN (ePDF): 978-0-8311-9452-9

Editorial Director: Judy Bass
Project Editor: Francis J. Donegan
Copy Editor: Judy Duguid
Compositor: Paradigm Data Services (P) Ltd., Chandigarh
Cover Designer: Janet Romano-Murray

industrialpress.com
ebooks.industrialpress.com

10 9 8 7 6 5 4 3 2 1

To my family

Contents

Introduction xvii

CHAPTER 1

Basics of Mechanical Engineering Calculations and Design 1

1-1. Mechanical Engineering Practice 1
1-2. Basics of Engineering Design 2
1-2.1. Engineering Calculations of Machine Elements 5
1-3. Plane Stresses 7
1-3.1. Reactions in Supports, Shear Forces, Bending Moments, and Torque Diagrams 7
Example 1-3.1a: Shear forces and bending moments caused by an external force. 9
Example 1-3.1b: Shear forces and bending moments caused by a moment. 12
Example 1-3.1c: Torque diagram. 14
1-4. Types of Stresses and Their Calculations 16
1-4.1. Normal and Tangential Stresses 16
1-4.2. Bending Stresses 17
1-4.3. Shear Stresses at Bending 20
1-4.4. Torsional Stresses 20
1-4.5. Bending and Shear Stresses of a Rectangular Bar 21
1-4.6. Contact Stresses 22
1-4.7. Bearing Stresses 24
1-4.8. Units of Stresses 25
1-4.9. The Safety Factor 25
1-5. Principal Stresses 26
1-6. Failure Theories 28
1-6.1. The Theory of Maximum Normal Stresses 29
1-6.2. The Theory of Maximum Strain 29
1-6.3. The Theory of Maximum Shear Stresses 29
1-6.4. The Theory of Maximum Distortion Energy 30

1-6.5. The Modified Mohr Theory of Failure. 31
Example 1-6.1a: Round bar mounted in a fixed support. 31
Example1-6.1b: Round bar in two bearings. 33
Example 1-6.1c: A shaft carrying certain elements is mounted on two bearings. 36
1-7. Assemblies and Prototypes. 39
1-8. Fatigue Phenomenon of Materials 40
1-8.1. Fatigue Testing 41
1-8.2. The Endurance Limit. 44
1-9. Factors Influencing the Endurance Limit 45
1-9.1. The Surface Finish Factor k_a 46
1-9.2. The Size Factor k_b 47
1-9.3. The Load Factor k_c 48
1-9.4. The Temperature Factor k_d 48
1-9.5. Miscellaneous-Effects Factor k_e 48
Example 1-9.1a: Calculating endurance limits. 49
Example 1-9.1b: Determining safety factors. 49
Example 1-9.1c: Calculating the safety factor for a shaft in plane stress state conditions. 50
1-10. Stress Concentration. 50
Example 1-10.1a: Stress concentration. 52
Example 1-10.1b: Multiple stress concentration raisers. 53
1-11. Types of Variable Stress Cycles 53
1-11.1. Calculating the Mean Stress and Stress Amplitude for a Completely Reversed Cycle. 56
1-11.2. Calculating the Mean Stress and Stress Amplitude for a Repeated Cycle 56
1-11.3. Calculating the Mean Stress and Stress Amplitude for a Random Cycle 56
1-12. Effect of the Mean Stress on the Endurance Limit .56
1-12.1. Analysis of the Modified Goodman Line 60
Example 1-12.1a: Determining the safety factor for a repeated cycle. 63
Example 1-12.1b: Determining the diameter of a shaft for a random cycle of axial loading. 64
Example 1-12.1c: Determining the diameter of a bar for a random cycle of bending loading. 65
1-13. Finite Life Cycle. 65
Example 1-13.1a: Calculating the fatigue strength of a specimen for a finite life cycle (completely reversed cycle). 67
Example 1-13.1b: Calculating the fatigue strength of an axle for a finite life cycle (completely reversed cycle). 67

Example 1-13.1c: Calculating the fatigue strength of a shaft for a finite life cycle (completely reversed and repeated cycles). 68
Chapter 1. Problems 69

CHAPTER 2
Bearings 73
2-1. Sliding Bearings 73
2-1.1. Materials for Sliding Bearings 74
2-1.2. Sliding Friction in Journal Bearings 75
2-2. The *PV* Factor for Sliding Bearings 78
Example 2-2.1a: Selecting the material for a sliding bearing. 79
Example 2-2.1b: Determining the material for a sliding bearing and the maximum values of the forces applied to the shaft. 79
2-3. Rolling Bearings 80
2-3.1. Types of Rolling Bearings 81
2-3.2. Rolling Bearing Materials, Thermal Treatment, and Standardization 88
2-4. The Role of the Bore Diameter and Service Life in Selecting a Bearing 88
2-4.1. Rolling Bearing Life Ratings 89
2-4.2. Dynamic Load Ratings 90
2-4.3. Static Load Ratings 92
2-4.4. Calculating the Service Life for Rolling Bearings Based on Dynamic Load Ratings 92
Example 2-4.1a: Selecting a radial ball bearing. 95
2-5. Design Considerations 96
Chapter 2. Problems 99

CHAPTER 3
Shafts and Related Elements 101
3-1. Analysis of the Static Plane Stress State 103
3-2. Engineering Calculations of Shafts 110
3-3. Shaft Calculations for Static Stresses 111
3-3.1. Design Calculations for Unidirectional Stress State 111
3-3.2. Design Calculations of a Shaft Subjected to Bending and Torsion 112
Example 3-3.2a: Design calculations of a shaft subjected to torsion and bending. 114
3-3.3. Verifying Calculations for Static Stresses 116
3-4. Calculations of Variable Stresses in a Unidirectional Stress State 117

3-5. Calculations of Variable Stresses at a Plane Stress State 118
3-5.1. Alternating Bending and Steady Torsion 118
Example 3-5.1a: Calculations of a shaft subjected to alternating bending and steady torsion (continuation of example 3-3.2a). 119
3-5.2. Alternating Bending and Alternating Torsion 120
Example 3-5.1a: Calculating the diameter of a shaft subjected to variable bending and torsional stresses. 121
3-6. Deflection and Angle of Rotation of Shafts 122
3-7. Critical Speed of Shafts 128
3-8. Keys and Couplings 133
3-8.1. Keys 133
3-8.2. Calculations of Keys 134
3-8.3. Couplings 135
3-9. Design Project 136
Chapter 3. Problems 144

CHAPTER 4

Belt Transmissions 149

4-1. Basic Parameters of Belt Transmissions 150
Example 4-1.1a: Calculating basic parameters 152
4-1.1. Characteristics of Belt Transmissions 153
4-2. Types of Belts 153
4-2.1. Flat Belts 154
4-2.2. V-Belts 155
4-2.3. Tensioners 158
4-3. Forces Acting on Belt Transmissions 158
4-4. Stresses in Belts 165
Example 4-4.1a: Calculating the forces applied to the belt and the shaft, and the maximum stress acting in the belt. 166
4-5. Design of Flat-Belt Transmissions 168
4-5.1. Design of Flat-Belt Transmissions Made From Common Materials 168
Example 4-5.1a: Determining the parameters of a leather flat-belt drive. 169
4-5.2. Design of Flat-Belt Transmissions Using Synthetic Materials Reinforced by Steel Cords 169
Example 4-5.2a: Design of an open flat-belt drive using a polyurethane belt reinforced by steel cords. 170
4-6. Selecting V-belts for Belt Transmissions 171
4-7. Design Project 172
Chapter 4. Problems 181

CHAPTER 5

Basics of Gearing and Cylindrical Gears 183

5-1. Basics of Teeth Engagement.......................................185
5-2. Structure and Parameters of Meshing Gears..........................187
5-2.1. Basic Geometric Parameters of Spur Gears 189
5-3. Contact Ratio, Interference, and Minimum Number of Teeth 193
5-4. Gear Manufacturing Methods.......................................195
5-4.1. Single-Tooth Milling 195
5-4.2. Teeth Generating Method.....................................195
5-5. Gear Trains...198
5-5.1. Simple Gear Train...198
5-5.2. Compound Gear Trains..200
5-5.3. Planetary, or Epicyclic, Gear Train.........................201
Example 5-5.3a: Planetary gear train calculations. 204
5-6. Gear Materials 204
5-7. Engineering Calculations of Gears 204
5-7.1. Forces in Spur Gear Engagement..............................207
5-7.2. Bending Stresses in Spur Gears Subjected to Static Loading 209
5-8. Verifying Calculations for Bending Stresses in Spur Gears Subjected to Variable Loading..212
5-8.1. Verifying Calculations of Bending Stresses for Repeated Cycles....213
Example 5-8.1a: Verifying calculations for a repeated cycle of bending stresses for a pitch system. 214
Example 5-8.1b: Verifying calculations for a repeated cycle of bending stresses in a module system.................214
5-8.2. Verifying Calculations of Bending Stresses for a Completely Reversed Cycle 215
Example 5-8.2a: Verifying bending stress calculations for a completely reversed cycle of a pitch system. 215
Example 5-8.2b: Verifying bending stress calculations for a completely reversed cycle for a module system. 216
5-9. Design Calculations for Bending Stresses 216
5-9.1. Design Calculations for Bending Stresses for Repeated Cycles...218
Example 5-9.1a: Design calculation of pitch system gears for bending stresses for a repeated cycle. 218
Example 5-9.1b: Design calculations of module system gears for bending stresses for a repeated cycle. 219
5-9.2. Design Calculations for Bending Stresses for Completely Reversed Cycles 220
Example 5-9.2a: Design calculations of pitch system gears for bending stresses for a completely reversed cycle. 220
Example 5-9.2b: Design calculations of module system gears for bending stresses for a completely reversed cycle. ...221

5-10. Surface Durability Calculations for Spur Gears 221
5-10.1. Verifying Calculations of Surface Durability 222
Example 5-10.1a: Verifying calculations of contact stresses for pitch system gears. 222
Example 5-10.1b: Verifying calculations of contact stresses for module system gears. 223
5-10.2. Design Calculations for Surface Durability 224
Example 5-10.2a: Design calculations of spur gear sets for surface durability. 225
Example 5-10.2b: Design calculations based on surface durability and verifying calculations for bending stresses. 225
5-11. Parallel Helical Gears 227
5-11.1. Geometry of Parallel Helical Gears 228
5-11.2. Forces in Parallel Helical Gears 230
5-12. Bending Stresses in Helical Gears 231
5-13. Verifying Calculations of Bending Stresses for Helical Gears 232
5-13.1. Verifying Calculations of Bending Stresses for Helical Gears Subjected to a Repeated Cycle 232
Example 5-13.1a: Verifying calculations of bending stresses for pitch system helical gears subjected to a repeated cycle. 233
Example 5-13.1b: Verifying calculations of bending stresses for a module system subjected to a repeated cycle. . . .233
5-13.2. Verifying Calculations of Bending Stresses for Helical Gears Subjected to a Completely Reversed Loading Cycle 234
Example 5-13.2a: Calculating the bending stresses and selecting the materials for a set of helical gears subjected to a completely reversed loading cycle. 234
Example 5-13.2b: Verifying bending stresses calculations for a module system helical gear for a completely reversed cycle. 235
5-14. Design Calculations for Bending Stresses of Helical Gears. 236
5-14.1. Design Calculations Based on Bending Stresses for a Repeated Cycle 236
Example 5-14.1a: Design calculations of a pitch system helical gear for a repeated bending cycle. 236
Example 5-14.1b: Design calculations of a module system helical gear for a repeated cycle. .237
5-14.2. Design Calculations of Helical Gears for a Completely Reversed Bending Cycle 237
Example 5-14.2a: Design calculations of a set of helical gears subjected to a completely reversed loading cycle. 238

5.15. Surface Durability, Verifying, and Design Calculations for Parallel Helical Gears. 238
Example 5-15.1a: Verifying calculations of contact stresses and selecting materials for pitch system helical gears. 239
Example 5-15.1b: Verifying calculations of contact stresses for module system helical gears. 239
Example 5-15.1c: Design calculations of contact stresses for module system helical gears. 240
5-16. Design Project 241
Chapter 5. Problems 249

CHAPTER 6

Bevel Gears 251

6-1. Bevel Gears: Their Geometry and Acting Forces 251
6-2. Bending Stress Calculations for Bevel Gears. 255
6-2.1. Verifying Calculations for a Repeated Cycle of Bending Stresses in Bevel Gears. 257
Example 6-2.1a: Verifying calculations for a repeated cycle of bending stresses for a pitch system bevel pinion. 257
6-2.2. Verifying Calculations for a Completely Reversed Cycle of Bending Stresses of Bevel Gears. 258
Example 6-2.2a: Verifying calculations for a completely reversed cycle of bending stresses for module system bevel gears. 258
6-3. Design Calculations of Bevel Gears for Bending Stresses 259
6-3.1. Design Calculations of Bevel Gears for a Repeated Cycle of Bending Stresses. 259
Example 6-3.1a: Design calculations of a pitch system bevel pinion for a repeated cycle of bending stresses. 260
6-3.2. Design Calculations of Bevel Gears for a Completely Reversed Cycle of Bending Stresses 261
Example 6-3.2a: Design calculations of a module system bevel gear for a completely reversed cycle of bending stresses. 261
6-4. Surface Durability of Bevel Gears 262
6-4.1. Verifying Calculations of Bevel Gears Based on Surface Durability 262
Example 6-4.1a: Verifying calculations of contact stresses for module system bevel gears. 263
6-4.2. Design Calculations of Bevel Gears Based on Surface Durability 263
Example 6-4.2a: Design calculations of bevel gears based on surface durability. 264

6-5. Design Project 265
Chapter 6. Problems 273

CHAPTER 7

Worms and Worm Gears 275

7-1. Worms and Worm Gears and Their Geometry 275
7-2. Materials and Manufacturing of Worms and Worm Gears 286
7-3. Efficiency and Heat Calculations for Worm Gear Drives 286
7-4. Forces Acting on Worm Gear Sets 289
7-5. Bending Stress Calculations for Worm Gears 290
7-5.1. Verifying Worm Gear Bending Stress Calculations 292
Example 7-5.1a: Verifying calculations of bending stresses for an open worm gear set. 292
7-5.2. Design Calculations of Worm Gear Sets Based on Bending Stresses 292
Example 7-5.2a: Design calculations of a worm gear set with regard to bending stresses. 293
7-6. Calculations of Worm Gear Surface Durability 294
7-6.1. Verifying Calculations of Worm Gears that Experience Contact Stresses 296
Example 7-6.1a: Verifying calculations of a worm gear that experiences contact stresses. 296
7-6.2. Design Calculations of Worm Gears that Experience Contact Stresses 296
Example 7-6.2a: Design calculations of a worm gear set that experiences contact stresses 297
7-7. Design Project 298
Chapter 7. Problems 307

CHAPTER 8

Power Screw Drives 309

8-1. Power Screw Mechanics 309
8-2. Power Screw Drive Calculations 316
8-2.1. Bearing Stress Calculations for Power Screw Drives and Determining Power Screw Diameters 317
Example 8-2.1a: Determining the diameter of a power screw. 319
8-2.2. Bending and Shear Stresses in the Threads of Nuts. 319
Example 8-2.2a: Verifying calculations of stresses in the threads 320
8-2.3. Calculating the Compression and Torsion Stresses and Verifying the Stability of Power Screws. 321
Example 8-2.3a: Determining the stability safety factor. 322

8-3. Design Project 323
Chapter 8. Problems 326

CHAPTER 9

Threaded Joints 327

9-1. Thread Profiles 328
9-1.1 Bolts, Screws, Studs, Nuts. 330
9-2. Characteristics of Fasteners and Threaded Joints 331
9-2.1. Allowable Stresses for Threaded Fasteners 332
9-2.2. Stiffness Coefficients of Joint Elements 332
9-3. Calculations of Typical Threaded Joints. 335
9-3.1. Fasteners Subjected to Tensile Stress 336
Example 9-3.1a: Calculating screw diameter. 337
9-3.2. Calculating the Required Preload Torque. 337
9-3.3. Calculations of Preloaded Joints Subjected to External Forces 338
9-3.4. Characteristics of Preload Forces. 342
9-4. Verifying Calculations of Preloaded Threaded Joints Subjected to Static External Loading. 343
9-4.1. Joint Tightness Factor, Preload Torque, and Equivalent Stress. 343
Example 9-4.1a: Verifying calculations of a preloaded threaded joint subjected to static loading. .345
9-4.2. Design Calculations of Preloaded Joints Subjected to Static Loading 346
Example 9-4.2a: Determining the diameter of the bolt for a preloaded joint subjected to static loading. 347
9-5. Preloaded Threaded Joints Subjected to Variable External Loading 348
9-5.1. Verifying Calculations of Preloaded Joints Subjected to Variable External Loading. 349
Example 9-5.1a: Verifying calculations of a preloaded joint subjected to variable external forces. 349
9-5.2. Design Calculations of Preloaded Joints Subjected to Variable Loading. 351
9-6. Threaded Joints Subjected to Shear Forces 352
9-6.1. Bolts Installed in the Holes with an Interference Fit. 352
9-6.2. Preloaded Bolts Installed with a Clearance Fit. 353
Example 9-6.1a: Calculating a threaded joint subjected to shear stresses. 355
9-7. Design Project 355
Chapter 9. Problems 359

CHAPTER 10

Riveted Joints .361

10-1. Types of Rivets, Riveted Joints, and Riveting Processes. 361
10-2. Calculations of Riveted Joints 362
10-2.1. Verifying Stress Calculations of Rivets 364
10-2.2. Design Calculations of Rivets. 364
10-2-3. Calculating the Stresses in the Plates 365
10-3. Calculations of a Group of Rivets in a Joint Subjected to Eccentric Loading 365
Example 10-3.1a: A riveted joint subjected to an eccentric inclined force. 369
Chapter 10. Problems 372

CHAPTER 11

Welded Joints .375

11-1. Types of Welded Joints 376
11-2. Strength Calculations of Welded Joints Subjected to Static Loading 378
11-2.1. Strength Calculations of Butt Joints. 378
11-2.2. Strength Calculations of Fillet Joints. 379
Example 11-2.1a: Calculating the width of a welded butt joint subjected to static loading. 382
Example 11-2.1b: Calculating the thickness of a welded butt joint subjected to static loading. 383
Example 11-2.1c: Calculating the width of the weld of a fillet lap joint subjected to static shear forces. 383
11-2.3. Fillet Joint of a Round Bar Subjected to a Combined Loading 384
11-3. Fatigue Calculations of Welded Joints 385
Example 11-3.1a: Welded butt joint subjected to variable loading. 386
Example 11-3.1b: Determining the width of a fillet lap joint subjected to a completely reversed loading cycle. 386
11-4. Fillet Lap Joints Subjected to an Eccentric Force 387
Chapter 11. Problems 391

CHAPTER 12

Springs. 395

12-1. Types of Springs. 395
12-2. Parameters and Basic Calculations of Helical Springs. 396
12-2.1. Parameters of Helical Springs. 397
12-2.2. Stresses in Helical Springs Subjected to Static Loading 400
12-3. Calculations of Springs Subjected to Variable Stresses 401

12-4. Stiffness Coefficient (Spring Rate) of Springs 402
12-5. Spring Materials and Allowable Stresses 404
12-6. Design of Helical Coil Springs Subjected to Static Loading 405
12-7. Design of Helical Coil Springs Subjected to Variable Loading—Fatigue Calculations 406
12-8. Impact Loading of Helical Compression Springs 408
Example 12-8.1a: Design of a helical compression spring subjected to impact loading. 410
Chapter 12. Problems 413

APPENDIXES

Appendix A-1 415
Appendix A-1-1..415
Appendix A-1-2..416
Appendix A-1-3..416
Appendix A-1-4..417
Appendix A-1-5..419
Appendix A-2...423
Appendix A-2-1..423
Appendix A-2-2..423
Appendix A-2-3..424
Appendix A-2-4..425
Appendix A-3...426
Appendix A-3-1..426
Appendix A-3-2..426
Appendix A-4...426
Appendix A-4-1..426
Appendix A-4-2..427
Appendix A-4-3..428
Appendix A-4-4..428
Appendix A-4-5..429
Appendix A-4-6..430
Appendix A-4-7..431
Appendix A-4-8..431
Appendix A-4-9..432
Appendix A-4-10...432
Appendix A-5 433
Appendix A-5-1..433
Appendix A-5-2..433
Appendix A-6 434
Appendix A-6-1..434
Appendix A-6-2..435
Appendix A-6-3..436

Appendix A-7 437
    Appendix A-7-1 437
    Appendix A-7-2 438
Appendix A-8 438
    Appendix A-8-1 438
Appendix A-9 438
    Appendix A-9-1 438
    Appendix A-9-2 439

Bibliography 441
Answers 443
Index 449

Introduction

This textbook offers fundamental studies of mechanical engineering design calculations and considerations that will enable students to successfully meet the challenges of the contemporary industrial world. It contains a comprehensive discussion of the following distinct aspects of machine design:

1) Verifying calculations to determine whether the specifications of the machine elements in an assembly fit its operational requirements.
2) Design calculations to determine the required specifications for the machine elements in a new assembly under development.
3) Adjustments for engineering calculations to account for misalignment, non-uniform stress distribution, and other consequences of interactions between machine elements.
4) Design considerations for mechanical engineering systems to allow for adjustment, assembly, disassembly, maintenance, etc.

Basically, mechanical engineering systems are composed of two groups of machine elements: machine elements for general applications and machine elements for specific applications. Machine elements belonging to the group for general applications are present in all mechanical engineering systems. Examples of these elements include screws, nuts, gears, springs, bearings, etc. Machine elements for specific applications, such as camshafts, crankshafts, pistons, valves, etc., are used exclusively in certain types of systems. There is a limited number of types of machine elements for general applications, making it possible to cover them in one course, which is the focus of this text. Dedicated courses exist for machine elements for specific applications, but familiarity with machine elements for general applications is a prerequisite.

The vast majority of mechanical engineering design activities are associated with developing new assemblies composed of machine elements for general applications.

Machine Design is the last in the sequence of fundamental courses in Mechanical Engineering and related programs. This course combines the principles of the core prerequisites with a variety of applied analytical and experimental studies that represent the most important sources of knowledge required for a successful career in mechanical engineering design. It is vital to keep this textbook during your entire engineering career.

Engineering calculations represent one of the most important aspects of the engineering design process and are discussed at length in this book. Since machine elements for general applications have conventional geometric shapes that are based on extensive industrial experience, they normally do not require change. The calculations an engineer performs are intended to determine quantitative geometric characteristics associated with mechanical strength of certain typical parameters of the elements such as diameters of shafts (bolts, rivets), pitches, and modules of teeth of gears, etc.

In general, all other conditions being equal, the strength of an isolated body is proportional to its geometric characteristics, such as cross-sectional area or moment of inertia. However, this is often not the case with machine elements, as they generally interact with other elements and cannot be considered as isolated. For example, due to inevitable misalignment of the meshing teeth of a pair of gears, the calculated strength of a tooth in isolation might be much higher than the actual strength of the same tooth when engaged with another tooth. Therefore, any changes to the geometry of these teeth should comply with appropriate limitations established by experimental data. Arbitrary changes of the geometric characteristics of teeth are not acceptable because the analytically derived formulas for these cases cannot account for things like the misalignment of the teeth. Meshing gears is just one of many examples of complex interactions between machine elements.

It is reasonable to assume that machine elements that are off-the-shelf products have been designed according to appropriate methodologies. Thus, if an existing machine element fails under regular operating conditions, it is justifiable to carry out the corresponding verifying engineering calculations in order to determine the actual stresses acting in the element. The results of these calculations could be helpful for solving the problem associated with the failure of the particular element by replacing it with an identical element of stronger material or replacing the interacting pair of elements with a pair with increased values of the corresponding parameters.

Engineering activities very often include development of new assemblies for which the operational requirements are predetermined. Some textbooks recommend using verifying calculations in order to select appropriate off-the-shelf machine elements for the new assembly. Usually, the engineer selects from the corresponding catalogs the machine elements and applies the verifying calculations to determine whether the geometric characteristics of these elements fit the requirements of the operational conditions. It is not a straightforward method of design of a new assembly. This design process includes several iterations, but more importantly, in general, the arbitrary selection of components cannot result in a relatively optimized assembly that is characterized by adequate weight, overall dimensions, etc.

This textbook offers not only the calculations to verify the specifications of a machine element in a new assembly, but also the design calculations for determining the required specifications in the first place. The use of design calculations directly determines the most appropriate specifications of the machine elements that fit the new assembly. These straightforward design methodologies result in optimized assemblies with respect to the above mentioned parameters. The required formulas for verifying and design calculations of machine elements for general applications are presented in this book in an appropriate way, taking into consideration the variety of loading characteristics.

Gear transmissions represent a significant part of the machine design course. The majority of existing textbooks offer only verifying calculations for gears. These calculations are based on the methodologies of the American Gear Manufacturing Association that emphasize a detailed approach to determining the values of the empiric factors that modify the fundamental analytical expressions. The design of new transmissions in existing books is based predominately on verifying calculations.

This textbook offers verifying calculations for existing gears as well as direct design calculations of gears for new transmissions. These verifying and design calculations include methodologies for determining the stresses in the teeth subjected to repeated cycles of loading where $\sigma_{min} = 0$ as well as for completely reversed cycles where $|\sigma_{max}| = |\sigma_{min}|$. In this book, the values of the typical empiric modifying factors for the calculations of gears as machine elements for general applications are accepted according to the recommendations that are available in corresponding published sources. In cases when the designed transmission is intended for mass production, it is justifiable to correct the obtained values of the geometric characteristics of the gears from the design calculations by applying the methodologies of verifying calculations from the American Gear Manufacturing Association or similar entities.

Creating assemblies composed of machine elements requires design skills that engineers acquire and perfect during their practical work. The design of assemblies is associated with numerous requirements, criteria, conditions, and a variety of other aspects that represent design considerations. This book addresses some typical design considerations that are related to the vast majority of assemblies. This makes it easier for graduates to enter into the world of mechanical engineering design.

Engineers use reference handbooks for obtaining supplemental information for their design projects. Accordingly, this textbook does not include this supplemental material or detailed reviews of prerequisite course work. This significantly reduces the page count of the book, allowing it to be more affordable. However, the appendixes of this book do contain sufficient material for solving the examples and the homework problems.

This textbook presents several assembly design projects that demonstrate the typical design steps for the development of an assembly (mechanical system). Corresponding design considerations are explained during the descriptions of these assemblies. In order to present these design projects in a rational way, it was necessary to arrange the structure of the book so that the complexity of the projects increases from chapter to chapter.

This textbook presents machine elements in a certain sequence, where almost every chapter concludes with a relevant design project based on the topics learned up to that point. The design project in each chapter builds upon the last, becoming more and more complex as it incorporates additional design considerations. The sequence of chapters is as follows:

1) Basics of Mechanical Engineering Calculations and Design
2) Bearings
3) Shafts and Related Elements
4) Belt Transmissions
5) Basics of Gearing and Cylindrical Gears
6) Bevel Gears

7) Worms and Worm Gears
8) Power Screw Drives
9) Threaded Joints
10) Riveted Joints
11) Welded Joints
12) Springs

In many engineering programs, a junior design project course is offered in parallel with the second semester of the Machine Design course. The arrangement of chapters in this text allows students to learn their junior project prerequisites in the first semester, preparing them to start working on the project at the beginning of the second semester.

The stress and strength calculations in this book, as in other books, are based on the rigorous analytical methodologies presented in the course of Strength of Materials (Mechanics of Materials) that are rigorously developed predominately for unidirectional static loading. However, the majority of machine elements for general applications are subjected to two-dimensional variable loading. In order to evaluate stresses acting in material subjected to two- or three-dimensional loading, the course of Strength of Materials introduces so-called equivalent stresses that supposedly reflect the actual state of stresses in the material. These hypothetical equivalent stresses can be computed based on the so-called theories of failure. However, as it is well known, the theories of failure are actually just hypotheses; they are not theories. In fact, equivalent stresses represent just a hypothetical concept. Moreover, we cannot experimentally measure the equivalent stresses and, consequently, cannot experimentally justify these theories of failure. Because two-dimensional loading is characterized by two independent loading factors acting in different directions, the attempt of conducting verifying experimental investigations is associated with an endless combination of pairs of these two independent loading factors. Such experiments cannot be carried out in a finite period. This is the reason why the theories of failure cannot be justified experimentally. The applicability of the theories of failure is based on the statistical methods that utilize corresponding actual data for obtaining appropriate correlations that indirectly justify or reject these hypotheses.

Variable loading causes a so-called fatigue phenomenon in the materials that significantly reduces the strength of the material. Fatigue causes changes of certain mechanical properties of the material, resulting in the failure of an element at stresses that are significantly smaller than the yield strength. The mechanism of the material distortion under variable loading is only partially understood, and currently there is no an analytical method for performing the appropriate engineering calculations with regard to machine elements. There is a certain similarity between the problem of two-dimensional loading and the variable loading that is also characterized by two independent parameters: the maximum and minimum stresses. For the same reasons that apply to two-dimensional loading, it is impossible to carry out appropriate experimental investigations for variable loading that would allow justifying corresponding hypotheses to obtain analytical expressions related to the fatigue strength characteristics of machine elements. Similar to theories of failure, in the study of the fatigue phenomenon are utilized criteria of fatigue failure. These criteria also represent experimentally unverifiable hypotheses and are also indirectly justified by statistical methods involving related correlations. In addition to criteria of fatigue failure, the fatigue related engineering calculations include empiric factors that modify the theoretical formulas for stress calculations derived for static loading.

In this book, the graphical methods for determining certain characteristics related to engineering calculations of the fatigue phenomenon are replaced with corresponding analytical expressions that increase the accuracy of the results and eliminate the need for drawing accessories.

Due to the two-dimensional loading and to the fatigue phenomenon, it might be recommended to apply in the engineering calculations elevated values for safety factors with regard to the mechanical strength of the machine elements for general applications. The increase of these factors results in the increase of the strength of the machine elements and in the weight of the assemblies. However, in most real-life cases, the newly designed assemblies are produced in a very limited number, and the increased strength of the machine elements eliminates the necessity in experimental testing that significantly exceeds the expenses associated with the increased weight of the assembly.

Please note that this text contains a limited number of homework problems. There are 385 additional problems associated with this book, and they are available on the publisher's website: http://www.machinedesignea.com.

CHAPTER 1

Basics of Mechanical Engineering Calculations and Design

Engineering provides us with progressing technologies that play a significant role in our lives. Mechanical engineering design converts exciting ideas into reality. Successful design projects lead to the advanced systems and machines that enrich our lives.

The mechanical engineering design process is based on the interaction of many studies that result in creating engineering documentation for manufacturing machines, systems, and other products. Mechanical engineering practitioners working on design projects apply their knowledge and skill to create the documentation that comprises the necessary engineering calculations and corresponding drawings.

1-1. Mechanical Engineering Practice

Mechanical engineers work in fields that involve the development, design, and improvement of engineering systems intended to perform at the highest efficiency. In order to achieve the desired performance of a system, engineers must have a deep understanding of the relationships among the parameters of the system, their role in the functioning of the system, and the influence of these parameters on one another. This control of the parameters of a system is based on corresponding analytical and experimental investigations. The results of these investigations represent the foundation for creating new and modifying existing methodologies of engineering calculations. Engineers continuously implement these methodologies in real-life projects. Mechanical engineering practitioners successfully bring to life new and more sophisticated systems, machines, and equipment.

Another significant field of mechanical engineering activities involves the education of new generations of engineers. Without sharing their experience with the students, the educational programs would not achieve their effectiveness. Due to many co-operative educational programs and internships, students enhance their engineering skills under the supervision of highly qualified mechanical engineers, while getting the real-life training in the industry.

Many mechanical engineers are involved in manufacturing enterprises associated with metal working processes, including the assembly and testing of the products and systems. They design experimental stands, testing rigs, jigs, fixtures, attachments, special measuring tools, and the like. They are in charge of the development and design of special conveyers for moving work pieces from point to point in the production process. These engineers develop and design pneumatic and hydraulic equipment for handling work pieces and materials. They design semi-automatic and automatic manufacturing lines and robotic systems for fabrication and assembly of numerous machines and products. Some of the fields that mechanical engineers are involved with include military armament, aircraft, automotive, railroad, metal working, construction, ship building, metallurgy, steel production, chemistry, crude oil and natural gas, agricultural equipment, medical equipment, home appliances, and many others. Numerous mechanical engineers are working in the computer industry. It is, probably, very difficult to find an industrial field where mechanical engineers do not play a significant role.

A large number of mechanical engineers supervise the operating and servicing of these technologies and keep them in working order. Many engineers are involved in promoting and marketing all kind of mechanical engineering products and systems. Mechanical engineers occupy a significant number of managerial positions in the industry.

1-2. Basics of Engineering Design

The process of mechanical engineering design involves a variety of interrelated sciences and technologies. Design is a very rewarding and, at the same time, a very challenging field for engineers. It is important to realize that design problems can have more than one adequate solution. It would not be surprising if two different designers in the same industrial environment working on the same engineering problem came up with two different design solutions that adequately solve the problem.

That does not mean that these two solutions are equivalent because there could be important differences between these two acceptable solutions. This situation raises the question: what criteria makes one solution more acceptable than another? It is a complex issue, and there is no simple nor short answer to this question. Not going into the details, the answer to this question, in the opinion of the author, is which of these solutions is more based on solid knowledge. The keyword here is knowledge. This is what this textbook offers. Usually, the engineer generates a variety of solutions and upon analysis makes a final decision.

Design is one of the most important aspects of mechanical engineering practice, and it is worthwhile to become acquainted with the procedures related to the design process of a mechanical system (assembly). Here is a possible scenario that reflects, to a certain extent, the process of development and design of a mechanical engineering system (assembly).

- The process starts with the recognition of a mechanical engineering design problem.
- An engineer develops several versions of systems that could be applicable to the solution of the problem. Normally, the engineer makes sketches of these versions. It should be clear that these versions in the vast majority of cases do not represent inventions, and this should not by any counts be perceived as some kind of a disadvantage of design solutions. An engineer is obligated to develop the solution using as much existing knowledge as possible. An engineer is not obligated to come up with new patentable ideas when solving a problem.

- Normally, the versions mentioned above resemble features of existing systems, assemblies of machine elements, and other products that have successfully passed the test of time. The engineer analyzes these versions and very often discusses them with his or her colleagues.
- This leads to the acceptance of a modified version that combines features from a number of these versions.
- The engineer performs functional calculations in order to determine the parameters of the energy source (electrical motor or internal combustion engine) that is required to power the system.
- During the next step, he or she calculates the parameters of the involved machine elements.
- The engineer designs the general layout of the assembly, complying with all requirements of the standards for engineering documentation.
- This general layout of the assembly goes to the junior engineers who produce the detail drawings of the machine elements.

Obviously, this is just an approximate scenario, but it helps to visualize the design process. The detail drawings represent the engineering documentation that is necessary for manufacturing the machine elements of the assembly (system). Actually, this assembly is the embodiment of the system intended to solve the problem.

Assemblies consist of machine elements. There are thousands upon thousands of different machine elements. It is reasonable to ask if it is possible to describe the variety of engineering calculations and drawings of this enormous number of machine elements in one book. Obviously, this is impossible. However, this course exists in all mechanical and related engineering programs, and the corresponding textbooks exist as well.

The analysis of existing machine elements shows that they belong to two different groups: machine elements for general applications and elements for specific applications. Machine elements such as bolts, nuts, pins, gears, shafts, couplings, bearings, rivets, springs, welds, etc., belong to the group of elements for general applications. The number of elements for general applications is limited to approximately a dozen and, obviously, it is possible to describe in one book the corresponding engineering calculations and the related design features of these elements. Most machine elements for general applications are off-the-shelf products that are not associated with any particular machine or device. (Although a weld is not technically an element because it does not exist by itself but only as a welded joint. Similarly, a ball bearing is not an element, it is a unit consisting of several elements. However, welds and bearings are included in the study of machine elements because they are widely used in machines and other products.)

Some types of body components, such as bearing housings, casings, forgings, and others, are considered elements for general applications; however, they are not included in machine elements and assemblies design courses. The reason is we do not have established methodologies for engineering calculations of the strength of these types of elements. The methods of finite elements analysis provide certain opportunities to perform stress analysis of these body shapes; however, it is not a trivial task. It is important to state that the accumulated experience related to these body elements indicates that the stress and deflection calculations of these elements are unnecessary. Actually, their strength and stiffness by far exceed the required values, while the reduction of their dimensions is not acceptable or not achievable due to the design considerations or manufacturing limitations. Therefore, production methods place limits on the minimum thickness of the walls or ribs. Usually, the walls and ribs of body-type elements that have the minimum achievable thickness possess excessive strength and stiffness. This makes it clear that engineers account for the capabilities of the existing manufacturing methods for the corresponding machine elements and accept the fact that the body elements are somewhat overweighed.

Machine elements, such as crankshafts, camshafts, pistons, valves, connecting rods, and others, belong to the group for specific applications. However, the mechanical systems comprising a certain set of machine elements for specific applications necessarily include many machine elements for general applications. Some of the machine elements for specific applications are also off-the-shelf products; however, in order to get the needed elements (usually called parts), it is important to provide information explaining where these elements will be used (model of the machine, year of manufacturing, etc.).

Since the course Machine Design Elements and Assemblies deals with the group of machine elements for general applications, it is reasonable to ask if the knowledge of this course is sufficient to practice mechanical engineering design in the wide variety of different mechanical engineering environments. The answer is definitively positive, and there are no exceptions. Actually, an assembly or a system that comprises just machine elements of specific applications does not exist. For example, an internal combustion engine, an air compressor, a hydraulic press, and other specialized machines and systems comprise a certain number of elements for specific applications; however, the vast majority of the machine elements of these systems are general application elements.

The machine elements for general applications could be assembled in numerous combinations, creating a variety of mechanical systems. Engineering practitioners are involved in development and design, or improvement, of these systems. These activities represent the real essence of mechanical engineering design. The course of Machine Design Elements and Assemblies is the only course in the mechanical and related engineering programs that directly addresses the aspects of designing machine elements and assemblies.

The design process consists of generating engineering solutions, performing engineering calculations and drawings. Sometimes, testing of a mechanical system is a part of the design process. In fact, the acquired knowledge from this course is very helpful in solving the design problems associated with machine elements of specific applications. Finally, this course, along with other engineering courses of the mechanical engineering and related educational programs, prepares graduates to become familiar with the peculiarities of the calculations and design of machine elements for specific applications.

Here are some of the requirements that should be accounted for in the design processes of machine elements for general applications, as well as assemblies of these elements.

The basic requirement of a machine element is to possess sufficient strength to withstand the action of the loading factors applied to it during the operation of the assembly (system). Obviously, that to possess the required ability to interact with the other elements of the assembly is very important, but if the machine element fails due to a lack of strength, the other factors become secondary. In other words, the machine element should first possess the adequate strength and stiffness. It is important to emphasize that each element for general applications has its typical shape and is made of certain types of materials.

The engineering calculations of machine elements are focused on determining the appropriate geometric parameters (the dimensions of the cross-sectional areas, diameters, etc.) of the machine elements. These parameters depend on the characteristics of the loading factors applied to the machine elements. Thus, when the engineer is designing a machine element, the main issue is to determine the geometric parameters that will provide the required strength of this element. This text presents the appropriate methodologies for calculating the strength of the machine elements for general applications. These methodologies include the selection of the types of engineering materials for each element and address the characteristics of the typical shapes of these elements and related design considerations.

It should be clear that the design requirements of machine elements are different from the requirements of the assemblies. The assembly, which is actually a mechanical system or a machine, is characterized by its performance. The assembly also provides the appropriate conditions for the adequate functionality of the elements in the assembly. These conditions comprise the lubricating, adjustments, temperature compensations, assembling and disassembling of the elements, etc. The assembly drawing should show that all of these design considerations are accounted for and incorporated.

This textbook offers detailed information related to the design of machine elements of general applications and their assemblies.

1-2.1. Engineering Calculations of Machine Elements

The design of machine elements is based on numerous factors, such as strength, deflection, safety, manufacturability, weight, size, material, cost, and many others. But the most important factor is the strength of the machine element. It should be mentioned that strength and deflection are interrelated, and determining deflections is a part of strength calculations.

As it is mentioned above, when an element fails due to lack of strength all other considerations do not matter. If the element possesses the required strength all other design considerations become relevant. Therefore, this book focuses on engineering calculations associated with the strength of machine elements.

It is important to understand the role, purpose, and the special distinctions of engineering calculations of machine elements for general applications. These calculations are based on principles that are presented in many courses, such as Calculus, Statics, Strength of Materials, and other related studies.

The principles discussed in Strength of Materials deal with the fundamentals of the theory of elasticity that applies rigorous analytical methods of research for obtaining expressions that reveal internal stresses and cubic strain of deformable rigid bodies subjected to static loading. These principles are applicable only in cases where the deformations do not exceed the elasticity limits of the materials of these bodies. Thus, the purpose of stress calculations in the course of Strength of Materials is to reveal the state of the stresses in a particular rigid body subjected to the known magnitude of the applied static load, while the values of its geometric characteristics are also known. The analytical methodologies that reveal the state of stresses in deformable rigid bodies represent a certain research procedure. Thus, the purpose of the calculations in Strength of Materials consists in revealing the state of the stresses in the material, and they represent a research process.

Design is associated with creativity. The design process of assemblies of machine elements consists of activities that result in the creation of products that did not exist before. It does not necessarily mean that the conceptual design did not exist before, but it may mean that some quantitative characteristics of the new design are different from what existed before. The implementation of these differences requires a creative approach. Consider a very simplified case where it is required to increase the input load of a certain assembly. This is associated with design calculations that represent a part of the creative process. In this case, some of the machine elements should be redesigned in order to increase their load capacity. Unlike Strength of Materials calculations, where the initial data consists of two known parameters—the input load and the geometric characteristics—in Machine Design Elements and Assemblies, the input data consists of just one known parameter: the applied load, while the geometric characteristics need to be determined. In other words, a new machine element having the required geometric characteristics needs to be created.

It should be mentioned that sometimes engineering practitioners have to determine the state of stresses in certain elements. Assume that an element of an existing system failed due to its strength deficiency. Usually, in this case engineers perform calculations of the stress of the machine element in order to verify that the geometric characteristics correspond to the particular requirements related to this element. The results may show that the failure occurred due to an overload or that the original design was not accurate. These types of engineering calculations are considered verifying calculations. In this book we present two types of engineering calculations: design calculations and verifying calculations.

As mentioned above, the course of Strength of Materials predominantly deals with static stresses. However, the majority of machine elements for general applications is subjected to variable stresses. The variable stresses cause fatigue phenomena in the material, resulting in the failure of the elements under stress that would be considered save for static loading. Therefore, the available formulas for calculating static stresses cannot be directly applied to the machine elements subjected to variable stresses. Currently there are no analytically established methodologies for calculating the required strength of the machine elements subjected to the action of variable stresses. The existing methodologies for calculating the required strength are based on applying empiric modifying factors to the corresponding formulas derived for calculating static loading. All of these methodologies are based on very similar approaches and contain similar modifying factors that may have some differences in their values. The corresponding methodologies related to variable loading accepted in this book are adopted from many published sources.

It should be mentioned that there are different design considerations when it comes to the geometric characteristics and materials for machine elements for individual as opposed to mass production. For individual production, the relatively excessive weight of the element plays an insignificant role. It is justifiable to use less expensive materials and increase the strength of the element by increasing its weight and dimensions, avoiding testing expenses. However, in mass production the cost of manufacturing a single component is significantly less than in individual production. Therefore, in mass production there is a strong tendency to minimize the weight and dimensions of machine elements by using stronger materials and applying lesser, but sufficient, margins of strength. This results in a significant decrease in the cost of the elements and an increase in the effectiveness of the operation of the system. Extensive testing of the machine elements is a common practice in cases of mass production.

As mentioned above, the course of Machine Design Elements and Assemblies is based on the fundamentals of the studies of Engineering Drawing, Calculus, Statics, Dynamics, Strength of Materials (Mechanics of Materials), and others. The engineering calculations of machine elements described in this book consist of a combination of the principles of the mentioned studies with numerous specific modifying empiric factors associated with the fatigue phenomena. It should be stressed that the operational conditions of these elements in the majority of cases are characterized by variable stresses causing fatigue of the materials. Because machine elements are predominantly subjected to biaxial loading, it is necessary to apply certain theories of failure in order to determine the corresponding equivalent stresses in the elements. The forces applied to these elements are usually concentrated and directed parallel to the coordinate axes. The forces of gravity of the machine elements are usually insignificant in comparison with the applied forces. The mechanical engineering systems considered in this course represent statically determinate systems. Therefore, this course does not address:

- **a)** three-dimensional stress analysis
- **b)** distributed forces
- **c)** statically indeterminate systems
- **d)** forces of gravity of the machine elements

1-3. Plane Stresses

This section presents a brief review of some fundamentals of Statics and Strength of Materials (Mechanics of Materials) that are extremely important for the strength calculations of machine elements. The goal is not simply to serve as a reminder of the basic definitions, principles, and methods of stress calculations, but also to highlight the use of universal methodologies of the corresponding calculations. These methodologies help determine the reactions in the supports and to plotting shear forces, bending moments, and torque diagrams.

1-3.1. Reactions in Supports, Shear Forces, Bending Moments, and Torque Diagrams

Stress and strength calculations are based on the same fundamentals. Stress calculations determine the state of stresses of the rigid body (machine element) without any regard to the mechanical properties of the material of the body. In other words, stress calculations indicate the state of stresses in the body without any regard to its failure. In mechanical engineering, the failure of a machine element is associated with developing residual deformations in the element, damage of the machine element, or with developing pitting on its surface (pitting is addressed in the next and other chapters). However, strength calculations help prevent the failure of the machine element. This is achieved by accounting for the strength of the material that the element is made from. Nevertheless, the methodologies of stress and strength calculations use the same loading factors: shear forces, bending moments, and torques.

In order to avoid some possible confusion, it should become clear that the above mentioned loading factors are by nature the internal loading factors that represent the reactions of the material of the machine element to the applied external loading factors. The external loading factors come from the energy sources of the machine (assembly). These factors are external forces, external moments, and external torques. The characteristics of the external loading factors are determined from the calculations associated with the operation of the system.

It is important to remember the basic procedures for determining the internal loading factors. Many mechanical engineering systems (assemblies) comprise shafts and axles that carry rotating machine elements, such as pulleys, gears, etc. Shafts transmit torques, while axles do not. Both the shafts and axles with their supports (bearings) represent statically determinate systems and resemble round bars (rods). For example, due to the external loading factors applied to the shafts or axles, the supports react to these loading factors. The combined action of the reactions and the external loading factors results in occurrence of the shear forces, bending moments, and torques that act inside of these round bars. The action of these loading factors leads to the situation that some of the cross-sectional areas of the elements become more loaded than others and, consequently, more stressed than others. Obviously, the strength calculations should be performed for the most stressed areas. In order to determine the locations of the most loaded cross-sectional areas in such machine elements as axles and shafts, it is necessary to construct diagrams of shear forces, bending moments, and toques. The methodologies presented below determine the reactions in the supports and allow plotting the corresponding diagrams. To make these methodologies universal for determining the reactions, the shear forces, the bending moments, and torques, as well as plotting their diagrams, we accept the following rules:

1) We will always go from the left-hand end of the round bar toward the right.
2) An external or internal force is positive if it has the same direction as the corresponding coordinate axis, and negative if the force is in the other direction.
3) An external or internal moment or toque is positive if it rotates the round bar clockwise, and negative if it rotates the bar counterclockwise.
4) The reactions of the supports shown on the corresponding free-body diagrams are assumed to be positive. If the calculations show that the reaction is negative, we will not change its direction on the free-body diagram; however, we will account for the negative sign (−) of this reaction in the corresponding calculations.

Determining the reactions in the supports of the round bars is based on the equations of Statics for bi-axial loading:

$$\sum F_x = 0 \tag{1-3.1}$$

$$\sum F_y = 0 \tag{1-3.2}$$

$$\sum M = 0 \tag{1-3.3}$$

where $\sum F_x$ and $\sum F_y$ represent the algebraic sums of the projections of all forces on the coordinate axes x and y respectively, and $\sum M$ is the sum of all moments about any point in the plane x, y.

The methodology of plotting the diagrams is based on the method of successive imaginary cuts of the round bar into intervals placing the first cut between the left-hand end of the round bar and first loading factor. The idea is that by throwing away the right-hand side of the round bar and analyzing the equilibrium of the remaining left-hand part, it is possible to determine the internal loading factors.

Examples demonstrating the procedures of determining the reactions, the internal loading factors, and plotting the corresponding diagrams are presented below.

Example 1-3.1a: Shear forces and bending moments caused by an external force.
Figure 1-3.1a shows a bar AB mounted in supports A and B, while being loaded by a force of $6F$. We need to plot the shear forces and bending moments diagrams and determine the most loaded cross-sectional area of the bar.

Solution:
This example demonstrates the methodology of plotting the shear forces and bending moments diagrams when a bar is loaded by a force. The diagrams determine the absolute maximum values of the loading factors and the location of the most loaded cross-sectional area of the bar.

We begin by determining the reactions in the supports. Figure 1-3.1b shows the free-body diagram of the bar AB with the reactions of the supports R_A and R_B.

According to the rules mentioned above, it is assumed that these reactions are positive. If the calculations contradict this assumption, we do not change the directions of the reactions in the free-body diagram. We just account for the negative sign of the reaction in the subsequent calculations.

Based on equation (1-3.3) we can determine the sum of moments about point B in the following way:

$\sum M_B = 0$, therefore, $R_A 3a - 6Fa = 0$. Solving this equation for R_A, we obtain

$$R_A = 2F \tag{1-3.4}$$

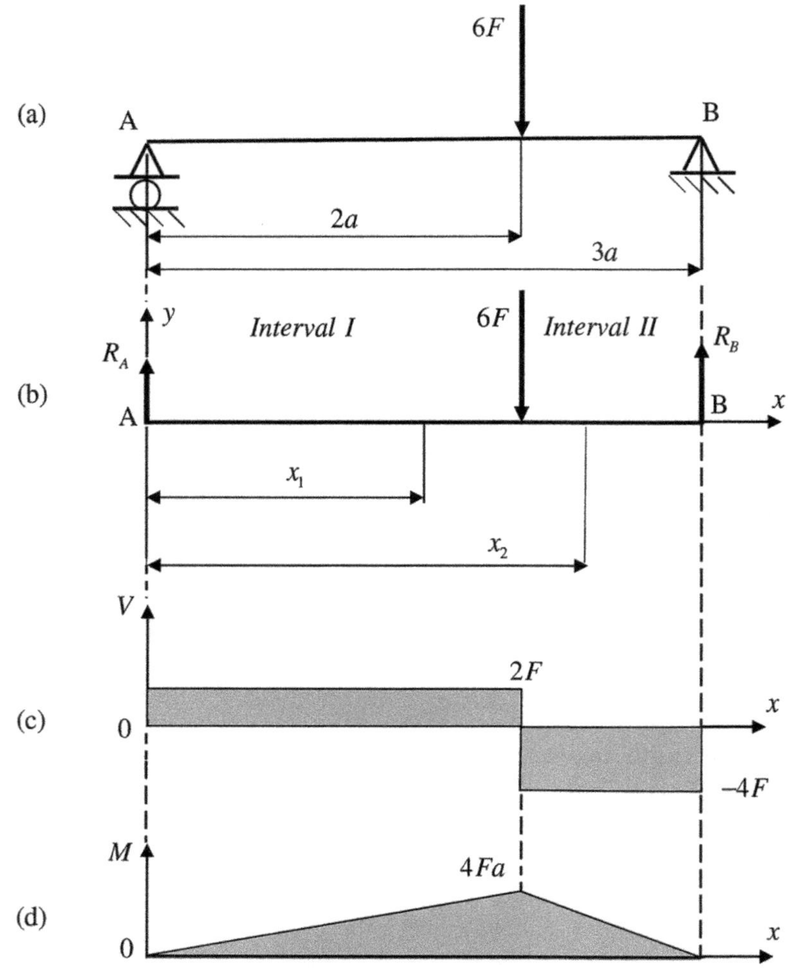

Figure 1-3.1. Free-body diagram for a bar loaded by an external force, shear forces, and bending moments.

Using equation (1-3.2), we determine the reaction in the support B. We may write $\sum F_y = 0$. This equation yields $R_A - F + R_B = 0$. Solving this equation for R_B, we have $R_B = R_A - F$. Substituting the value of R_A from equation (1-3.4), we obtain

$$R_B = 4F \tag{1-3.5}$$

Using the method of cuts described above, we divide the bar into two intervals *I* and *II*, as shown on the free-body diagram in Figure 1-3.1b. We make the first cut on the first interval at the distance of x_1 from the point A. Figure 1-3.1b shows that the abscissa x_1 may change its value from zero to $2a$, therefore, we may write

$$0 \le x_1 \le 2a \tag{1-3.6}$$

The next cut we make on the second interval in the point with the abscissa x_2 that changes its values from $2a$ to $3a$, and, consequently, we have

$$2a \le x_2 \le 3a \tag{1-3.7}$$

We begin with the analysis of the shear forces. According to the method of cuts, we consider just the remaining part of the bar that is located on the left side of the cut. The part of the bar on the right-hand side of the cut is thrown away. Figure 1-3.1b shows that on the remaining part of the bar at the first interval is subjected to the action of the reaction R_A. Therefore, the shear force V_I on the first interval can be determined from the following equation: $V_I = R_A$. Substituting the value of R_A from equation (1-3.4) we obtain

$$V_I = 2F \tag{1-3.8}$$

Similarly, we determine the shear force V_{II} on the second interval. Considering the remaining left-hand part of the second interval, we may write $V_{II} = R_A - 6F$, and substituting here the value of R_A, we obtain

$$V_{II} = -4F \tag{1-3.9}$$

Because equations (1-3.8) and (1-3.9) have constant values on their right-hand sides, the shear-forces diagrams for the first and second intervals represent straight lines parallel to axis x, as shown in Figure 1-3.1c. In order to close the shear-forces diagram, we need to apply at the point of the right-hand second interval a positive force of $4F$ (opposite to the shear force $V_{II} = -4F$). This force represents the vertical reaction of the support $R_B = 4F$.

Note as shown in Figure 1-3.1c at the point where the force $6F$ is applied (at the distance of $2a$ from point A), we have an abrupt jump of the shear-forces diagram. The value of the jump equals the absolute value of the applied force $|6F|$.

Now we determine the bending moments. Actually, the bending moment M_I on the first interval equals the moment produced by the reaction R_A about the point of the cut at the distance x_1 from point A. Thus, we may write $M_I = R_A x_1$. Substituting here the value of R_A, we have

$$M_I = 2Fx_1 \tag{1-3.10}$$

Equation (1-3.10) shows that on the first interval, the bending moment represents a linear function of the argument x_1. Therefore, the graph of this function is an inclined straight line that is determined by coordinates of two points. Combining expression (1-3.6) with equation (1-3.10), we obtain that when $x_1 = 0$ then $M_I = 0$, and when $x_1 = 2a$ then $M_I = 4Fa$. Thus, the inclined line connecting these two points (0 and $4Fa$) in Figure 1-3.1d represents the part of the bending-moment diagram that describes the behavior of the bending moment on the first interval. The bending moment M_{II} on the second interval, as shown in Figure 1-3.1b, equals the algebraic sum of moments about the point of the cut at the distance of x_2 from the point A. Therefore, we may write $M_{II} = R_A x_2 - 6F(x_2 - 2a)$, or substituting here the value of R_A, we obtain

$$M_{II} = 2Fx_2 - 6F(x_2 - a) \qquad (1\text{-}3.11)$$

Equation (1-3.11) shows that the bending moment on the second interval represents a linear function of the argument x_2. Substituting in consecutive order into equation (1-3.11) the values of this argument according to expression (1-3.7), we determine the two coordinates that are needed to draw the inclined line describing the behavior of the bending moment on the second interval. Thus, we may write when $x_2 = 2a$ then $M_{II} = 4Fa$ (this ordinate coincides with the coordinate at the end of the first interval), and when $x_2 = 3a$ then $M_{II} = 0$. The inclined line connecting the points $4Fa$ and 0 (on the right-hand side of the diagram) in Figure 1-3.1d represents the part of the diagram that describes the behavior of the bending moment on the second interval. Now, analyzing the diagrams of the shear forces and of the bending moments, it can be seen that the most loaded cross-sectional area is located in the point removed by $2a$ from point A. At this point, we have the maximum values of the shear force $V_{max} = |4F|$ and the bending moment $M_{max} = |4Fa|$.

Example 1-3.1b: Shear forces and bending moments caused by a moment.
Figure 1-3.2a shows a bar mounted on two supports, points A and B, and loaded by a moment $M = 4Fa$. It is necessary to determine the maximum values of the shear force and bending moment, and the location of the most loaded cross-sectional area of the bar.

Solution:
We begin by determining the reactions in the supports. Referring to the free-body diagram shown in Figure 1-3.2b and using formula (1-3.3), we compose the equation of the sum of moments about point B $\sum M_B = 0$. This equation yields $R_A 4a - M = 0$. Because $M = 4Fa$, we may write $R_A 4a - 4Fa = 0$. Solving this equation for R_A, we obtain

$$R_A = F \qquad (1\text{-}3.12)$$

In order to determine the reaction R_B, we use Figure 1-3.2b and apply equation (1-3.2) $\sum F_y = 0$. Hence, $R_A + R_B = 0$. Substituting the value of R_A from equation (1-3.12), we obtain

$$R_B = -F \qquad (1\text{-}3.13)$$

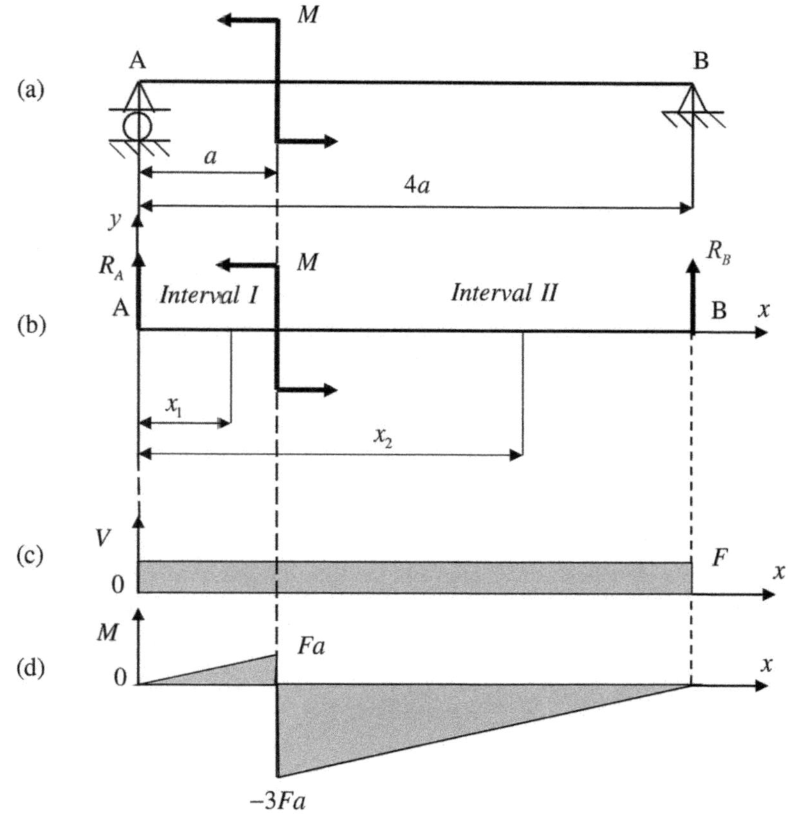

Figure 1-3.2. Free-body diagram for a bar loaded by an external moment, including shear-forces and bending-moment diagrams.

Dividing the bar into two intervals, as shown in Figure 1-3.2b, and making corresponding cuts, we determine the shear forces and bending moments acting on the these intervals. The shear force V_I on the first interval is expressed by the following equation: $V_I = R_A$. Substituting the value of R_A, we obtain

$$V_I = F \tag{1-3.14}$$

Similarly, we determine the shear force V_{II} on the second interval $V_{II} = F$. Replacing R_A with its value, we have

$$V_{II} = F \tag{1-3.15}$$

Equations (1-3.14) and (1-3.15) contain the same constant value on their right-hand sides, therefore, these equations describe the same line that is parallel to the axis x. This line having the ordinate that equals to F is shown in Figure 1-3.2c.

Based on Figure 1-3.2b, we determine the bending moments acting on the first and second intervals. Considering the moment of the reaction R_A about the point of the cut at the distance x_1 from the point A, we calculate the bending moment M_I acting on the first interval. Hence, $M_I = R_A x_1$. Replacing the reaction R_A by its value, we obtain

$$M_I = Fx_1 \tag{1-3.16}$$

Equation (1-3.16) represents a linear function the graph of which is determined by coordinates of two points. According to Figure 1-3.2b, the first interval is limited on its left-hand side by zero and on the right-hand side by a. Therefore, we may write

$$0 \le x_1 \le a \tag{1-3.17}$$

Combining equation (1-3.16) with expression (1-3.17), we determine that when $x_1 = 0$ then $M_I = 0$, while when $x_1 = a$ then $M_I = Fa$. In Figure 1-3.2d, the inclined line connecting the two points with the ordinates 0 and Fa represents the part of the bending-moments diagram describing the behavior of the bending moment on the first interval.

According to Figure 1-3.2b, we determine the bending moment M_{II} acting on the second interval as $M_{II} = R_A x_2 - M$, or substituting the corresponding values, we obtain

$$M_{II} = Fx_2 - 4Fa \tag{1-3.18}$$

Figure 1-3.2b shows that the second interval is characterized by the following limits: on the left-hand side the limit is a, while on the right-hand side the limit is $4a$. Thus, we may write

$$a \le x_2 \le 4a \tag{1-3.19}$$

Combining equations (1-3.18) and (1-3-19), we have that when $x_2 = a$ then $M_{II} = -3Fa$; when $x_2 = 4a$ then $M_{II} = 0$. The line in Figure 1-3.2d that connects the points with the ordinates $-3Fa$ and 0 characterizes the behavior of the bending moment on the second interval. It should be noted, as shown in Figure 1-3.2d, that at the point where the external moment is applied there occurs an abrupt jump of the bending moments diagram. This jump is equal to the absolute value of the applied moment $|4Fa|$.

Analyzing the shear-forces diagram in Figure 1-3.2c as well as the bending moments diagram in Figure 1-3.2d, it becomes clear that the most loaded cross-sectional area of the bar is located on the distance of a from the point A. At this point, the maximum values of the shear force and bending moment are $V_{max} = |F|$ and $M_{max} = |3Fa|$.

Example 1-3.1c: Torque diagram.

Figure 1-3.3a represents a round bar mounted in bearings A and B and loaded by torques T_1, T_2, and T_3. We must determine the location of the most loaded cross-sectional area of the bar and the absolute maximum value of the torque applied to this area.

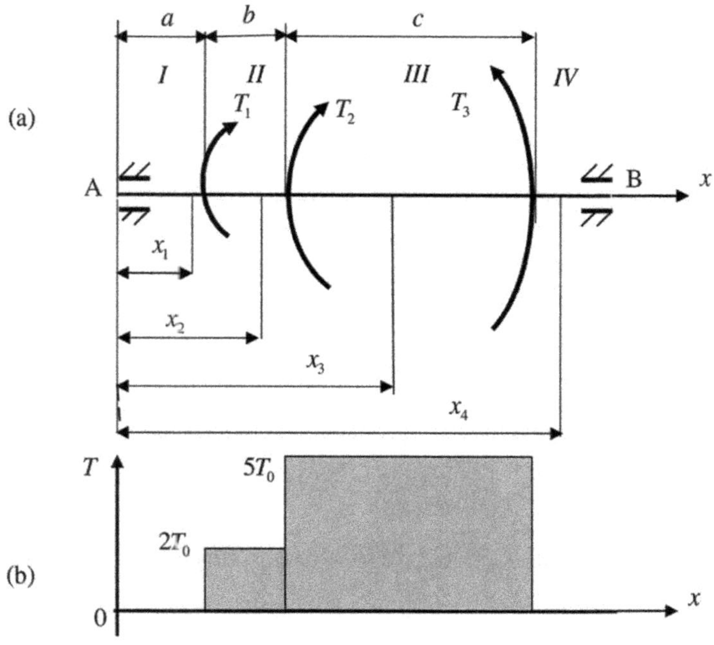

Figure 1-3.3. Torque diagram.

Solution:

Obviously, torques do not cause radial reactions in the supports, and there is no need to draw a free-body diagram.

It is important to emphasize that when a bar is rotating at a constant angular velocity the algebraic sum of the torques applied to the bar equals zero; therefore, we may write

$$\sum T = 0 \tag{1-3.20}$$

The methodology of plotting the torque diagrams is also based on the method of cuts, and it is very similar to constructing shear-forces diagrams. Figure 1-3.3a shows the corresponding distances a, b, and c between the torques and indicates that the bar is divided into four intervals. The values of the applied torques are $T_1 = 2T_0$, $T_2 = 3T_0$, and $T_3 = 5T_0$. The abscissas of the cuts x_1, x_2, x_3, and x_4 are shown in Figure 1-3.3a. The abscissa x_1 indicates the location of the cut on the first interval that is not loaded by any torque. Therefore, the torque on the first interval is $T_I = 0$. The next cut is done on the second interval at the distance x_2 from the bearing A. We throw away the right-hand side of the bar and consider the equilibrium of the remaining part of this bar. Referring to Figure 1-3.3a, we can see that the remaining part of the bar will be in equilibrium if we apply to the cross-section having the abscissa x_2 a torque $T_{II} = T_1 = 2T_0$. The cut on the third interval is made at the distance x_3 from bearing A. And again, considering the equilibrium of the remaining part of the bar on the third interval, we may write $T_{III} = T_2 + T_3$.

Therefore, we obtain $T_{III} = 2T_0 + 3T_0 = 5T_0$. The fourth cut is made on the fourth interval at the distance of x_4 from the bearing A. There are no torques on the fourth interval. Actually, torque T_3 is directed opposite to torques T_1 and T_2 and must be equal to the sum of these torques in order to assure that the system complies with the condition stated by equation (1-3.20). In other words, T_3 balances the combined action of torques T_1 and T_2. Based on these considerations, we write the equation for determining the value of the torque on the fourth interval as $T_{IV} = T_1 + T_2 - T_3$. Substituting the corresponding values, we have $T_{IV} = 2T_0 + 3T_0 - 5T_0 = 0$. Indeed, it is expected that the sum of all the torques applied to this bar equals zero. The equations describing the behavior of the torques on the second and third intervals show that they represent constant values and correspond to segments of straight lines parallel to the axis x, as shown in Figure 1-3.2b.

According to the torque diagram presented in this figure, the maximum value of the torque applied to the bar is $T_{max} = |5T_0|$, and the entire third interval of the bar is subjected to the action of this torque.

1-4. Types of Stresses and Their Calculations

The mechanical properties of materials describe their ability to withstood load. Stresses imposed on the materials reveal the strength of the materials. The critical state of a material occurs when the stresses approach the limit of the strength of the material, causing a failure. Machine elements in operational conditions are subjected to several types of stresses: normal, tangential, and surface stresses. Surface stresses can be subdivided in contact and bearing stresses. All of these types of stresses are considered below.

1-4.1. Normal and Tangential Stresses

The discussions of stresses and deformations are presented in the course of Strength of Materials (Mechanics of Materials), but it is a good idea to review the basic definitions and corresponding formulas.

There are two basic types of stresses: normal and tangential. Normal stresses, usually denoted by the Greek letter σ, act perpendicularly to the cross-sectional area of the material and are parallel to the applied forces. Normal stresses that cause tension are considered positive; normal stresses that cause compression are perceived as negative. Many brittle materials (like cast iron) have higher strength at compression than at tension. Ductile steels exhibit the same mechanical properties during tension and compression testing.

The tension or the compression stress can be determined using formula

$$\sigma = \frac{F}{A} \tag{1-4.1}$$

where F is the applied force, and A is the cross-sectional area.

Tension and compression stresses have different influences on the internal structure of the material. In a simplistic way, the differences can be described as follows: tension stresses cause an elongation of the material, decreasing its cross-sectional area; compression stresses cause the shrinking, or compression, of the material, expanding its cross-sectional area.

The normal or axial strain ε represents the ratio between the absolute deformation δ of a straight bar and its original length l:

$$\varepsilon = \frac{\delta}{l} \tag{1-4.2}$$

The relationship between the normal stress and normal axial strain is shown in Hooke's Law:

$$\sigma = \varepsilon E \tag{1-4.3}$$

where E is the modulus of elasticity having a constant value for the same type of materials.

The Poisson's ratio ν represents the ratio between the lateral strain ε' and the axial strain:

$$\nu = \frac{\varepsilon'}{\varepsilon} \tag{1-4.4}$$

Tangential stresses, often referred to as shear stresses, act on the cross-sectional area of the material and are caused by forces that are directed parallel with this cross-sectional area. The tangential (shear) stresses are usually denoted by the Greek letter τ and can be determined from the following formula:

$$\tau = \frac{F}{A} \tag{1-4.5}$$

where F the force acting tangential to the cross-sectional area A. The changes of the internal structure of the material do not depend on the direction of shear stresses.

Shear strain γ is an angle that represents the deviation of the shape during shear deformation and the relationship between them reads

$$\tau = \gamma G \tag{1-4.6}$$

where G is the shear modulus of elasticity or modulus of rigidity, and it also has a constant value for the same type of materials. The relationship between these two moduli reads

$$G = \frac{E}{2(1+\nu)} \tag{1-4.7}$$

1-4.2. Bending Stresses

The so-called bending stresses are actually normal stresses caused by bending moments. In the vast majority of bending, the normal stresses are accompanied by shear stresses. In the cases of pure bending, there are no shear forces and, consequently, no shear stresses. Pure bending plays a very important role in certain methods of testing materials. The following example explains some peculiarities related to pure bending. Figure 1-4.1a shows a bar loaded by two equal forces F at points A and B.

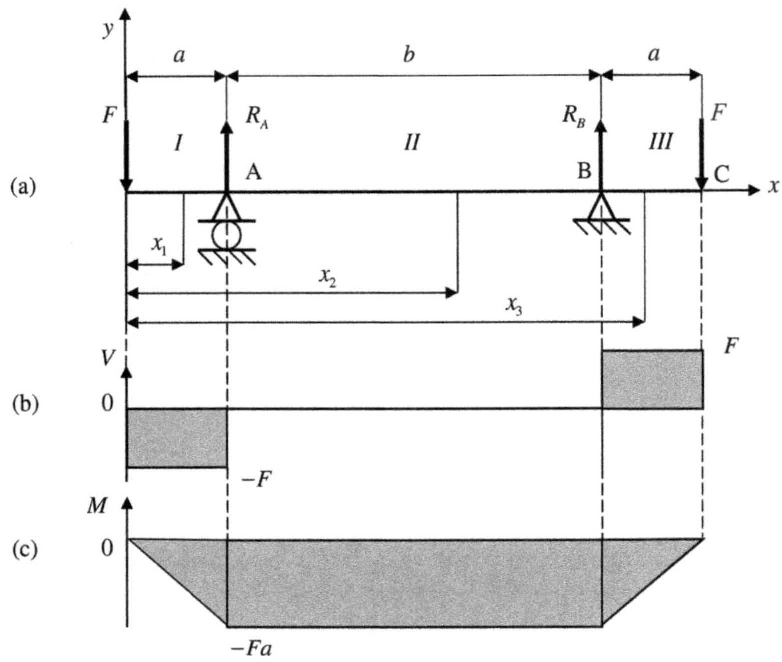

Figure 1-4.1. Pure bending.

The reactions in the supports, the distances between the forces, the three intervals, and the abscissas of the cuts on the corresponding intervals of the bar are presented in this figure. In order to determine the reactions in the supports, we recall equation (1-3.2) that produces $-F + R_A + R_B - F = 0$. Hence, $R_A + R_B = 2F$; therefore, we may write

$$R_B = 2F - R_A \tag{1-4.8}$$

Based on equation (1-3.3), we compose the equation for the sum of the moments about point C: $-F(a + b + a) + R_A(b - a) + R_B a = 0$. Combining this equation with equation (1-4.8) we obtain

$$R_A = F \tag{1-4.9}$$

Solving equations (1-4.8) and (1-4.9) we have

$$R_B = F \tag{1-4.10}$$

Computing the shear force on the first interval, we may write

$$V_I = -F \tag{1-4.11}$$

The shear force on second interval is calculated from the following equation: $V_{II} = -F + R_A$, or substituting the value of R_A from equation (1-4.9) we obtain

$$V_{II} = 0 \tag{1-4.12}$$

The equation for the shear force on the third interval reads $V_{III} = -F + R_A + R_B$, or substituting the values of R_A and R_B, we may write

$$V_{III} = F \tag{1-4.13}$$

Figure 1-4.1b shows the shear-forces diagrams according to the equations (1-4.11), (1-4.12), and (1-4.13). As indicated in the figure, the shear force on the second interval equals zero.

Computing the bending moment on the first interval, we write $M_I = -Fx_1$.

Because $0 \le x_1 \le a$, when $x_1 = 0$ then $M_I = 0$, and when $x_1 = a$ then

$$M_I = -Fa \tag{1-4.14}$$

On the second interval, the bending moment can be determined from the following equation: $M_{II} = -Fx_2 + R_A(x_2 - a)$. Figure 1-4.1a shows the limits of the second interval, hence $a \le x_2 \le a + b$. Therefore, we may write that when $x_2 = a$ then $M_{II} = -Fa$ that is the same as at the end of the first interval, and when $x_2 = a + b$ then bending moment still equals

$$M_I = -Fa \tag{1-4.15}$$

Equations (1-4.14) and (1-4.15) show that in the beginning and the end of the second interval the bending moments have the same values, therefore, the bending moment on the second interval has a constant value. In order to calculate the bending moment on the third interval, we write $M_{III} = -Fx_3 + R_A(x_3 - a) + R_B(x_3 - a - b)$, while the limits on the third interval are $a + b \le x_3 \le a + b + a$. Accounting for these limits for x_3, we obtain that when $x_3 = a + b$ then $M_{III} = -Fa$, and when $x_3 = a + b + a$ then

$$M_{III} = 0 \tag{1-4.16}$$

The bending-moments diagram corresponding to the equations (1-4.14), (1-4.15), and (1-4.16) is plotted in Figure 1-4.1c. The diagrams in Figures 1-4.1b and 1-4.1c indicate that there is no shear force on the second interval, and the bending moment on this interval represents a constant value. Therefore, the second interval of the bar is subjected to pure bending.

The distinction of pure bending allows for conducting tests on the specimen that clarify the influence of the bending stresses on the strength of material, without the presence of shear forces. The bending stresses produced during the bending process of a round bar are determined from the flexure formula:

$$\sigma_{max} = \frac{M_{max} c}{I} \tag{1-4.17}$$

where M_{max} is the absolute maximum value of the bending moment taken from the bending moments diagram; c is the distance between the center of gravity of the cross-sectional area and its periphery, actually $c = 0.5d$; while d is the diameter of the bar; and I is the moment of inertia of the cross-sectional area.

The moment of inertia of a circular cross-sectional area is determined from the following formula:

$$I = \frac{\pi d^4}{64} \tag{1-4.18}$$

Figure 1-4.2a shows the diagram of the bending stress that has its absolute maximum values at the periphery of the cross-sectional area, while it equals zero in the center of gravity of this area.

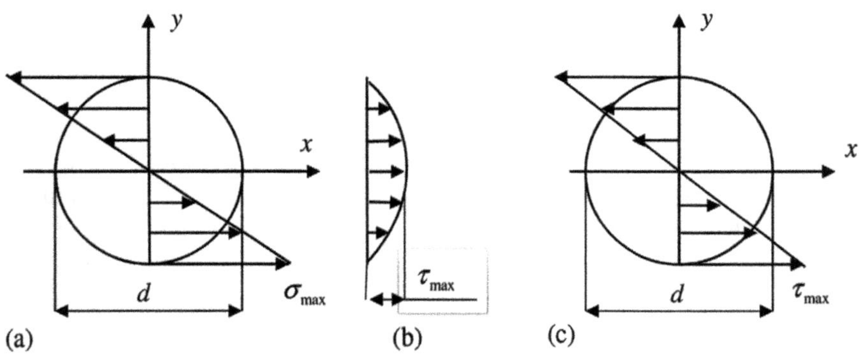

Figure 1-4.2. Diagrams of bending, shear, and torsional stresses.

1-4.3. Shear Stresses at Bending

In a general case of bending, the material experiences a combination of normal and shear stresses. When considering shear stresses, the absolute maximum value of the shear stress for a round bar is calculated from the shear formula:

$$\tau_{max} = \frac{4V_{max}}{3A} \tag{1-4.19}$$

where V_{max} is the absolute maximum value of the shear force taken from the shear-forces diagram and A is the cross-sectional area of a round bar. The shear-stress diagram presented in Figure 1-4.2b shows that the maximum shear stress is acting on the center of gravity of the cross-sectional area, while on the periphery of this area the stress equals zero.

1-4.4. Torsional Stresses

Torsional stresses are actually shear stresses, and for a round bar the maximum value of the stress τ_{max} is determined from the torsion formula:

$$\tau_{\max} = \frac{T_{\max} c}{J} \tag{1-4.20}$$

where $T_{\max}$ is the maximum value of the torque taken from the torque diagram and J is the polar moment of inertia of the cross-sectional area. For a round bar, the formula for calculating the polar moment of inertia reads

$$J = \frac{\pi d^4}{32} \tag{1-4.21}$$

The diagram of torsional stresses is presented in figure 1-4.2c that shows that the torsional stresses like the bending stresses have their maximum values on the periphery of the cross-sectional area, while in the center of gravity of the area they equal zero.

1-4.5. Bending and Shear Stresses of a Rectangular Bar

In case of bending of a rectangular bar, the flexure formula reads

$$\sigma_x = \frac{M_x c_y}{I'_x} \tag{1-4.22}$$

where M_x is the bending moment causing the rotation of the section around the axis x, $c_y = 0.5h$, while h is measured along axis y and represents the height of the rectangular area, and I'_x is the moment of inertia of the rectangular area around axis x, as it is shown in Figure 1-4.3. The moment of inertia I'_x is calculated from the following formula:

$$I'_x = \frac{bh^3}{12} \tag{1-4.23}$$

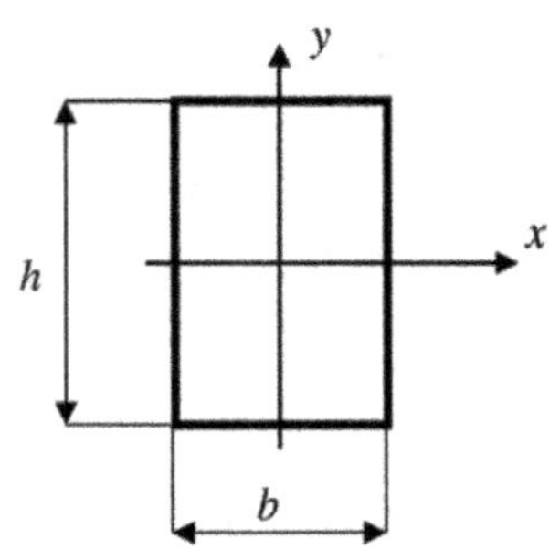

Figure 1-4.3. Axes and parameters of the rectangular area.

In cases when the bending moment M_y is rotating the section around axis y, the formula for the bending stress reads

$$\sigma_y = \frac{M_y c_x}{I'_y} \tag{1-4.24}$$

where $c_x = 0.5b$, while b is measured along the axis y and represents the width of the rectangular area, and I'_y is the moment of inertia of the rectangular area around axis y. The formula for the moment of inertia I'_y is

$$I'_y = \frac{hb^3}{12} \tag{1-4.25}$$

The shear stresses in a rectangular bar are calculated using the shear formula that reads

$$\tau'_{\max} = \frac{3V_{\max}}{A'} \tag{1-4.26}$$

where $V_{\max}$ is the maximum shear force and A' is the area of the rectangular bar.

1-4.6. Contact Stresses

Contact stresses are surface stresses that occur when a pair of bodies is brought together, and they initially touch in a point or in a line before being compressed to each other. The interaction between certain machine elements could be characterized by the contact of a pair of surfaces having either identical or different shapes. For example, in an external engagement of a pair of gears the contact of the teeth can be described as a contact of two convex cylindrical surfaces. In an engagement of a gear with a rack, the contact occurs between a convex and flat surface. The interaction between the roller and the outer ring of a roller bearing is characterized by the contact of convex and concave surfaces. When loading is absent, the contact occurs along a line. For a ball bearing in the absence of loading, the ball contacts the inner and outer rings at a point.

When the working load is applied, the materials in the zones of contact become deformed, and the lines become flat strips, while dots become circles. Moreover, the contact zones during the operation of these elements constantly move, and the surfaces become cyclically stressed and unstressed. After a sufficient number of stress cycles, the contacting surfaces become covered with pits, or cavities, of missing material. These cavities are filled in with solidified oil that further cuts out minuscule "scales" of material, and the surface becomes progressively destroyed. Machine elements affected by pitting should be replaced immediately.

Contact stresses have been analyzed and investigated extensively. But certain aspects of these stresses are still being researched. In the nineteenth-century, a German scientist H. Hertz (1881) carried out an analytical investigation of contact stresses that came to be called Hertzian stresses.

Figure 1-4.4a shows the contact of convex cylinders having radii r_1 and r_2 and the length l (not shown in the figure), or two spheres having radii r_1 and r_2. Figure 1-4.4b shows the contact of convex and concave cylinders having radii r_1 and r_2. Under the forces F the cylinders deform and create a contact area having the shape of a strip with the width of b and length l, while the contact area of the spheres has a circular area with the diameter of $2a$. The radius a is determined from the following formula:

$$a = 1.144\sqrt[3]{\frac{F[E_1(1-\nu_2^2)+E_2(1-\nu_1^2)]r_1 r_2}{2E_1 E_2(r_2 + r_1)}} \tag{1-4.27}$$

(a)

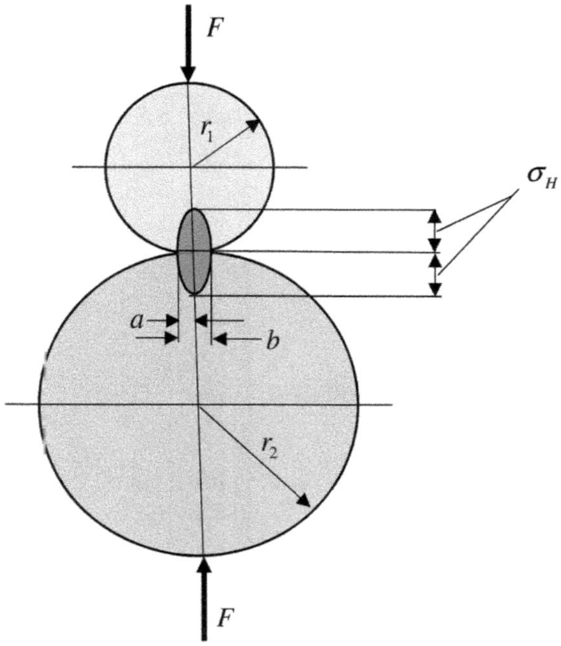

(b)

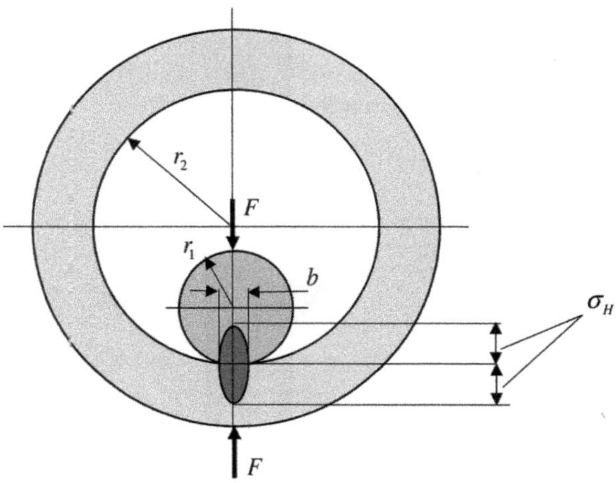

Figure 1-4.4. Hertzian stresses.

where E_1 and E_2 are the moduli of elasticity of the materials involved, while v_1 and v_2 are the Poisson's ratios. For steel spheres having $\nu = 0.3$, the maximum contact stresses in thespheres are calculated from the formula $\sigma_H = \dfrac{1.5F}{\pi a^2}$. Combining this formula with expression (1-4.27), we obtain

$$\sigma_H = 0.388\sqrt[3]{\frac{4FE_1^2E_2^2(r_2 + r_1)^2}{r_1^2r_2^2(E_1 + E_2)^2}} \tag{1-4.28}$$

In cases of contact of two cylinders having parallel axes and concave or convex contacting surfaces, the width of the strip can be determined from the formula

$$b = 1.593\sqrt{\frac{F[E_1(1-\nu_2^2)+E_2(1-\nu_1^2)]r_1r_2}{2lE_1E_2(r_2 \pm r_1)}} \tag{1-4.29}$$

If these two cylinders are made of steel, the maximum contact stress is calculated from the formula $\sigma_H = 1.274\frac{F}{lb}$. Using this formula with equation (1-4.29), and accounting that $\nu_1 = \nu_2 = 0.3$, we obtain

$$\sigma_H = 0.418\sqrt{\frac{FE(r_2 \pm r_1)}{2lr_1r_2}} \tag{1-4.30}$$

In the formulas (1-4.29) and (1-4.30) the plus sign (+) relates to the contact of convex surfaces (Figure 1-4.4a), while minus sign (–) relates to a combination of convex-concave surfaces (Figure 1-4.4b), where r_1 and r_2 are the corresponding radii of curvature.

1-4.7. Bearing Stresses

Bearing stresses represent another type of surface stress that occurs on the contacting surfaces of two bodies that are pressed against one another. Figure 1-4.5 shows a riveted joint connecting two plates having thicknesses a and b, while $a < b$.

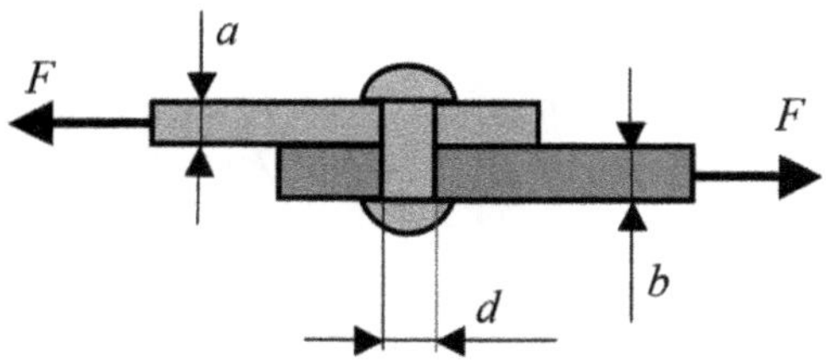

Figure 1-4.5. Bearing stresses.

The plates are loaded by forces F that develop bearing stresses on the surfaces of the holes in both plates as well as on the cylindrical surface of the rivet. In order to calculate the maximum bearing stress, we should consider the interaction of the rivet with the thinnest plate. The bearing stress for the rivet and the plate can be determined from the formula

$$\sigma_{br} = \frac{F}{ad} \tag{1-4.31}$$

where σ_{br} is the bearing stress. This formula shows that the bearing area ad represents the product of multiplying the diameter of the rivet (or the hole in the plate) by the thickness of the plate.

1-4.8. *Units of Stresses*

The units of stresses in the U. S. Customary System are psi (pound per square inch) and ksi (kilo-pounds per square inch). In the International System of Units (SI), the unit of stress is a pascal (Pa) that represents a newton (N) per square meter (m^2). It is more practical to use MPa (mega pascal) that equals to $\frac{N}{mm^2}$, for example, $100\ MPa = 100\frac{N}{mm^2}$.

1-4.9. *The Safety Factor*

Machine elements are predominantly made of steels. The American Society for Testing Materials (ASTM) established standards for testing steels. These standards require vendors to provide information regarding yield strength (S_y) and ultimate strength (S_u) based on testing the tensile strength of steels. Strength calculations of machine elements are built on these two characteristics.

Because residual deformations should not occur in machine elements, the stresses acting on these elements should be less than the yield stress. Considering both the imperfections of the materials and the numerous approximations accepted in the engineering calculations, it is clear that machine elements should possess a certain margin of strength that would compensate for the inevitable errors in the stress calculations. This is achieved by incorporating in the design process a factor that will ensure that the actual strength of the element exceeds the stress imposed on it. This factor is called the safety factor and represents the ratio between the strength of the material and the calculated (nominal) stress acting on the machine element. This ratio defines the safety factor in the following way:

$$safety\ factor = \frac{strength\ of\ the\ material}{stress\ of\ the\ element} > 1$$

Thus, the safety factor depends on two parameters: the strength of the material and the actual stress acting on the material. Let us briefly analyze the accuracy of the values of these two parameters starting with the strength of the material. Machine elements for general application are, for the most part, made from steels the mechanical properties of which and the allowable deviations of their values are adequately presented in published sources. Therefore, the available information regarding the strength of materials inherently contains errors.

Stresses acting on the elements are determined during engineering calculations based on theoretical fundamentals containing assumptions that decrease the accuracy of the calculated results. On top of this, very often the theoretical formulas have to be modified by empiric factors that further decrease the accuracy of the calculations. In addition, the structure of the analytical expressions for determining stresses consists of two components: the geometric characteristics and the loading factors. The geometric characteristics can be determined (measured) with an adequate level of accuracy; however, determining the characteristics of the loading factors is associated with several complications that definitely result in some degree of inaccuracy. Therefore, all these considerations should be accounted for in assigning safety factors that compensate for these inaccuracies, providing an additional margin of strength in order to account for other unspecified factors, such as a an accidental overload or a severe defect in the material structure, etc. It may be recommended, before finalizing the decision regarding the value of the safety factor, to find out the appropriate information related to the safety factors in similar systems.

Denoting the safety factor by n, we may write

$$n = \frac{S}{\sigma} \tag{1-4.32}$$

where S is the strength of the material of the machine element, and σ is the stress in this element.

Usually for normal stresses, the safety factor n_σ is defined as

$$n_\sigma = \frac{S_y}{\sigma} \tag{1-4.33}$$

where S_y or S_{yt} is the tensile yield strength. For shear stresses, the safety factor n_τ is determined in a similar way:

$$n_\tau = \frac{S_{sy}}{\tau} \tag{1-4.34}$$

where S_{sy} is the shear yield strength. Obviously, $n > 1$ and the margin of safety is $n - 1$.

The value of the safety factor depends on many considerations, and it should be carefully determined for each design assignment. It should be clear, the higher the safety factor, the higher the cost of the element. The minimum values of the safety factor are about 1.2. In some cases, the safety factor could equal 10 or more.

The methodology of strength calculations comprises the same steps as the methodology of stress calculations, while in addition it includes steps related to material selection and establishing the safety factor. In conclusion, it should be stressed that assigning of safety factors is an extremely important issue and should be made based on solid considerations.

Based on the definition of safety, the concept of an allowable stress that represents the ratio between the yield strength (or ultimate strength) and the safety factor can be introduced by the following formula:

$$\sigma_{all} = \frac{S_y}{n} \tag{1-4.35}$$

where σ_{all} is the allowable stress.

1-5. Principal Stresses

In the vast majority of cases, machine elements are subjected to biaxial loading that is characterized by normal and shear stresses acting on each of four planes of an infinitesimal cube (or rectangular parallelepiped) that is cut off from the material. It is acceptable to simulate this cube in a shape that allows analyzing the stresses acting on its four plates. Figure 1-5.1a represents this cube, where σ_x and σ_y are the normal stresses acting along axes x and y, respectively.

The shear stresses τ acting in each plate have double subscripts. For instance, for the stress τ_{xy} the first subscript indicates that the stress acts perpendicularly to the axis x, while the second subscript shows the direction of this stress. The values of all these shear stresses are equal to each other. Bending moments, shear forces, and torques cause these normal and shear stresses.

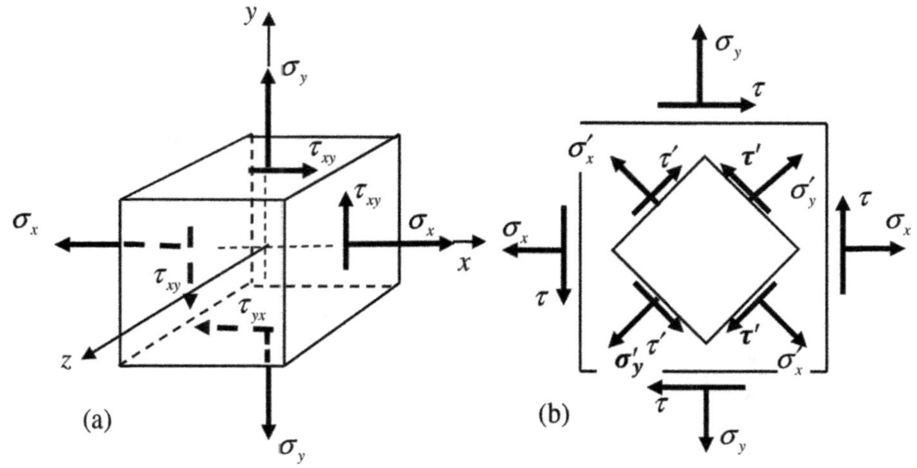

Figure 1-5.1. Plane stress state.

It is acceptable to represent the plane stress state in a two-dimensional view as it is shown in Figure 1-5.1b where the shear stresses are denoted by τ. This figure shows an element that is cut off from the cube and rotated around the axis z on a certain angle. The stresses acting on the plates of this element are represented by σ'_x, σ'_x, and τ'. It is known from the course of Strength of Materials that the most stressful state of this element occurs when the normal stresses obtain their extreme values and the shear stresses disappear (they become equal to zero). The extreme values of the normal stresses are called principal stresses, and they are always denoted by σ_1 and σ_2, where $\sigma_1 > \sigma_2$. The principal stresses are shown in Figure 1-5.2 and can be calculated using the following equations:

$$\sigma_1 = \frac{\sigma_x + \sigma_y}{2} + \sqrt{\left(\frac{\sigma_x - \sigma_y}{2}\right)^2 + \tau_{xy}^2} \tag{1-5.1}$$

and

$$\sigma_2 = \frac{\sigma_x + \sigma_y}{2} - \sqrt{\left(\frac{\sigma_x - \sigma_y}{2}\right)^2 + \tau_{xy}^2} \tag{1-5.2}$$

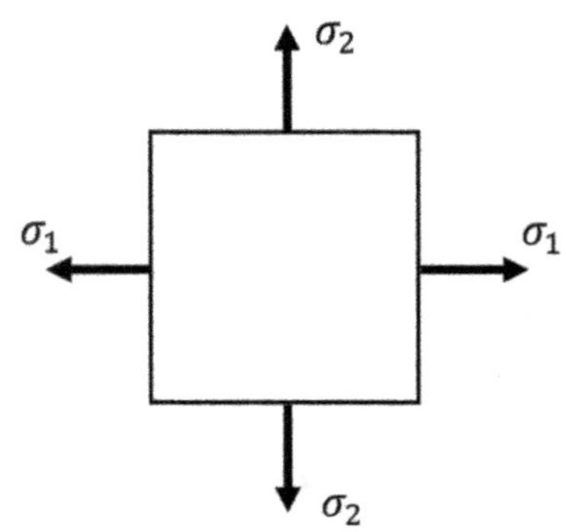

Figure 1-5.2. Principal stresses.

During the stress calculations of the machine elements it often happens that $\sigma_y = 0$. Including this in equations (1-5.1) and (1-5.2) and replacing σ_x with σ and τ_{xy} with τ, we have

$$\sigma_1 = \frac{\sigma}{2} + \sqrt{\frac{\sigma^2}{4} + \tau^2} \tag{1-5.3}$$

$$\sigma_2 = \frac{\sigma}{2} - \sqrt{\frac{\sigma^2}{4} + \tau^2} \tag{1-5.4}$$

In order to evaluate the strength of a machine element or another body subjected to the action of the principal stresses σ_1 and σ_2, which actually represent the extreme stresses, they should be compared with the appropriate characteristics of the material. In other words, in order to evaluate the strength of a machine element, we need to determine its safety factor, which represents the ratio between the strength of the material and the stress, while it is just one stress and not two stresses, σ_1 and σ_2, as in the case of a plane stress state. In addition, the only available strength characteristics of the materials are the results of axial (one directional) tensile testing. The situation becomes much more complicated after realizing that it is practically impossible to carry out any systematic biaxial testing of materials because the number of combinations of the values of σ_1 and σ_2 is, actually, endless. It is an unsolvable problem to obtain the appropriate characteristics of materials attempting to carry out two or three directional testing. Therefore, it becomes necessary to find a way to use the results of the axial tensile testing for evaluating the biaxial stress state. This problem has been known for many years; however, a strict theoretical solution of the problem was not yet found until now. Numerous hypotheses were suggested that consist of creating one expression that comprises σ_1 and σ_2 for calculating an equivalent normal stress σ_{eq} that characterizes the biaxial stress state. This equivalent stress then could be compared with the yield strength (or ultimate strength) of the material (keeping in mind that the values of yield or ultimate strength were measured during one directional tensile testing). Accounting for the safety factor, this hypothetical expression could have the following shape:

$$\sigma_{eq} = f(\sigma_1, \sigma_2) \le \frac{S_y}{n} \tag{1-5.5}$$

These hypotheses are called failure theories, and the most applicable of them are discussed in the next section.

1-6. Failure Theories

The hypotheses of failure represent attempts to predict the failure of materials based on the factors associated with the stresses and deformation of the materials. Hence, it may be expected that the failure occurs due to the maximum normal stress acting in one of the two directions. It may be assumed that the maximum shear stresses or the maximum elongation of the material is causing the failure.

Other hypotheses of failure are built on considerations associated with the energy of deformation of the materials. There are software programs that analyze various combinations of factors related to the failure of materials. However, the accuracy of the results obtained by using these programs is not essentially higher than the results based on the relatively simple formulas of the conventional failure theories presented below. Traditionally, we consider five theories, three of which are in use by engineers.

It should be stressed that all of these theories of failure are not theories; they are just assumptions that like all assumptions in research investigations reflect the results of experimental investigations.

1-6.1. *The Theory of Maximum Normal Stresses*

The maximum normal stresses theory is historically the first and simplest hypothesis of material failure. It suggests that failure occurs when the absolute maximum value of the normal stresses exceed the strength of the material. This theory is also called the first theory of failure and represents an historical interest. The first theory of failure can be expressed in the following way:

$$\sigma_{eq} = \sigma_1 \leq \frac{S_y}{n} \tag{1-6.1}$$

This theory does not take into the consideration the normal stress σ_2. In some cases associated with testing of brittle steels, this theory gives satisfactory results. However, the results of extensive experimental testing of this theory do not support this theory. Therefore, the first theory of failure is not accepted in the engineering practice.

1-6.2. *The Theory of Maximum Strain*

The maximum strain theory is known as the second theory of failure, and it is presented by the following expression:

$$\sigma_{eq} = \sigma_1 - \nu\sigma_2 \leq \frac{S_y}{n} \tag{1-6.2}$$

where ν is the Poisson's ratio. This theory is supported by the results of testing brittle steels; however, it contradicts the experimental data related to ductile steels. This theory does not have practical applications in the engineering strength calculations.

1-6.3. *The Theory of Maximum Shear Stresses*

The theory of maximum shear stresses is the third theory of failure, and it is based on the assumption that maximum shear stresses play the dominant role in the failure of materials. It should be noted that in compression testing the surface of the fracture of the specimen is directed approximately at an angle of 45° to the direction of the compression (to the axis). Considering that the corresponding calculations show that at this angle the shear stresses approach their maximum, it becomes reasonable to assume that the maximum shear stresses are the reason for failure. The maximum shear stress τ_{max} is calculated using the following formula:

$$\tau_{max} = \frac{\sigma_1 - \sigma_2}{2} \tag{1-6.3}$$

This implies that $\tau_{max} \le 0.5S_y$ and $S_{sy} = 0.5S_y$. Therefore, replacing in formula (1-6.3) τ_{max} with σ_{eq}, we may state the third theory of failure with the following expression:

$$\sigma_{eq} = \frac{\sigma_1 - \sigma_2}{2} \le \frac{S_{sy}}{n} \tag{1-6.4}$$

Considering that according to this theory $S_y = 2S_{sy}$, we may rewrite expression (1-6.4) as follows:

$$\sigma_{eq} = \sigma_1 - \sigma_2 \le \frac{S_y}{n} \tag{1-6.5}$$

Substituting into expression (1-6.5) the values of σ_1 and σ_2 according to equations (1-5.1) and (1-5.2) and performing the conventional procedures, we obtain

$$\sigma_{eq} = \sqrt{(\sigma_x - \sigma_y)^2 + 4\tau^2} \le \frac{S_y}{n} \tag{1-6.6}$$

In cases when $\sigma_y = 0$, the expression for the theory of the maximum shear stresses reads

$$\sigma_{eq} = \sqrt{\sigma^2 + 4\tau^2} \le \frac{S_y}{n} \tag{1-6.7}$$

The theory of the maximum shear stresses is supported with sufficient accuracy by the experimental data related to ductile and brittle steels. Sometimes the use of this theory for calculating the parameters of the cross-sectional areas (diameters, thickness, etc.) of machine elements results in some excessive strength of this element. It should be noted that this theory is applicable for materials that exhibit the same mechanical properties in tensile and compression testing. The vast majority of steels possess these properties.

This theory of failure is widely used in industries that utilize steels.

1-6.4. The Theory of Maximum Distortion Energy

The theory of the maximum distortion energy represents the fourth failure theory, and it is called the von Mises theory. This theory is based on a concept associated with the potential energy of material deformation resulting in distortion of the shape of the material. The analytical expression of the concept of the maximum distortion energy reads

$$\sigma_{eq} = \sqrt{\sigma_1^2 + \sigma_2^1 - \sigma_1\sigma_2} \le \frac{S_y}{n} \tag{1-6.8}$$

Combining expression (1-6.7) with equations (1-5.1) and (1-5.2), we may write

$$\sigma_{eq} = \sqrt{\left(\frac{\sigma_x + \sigma_y}{2}\right)^2 + 3\left(\frac{\sigma_x - \sigma_y}{2}\right)^2 + 3\tau^2} \le \frac{S_y}{n} \tag{1-6.9}$$

In cases when $\sigma_y = 0$, expression (1-6.8) becomes the following:

$$\sigma_{eq} = \sqrt{\sigma^2 + 3\tau^2} \le \frac{S_y}{n} \tag{1-6.10}$$

Assuming that $\sigma = 0$ and $n = 1$, we obtain $\tau_{max} \le 0.577 S_y$ compared to the third theory, or theory of maximum shear stresses, where $\tau_{max} \le 0.5 S_y$.

The von Mises theory is applicable to ductile steels that have the same mechanical properties at tensile and compression testing. This theory in comparison with the third theory results in somewhat smaller dimensions of the cross-sectional areas of the machine elements. The fourth theory is also widely used in design practice throughout the industry. It is justifiable to apply the third and fourth theories of failure to the strength calculations of any machine elements; however, keep in mind that the third theory is considered a more conservative theory compared with the fourth theory. In other words, the third theory results in solutions associated with excessive strength of the element compared to the fourth theory. This textbook during discussions of certain machine elements provides recommendations related to the selection of the most applicable failure theory for the particular element.

1-6.5. The Modified Mohr Theory of Failure

This theory is known as the fifth theory of failure. The modified Mohr theory of failure can be used for brittle materials, such as cast iron, that are characterized by differences of their mechanical properties depending on whether they are undergoing tensile or compression testing. Actually, this theory represents a modification of the third theory and is expressed by the following:

$$\sigma_{eq} = \sigma_1 - \frac{S_y}{S_{cy}} \sigma_2 \le \frac{S_y}{n} \tag{1-6.11}$$

where *Scy* is the yield strength of the material in compression testing.

Below are presented a few examples of using the most applicable theories of failure.

Example 1-6.1a: Round bar mounted in a fixed support.

Figure 1-6.1a shows a round bar mounted in a fixed support at point A. The bar is loaded by a moment $M_0 = 600$ lb-in. and a torque $T_0 = 400$ lb-in. The distance between the moment and the torque is $a = 3$ in., and the length of the bar is $l = 6$ in. The bar is made of annealed AISI 1015 steel having the yield strength $S_y = 41, 250$ psi.

Using the third theory of failure, it is necessary to determine the diameter of the bar if the safety factor is $n = 2.5$.

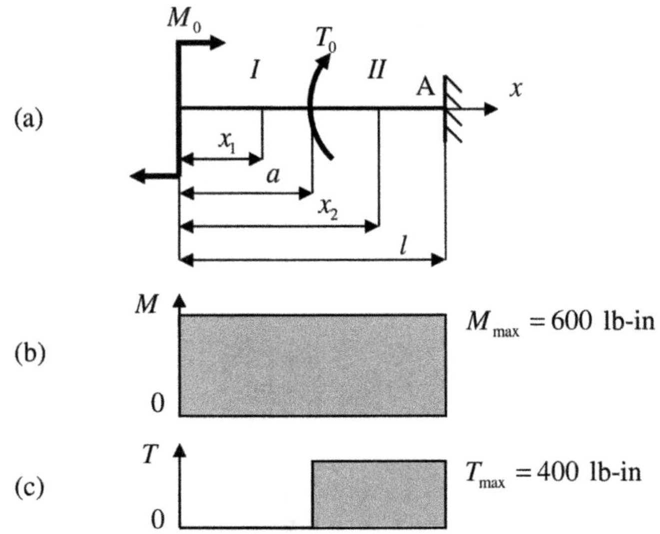

Figure 1-6.1. Bending moments and toque diagrams for a bar in a fixed support.

Solution:

In this case, we do not have on the left-hand side of the bar any unknown loading factors; therefore, we may start with determining the bending moments. Dividing the bar into two intervals, as shown in Figure 1-6.1a, and making cuts having abscissas x_1 and x_2, we may write $0 \leq x_1 \leq a$ and $a \geq x_2 \leq l$. Composing the equation for the bending moments on the first interval, we write $M_I = M_0 = 600$ lb-in. This equation shows that the bending moment on the first interval represents a constant value and does not depend on the abscissa. The equation for the bending moment on the second interval reads: $M_{II} = M_0$. So, the same bending moment is acting on the first and second intervals, and the diagram represents a segment of a straight line parallel to the axis. Actually, the maximum bending moment $M_{max} = 600$ lb-in. Therefore, in this case, we have a pure bending and, consequently, the shear forces are absent.

The diagram of the bending moments acting on the first and second intervals is shown in Figure 1-6.1b. The equation for the torque on the first interval reads: $T_1 = 0$. For the second interval, the equation for the torque is the following: $T_{II} = T_0 = 400$ lb-in. Therefore, we may write $T_{max} = 400$ lb-in.

Figure 1-6.1c shows the torque diagram. It should be mentioned that the reactions in the support have the opposite values of the corresponding loading factors shown in the right-hand side of the diagrams: the bending moment in the support is –600 lb-in. and the torque is *–400* lb-in.

Now we have to determine the bending and torsional stresses. Based on formulas (1-4.17) and (1-4.18) and considering that $c = 0.5d$, we have $\sigma_{max} = \frac{32M_{max}}{\pi d^3}$. Substituting the value of M_{max}, we have $\sigma_{max} = \frac{6{,}115}{d^3}$ psi. Using formulas (1-4.20) and (1-4.21), we obtain $\tau_{max} = \frac{16T_{max}}{\pi d^3}$. Accounting for the value of T_{max}, we write $\tau_{max} = \frac{2{,}032}{d^3}$ psi.

We determined the maximum values of the bending and torsional stresses acting on the most loaded cross-sectional area having the abscissa a. It is important to underline that in this case the maximum bending stresses and the maximum torsional stresses are acting simultaneously on one point of the cross-sectional area. In order to evaluate the stress state in this point, we have to apply a certain theory of failure. It should be mentioned that in applying the theories of failure we accept that $\sigma_{max} = \sigma$ and $\tau_{max} = \tau$. For this case, we use the third theory of failure that is represented by expression (1-6.7). Using the values of the bending and torque stresses, we may write $\sigma_{eq} = \sqrt{\left(\frac{6,115}{d^3}\right)^2 + 4\left(\frac{2,032}{d^3}\right)^2} \le \frac{S_y}{n}$. Substituting into this expression the values of the yield strength $S_y = 41,250$ psi and the safety factor $n = 2.5$ and applying some obvious actions, we have $\sigma_{eq} = \frac{7,342}{d^3} = \frac{41,250}{2.5} = 16,500$ psi. Solving this equation for the diameter d, we find $d = \sqrt[3]{\frac{7,342}{16,500}} = 0.76$ in.

Example1-6.1b: Round bar in two bearings.

Figure 1-6.2a shows a round bar rotating on bearings A and B. The bar has a diameter $d = 25$ mm. The bar is loaded by a force $F_1 = 3,000$ N, a moment $M_1 = 250$ N·m, and torques $T_1 = 200$ N·m, $T_2 = 100$ N·m, and $T_3 = 300$ N·m. The distances between the loads as well as the length of the bar are indicated in the figure. Based on the third theory of failure, it is necessary to select an appropriate steel for the bar. The safety factor $n = 2$ and $a = 35$ mm.

Solution:

Figure 1-6.2b shows the free-body diagram of the bar indicating the reactions in the supports R_A and R_B. Recalling the equation (1-3.3), we write the sum of the moments about point B. Hence, $-F_1 3a + R_A 2a + M_1 = 0$. Substituting into this equation the corresponding values, we have $-3,000 \cdot 3 \cdot 35 + R_A\, 2 \cdot 35 + 250,000 = 0$. Solving this equation for R_A, we obtain $R_A = \frac{315,000 - 250,000}{2 \cdot 35} = 929$ N. Based on the equation (1-3.2), we determine the reaction R_B. Hence, $-F_1 + R_A + R_B = 0$, therefore, $R_B = -3,000 + 929 = 2,071$ N.

In order to plot the shear-forces, bending-moments, and torques diagrams we cut the bar on the free-body diagram into three intervals as shown in Figure 1-6.2b. The locations of cuts are indicated by the abscissas x_1, x_2, and x_3. Based on the free-body diagram, we determine the limits of the intervals. Hence, the limits for the first, second, and third intervals are characterized by the following expressions: $0 \le x_1 \le a$, $a \ge x_2 \le 2a$, and $2a \le x_3 \le 3a$, respectively.

The equation for the shear forces on the first interval reads $V_I = -F_1 = -3,000$ N. The shear forces acting on the second interval are $V_{II} = -F_1 + R_A = -3,000 + 929 = -2,071$ N. The equation for the shear forces on the third interval is identical to the equation on the second interval; therefore, we have $V_{III} = -2,071$ N. Figure 1-6.2c shows the diagram of the shear forces.

The equation of the bending moments on the first interval reads $M_I = -F_1 x_1$. According to the limits indicated above of the intervals, we have that when $x_1 = 0$ then $M_1 = 0$, and when $x_1 = a$ then $M_I = -3,000 \cdot 35 = -105,500$ N·mm. Composing the equation of bending moments for the second interval, we write $M_{II} = -F_1 x_2 + R_A(x_2 - a)$. Accounting for the limits for this interval, we have that when $x_2 = a = 35$ mm then $M_{II} = -105,500$ N·mm, and when $x_2 = 2a$ then $M_{II} = -F_1 2a + R_A(2a - a)$.

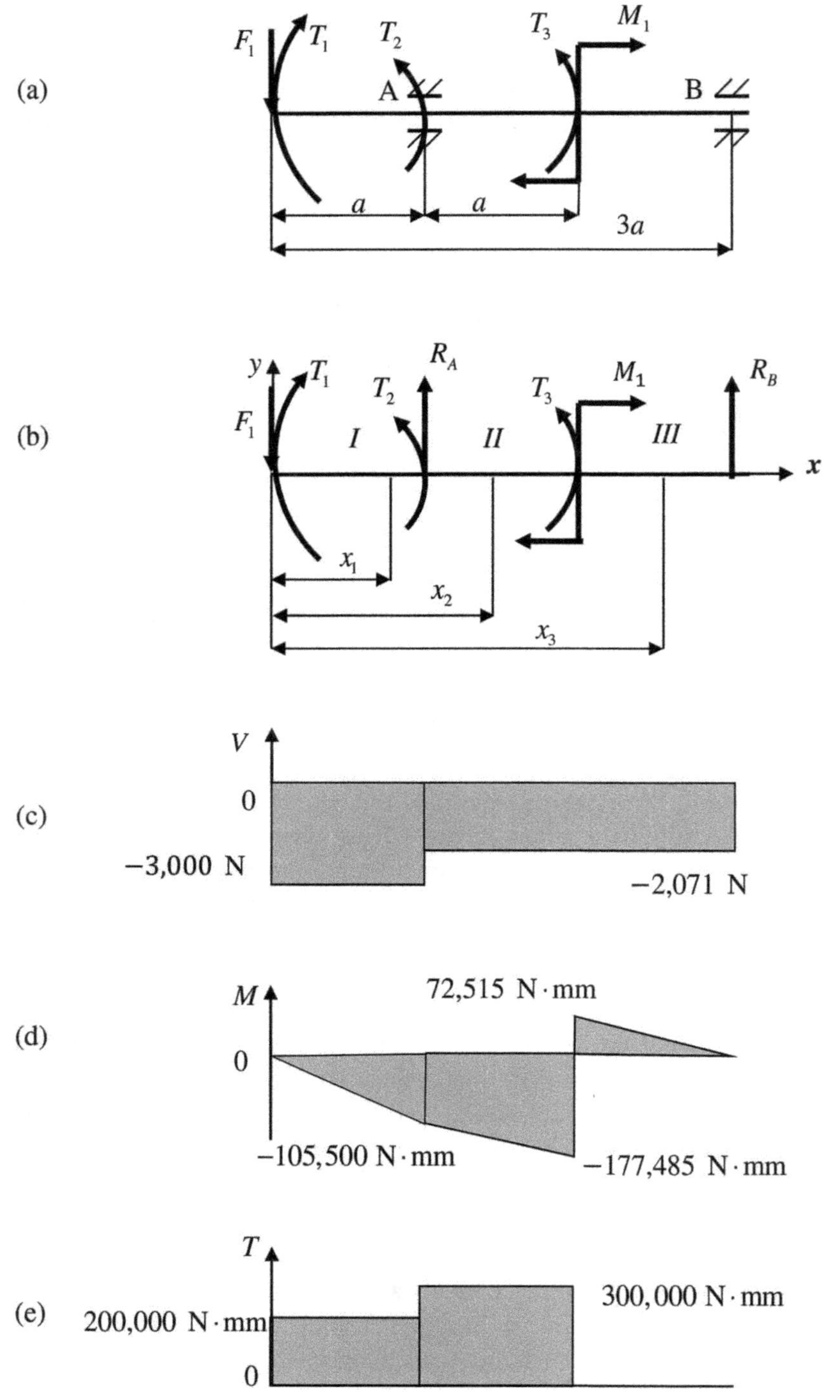

Figure 1-6.2. Round bar in two bearings. Free-body, shear-forces, bending-moments, and torque diagrams.

Substituting the corresponding values, we obtain $M_{II} = -3{,}000 \cdot 2 \cdot 35 + 929(2 \cdot 35 - 35) = -177{,}485$ N·mm. The bending moments on the third interval are determined by the following equation: $M_{III} = -F_1x_3 + R_A(x_3 - a) + M_1$.

Taking into consideration the limits for this interval, we may write that when $x_3 = 2a$ then $M_{III} = -F_12a + R_A(2a - a) + M_1$. Substituting the corresponding values, we obtain $M_{III} = -3{,}000 \cdot 2 \cdot 35 + 250{,}000 = 72{,}515$ N·mm; and when $x_3 = 3a$ then $M_{III} = -3{,}000 \cdot 2 \cdot 35 + 929(3 \cdot 35 - 35) + 250{,}000 = 0$. Figure 1-6.3d shows the diagram of bending moments based on the above calculations.

The equation for the torque on the first interval reads $T_I = T_1 = 200{,}000$ N ·mm. On the second interval, the torque is calculated from the following equation: $T_{II} = T_1 + T_2$. Substituting the corresponding values, we may write that $T_{II} = 200{,}000 + 100{,}000 = 300{,}000$ N·mm. For the third interval, the equation of the torque reads $T_{III} = T_1 + T_2 - T_3 = 200{,}000 + 100{,}000 - 300{,}000 = 0$. The Figure 1-6.2e shows the torque diagram according to the calculations presented above.

We plotted the diagrams of shear forces, bending moments, and torques, and we can calculate the corresponding stresses. Thus, the maximum value of the shear force is $V_{max} = |3.000|$ N.

Denoting the maximum shear stress caused by the shear force by τ_F and applying formula (1-4.19), we calculate $\tau_F = \dfrac{4 \cdot 3{,}000 \cdot 4}{3\pi 25^2} = 8.153 \dfrac{\text{N}}{\text{mm}^2}$ (8.253 MPa).

The bending moments and the torque diagrams indicate that the maximum values of the bending moment and torque acting in the right-hand side of the second interval and they respectively are: $M_{max} = |177{,}485|$ N·mm and $T_{max} = |300{,}000|$ N·mm. Using formula (1-4.18), we determine the moment of inertia: $I = \dfrac{\pi 25^4}{64} = 19{,}165 \text{ mm}^4$. Applying this result to formula (1-4.17) and accounting that $c = 0.5d$, we calculate the maximum bending stress $\sigma_{max} = \dfrac{177{,}485 \cdot 0.5 \cdot 25}{19{,}165} = 115{,}8 \dfrac{\text{N}}{\text{mm}^2}$ (115.8 MPa). Recalling formula (1-4.21), we determine the polar moment of inertia: $J = \dfrac{\pi 25^4}{32} = 38{,}330 \text{ mm}^4$. Substituting into formula (1-4.20) the polar moment of inertia from formula (1-4.21), we calculate the maximum torsional stress: $\tau_{max} = \dfrac{300{,}000 \cdot 0.5 \cdot 25}{38{,}330} = 97{,}834 \dfrac{\text{N}}{\text{mm}^2}$ (97,834 MPa).

Thus, we determined the maximum values of stresses acting in corresponding cross-sectional areas of the bar. It is important to highlight that the maximum value of the shear stress resulting from the shear force is applied to the center of gravity of the cross-sectional area. However, the maximum torsional stress occurs at all points of the periphery of this area, while in the center of this area the torsional stress equals zero. Consequently, these two stresses cannot be considered together in evaluating the stress state at any certain point of the cross-sectional area. However, the maximum bending and torsional stresses act simultaneously in each point of the periphery of this cross-sectional area. In order to evaluate the stress state in this case, we have to apply an appropriate theory of failure.

Using the expression (1-6.7) that represents the third theory of failure, we may write $\sigma_{eq} = \sqrt{115.8^2 + 4 \cdot 97.834^2} \le \dfrac{S_y}{n}$, or $227.4 \le \dfrac{S_y}{n}$. Considering that $n = 2$, we have $S_y \ge 227.4 \cdot 2 = 454.8 \dfrac{\text{N}}{\text{mm}^2}$ (MPa). According to the table of the mechanical properties of steels presented in Appendix A-1-4, we select the AISI 1060 as rolled steel that has $S_y = 483$ MPa.

Taking the most conservative approach, we determine $S_{sy} = 0.5S_y = 0.5.414 = 207$ MPa. The maximum shear stress caused by the shear force is $\tau_F = 8.253$ MPa which is significantly less than the computed value of S_{sy} and, obviously, does not represent any strength problem.

Example 1-6.1c: A shaft carrying certain elements is mounted on two bearings.
Figure 1-6.3a shows a shaft supported by bearings A and B. The shaft carries certain elements exerting different types of loads on the shaft. The segment of a straight line CD represents a machine element rigidly connected to the shaft and is loaded by forces F_2, F_3, and F_4, while the forces F_2 and F_4 act in the vertical plane, the force F_3 acts in the horizontal plane.

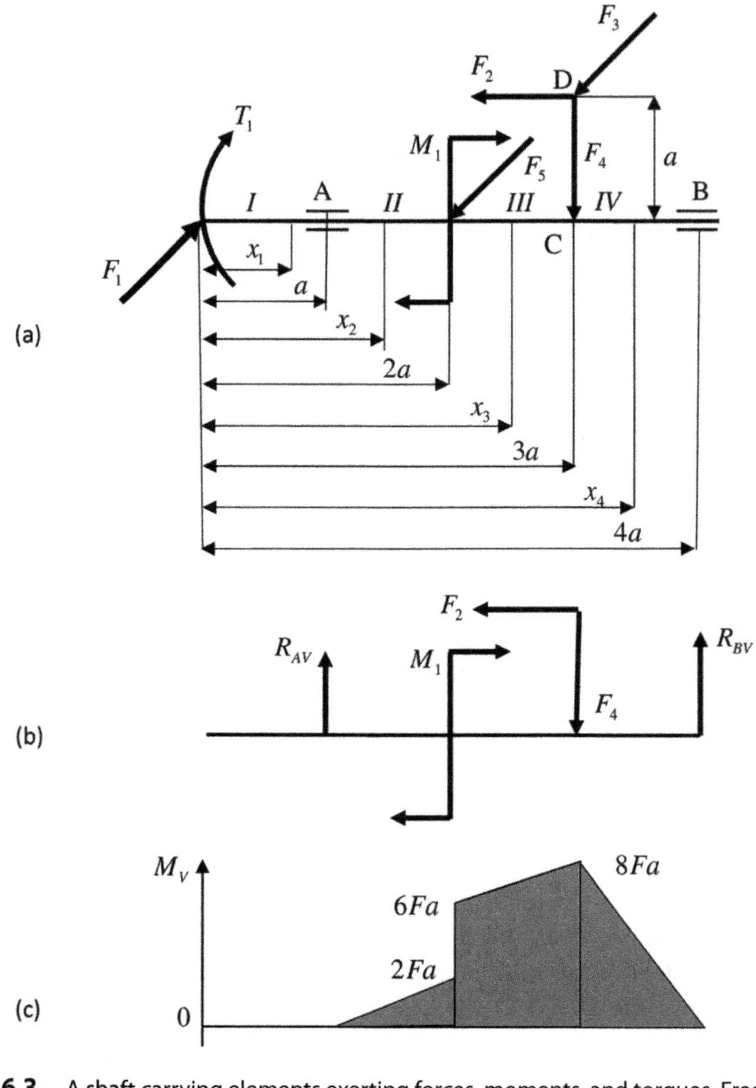

Figure 1-6.3 A shaft carrying elements exerting forces, moments, and torques. Free-body and bending-moments diagrams in the vertical plane.

According to Figure 1-6.3a, the force F_2 develops an external moment that is equal to F_2a and acts in the vertical plane. The force F_3 creates a torque that is equal to F_3a, and it is balanced by the torque T_1. The forces F_1 and F_5 act in the horizontal plane.

Figure 1-6.4 is a continuation of the figure 1-6.3 and shows the diagrams in the horizontal plane.

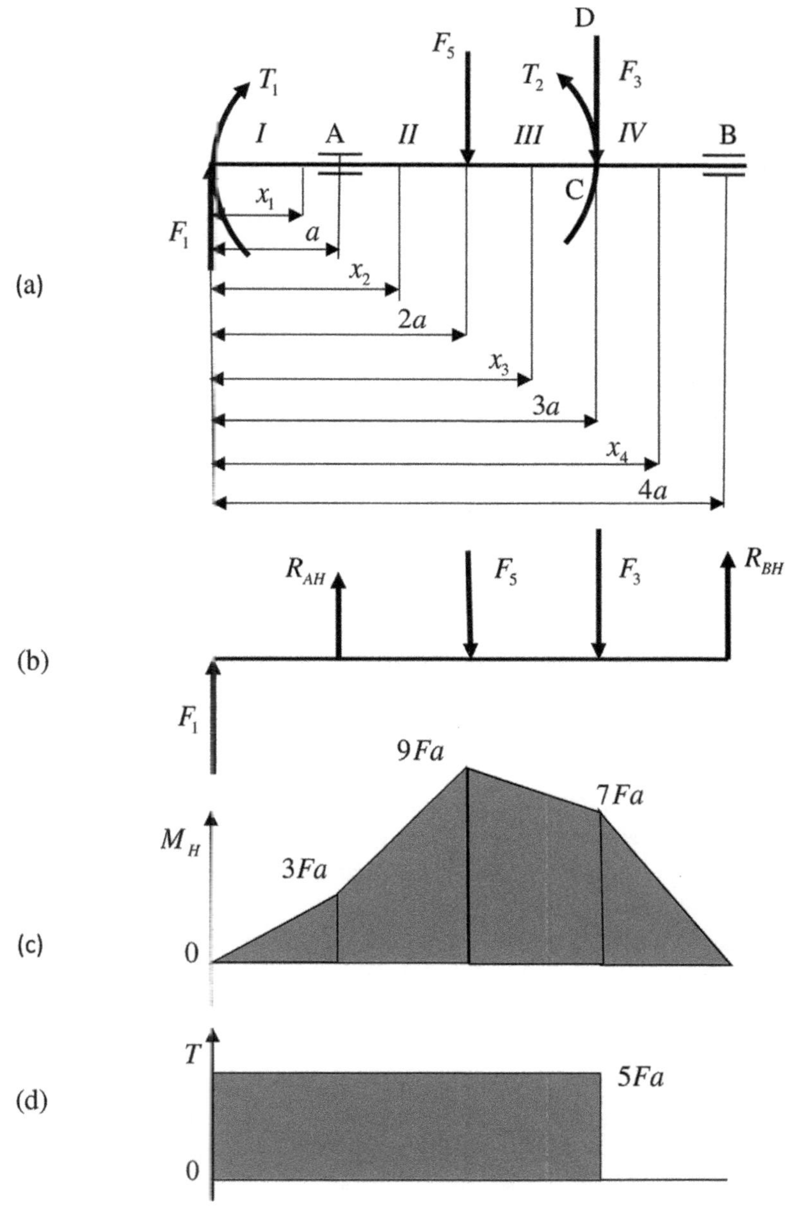

Figure 1-6.4. A shaft carrying elements exerting forces, moments, and torques. Free-body and bending-moments diagrams in the horizontal plane.

It is necessary to determine, according to the fourth theory of failure, the equivalent stress in the shaft having a diameter d, while the corresponding values of the loads are $F_1 = 3F$, $F_2 = 6F$, $F_3 = 5F$, $F_4 = 4F$, $F_5 = 8F$, $M_1 = 4Fa$, and $T_1 = 5Fa$.

Solution:

The solution of the problem requires appropriate calculations related to the vertical and horizontal planes. It should be mentioned that the solution does not include the analysis of the shear forces since the shear stresses caused by these forces cannot be accounted for in the theories of failure.

Figure 1-6.3b shows the free-body diagram in the vertical plane. Based on equation (1-3.3), we compose the sum of moments about point B in order to determine the reaction R_{AV}. Hence, $R_{AV}\,3a + M_1 - F_2a - F_4a = 0$. Substituting into this equation the corresponding values and solving it for R_{AV}, we obtain that $R_{AV} = 2F$. Applying equation (1-3.2), we calculate that $R_{AV} - F_4 + R_{BV} = 0$. Therefore, $R_{BV} = 2F$. In order to determine the bending moments acting in the shaft, we divide the shaft into four intervals and make corresponding cuts as shown in Figure 1-6.3a. According to these cuts, we determine that on the first interval $M_i = 0$. The bending moment on the second interval is calculated from the following equation: $M_{II} = R_{AV}(x_2 - a)$. The limits of x_2 are $a \le x_2 \le 2a$. Thus, when $x_2 = a$ then $M_{II} = 0$, and when $x_2 = 2a$ then $M_{II} = 2Fa$. On the third interval the equation for bending moments reads $M_{III} = R_{AV}(x_3 - a) + M_1$, while limits of this interval are $2a \le x_3 \le 3a$. Hence, when $x_3 = 2a$ then $M_{III} = 2Fa + 4a = 6Fa$, and when $x_3 = 3a$ then $M_{III} = 4Fa + 4Fa = 8Fa$. The equation for calculating the bending moments on the fourth interval is $M_{IV} = R_{AV}\,(x_4 - a) + M_1 - F_2a - F_4(x_4 - 3a)$, while the limits of this interval are $3a \le x_4 \le 4a$. When $x_4 = 3a$ then $M_{IV} = R_{AV}(3a - a) + M_1 - F_2a - F_4(3a - 3a) = 7Fa$, and when $x_4 = 4a$ then

$$M_{IV} = R_{AV}\,(4a - a) + M_1 - F_2a - F_4(4a - 3a) = 6Fa + 4Fa - 6Fa - 4Fa = 0.$$

Figure 1-6.3c shows the diagram of the bending moments M_V acting in the vertical plane.

Now we determine the bending moments acting in the horizontal plane. Figure 1-6.4a shows the loading factors applied to the shaft in the horizontal plane. The free-body diagram in the horizontal plane is presented in Figure 1-6.4b that allows in accordance with equation (1-3.3) to compose the sum of moments about point B and determine the reaction R_{AH}. Hence, we write $F_14a + R_{AH}3\text{a} - F_52a - F_3a = 0$. Solving this equation for R_{AH}, we obtain that $R_{AH} = 3F$. Using equation (1-3.2), we determine the reaction R_{BH}. Hence, we write $F_1 + R_{AH} - F_5 - F_3 + R_{BH} = 0$. This equation yields $R_{BH} = 7F$.

Figure 1-6.4a shows that the shaft is divided in four intervals that allow us to determine the internal bending moments. The bending moments acting on the first interval can be calculated from the following equation: $M_I = F_1x_1$, while $0 \le x_1 \le a$. Hence, when $x_1 = 0$ then $M_I = 0$, and when $x_1 = a$ then $M_1 = F_1a = 3Fa$. For the second interval, the equation for calculating the bending moments reads $M_{II} = F_1x_2 + R_{AH}(x_2 - a)$. The limits of this interval are $a \le x_2 \le 2a$. Thus, when $x_2 = a$ then $M_{II} = 3Fa$, and when $x_2 = 2a$ then $M_{II} = F_12a + R_{AH}\,a = 6Fa + 3Fa = 9Fa$. The bending moments on the third interval are calculated from the following equation:

$M_{III} = F_1x_3 + R_{AH}(x_3 - a) - F_3(x_3 - 2a)$, while the limits of the third interval are $2a \le x_3 \le 3a$. Hence, when $x_3 = 2a$ then $M_{III} = 9Fa$, and when $x_3 = 3a$ then $M_{III} = F_13a + R_{AH}(3a - a) - F_3(3a - 2a) = 7Fa$.

The equation for the bending moments on the fourth interval is $M_{IV} = F_1x_4 + R_{AH}(x_4 - a) - F_5(x_4 - 2a) - F_3(x_4 - a)$, and the limits of the interval are $3a \le x_4 \le 4a$. Thus, when $x_4 = 3a$ then $M_{IV} = 5Fa$, and when $x_4 = 4a$ then $M_{IV} = F_14a + R_{AH}(4a - a) - F_5(4a - 2a) - F_3(4a - 3a) = 0$.

Figure 1-6.4c represents the bending-moments diagram in the horizontal plane.

Figures 1-6.3a and 1-6.4a indicate that the torque T_1 is acting on the first, second, and third intervals, and Figure 1-6.4d shows that this torque is balanced by torque T_2. Hence, $T_1 + T_2 = 0$ and on the first interval $T_I = T_1 = F_3a = 5Fa$. The same torque is acting on the second and third intervals, while $T_{IV} = 0$. The torque diagram is shown in Figure 1-6.4d.

In general, the bending moment acting in a certain cross-sectional area is the result of the vector summation of the corresponding moment vectors in vertical and horizontal planes. In our case, it is not obvious which cross-sectional area is the most loaded. We need to determine the bending moments acting at the end of the second and third intervals and figure out which one is bigger. Hence, $M'_{II} = \sqrt{(6Fa)^2 + (9Fa)^2} = 10.82Fa$ and $M'_{III} = \sqrt{(8Fa)^2 + (7Fa)^2} = 10.63Fa$, since $M'_{II} > M'_{III}$, we take $M_{\max} = 10.82\ Fa$. According to the diagram in Figure 1-6.4d, $T_{\max} = 5Fa$. Based on formulas (1-4.17) and (1-4.18) we calculate $\sigma_{\max} = \dfrac{32 \cdot 10.82Fa}{\pi d^3} = \dfrac{110.27Fa}{d^3}$. Using formulas (1-4.20) and (1-4.21), we determine $\tau_{\max} = \dfrac{16 \cdot 5Fa}{\pi d^3} = \dfrac{25.48Fa}{d^3}$. Applying expression (1-6.10), we determine the equivalent stress according to the fourth theory of failure $\sigma_{eq} = \sqrt{\left(\dfrac{110.27Fa}{d^3}\right)^2 + 3\left(\dfrac{25.48Fa}{d^3}\right)^2} = \dfrac{118.8Fa}{d^3}$.

Taking $F = 2{,}000$ N, $a = 40$ mm, and $d = 25$ mm, we obtain

$$\sigma_{eq} = \frac{118.8Fa}{d^3} = \frac{118.8 \cdot 2{,}000 \cdot 40}{25^3} = 608.26 \frac{\text{N}}{\text{mm}^2}\ (\text{MPa})$$

1-7. Assemblies and Prototypes

New mechanical systems (assemblies) are often prototypes of products that will be commercialized. The main concern during the design step of these assemblies is making sure the product provides the required functionality. If the system is intended to be used for the inside needs of an enterprise, it will be manufactured as single or several assemblies at the lowest possible cost by using ordinary carbon steels and applying conventional manufacturing processes. In these cases the machine elements are normally fabricated using universal metal cutting equipment. Usually, it is desirable to avoid polishing, honing, grinding, and other costly manufacturing operations. Quenching and carburizing as well as other thermal treatment procedures of steel are applied in cases of extreme necessity. Hence, in cases of a very limited production, the elements should possess sufficient strength, while an assembly that is reasonably overweight does not represent a problem in terms of space, cost, or handling.

In these cases, it is preferable to use elevated safety factors in the engineering calculations in order to eliminate testing of the elements with regard to their strength. The analysis of the plane stress state of a shaft caused by bending and torsion shows that the difference between the equivalent stresses calculated according to the third and fourth theories of failure averages about 11%. It is clear that, usually, the safety factors compensate for this difference with a significant margin of safety. Therefore, the design of the machine elements of a new assembly that will be manufactured in the quantity of a single or several pieces should be focused on providing sufficient strength of the elements in order to avoid, as it is mentioned above, costly and time-consuming testing. However, in most situations the testing with regard to the functionality of the assembly is inevitable.

In case of mass production, the same considerations are applicable to the preliminary stage of development of a prototype of a product. During this stage, the main concern is associated with the functionality of the prototype, while the strength of the machine elements should be guaranteed by using elevated safety factors. The preliminary testing is intended to reveal some shortcomings of the concept of the design. Usually it takes a few attempts to finalize the preliminary design of prototypes. When the conceptual design is finalized, a pilot group of prototypes is manufactured applying the most appropriate materials and fabrication methods. The pilot group of prototypes usually undergoes intensive testing with regard to the strength of the machine elements and functionality of the assembly. Obviously, appropriate corrections in the dimensions of the machine elements as well as in the safety factors could by applied as a result of testing.

Obviously, the engineering calculations provide a reliable basis to determine the geometric parameters of the corresponding machine elements for the new or improved assembly.

Summarizing, design of mechanical engineering systems is a process where the strength calculations of machine elements represent an extremely important stage. The existing methodologies of strength calculations provide adequate solutions that satisfy the real-life requirements of the machine elements for general applications.

The procedures described above reflect the general approach related to assemblies and prototypes. Obviously, in real-life conditions there are numerous scenarios possible that differ from what is described above.

1-8. Fatigue Phenomenon of Materials

Stress and strength calculations are developed for loading factors having constant values, or in other words, for static loading. In addition, in the past these methodologies were applied to the strength calculations of machine elements without considering any role of variable loading. However, repeated observations have shown that some machine elements, such as railroad axles, designed in accordance with these fundamentals suddenly failed, becoming fractured in two pieces. Moreover, this happened without any visible signs (like necking) that usually precede the failure of machine elements made of steel. What was more important is that the actual stresses acting in these failed axels were significantly lower than the yield strength limit of the material.

The railroad axles are subjected to bending stresses that, as we know, reach their maximum values in the peripheral fibers of the cross-sectional area. It is clear that during one revolution of the axle each fiber changes its stress from zero to maximum, and from maximum to zero, and then to minimum (having the same absolute value as the maximum), and then to zero. Obviously, the stresses in the axles are variable and cannot be characterized as static (constant) stresses.

These observations stimulated intensive testing of axles in laboratory conditions. The experimental investigations of railroad axles started in Europe in the beginning of the twentieth-century, resulting in the confirmation of the severely destructive role of variable stresses on the strength of materials. The sudden failure of materials subjected to variable stresses that are significantly lower than the yield strength of this material is known as the phenomenon of fatigue of materials. The assumptions explaining this phenomenon suggest that due to the imperfections in the microstructures of the materials, there are microscopic cracks inside the material. In case of static stresses, these cracks remain almost unchanged. However, under variable stresses in certain cross-sectional areas of the material, there begins a process of nucleation of one or more of these cracks that have a tendency to grow. These cracks propagate and gradually damage the cross-sectional area. When the crack reaches the periphery of the cross-sectional area, it significantly weakens the material, and after a time, the material suddenly fractures.

It is reasonable to ask if it is possible to produce a material without these imperfections and, therefore, eliminate the fatigue phenomenon. Another question that may arise: are the micro cracks generated by the variable stresses in the parts of material that are free from imperfections? To my knowledge, there are currently no answers to these questions. However, there is undeniable evidence that in many cases both surfaces of the fractured pieces are relatively flat and are ground against each other. And just a part of the area of the fractured pieces has a rough crystalline structure. It is important to note that fractured surfaces are closely perpendicular to the axis of rotation of the axle. All this suggests that the ground parts of the surfaces were developed some time ago, while the crystalline parts of surfaces are related to the moment of fracture. The nature of the described fracture lets us assume that the fracture is associated with a crack that gradually grows along the shortest distance to the periphery.

Fatigue failure played a significant role in numerous accidents associated with mechanical engineering systems. Numerous investigations of the engineering and scientific communities were devoted to the study of the fatigue phenomenon with the goal to determine the conditions under which the fatigue failure could be prevented. Fortunately, the investigations revealed that for some materials, such as steel and others, fatigue failure is preventable. Because of the remarkable achievements of the scientists and engineering practitioners who worked on the problem, many of the fatigue failures of the past no longer occur. The engineering calculations related to preventing fatigue failure are presented below.

1-8.1. Fatigue Testing

Fatigue testing of materials, specifically of steels, became a standard procedure that was developed by the American Society for Testing and Materials (ASTM). The conventional methods of steel testing are the axial and bending loadings of corresponding specimens. During these tests, the dedicated specimens are subjected to variable stresses where their absolute values of the maximum and minimum stresses are equal. It is important to emphasize that the methods of bending testing are based on imposing pure bending into the specimen excluding any presence of shear stresses. Therefore, as it was mentioned earlier, due to the phenomenon of pure bending, it is possible to reveal the characteristics of bending loading without interference of shear forces. Figure 1-8.1 shows a standard specimen for bending testing of steel. The surface of the specimen is polished. The laboratory conditions of the testing procedure are controlled in order to satisfy certain requirements.

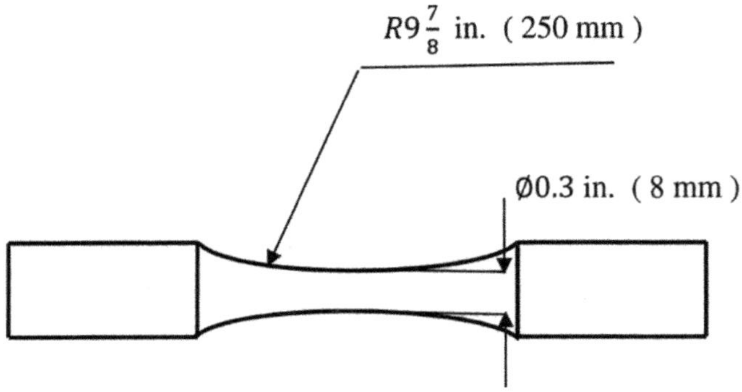

Figure 1-8.1. Fatigue bending test specimen.

The bending testing machine controls the bending stress in the specimen. During a single test, the machine loads the specimen with a certain alternating stress and counts the number of revolutions or stress cycles that the specimen can endure before fracturing. The results of the corresponding tests allow building graphs expressing the dependence between alternating stresses and the number of cycles at which the specimen failed. The analysis of these graphs provides the basics for creating calculations that address the fatigue phenomenon. Figure 1-8.2 shows a typical graph that resembles the actual graphs expressing the relationship between the alternating stresses at pure bending and the number of cycles that the specimen endures before failing.

The alternating bending stresses S_a are plotted on the ordinate axis, while the abscissas represent the numbers of corresponding cycles (revolutions) N. The number of cycles symbolizes the duration of life of the specimen or machine element. It became acceptable to use the word "life" as a synonym for the number of cycles in association with the fatigue phenomenon. The words "life cycle" are associated with number of revolutions (cycles) that a specimen or machine element endures with no failure. Obviously, if the alternating stress could have the value of the ultimate strength S_u (or S_{ut}) then the specimen would fracture immediately (at zero revolutions). This is shown in Figure 1-8.2, where at point A the alternating stress has the value of the ultimate strength S_u (S_{ut}) and the abscissa of this point equals zero. The curved line of graph ABC shows that decreasing values of the alternating stresses S_1, S_2, and S_3 leads to the increasing of the number of cycles N_1, N_2, and N_3 at which the specimens become fractured. According to Figure 1-8.2, the typical graph describing the relationship between the number of cycles and fatigue strength starting from a certain point (point B) having the abscissa equal

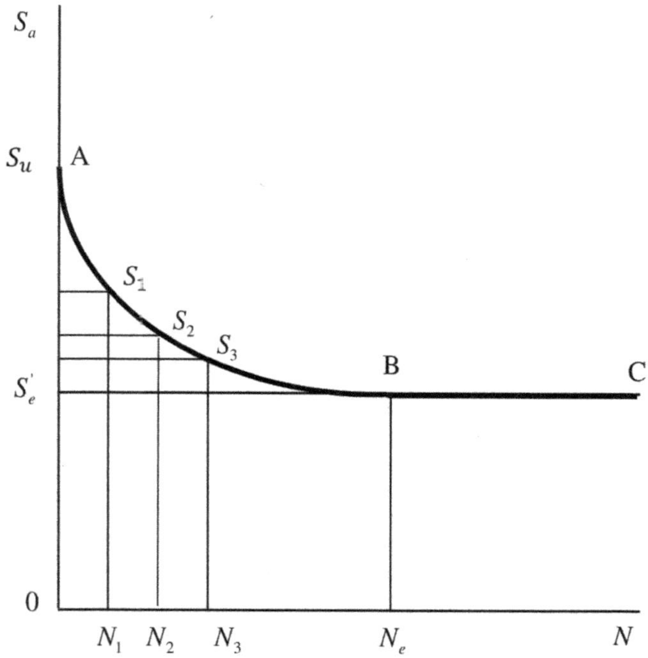

Figure 1-8.2. Typical graph approximating the results of fatigue testing of alternating bending stresses S_a versus the number of cycles N.

to N_e continues as a parallel line to the abscissa axis. So, the segment BC indicates that starting from point B increasing of the number of cycles does not result in fracturing of the specimen. This means that for this specimen there exists a certain level of the alternating stresses that if being loaded below this level this specimen will never fail. Hence, the value of the alternating stress below which the specimen will never fail regardless of the number of the loading cycles is called the endurance limit and is denoted by S_e', as shown in Figure 1-8.2. The prime mark at the notation letter S of the endurance limit indicates that the experimentally obtained value of this limit belongs just to this particular specimen, not to any machine element that is made of the same material as this specimen. As we know, the results of ordinary tensile testing of a steel specimen are applicable to any machine element made of the same steel. However, the fatigue phenomenon in this regard is very different. The experimental investigations revealed that the endurance limit depends on a variety of factors that are not limited to the mechanical properties of the material, but include characteristics or factors that are related to actual machine elements and types of their dynamic loading. These factors are discussed below.

Nevertheless, the existence of an endurance limit presents numerous opportunities for developing appropriate methodologies that indicate the ways to prevent fatigue failure.

It is interesting to note that the investigations revealed that materials possess a specific type of "memory." They "remember" the number of loading cycles that they were subjected to; however, they are not "sensitive" to the frequency of these cycles. They are sensitive to the maximum and minimum values of the stresses during the cycle; however, they do not "pay attention" to the intermediate values of the stresses. They "memorize" the number of the stress changes regardless of the order they were imposed. For example, if a shaft in an assembly is loaded just by a torque and is rotating all the time in the same direction, the shaft is subjected, actually, to steady state (static) loading (the torsional stress in the shaft represents a static stress). However, if the shaft periodically works in the reverse mode, then the material "memorizes" that it was stressed in both directions, and in this case, the shaft is subjected to the fatigue phenomenon.

It is accepted to divide the spectrum of loading cycles shown in Figure 1-8.2 into two categories: the low life cycle, which includes the range from zero to 10^3 cycles; and the high life cycle, which starts at 10^3 cycles and goes up to the end of the right-hand side of the graph (infinity). According to some authors, the low life cycle is characterized by a reduced ultimate strength, and at 10^3 cycles the ultimate strength is lowered to the level of about 90%. However, it is acceptable in most cases to perform the corresponding strength calculations for the low life cycle based on the methodologies of static loading; in other words, to consider that the ultimate strength is not lowered. As several authors suggest, the range between 10^3 and 10^6 cycles is considered the finite life cycle, and from 10^6 to the right-hand end of the graph (infinity) is the infinite life cycle. It is obvious that the required duration of life of some machine elements, even those that belong to the same mechanical system, may significantly differ from each other. For example, during its life, the door axle of a car will be subjected to the number of cycles that is much less than 10^6, while a water pump shaft will be subjected to the number of cycles that exceeds 10^6 by far. Therefore, in the same machine the elements could be characterized by the low life, the finite life, and the infinite life cycles. It would be unjustifiable to calculate the strength of all these machine elements, applying to them the same most conservative methodology of strength calculations. For each of these life cycles there should be corresponding methodologies for strength calculations. Hence, the methodologies of engineering calculations for all of these types of cycles are discussed in detail in this book.

1-8.2. The Endurance Limit

The study of the fatigue phenomenon is far from complete. Some of the existing data are already on solid ground, while others are still in progress. It would be safe to say that these types of experimental studies are always in progress. With respect to the endurance limit of ductile steel specimens subjected to pure bending, the accumulated data are consistent and represent a reliable basis to build appropriate engineering calculations. Numerous experimental investigations with steel specimens subjected to alternating stresses at bending revealed that the endurance limits for these specimens are in the range between 10^6 and 10^7 loading cycles. For machine elements made from steel, several published sources recommend using the following expression to estimate the values of the endurance limits of the specimens alternating bending:

$$S'_e = \begin{cases} 0.5S_{ut} \text{ for } S_{ut} \le 200 \text{ ksi (1,400 MPa)} \\ 100 \text{ ksi (700 MPa) for } S_{ut} > 200 \text{ ksi (1,400 MPa)} \end{cases} \tag{1-8.1}$$

It should be clear that the alternating loading causes the same absolute values of the maximum and minimum stresses. For the low and medium carbon steels that are usually used for machine elements for general applications, the yield strength averages about $0.7S_{ut}$, hence, the endurance limit is about $S'_e = 0.714S_{yt}$. Therefore, while using these values of endurance limits according to expression (1-8.1), it is necessary to realize that the nature of testing of the fatigue characteristics is much more complicated than the tensile testing, and it is justifiable to apply increased safety factors in fatigue calculations in comparison with the calculations for static loading.

Not all engineering materials possess definitive endurance limits; many of the materials will eventually fail due to the fatigue phenomenon. These materials include most non-ferrous metals, plastics, and others. For general information, we may note that some authors suggest that some aluminum alloys and bronze possess certain fatigue strength that in some way resembles the endurance limit. Appropriate experimental investigations allow determining the duration of life (in terms of loading cycles) of the machine elements made of materials that do not exhibit a definite endurance limit. Based on the appropriate strength calculations, it is possible to design machine elements made of these materials for a required life duration.

1-9. Factors Influencing the Endurance Limit

The endurance limit of an actual machine element differs significantly from the endurance limit of a specimen made of the same material. The prime mark assigned to the endurance limits S'_e of specimens emphasizes the differences between the endurance limits of specimens and the endurance limits of actual machine elements that are denoted by S_e (without a prime mark). In static loading, the yield strength S_y as well as the ultimate strength S_u of the material does not depend on the peculiarities of a particular machine element. The characteristics of strength obtained for the specimen during its tensile testing remain the same for the machine element made of the same material. However, in cases of dynamic loading that impose variable stresses on the element, the material of this element experiences, in most cases, a decline in its strength compared with the strength of the specimen. Some factors that do not play any role during the static loading of an element become active when the element is subjected to dynamic loading, and they reduce the endurance limit of this element.

Numerous experimental investigations related to the fatigue phenomenon provide a significant amount of data related to machine elements for general applications. An essential amount of fatigue testing was conducted with round bars resembling shafts that actually represent one of the most widespread machine elements for general applications. When considering shafts as machine elements for general applications, we should note that, in most cases, they have a stepwise shape; they are machined from a certain group of steels; and they have a corresponding range of dimensions. With this in mind, it is possible to focus on data that are more appropriate for the shafts representing one of the machine elements for general applications. However, the fatigue phenomenon data related to the shafts is, for the most part, applicable for fatigue calculations of machine elements that have some structural similarities with round and rectangular bars.

Several factors influence the endurance limit of actual machine elements, including the surface finish of the element. The lower the rate of roughness of the surface the less the reduction of the endurance limit occurs. The surface of a standard specimen is polished, and it is accepted that this type of surface finish does not affect the level of fatigue strength. It is also observed that for the same surface finish the ultimate strength of the component significantly influences the endurance limit. So, the higher the ultimate strength of the element the lower is its endurance limit. However, this tendency is correct to a certain value of the ultimate strength for steel (200 ksi), while above this value the further increase of ultimate strength does not result in further decrease of the endurance limit. It is also established that the size of the cross-sectional areas of the elements compared to the size of a specimen plays a sensible role in the fatigue strength of the element. Hence, increasing the cross-sectional areas of the elements results in a corresponding decrease in their endurance limits. There exists a similar tendency in the behavior of some materials subjected to static loading. The type of dynamic loading is also among the factors that influence the endurance limit. We consider three types of dynamic loading: pure bending, axial (push-pull) loading, and pure torsion. Each of these loading types has a different influence on endurance limits.

The operational temperature of the elements and some other factors influence the endurance limit of actual elements.

All factors mentioned above can be accounted for in the following expression known as Marin's equation that modifies the endurance limit of a machine element:

$$S_e = k_a k_b k_c k_d k_e S_e' \tag{1-9.1}$$

where k_a is the surface finish factor, k_b is size factor, k_c is the load factor, k_d is the temperature factor, k_e is the miscellaneous-effects factor. All of these factors have values that are less than unity.

One more factor, namely, the stress concentration factor plays an important role in stress calculations of a machine element. Sometimes the inverse value of this factor is considered an endurance limit reduction factor and could modify equation (1-9.1). Actually, the fatigue stress concentration factor increases the acting stress in the machine element and, consequently, should be used as a corresponding correcting (increasing) coefficient related to the acting stress. However, in some cases it does not make a difference if the inverse value of the fatigue stress concentration factor was used to reduce the endurance limit or the fatigue stress concentration factor was used to increase the acting stress. The fatigue stress concentration factor is discussed below in a separate section in order to consider the different circumstances for accounting for this factor, depending on type of loading.

1-9.1. The Surface Finish Factor k_a

Numerous experimental investigations of round bars resembling shafts indicate that the endurance limit of the bar depends on the combination of its ultimate strength and the manufacturing method applied to the outside surface of the bar. The lower the quality of the surface finish and the higher the ultimate strength, the higher is the reduction rate of the endurance limit. It should be noted that shafts and other machine elements of general applications are manufactured of steels having the ultimate strength that rarely exceeds 200 ksi (1.400 MPa). In addition, the outside surfaces of the shafts and other machine elements for general applications are usually machined. The available experimental data allow us to estimate the surface finish factor for machined elements by using the following empiric expression:

$$k_a = \begin{cases} 1.0 & S_{ut} < 41\text{ ksi} \\ 1.0 - 0.00337(S_{ut} - 41) & 41\text{ ksi} \le S_{ut} < 100\text{ ksi} \\ 0.8 - 0.00140(S_{ut} - 100) & 100\text{ ksi} \le S_{ut} \le 200\text{ ksi} \\ 0.66 & S_{ut} > 200\text{ ksi} \\ 1.0 & S_{ut} < 280\text{ MPa} \\ 1.0 - 0.0004884(S_{ut} - 280) & 280\text{ MPa} \le S_{ut} < 700\text{ MPa} \\ 0.8 - 0.0002029(S_{ut} - 700) & 700\text{ MPa} \le S_{ut} \le 1{,}400\text{ MPa} \\ 0.66 & S_{ut} > 1{,}400\text{ MPa} \end{cases} \quad (1\text{-}9.2)$$

1-9.2. The Size Factor k_b

Engineering materials such as steel, cast iron, aluminum, and others demonstrate the decrease in the endurance limit of the element being subjected to all kinds of variable stresses, while the size of the element increases. Therefore, it is necessary to estimate the influence of the size of a machine element on its endurance limit. There are some corresponding approaches to account for the size of a component by using an appropriate size factor.

As shown in the Figure 1-8.1, the minimum diameter of a standard specimen is 0.3 in., or 8 mm; it is considered that the size factor equals unity for diameters of 0.3 in. (8 mm) or less. In some earlier published sources, it is suggested that for bending and torsion stresses the factor may be $k_b = 0.85$ for the range of diameters of round bars from 0.3 in. to 2.0 in., while for lager diameters this factor should be equal to 0.75, and for axial stresses the size factor equals unity. The later sources contain different empiric formulas for calculating the values of this factor for diameters up to 10 in. It is reasonable to assume that in the engineering practice associated with machine elements for general applications the diameters of shafts would rarely exceed 6.0 in. Therefore, this book considers empiric formulas dealing with the size factors for shafts and round bars with diameters up to 6.0 in. subjected to all kinds of variable stresses. These empiric formulas are presented below:

$$k_b = 1\begin{cases} 1.0 & d < 0.3\text{ in.} \\ 1.0 - 0.185(d - 0.3) & 0.3\text{ in.} \le d < 1.0\text{ in.} \\ 0.87 - 0.046(d - 1.0) & 1.0\text{ in.} \le d \le 6.0\text{ in} \\ 1.0 & d < 8\text{ mm} \\ 1.0 - 0.0073(d - 8) & 8.0\text{ mm} \le d < 25.4\text{ mm} \\ 0.87 - 0.018(d - 25.4) & 25{,}4\text{ mm} \le d \le 152\text{ mm} \end{cases} \quad (1\text{-}9.3)$$

where d is the diameter of a round bar.

It should be mentioned that in some cases it is necessary to determine the size factor for a component having a rectangular cross-sectional area. It is suggested for these cases to take as the equivalent diameter the smallest side of the perimeter of this area. Thus, if a rectangular bar, for example, has the cross-sectional area of 0.8·4.5 in^2 , than $k_b = 1.0 - 0.185(0.8 - 0.3) = 0.907$.

1-9.3. The Load Factor k_c

A round bar could be subjected to three types of variable stress that are characterized by maximum and minimum stresses per cycle, while in each type of stress the maximum and minimum stresses have the same absolute values. These types of variable stresses are bending, axial, and torsional. As mentioned above, conventional fatigue testing is associated with pure bending, and the endurance limit of the specimen is based on this testing. The endurance limit in this case can be presented by the following formula: $S_e' = k_c 0.5 S_{ut}$, where k_c is the loading factor. Consequently, for pure bending the loading factor $k_c = 1$.

Machine elements for general applications such as threaded fasteners could be subjected to axial dynamic loading. Several authors report that in a case of variable axial loading that generates the same absolute values of the maximum and minimum stresses in the element the loading factor is in the range from 0.7 to 0.9. Averaging these numbers we assign the loading factor $k_c = 0.8$.

With regard to the variable pure torsion, several authors consider that the loading factor may be taken in range from 0.577 to 0.59. We take for pure torsion loading factor $k_c = 0.577$.

However, in cases of combined action of variable bending and torsion, it is recommended to take the loading factor $k_c = 1$. Usually, the equivalent stress is calculated based on the fourth theory of failure.

Based on the considerations presented above, here is the expression for determining the loading factor depending on the type of loading:

$$k_c = \begin{cases} 1 & \text{for bending and combined bending and torsion} \\ 0.8 & \text{for axial} \\ 0.577 & \text{for pure torsion} \end{cases} \qquad (1\text{-}9.4)$$

1-9.4. The Temperature Factor k_d

For the operational conditions of machine elements characterized by temperatures below 400°F (204°C), the temperature factor equals unity. Actually, the operational conditions of machine elements for general applications are usually characterized by temperatures below 400°F and, consequently, we may accept $k_d = 1$.

1-9.5. Miscellaneous-Effects Factor k_e

It is known that there are additional factors that affect the actual endurance limit of machine elements. However, the quantitative characteristics of these factors are not available at this time and cannot be accounted for in the corresponding calculations. Therefore, in this course we take $k_e = 1$.

Because of the considerations related to the factors influencing the endurance limits of machine elements, we modify the equation (1-9.1):

$$S_e = k_a k_b k_c S_e' \tag{1-9.5}$$

Example 1-9.1a: Calculating endurance limits.
A shaft having the diameter d = 1.5 in. is made from AISI 1030 steel having S_{ut} = 80 ksi. We determine the endurance limit for alternating bending, axial loading, and pure torsion.

Solution:

1) Bending.
Using expressions (1-9.1), (1-9.2), (1-9.3), and (1-9.4), we determine $S_e' = 0.5S_{ut} = 0.5 \cdot 80 = 40$ ksi, $k_a = 0.869$, $k_b = 0.867$, and $k_c = 1$.
Based on the equation (1-9.5), we have $S_e = 0.869 \cdot 0.867 \cdot 1 \cdot 40 = 30.16$ ksi.
2) Axial loading.
The S_e', k_a, and k_b are the same as for bending. From expression (1-9.4), we determine $k_c = 0.8$. Therefore, we have $S_e = 0.869 \cdot 0.867 \cdot 0.8 \cdot 40 = 24.13$ ksi.
3) Pure torsion.
From the expression (1-9.1), we have $k_c = 0.577$. The rest is the same as in the previous cases. Thus, we write $S_e = 0.869 \cdot 0.867 \cdot 0.577 \cdot 40 = 17.40$ ksi.

Example 1-9.1b: Determining safety factors.
For the shaft in Example 1-9.1a, we calculate the safety factor using the following:

(1) bending stresses where the maximum bending moment M_{max} = 4,800 lb-in;
(2) normal stresses caused by axial loading characterized by the maximum axial load F_{max} = 17,800 lb;
(3) torsional stresses where the maximum torque T_{max} = 2,600 lb-in.

Solution:

1) Bending.
Using formulas (1-4.17) and (1-4.18), we calculate the bending stress:
$\sigma_{max} = \dfrac{64 \cdot 4{,}800 \cdot 0.5 \cdot 1.5}{\pi 1.5^4} = 14{,}490$ psi. Since the safety factor $n = \dfrac{S_e}{\sigma_{max}}$, we have
$n = \dfrac{30.16}{14.49} = 2.08$.
2) Axial loading.
Based on formula (1-4.1), we determine the normal stress
$\sigma_{max} = \dfrac{4 \cdot 17{,}800}{\pi 1.5^2} \dfrac{lb}{in^2}$ = 10,080 psi.
Therefore: $n = \dfrac{24.13}{10.08} = 2.40$.

3) Torsion.

Recalling formulas (1-4.20) and (1-4.21), we calculate the torsion stress as

$$\tau_{max} = \frac{32 \cdot 5{,}600 \cdot 0.5 \cdot 1.5}{\pi 1.5^4} = 8{,}460 \text{ psi}$$

Hence, $n = \frac{17.40}{8.46} = 2.057$.

Example 1-9.1c: Calculating the safety factor for a shaft in plane stress state conditions. For the same shaft from the previous examples and using the fourth theory of failure, we determine the safety factor if the shaft is simultaneously subjected to alternating bending M_{max} = 3,500 lb-in. and alternating torsion T_{max} = 3,100 lb-in.

Solution:

Contrary to the previous cases, the shaft is subjected to plane stress state conditions that are characterized by bending and torsional stresses. As emphasized above, the loading factor should be taken as for bending, therefore $k_c = 1$, while the endurance limit is the same as in previous cases, S_e = 30.16 ksi. Based on formulas (1-4.17) and (1-4.18), we determine that $\sigma_{max} = \frac{64 \cdot 3{,}500 \cdot 0.5 \cdot 1.5}{\pi 1.5^4} = 10{,}569 \text{ psi}$. Using formulas (1-4.20) and (1-4.21), we calculate

$$\tau_{max} = \frac{32 \cdot 3{,}100 \cdot 0.5 \cdot 1.5}{\pi 1.5^4} = 4{,}680 \text{ psi}$$

Modifying the expression (1-6.10) by replacing S_y with S_e, and σ and τ with σ_{max} and τ_{max} respectively, we write $\sigma_{eq} = \sqrt{\sigma_{max}^2 + 3\tau_{max}^2} = \frac{S_e}{n}$. Hence: $n = \frac{S_e}{\sigma_{eq}}$. Substituting the corresponding values, we have $\sigma_{eq} = \sqrt{10{,}569^2 + 3 \cdot 4{,}680^2} = 13{,}320 \text{ psi}$. Therefore,

$$n = \frac{30.16}{13{,}320} = 2.264$$

1-10. Stress Concentration

It is well known from the earlier studies (Strength of Materials, Mechanics of Materials) that in places of discontinuity of the geometric shape of the material a significantly increased stress in comparison with the nominal stress appears. The discontinuity in the shape can be characterized by the presence of a notch, by a step in the shaft diameters, by a radial hole or keyway in the shaft, and others. The local increase of the stress in the material due to the discontinuity of its shape represents the stress concentration, and the discontinuity by itself is known as a stress raiser. However, ductile and some brittle steels under static loading are practically not sensitive to stress concentration. In cases of dynamic loading, stress concentration becomes one of the most critical factors influencing the increase of the actual stresses.

Figure 1-10.1 presents a stepped shaft having diameters D and d and a fillet of radius r, while the shaft is subjected to an axial loading force F. This figure shows a diagram of normal stresses that are developed in the part of the shaft having the smaller diameter d. The normal stress σ_0 represents the nominal (calculated) stress that is equal $\sigma_0 = \frac{4F}{\pi d^2}$, while σ_{max} is the stress that acts on the periphery of the shaft caused by the discontinuity of the shape of the shaft. The stress σ_{max} represents the result of stress concentration. The stress σ_{max} can be determined experimentally or in some cases analytically. The ratio between σ_{max} and σ_0 is called the theoretical stress concentration factor K_t, thus

$$K_t = \frac{\sigma_{max}}{\sigma_0} \tag{1-10.1}$$

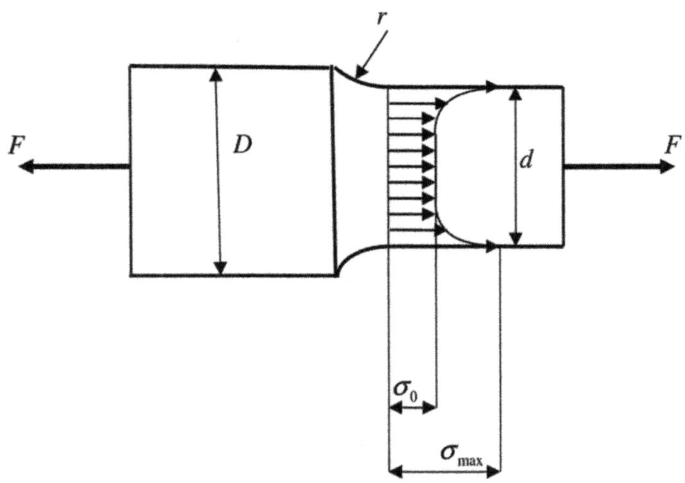

Figure 1-10.1. Example of stress concentration.

In Appendix A-1-5 there is a set of graphs that determine the values of K_t for particular cases. Fatigue stress concentration factor K_f is calculated from the following formula:

$$K_f = 1 + q(K_t - 1) \tag{1-10.2}$$

where q is the notch sensitivity factor that depends on the notch radius and ultimate strength of the material. For machine elements for general applications made of ductile steels, it is justifiable to accept that $q = 0.6$ keeping in mind the recommendations of applying elevated safety factors to these elements. Since $K_t > 1$, hence $K_f > 1$, and usually K_f is used as a multiplier to the actual stress of the element. Since the theoretical and, consequently, the fatigue stress concentration factors have different values for bending, torsion, and axial loading, it is more reasonable to account for the stress concentration phenomenon during the calculations of actual stresses. For example, if a shaft has a stress raiser and it is subjected to the action of alternating bending and alternating torsion, the fatigue stress concentrations factors for bending and torsion stresses should be determined, and these stresses should be increased by multiplying them by these factors respectively.

In addition, in cases when two or more raisers are acting simultaneously, determine the raiser that results in the larger fatigue stress concentration factor. The examples below show how to determine the stress concentration phenomenon in the stress calculations.

The graphs in the Appendix A-1-5 contain the corresponding drawings characterizing the discontinuity of the shape and the type of loading.

Example 1-10.1a: Stress concentration.

A stepped shaft having $D = 1.2$ in., $d = 1.0$ in., and $r = 0.1$ in. is made of AISI 1340 steel having $S_{ut} = 121{,}250$ ksi, and it is loaded by the following alternating loads: bending moment $M = 1{,}150$ lb-in. and torque $T = 900$ lb-in. We determine the safety factor of the shaft using the fourth theory of failure.

Solution:

Starting by determining the endurance limit , we calculate $k_a = 0.912$, $k_b = 0.87$, and $k_c = 1$. Thus, $S_e = 0.912 \cdot 0.87 \cdot 1 \cdot 0.5 \cdot 121{,}25 = 48.1$ ksi. In order to use the charts from Appendix A-1-5 for estimating the theoretical stress concentration factor K_t, it is necessary to determine the following ratios: $\frac{D}{d} = \frac{1.2}{1.0} = 1.2$ and $\frac{r}{d} = \frac{0.1}{1.0} = 0.1$. Referring to Appendix A-1-5 (Figure 3), we determine that for bending loading $K_t = 1.75$. Similarly, using Appendix A-1-5 (Figure 2), we found that for torsion loading $K_t = 1.35$. Taking the recommended value of the notch sensitivity factor $q = 0.6$ and using the equation (1-10.2), we calculate the fatigue stress concentration factors for bending $K_{f\,bend} = 1 + 0.6(1.75 - 1) = 1.45$ and for torsion $K_{f\,tor} = 1 + 0.6(1.35 - 1) = 1.21$. As we can see, the factors are different for bending and torsion and cannot be used to modify the calculated endurance limit. Therefore, the bending and torsional stresses should be increased according to their fatigue stress concentration factors.

Using formulas (1-4.17) and (1-4.18), we calculate the bending stress (it is not necessary to emphasize this as the maximum stress): $\sigma = \frac{64 \cdot 1{,}150 \cdot 0.5 \cdot 1.0}{\pi 1.0^4} = 11{,}720$ psi. Multiplying this result by the fatigue stress concentration factor for bending we determine the effect of stress concentration $\sigma_{\max} = \sigma K_{f\,bend} = 11{,}720 \cdot 1.45 = 16{,}994$ psi. Applying formulas (1-4.20) and (1-4.21), we determine the torsional stress $\tau = \frac{32 \cdot 900 \cdot 0.5 \cdot 1.0}{\pi 1.0^4} = 4{,}586$ psi. Similarly we increase the torsional stress: $\tau_{\max} = \tau K_{f\,tor} = 4{,}586 \cdot 1.21 = 5{,}549$ psi. Applying expression (1-6.10), we determine the equivalent stresses according to the fourth failure theory $\sigma_{eq} = \sqrt{16{,}994^2 + 3 \cdot 5{,}549^2} = 19{,}524$ psi. The safety factor can be calculated using the following formula: $n = \frac{S_e}{\sigma_{eq}} = \frac{48.1}{19{,}524} = 2{,}464$.

It may be noted that in case of alternating bending in combination with a steady state (static) torsion we should apply the fatigue stress concentration factor just to the bending stress. In this case, a stress concentration modifier k_f for the formula (1-9.5) can be used:

$$k_f = \frac{1}{K_f} \tag{1-10.3}$$

Therefore, for certain situations the formula (1-9.5) can be modified:

$$S_e = k_a k_b k_c k_f S'_e \tag{1-10.4}$$

Example 1-10.1b: Multiple stress concentration raisers.
Figure 1-10.2 shows a rotating stepped shaft having a hole for a pin. The shaft is loaded by a bending moment M and torque T. The rotation of the shaft is reversible. Determine the corresponding fatigue concentration factors if $\frac{D}{d}1.2, \frac{r}{d} = 0.05$, and $\frac{a}{d} = 0.25$.

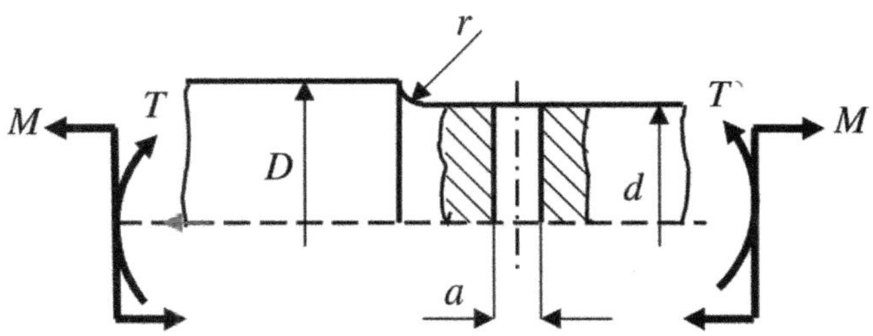

Figure 1-10.2. Multiple stress concentration raisers.

Solution:
Based on the Appendix A-1-5 (Figures 2 and 4) we determine the theoretical stress concentration factors for torsion due to the step $K_{tt1} = 1.6$ and due to the hole $K_{tt2} = 2.65$. Applying the notch sensitivity factor $q = 0.6$ and taking the larger theoretical stress concentration factor, we determine the torsion fatigue stress concentration factor $K_{ft} = 1 + 0.6(2.65 - 1) = 1.99$.

Now using Appendix A-1-5 (Figures 3 and 7), we determine the bending theoretical stress concentration factor due to the step $K_{tb1} = 1.90$ and due to the hole $K_{tb2} = 1.95$. Taking the larger theoretical stress concentration factor, we calculate the bending fatigue stress concentration factor $K_{fb} = 1 + 0.6(1.95 - 1) = 1.57$.

1-11. Types of Variable Stress Cycles

The variable (alternating) loading that is considered above has actually the same absolute values of the maximum and minimum stresses acting on the materials of the components. However, this is not always the case in the real-life operational conditions of machine elements. Very often, the maximum and minimum absolute values of stresses imposed on the elements differ from each other during the loading cycles mentioned above. The materials "remember" only the peaks of the stresses (the maximums and minimums), and they are not sensitive to the intermediate values of the stresses nor the frequencies of the peaks. All this allows us to consider three types of variable stresses and present them as graphs resembling harmonic curves, and reflecting the change of these stresses as functions of time. Hence, the variable stresses can be presented by the following types of stress cycles: completely reversed cycle, repeated cycle, and random cycle. Thus, Figure 1-11.1 represents corresponding graphs of these three basic types of dynamic stresses.

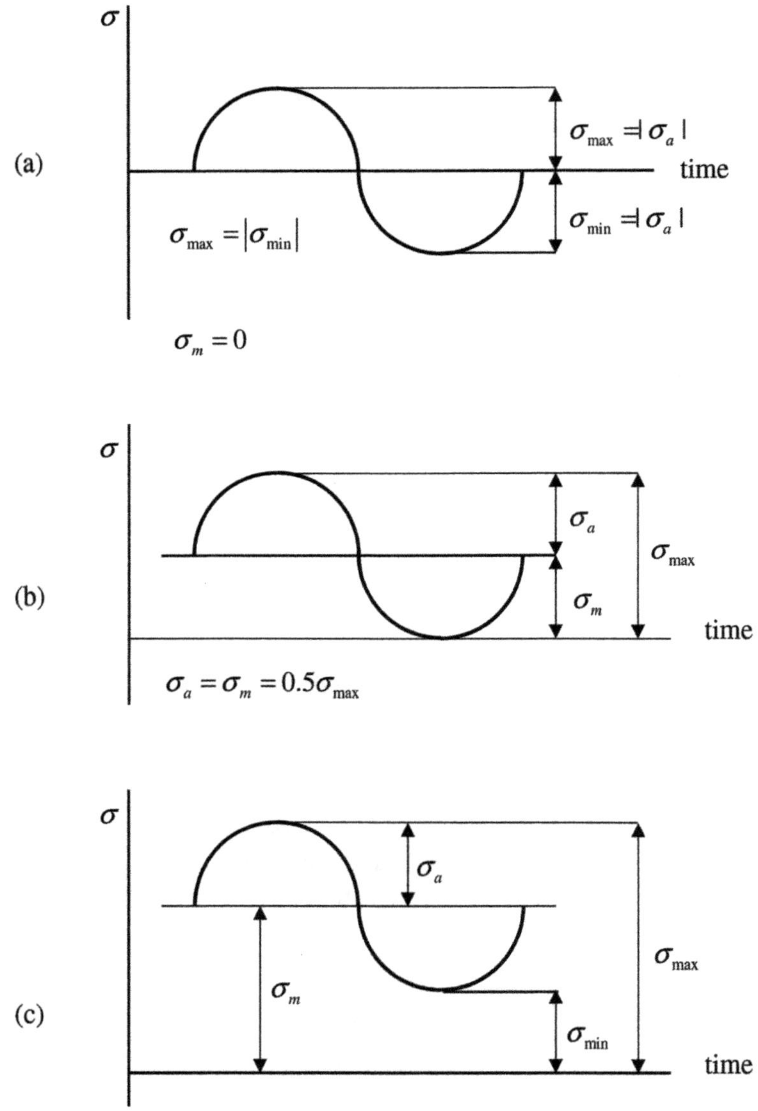

Figure 1-11.1 Basic types of variable stresses.

As mentioned above, the existing fatigue testing equipment develops alternating stresses having the same absolute values of their maximums and minimums in the specimens, while as it is mentioned above in the real-life operational conditions, the absolute values of the maximum and minimum stresses are not always equal to each other.

Figure 1-11.1a shows a graph of a completely reversed cycle that is also known as a reversed or symmetric cycle. This cycle is characterized by the same absolute values of the maximum and minimum stresses : $\sigma_{max} = |\sigma_{min}|$. The fatigue testing equipment executes this cycle for determining the endurance limits and the modifying factors for pure bending stress, for axial stress, and for pure torsional stress. Therefore, all information related to the fatigue phenomenon discussed above is obtained based on the completely reversed cycle.

Figure 1-11.1b shows a cycle in which the minimum stress equals zero, $\sigma_{min} = 0$. This cycle is known as a repeated cycle or as a fluctuating cycle. However, the word "fluctuating" is sometimes applied to any variable type of loading. A graph of a random cycle is shown in Figure 1-11.1c that is characterized by different values of the extreme peaks of stresses, hence, $\sigma_{max} \neq \sigma_{min} \neq 0$.

The maximum and minimum stresses as the characteristics of loading cycles do not reflect some important features of the cycles. In addition to the characteristics of maximum and minimum stresses, there are two more parameters of the cycles that allow for analyzing these cycles in a more adequate way. These parameters, or characteristics, are called the mean stress σ_m and stress amplitude σ_a that can be determined using the following formulas:

$$\sigma_m = \frac{\sigma_{max} + \sigma_{min}}{2} \tag{1-11.1}$$

and

$$\sigma_a = \frac{\sigma_{max} - \sigma_{min}}{2} \tag{1-11.2}$$

Based on formulas (1-11.1) and (1-11.2), we may write that the absolute values of the maximum and minimum stresses are

$$|\sigma_{max}| = |\sigma_m| + |\sigma_a| \tag{1-11.3}$$

and

$$|\sigma_{min}| = |\sigma_m - |\sigma_a| \tag{1-11.4}$$

Actually, the mean stress represents a static stress and the stress amplitude represents an alternating stress. Therefore, any cycle of variable stresses can be characterized by a superposition of a static stress σ_m and an alternative stress σ_a .

It should be noted that all these definitions and related formulas are completely applicable to tangential stresses, while τ_m and τ_a are, respectively, the mean shear stress and the shear stress amplitude.

Let us calculate the values of the mean stress and stress amplitude for each of these three cycles.

1-11.1. Calculating the Mean Stress and Stress Amplitude for a Completely Reversed Cycle

As shown in Figure 1-11.1, this cycle does not have a mean stress since $\sigma_{max} = |\sigma_{min}|$, and according to formula (1-11.1), we may write $\sigma_m = \dfrac{\sigma_{max} + (-\sigma_{min})}{2} = 0$. Based on formula (1-11.2) we determine $\sigma_a = \dfrac{\sigma_{max} - (-\sigma_{min})}{2} = \sigma_{max}$; therefore, the stress amplitude for this cycle equals the maximum stress.

The completely reversed cycle can be presented as a superposition of a zero mean stress and the stress amplitude, as shown in Figure 1-11.1a.

1-11.2. Calculating the Mean Stress and Stress Amplitude for a Repeated Cycle

This cycle represents a nonzero mean stress cycle. In a repeated cycle, the minimum stress equals zero, while $\sigma_m = \dfrac{\sigma_{max}}{2}$ and $\sigma_a = \dfrac{\sigma_{max}}{2}$. Figure 1-11.1b shows that this cycle can be characterized by the superposition of a constant stress σ_m and an alternative stress σ_a, while $\sigma_a = \sigma_m$.

1-11.3. Calculating the Mean Stress and Stress Amplitude for a Random Cycle

In a random cycle, the maximum and minimum values are determined upon their algebraic value. Therefore, the minimum stress may exceed the maximum stress based on their absolute values. This cycle and the repeated cycle represent the nonzero cycles. Let us consider an example, where $\sigma_{max} = 25$ psi and $\sigma_{min} = -95$ psi. Therefore, $\sigma_m = -35$ psi and $\sigma_a = 60$ psi. The maximum value of the stress in this case according to formula (1-11.3) is $|\sigma_{max}| = |35| + |65| = |95|$ psi.

Figure 1-11.1c illustrates a random cycle that represents a superposition of the mean stress σ_m and the stress amplitude σ_a. It should be clear that the characteristics of a random cycle could have positive as well as negative values.

As a general note, in cycles of variable stresses the stress amplitude cannot be equal to zero, while the mean stress can.

1-12. Effect of the Mean Stress on the Endurance Limit

As it was mentioned above, fatigue testing is carried out for alternating stresses having the same absolute values for their maximum and minimum stresses. In other words, all these tests were conducted based on a completely reversed cycle that is characterized by absence of the mean stress ($\sigma_m = 0$). However, in many real-life situations associated with fatigue, the mean stress is not equal to zero, such as in repeated and random cycles. The question arises: how do you determine the endurance limit of a specimen or a machine element when the mean stress exists?

In such cases, the experimental testing should involve two variables: the mean stress σ_m and the stress amplitude σ_a. This resembles the situation with the plane stress state that is caused by σ_1 and σ_2 and, as it was explained earlier, the testing would involve an endless number of pairs of stresses. It is impossible to carry out such testing in a finite interval of time. As mentioned above, the solution for the plane stress state analysis is based on applying justifiable assumptions that are supported by experimental data. Thus, the experimental way of determining the corresponding endurance limits of values σ_m and σ_a is not achievable for the same reason as for the pairs of σ_1 and σ_2. In order to estimate the endurance limits for different materials and for different pairs of σ_m and σ_a, some authors collected numerous actual data of fatigue failures associated with real cases. According to these data, corresponding points characterizing the pairs of σ_m and σ_a at which fatigue failure occurred where plotted on a system of coordinates where the stress amplitude σ_a (the ordinate) is presented as a function of the mean stress σ_m (the abscissa). Applying appropriate processing methods to the plotted points, different authors created diagrams characterizing the relationship between the stress amplitude σ_a and mean stress σ_m associated with fatigue failure. These diagrams are often used as criteria of fatigue strength.

Figure 1-12.1 shows the most common diagrams of fatigue criteria for the nonzero mean stress cycles. These diagrams are built on experimental data reflecting bending and axial loading cycles, resulting in positive values of normal stresses representing the mean stresses.

This figure represents the following three criteria of fatigue failure: the Soderberg line, the Goodman line, and the Gerber line. In order to comprehend the essence of the criteria of fatigue failure, we consider a random point A located on the intersection of the coordinates of the mean stress σ_m and the stress amplitude σ_a, representing the state of stresses in an infinitesimal cube of material of a machine element in conventional operational conditions. The purpose of the analysis of the random point A is to determine the value of the margin of safety or the safety factor

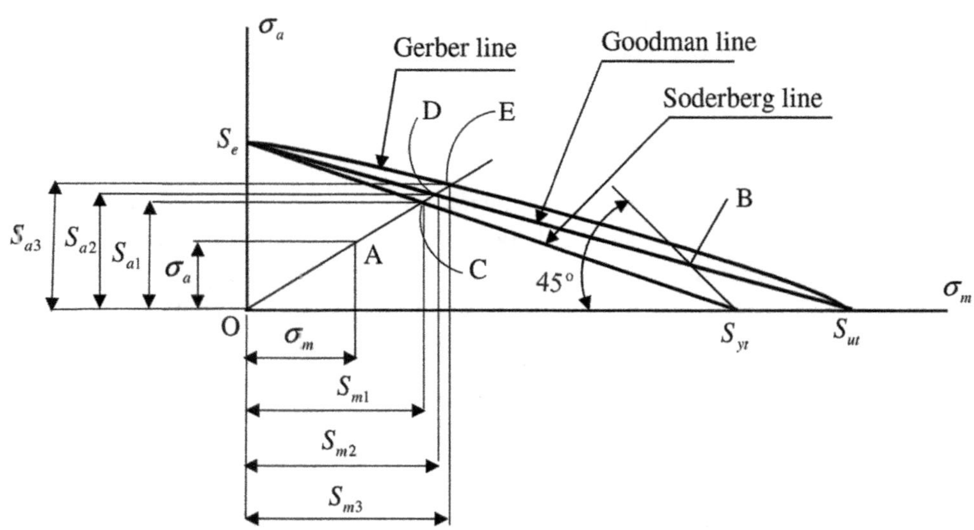

Figure 1-12.1. Fatigue strength diagrams for nonzero mean stress cycles.

of the strength of this element. The fatigue failure criteria are intended to provide the answer to this question.

We begin with the Soderberg line that connects the point corresponding to the approximated value of the endurance limit S_e on the ordinates axis with the point representing the yield strength S_{yt} on the abscissas axis. According to the Soderberg line, each stress state point on the diagram representing a certain pair of stresses σ_m and σ_a that is located in the triangular area contained between the Soderberg line and the coordinate axes is in a safe state stress zone regarding fatigue failure. In other words, the triangular area (S_eOS_{yt}) represents a safe zone free of fatigue failure. Obviously, the zone of failure is located above the Soderberg line. As we know, the safety factor can be determined from the following general formula: $n = \frac{S}{\sigma}$, where S is a certain strength limit and σ is the actual stress. However, in our case we have two actual stresses σ_m and σ_a, while the stress state of a machine element can be evaluated just by one safety factor. The appreciable characteristic of the criteria of fatigue failure consists in providing the possibility to determine one value of the safety factor that is appropriate for the two actual stresses σ_m and σ_a. Thus, in order to estimate the safety factor of the element subjected to the action of σ_m and σ_a, we extend the line OA to the intersection with the Soderberg line in point C. Any point on the line OC is within the safe zone with regard to fatigue failure. The line OA or OC as well as any extension of this line is called the load line. The mean stress S_{m1} and the stress amplitude S_{a1} represent the corresponding mean strength and amplitude strength limits of the machine element loaded by σ_m and σ_a. Therefore, the safety factor for this element is

$$n_1 = \frac{S_{m1}}{\sigma_m} = \frac{S_{a1}}{\sigma_a} \tag{1-12.1}$$

However, it is possible to consider additional options to evaluate the safety factors with regard to mean stress and stress amplitude. Figure 1-12.2 shows the additional fatigue strength limits S_m^* and S_a^* that relate to σ_m and σ_a.

According to these additional strength limits, two additional safety factors become available: $n_m^* = \frac{S_m^*}{\sigma_m}$ and $n_a^* = \frac{S_a^*}{\sigma_a}$. Obviously, $n_m \neq n_a$, this approach is associated with two different safety factors that separately evaluate the stress state of the machine element with regard to the mean stress (constant stress) and to the stress amplitude (alternating stress). This approach differs from the conventional approach that is based on a single safety factor. But more importantly, to the knowledge of the author this approach is not addressed in published sources and not used in engineering practice; however, it seems that this approach could have some value.

Analyzing the Goodman line (the initial version) in Figure 1-12.1, we see that a straight line connects the ordinate of the endurance limit S_e with the abscissa of the ultimate strength S_{ut}. According to the initial version of the Goodman line, the safe zone, or fatigue failure free zone, is in the triangular area ($S_e\ OS_{ut}$) contained by this line and the coordinate axes. Later, the Goodman line was combined with the yield stress line, shown in Figure 1-12.1, by a segment line inclined by an angle of 45° to the coordinate axes and having the abscissa S_{yt}. The resulting

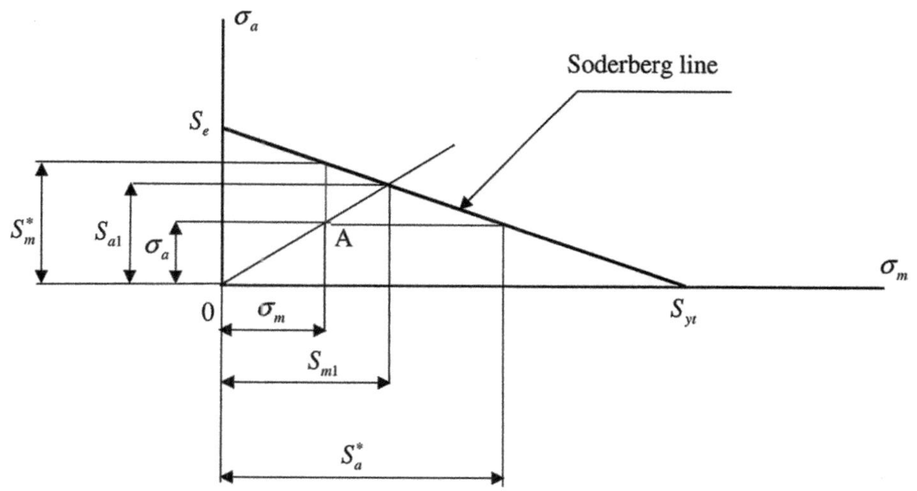

Figure 1-12.2. Additional fatigue strength limits.

broken line (S_eBS_{yt}) connecting the points of the endurance limit S_e , B, and the yield strength S_{yt} is called the Modified Goodman line. According to this line, the free of fatigue failure area is located between the coordinate axes and the broken line S_eBS_{yt} mentioned above. Extending the load line OA to the intersection with the Modified Goodman line in point D, we determine strength limits S_{m2} and S_{a2} that allow determining the safety factor of the machine element:

$$n_2 = \frac{S_{m2}}{\sigma_m} = \frac{S_{a2}}{\sigma_a} \tag{1-12.2}$$

Thus, according to the Modified Goodman line, the mean stress limit is S_{m2} and the stress amplitude limit is S_{a2} .

The Gerber criterion of fatigue failure, shown in Figure 1-12.1, represents a curved line (parabola) connecting the ordinate of the endurance limit S_e with the abscissa of the ultimate strength S_{ut}. According to the Gerber line, the free of fatigue failure area is located between the coordinate axes and the Gerber line. Extending the load line OA to the intersection with the Gerber line in point E, we determine the mean stress limit S_{m3} and the stress amplitude limit S_{a3} that allow us to calculate the safety factor for this case:

$$n_3 = \frac{S_{m3}}{\sigma_m} = \frac{S_{a3}}{\sigma_a} \tag{1-12.3}$$

Analyzing the diagrams in Figure 1-12.1 and the equations (1-12.1), (1-12.2), and (1-12.3), we see that $n_1 < n_2 < n_3$, which means that the approach associated with the Soderberg line results in design solutions for relatively smaller cross-sectional areas of the machine elements when compared with the other two criteria. The criteria of the Modified Goodman line and the Gerber line are considered more conservative, or in other words, they lead to stronger machine elements. The Modified Goodman line criterion is between the Soderberg line and Gerber line, and numerous authors report that the Modified Goodman line criterion is adequately supported by experimental data. This is the reason why in this book the Modified Goodman line is accepted as the criterion of fatigue strength. In addition, it should be stressed again that applying relatively higher safety factors in fatigue strength calculations compensates for the differences between the fatigue failure criteria.

1-12.1 Analysis of the Modified Goodman Line

The fatigue failure criterion based on the Modified Goodman line determines the stress amplitude limit S_a and the mean stress limit S_m based on algebraic formulas derived from the analysis of the geometric structure of the diagram presented in Figure 1-12.3.

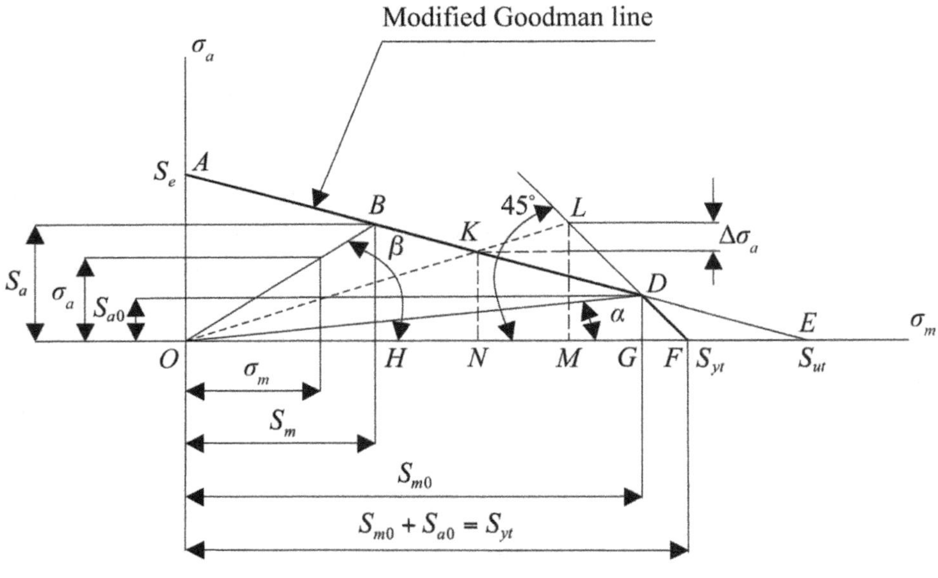

Figure 1-12.3. The Modified Goodman line diagram.

Based on the Modified Goodman line the methodology of solving nonzero mean stress fatigue problems consists of several steps that are presented below. It should be clear that the ratio between the stress amplitude limit S_a and the stress amplitude σ_a is similar to the ratio between the endurance limit S_e and the maximum stress σ_{max} for all stress or strength calculations in cases associated with the nonzero mean stress cycles. As for the completely reversed cycle, the safety factor $n_\sigma = \frac{S_e}{\sigma_{\max}}$, hence, for a nonzero mean stress cycle the safety factor is $n_\sigma = \frac{S_a}{\sigma_a}$. And since $\frac{S_a}{\sigma_a} = \frac{S_m}{\sigma_m}$, then $n_\sigma = \frac{S_a}{\sigma_a} = \frac{S_m}{\sigma_m}$.

Figure 1-12.3 shows that when the load line is inclined to the abscissa axis on the angle α and intersects with the point D that represents the intersection point of lines AF and LF, then $S_{m0} + S_{a0} = S_{yt}$, while the line segment $GF = S_{a0}$. However, it should be clear that in this case $\sigma_m + \sigma_a \leq S_{yt}$. This case reflects the situation when the stress amplitude is relatively small in comparison with the mean stress, while S_{m0} represents the maximum mean stress limit and S_{a0} represents the minimum stress amplitude limit. Hence, when the load line intersects the line segment DF, then the influence of the alternating stress on the strength of the machine element is very limited, and strength calculations of the corresponding machine element should be carried out based on the principles of static loading (without fatigue considerations). According to the triangle DOG, we may write $\tan\alpha = \dfrac{DG}{OG}$, $DG = GF = S_{a0}$, and $OG = OF - GF = S_{yt} - DG$. Considering the similarity of triangles AEO and DEG, we have $\dfrac{AO}{DG} = \dfrac{OE}{GE} = \dfrac{S_e}{DG} = \dfrac{S_{ut}}{S_{ut} - S_{yt} + DG}$. From this ratio we obtain that $DG = \dfrac{S_e(S_{ut} - S_{yt})}{S_{ut} - S_e}$. Based on expressions presented above, we may write

$$\tan\alpha = \frac{S_e(S_{ut} - S_{yt})}{S_{ut}(S_{yt} - S_e)} \tag{1-12.4}$$

Analyzing equation (1-12.4), it can be seen that tan α could be calculated as soon as the endurance limit is determined, while the angle α does not depend on actual values of the mean stress and stress amplitude.

Figure 1-12.3 shows that

$$\tan\beta = \frac{\sigma_a}{\sigma_m} \tag{1-12.5}$$

Thus, if for an actual machine element it occurs that $\beta < \alpha$ then the stress or strength calculations, as it is discussed above, should be based on the principles of static loading. It is acceptable at the beginning of the fatigue stress calculations to verify that, according to the first cycle rule of Langer, the sum of the mean stress and the stress amplitude does not exceed the yield strength of the material ($\sigma_a + \sigma_m < S_{yt}$). Based on Figure 1-12.3, it can be shown that when $\beta > \alpha$ the first cycle rule is confirmed. Consider the load line OK and its extension to the point L on the yield line. The angle between the line OL and abscissa axis is larger than α, while the sum of the mean stress limit (line segment ON) and the stress amplitude limit (line segment KL) is smaller by $\Delta\sigma_a$ than S_{yt} (since $LM = MF$). Therefore, while solving a problem using the Modified Goodman line, it is desirable during the initial steps of the solution to determine α and β, and if $\beta > \alpha$; we can proceed to the next steps of the calculations associated with the determining of S_m and S_a and there is no need to verify the rule of the first cycle.

It should be mentioned that for a repeated cycle $\tan\beta = \dfrac{\sigma_a}{\sigma_m} = 1.0$ and sometimes for the random cycle it is possible to determine the tan β based on the initial information regarding the input loads. This provides the possibility at the beginning of fatigue calculations to verify if $\beta > \alpha$ and conclude if it is justifiable to proceed applying the fatigue related methodologies.

In order to determine the stress amplitude limit S_a and the mean stress limit S_m, we consider the similarity of the triangles AOE and BHE that allows $\frac{BH}{OE-OH}=\frac{AO}{OE}$, therefore,

$$\frac{S_a}{S_{ut}-S_m}=\frac{S_e}{S_{ut}} \tag{1-12.6}$$

Accounting that $\frac{\sigma_a}{\sigma_m}=\frac{S_a}{S_m}$ and combining this expression with expression (1-12.6), we determine the mean stress limit and stress amplitude limit, respectively:

$$S_m=\frac{S_e S_{ut}\sigma_m}{S_{ut}\sigma_a+S_e\sigma_m} \tag{1-12.7}$$

and

$$S_a=\frac{S_e S_{ut}\sigma_a}{S_{ut}\sigma_a+S_e\sigma_m} \tag{1-12.8}$$

The last step of the calculations determines the safety factor:

$$n_\sigma=\frac{S_a}{\sigma_a}=\frac{S_m}{\sigma_m}=\frac{S_e S_{ut}}{S_{ut}\sigma_a+S_e\sigma_m} \tag{1-12.9}$$

For the nonzero cycles, we have to introduce the mean and amplitude values of the loading factors that are similar to mean and amplitude stresses. For bending moments we have

$$M_m=\frac{M_{\max}+M_{\min}}{2} \tag{1-12.10}$$

and

$$M_a=\frac{M_{\max}-M_{\min}}{2} \tag{1-12.11}$$

where M_m and M_a are the mean and amplitude values of the bending moments, respectively.

Similarly, for the torques we may write

$$T_m=\frac{T_{\max}+T_{\min}}{2} \tag{1-12.12}$$

and

$$T_a=\frac{T_{\max}-T_{\min}}{2} \tag{1-12.13}$$

where T_m and T_a are the mean and the amplitude values of the torques, respectively. Therefore, in the expressions for determining the stresses, the geometric parameters, or the safety factors we use the amplitude values of the loading factors, while the allowable stresses should be presented by the stress amplitude limit S_a from formula (1-12.8) to account for the required safety factor.

Since for a repeated cycle $\sigma_m = \sigma_a$, formulas (1-12.7) and (1-12.8) read

$$S_m = S_a = \frac{S_e S_{ut}}{S_{ut} + S_e} \tag{1-12.14}$$

In addition, it should be emphasized that the analysis of experimental data do not lead to the acceptance of a certain methodology of determining the stress amplitude limit and the mean stress limit for machine elements subjected to repeated or random cycles of tangential stresses, shear, or torsional stresses. Therefore, since the completely reversed cycle is the most destructive loading cycle, hence this cycle is accepted in this book for the fatigue related calculations of variable tangential stresses acting in machine elements.

Below are examples of using the Modified Goodman line for solving fatigue strength problems characterized by nonzero mean stress cycles.

Example 1-12.1a: Determining the safety factor for a repeated cycle.
Figure 1-12.4 shows a grooved axle having a diameter $D = 2.0$ in. and a groove diameter $d = 1.75$ in., and the notch radius $r = 0.125$ in. The axle is machined from AISI 1050 steel having $S_{ut} = 105{,}000$ psi and $S_{yt} = 60{,}000$ psi. It is loaded by a repeated bending moment $M = 10{,}000$ lb-in. We determine the safety factor.

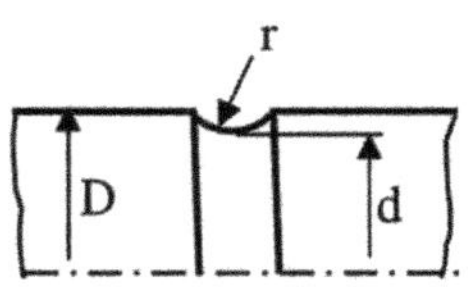

Figure 1-12.4. Grooved axle.

Solution:
First, we determine the endurance limit using the formula (1-10.4) $S_e = k_a k_b k_c k_f S_e'$. Based on expressions (1-9.2), (1-9.3), and (1-9.4) we calculate that $k_a = 0.793$, $k_b = 0.83$, and $k_c = 1$. Because $\frac{D}{d} = \frac{2.0}{1.75} = 1.14$ and $\frac{r}{d} = \frac{0.125}{1.75} = 0.07$, we determine from Appendix A-1-5 (Figure 5) that for our case the theoretical stress concentration factor $K_t = 2.0$. Using the equation (1-10.2) and considering that the notch sensitivity factor $q = 0.6$, we calculate the fatigue stress concentration factor $K_f = 1 + 0.6(2.0 - 1) = 1.6$ that allows using the formula (1-10.3) to write $k_f = \frac{1}{1.6} = 0.625$. Therefore, the endurance limit is $S_e = 0.793 \cdot 0.83 \cdot 1 \cdot 0.625 \cdot 0.5 \cdot 105 = 21.6$ ksi. Using the equation (1-12.4) we calculate $\tan\alpha = \frac{21.6(105-60)}{105(60-21.6)} = 0.241$. Since the axle is loaded by a repeated cycle, we may write $\tan\beta = \frac{\sigma_c}{\sigma_{ra}} = 1$. Therefore, $\tan\beta > \tan\alpha$ and we may continue to work on the problem.

Using the formulas (1-4.17) and (1-4.18), we calculate the maximum bending stress $\sigma_{max} = \frac{64 \cdot 9{,}000 \cdot 0.5 \cdot 1.75}{\pi 1.75^4} = 17{,}114$ psi. Since for a repeated cycle $\sigma_m = \sigma_a = \frac{\sigma_{max}}{2}$, we obtain $\sigma_m = \sigma_a = \frac{17{,}114}{2} = 8{,}557$ psi. Therefore, according to equation (1-12.14) we have $S_m = S_a = \frac{S_e S_{ut}}{S_{ut} + S_e} = \frac{21.6 \cdot 105}{105 + 21.6} = 17{,}915$ ksi. Hence, $n_\sigma = \frac{17{,}915}{8{,}557} = 2.09$.

Example 1-12.1b: Determining the diameter of a shaft for a random cycle of axial loading. A stepped shaft having the diameters ratio $\frac{D}{d} = 1.1$ and the ratio between the fillet radius and the smaller diameter $\frac{r}{d} = 0.05$ is made from AISI 1040 steel having S_{ut} = 620 MPa and S_{yt} = 415 MPa. The shaft is subjected to a static axial tensile force F = 160 kN and a repeated axial load F_{rep} = 210 kN. For a safety factor n_σ = 2, we determine the diameter of the shaft d.

Solution:

Since the diameter of the shaft is unknown, we assume that the size factor k_b = 0.8, keeping in mind that this factor could be verified when the diameter will be estimated. Using the expressions (1-9.2) and (1-9.4), we determine the surface and load factors, respectively, hence k_a = 0.833 and k_c = 0.8. From Appendix A-1-5 (Figure 1) we determine for our case the theoretical stress concentration factor K_t = 1.7. Using the formula (1-10.2) and considering that q = 0.6, we have $K_f = 1 + q(K_t - 1) = 1 + 0.6(1.7 - 1) = 1.42$. According to formula (1-10.3), we write $k_f = \frac{1}{1.42} = 0.7$. Therefore, the endurance limit equals $S_e = 0.833 \cdot 0.8 \cdot 0.8 \cdot 0.637 \cdot 0.5 \cdot 620 = 105.28$ MPa. At this time we cannot calculate σ_a and σ_m; however, we can determine the values of the minimum and maximum forces applied to the shaft. Hence, F_{min} = 160 kN, and F_{max} = 160 + 210 = 370 kN. Therefore, the mean value of the force is $F_m = \frac{370 + 160}{2} = 265$ kN and the force amplitude is $F_a = \frac{370 - 160}{2} = 105$ kN. Therefore, $\sigma_m = \frac{265 \cdot 10^3}{A}$ and $\sigma_a = \frac{105 \cdot 10^3}{A}$, where A is the cross-sectional area of the shaft. We can compute $\tan\beta = \frac{105A}{265A} = 0.396$. Now, based on formula (1-12.4) we calculate $\tan\alpha = \frac{105.28(620 - 415)}{620(415 - 105.28)} = 0.112$. Since $\beta > \alpha$, we may proceed with the calculations.

Using formula (1-12.8), we determine the value of the stress amplitude limit $S_a = \frac{105.28 \cdot 620 \cdot 105A}{105A \cdot 620 + 105.28 \cdot 265A} = 73.7$ MPa. In order to determine the diameter, we write $\sigma_a = \frac{105}{A} = \frac{4 \cdot 105 \cdot 10^3}{\pi d^2}$. For the safety factor $n_\sigma = \frac{S_a}{\sigma_a} = 2$ we obtain $2 = \frac{\pi 73.7 \cdot d^2}{4 \cdot 105 \cdot 10^3}$.

Solving this expression for the diameter, we have $d=\sqrt{\dfrac{2\cdot 4\cdot 105\cdot 10^3}{\pi 73.7}}=60.25$ mm. Verifying the size factor we determine that for this diameter k_b = 0.807. Comparing that with the assumed k_b = 0.8, it is clear that there is no need to recalculate the endurance limit.

Example 1-12.1c: Determining the diameter of a bar for a random cycle of bending loading.

A round bar machined of AISI 1022 steel having S_{ut} = 70 ksi and S_{yt} = 52 ksi is loaded by a random cycle characterized by the maximum bending moment M_{max} = 1.500 lb-ft and minimum bending moment M_{min} = −500 lb-ft. We determine the diameter of the bar taking the safety factor n_σ = 2.

Solution:

We start by determining the endurance limit. Since the diameter of the shaft is unknown, we assume that the size factor k_b = 0.85. This factor might be verified later based on the calculated diameter. Based on the expressions (1-9.2) and (1-9.4), we determine the surface and load factors: k_a = 0.902 and k_c = 1. The endurance limit is S_e = 0.902·0.85·1·0.5·70 = 26.83 ksi. Using formulas (1-12.10) and (1-12.11), we determine the values of the mean and amplitude bending moments, respectively: $M_m=\dfrac{1{,}500+(-500)}{2}=500$ lb-ft and $M_a=\dfrac{1{,}500-(-500)}{2}=1{,}000$ lb-ft.

Since $\dfrac{\sigma_a}{\sigma_m}=\dfrac{M_a}{M_m}$, we may write $\tan\beta=\dfrac{1{,}000}{500}=2$. We calculate $\tan\alpha=\dfrac{26.83(70-52)}{70(52-26.83)}=0.274$. Since $\tan\beta > \tan\alpha$ we may continue with the fatigue considerations.

Now combining formulas (1-4.17) and (1-4.18), we determine the values of σ_a and σ_m: $\sigma_a=\dfrac{32M_a}{\pi d^3}$ and $\sigma_m=\dfrac{32M_m}{\pi d^3}$. Then based on formula (1-12.8), we calculate the stress amplitude limit $S_a=\dfrac{26.83\cdot 70\cdot 32M_a\pi d^3}{70\cdot 32M_a\pi d^3+26.83\cdot 32M_m\pi d^3}=\dfrac{26.83\cdot 70\cdot 1{,}000}{70\cdot 1{,}000+26.83\cdot 500}=22.52$ ksi.

Accounting for the safety factor, we may write $n_\sigma=\dfrac{S_a}{\sigma_a}=2=\dfrac{22.52\cdot 10^3\pi d^3}{32M_a}$.

Solving this expression for d we calculate $d=\sqrt[3]{\dfrac{32nM_a}{\pi S_a}}=\sqrt[3]{\dfrac{32\cdot 2\cdot 10^3\cdot 12}{\pi 22.52\cdot 10^3}}=2.21$ in.

The recalculated size factor k_b = 0.814. The recalculated endurance limit is S_e = 25.694 ksi and S_a = 21.71 ksi. The value of the recalculated diameter is d = 2.24 in. Obviously, there was no need to perform this iteration.

1-13. Finite Life Cycle

The fatigue stress calculations described above are associated with the infinite life cycle. However, many machine elements execute fewer than 10^6 loading cycles during their life, and applying the strength calculation methods that are intended for infinite life are often not justified. It is desirable to adjust the strength of the elements to the required duration of their life.

The experimental graphs discussed above address the relationship between the fatigue strength and the number of loading cycles providing the possibility of determining the value of fatigue strength for the number of cycles that is below 10^6. Figure 1-13.1 shows that constructing the fatigue test curve (S_f, N) in a scale of common logarithms (logN, log S_f), the curve straightens in the range, approximately, between 10^3 and 10^6 cycles that actually represents the range of finite life cycle. This graph provides the possibility to derive an algebraic formula for calculating the fatigue strength S_f' for the finite life cycle of a specimen.

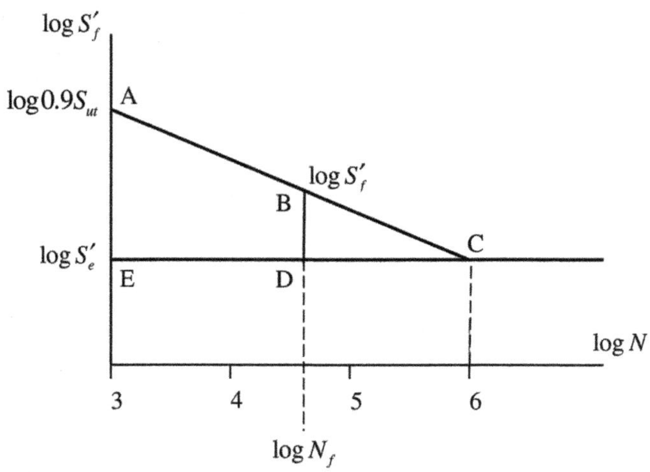

Figure 1-13.1. Fatigue strength diagram in the scale of common logarithms (log N, log S_f).

The ordinate axis of the graph in Figure 1-13.1 starts at the abscissa of log $10^3 = 3$. As mentioned earlier, experimental data indicate that at 10^3 cycles the fatigue strength S_f' drops to about $0.9S_{ut}$, as shown on the ordinate axis. Assume that it is necessary to determine the fatigue strength S_f' that corresponds to a certain number of cycles N_f, while $10^3 < N_f \leq 10^6$. Drawing a line perpendicular to the abscissa axis through the point representing the value of log N_f, we obtain a segment line DB for which point D represents the coordinate of log S_f' (fig. 1-13.1). The similarity of the triangles ACE and BCD allows composing the following proportion: $\frac{\text{AE}}{\text{BD}} = \frac{\text{EC}}{\text{DC}}$, where, AE = $\log 0.9S_{ut} - \log S_e'$, BD = $\log S_f' - \log S_e'$, EC = 6 – 3, and DC = 6 – log N_f. Substituting into this proportion the corresponding values, we obtain $\frac{\log 0.9S_{ut} - \log S_e'}{\log S_f' - \log S_e'} = \frac{6-3}{6-\log N_f}$. Applying conventional algebra, we derive the following equation for calculating the fatigue strength of a specimen for the finite life cycle:

$$\log S_f' = \log 0.9S_{ut}\left(2 - \frac{1}{3}\log N_f\right) + \log S_e'\left(\frac{1}{3}\log N_f - 1\right) \tag{1-13.1}$$

Denoting

$$s = \log 0.9S_{ut}\left(2 - \frac{1}{3}\log N_f\right) + \log S_e'\left(\frac{1}{3}\log N_f - 1\right) \tag{1-13.2}$$

we calculate the fatigue strength for a finite life cycle from the following formula:

$$S'_f = 10^s \qquad (1\text{-}13.3)$$

The following examples demonstrate the steps related to using this formula.

Example 1-13.1a: Calculating the fatigue strength of a specimen for a finite life cycle (completely reversed cycle).

We determine the fatigue strength of a specimen made of AISI 1022 steel having the ultimate strength of 70 ksi and subjected to 74,000 cycles of a completely reversed bending cycle.

Solution:

The diameter and surface finish of the specimen are in accordance with the corresponding standards; therefore its endurance limit is $S'_e = 0.5S_{ut} = 0.5{\cdot}70 = 35$ ksi.

First we determine the common logarithms of the given parameters: log 0.9·70 = 1.799, log 35 = 1.544, log 74,000 = 4.869. Substituting these values into equation (1-13.2), we determine the exponent s: $s = 1.799\left(2 - \frac{1}{3}\cdot 4.869\right) + 1.544\left(\frac{1}{3}4.869 - 1\right) = 1.64$. Based on the formula (1-13.3), we calculate the fatigue strength for the specimen: $S'_f = 10^{1.64} = 43.65$ ksi.

In the case of an actual machine element, the fatigue strength for the finite life cycle is calculated similar to the infinite life cycle using the same modifying factors as in cases of infinite life, hence, $S_f = k_a k_b k_c S'_f$, and if applicable, the stress concentration factor K_f should be accounted for.

Example 1-13.1b: Calculating the fatigue strength of an axle for a finite life cycle (completely reversed cycle).

An axle having a diameter of 15 mm is made of steel characterized by ultimate strength of $S_{ut} = 360$ MPa. We determine the fatigue strength of the axle that is subjected to 550,000 cycles of completely reversed bending.

Solution:

Using expressions (1-9.2), (1-9.3), and (1-9.4), we determine respectively $k_a = 0.96$, $k_b = 0.95$, and $k_c = 1$. Hence, $S_e = 0.96\cdot 0.95\cdot 1\cdot 0.5\cdot 360 = 164.16$ MPa.

Now we determine the common logarithms of the values of the ultimate strength, endurance limit, and number of cycles. Hence, log 0.9·360 = 2.51, log 164.16 = 2.215, log 550,000 = 5.74

Substituting these values into equation (1-13.2), we determine the exponent s:

$$s = 2.51\left(2 - \frac{1}{3}\cdot 5.74\right) + 2.215\left(\frac{1}{3}5.74 - 1\right) = 2.24.$$

Using formula (1-13.3), we calculate the fatigue strength:

$$S_f = 10^{2.24} = 173.9 \text{ MPa}.$$

Example 1-13.1c: Calculating the fatigue strength of a shaft for a finite life cycle (completely reversed and repeated cycles).

A shaft is made of steel having the ultimate strength of $S_{ut} = 147$ ksi. The diameter of the shaft is $d = 3.0$ in. The finite life of the shaft is 400,000 cycles. We determine the fatigue strength of the shaft for the completely reversed and repeated bending loading cycles.

Solution:

Applying expressions (1-9.2), (1-9.3), and (1-9.4), we determine $k_a = 0.734$, $k_b = 0.778$, and $k_c = 1$. Hence, $S_e = 0.734 \cdot 0.778 \cdot 1 \cdot 0.5 \cdot 147 = 41.9$ ksi.

The common logarithms of the determined values of the ultimate strength, endurance limit, and number of cycles are $\log 0.9 \cdot 147 = 2.12$, $\log 41.9 = 1.622$, $\log 400{,}000 = 5.6$.

Substituting these calculated values into equation (1-13.2), we determine the exponent s: $s = 2.12\left(2 - \frac{1}{3} \cdot 5.6\right) + 1.622\left(\frac{1}{3}\, 5.6 - 1\right) = 1.69$. Applying formula (1-13.3), we calculate the fatigue strength for the completely reversed cycle $S_f = 10^{1.69} = 48.98$ ksi.

The fatigue strength for the repeated or random cycles is associated with the mean stress limit S_m and stress amplitude limit S_a that can be determined from formulas (1-12.7) and (1-12.8).

Similarly, denoting by S_{mf} and S_{af} the mean stress limit and stress amplitude limit for the finite life of repeated and random cycles, we may write

$$S_{mf} = \frac{S_f S_{ut} \sigma_m}{S_{ut}\sigma_a + S_f \sigma_m} \tag{1-13.4}$$

and

$$S_{mf} = \frac{S_f S_{ut} \sigma_m}{S_{ut}\sigma_a + S_f \sigma_m} \tag{1-13.5}$$

As it is discussed earlier, in a repeated cycle $\sigma_a = \sigma_m$ and $S_a = S_m$, therefore, based on formulas (1-13.4) and (1-13.5), we obtain

$$S_{mf} = S_{af} = \frac{S_f S_{ut}}{S_{ut} + S_f} \tag{1-13.6}$$

Using this formula for our case, we have $S_{af} = \frac{48.98 \cdot 147}{147 + 48.98} = 36.74$ ksi. Thus, the fatigue stress amplitude limit for the finite repeated cycle in our case equals to $S_{af} = 36.74$ ksi.

In the case of a random cycle, the fatigue stress for finite life can be calculated from the formulas (1-13.4) or (1-13.5), however, the σ_a and σ_m should be determined first. It should be noted that in order to determine the safety factor for repeated and random cycles for the finite life it is sufficient to deal with the pair σ_a and S_{af}.

Chapter 1. Problems

1-01. Determine the maximum shear force and bending moment acting in the beam shown in Figure 1p-01, while $F_1 = F$, $F_2 = 2F$, $l_1 = a$, and $l_2 = 4a$.

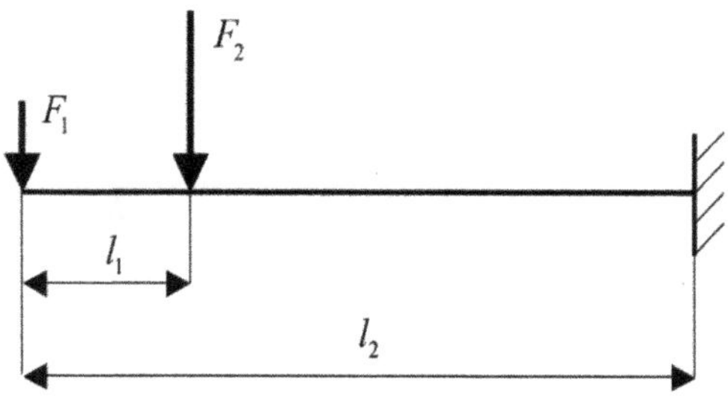

Figure 1p-01

1-02. Determine the maximum shear force and bending moment acting in the beam shown in Figure 1p-02, while $F_1 = F$, $F_2 = 2F$, $l_1 = a$, $l_2 = 2a$, and $l_3 = 3a$.

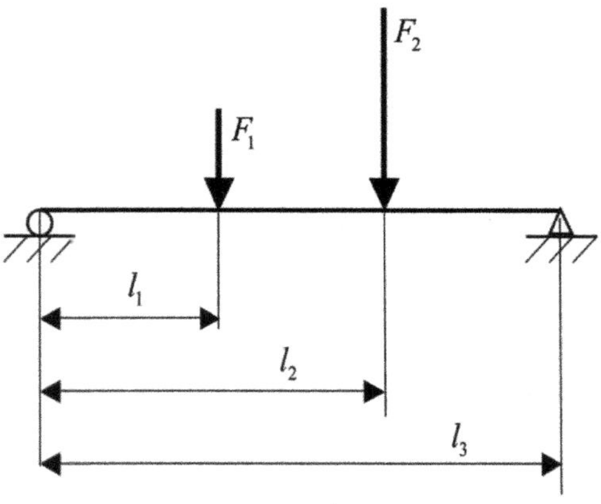

Figure 1p-02

1-03. Determine the maximum shear force and bending moment acting on the beam shown in Figure 1p-03, considering that the force F_2 is acting in the horizontal plane, while $F_1 = F$, $F_2 = 2F$, $l_1 = a$, $l_2 = 2a$, and $l_3 = 3a$.

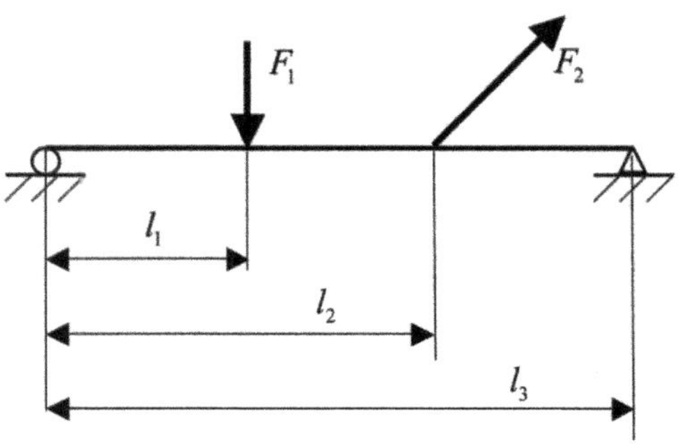

Figure 1p-03

1-04. Determine the maximum shear force and bending moment acting on the beam shown in Figure 1p-04, considering that the force F_2 is acting in the horizontal plane, while $F_1 = F$, $F_2 = 2F$, $F_3 = 3F$, $l_1 = a$, $l_2 = 2a$, $l_3 = 3a$, and $l_4 = 4a$.

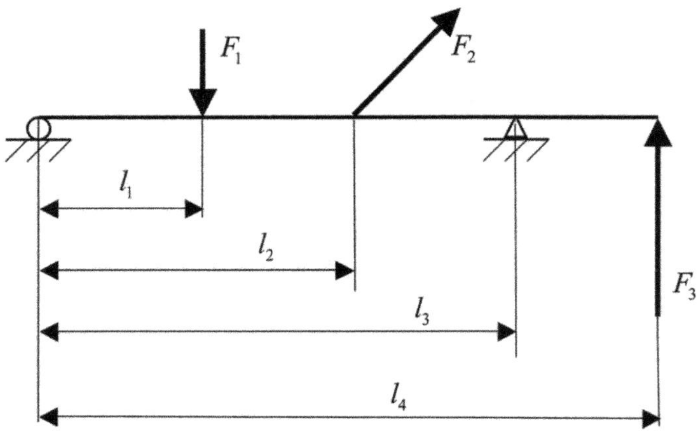

Figure 1p-04

1-05. Determine the maximum shear forces and bending moments acting on the beam shown in Figure 1p-05, while $M_1 = 2Fa$, $M_2 = 4Fa$, $F_1 = 3F$, $F_2 = 5F$, $l_1 = a$, $l_2 = 2a$, and $l_3 = 4.5a$. The force F_1 is acting in the horizontal plane.

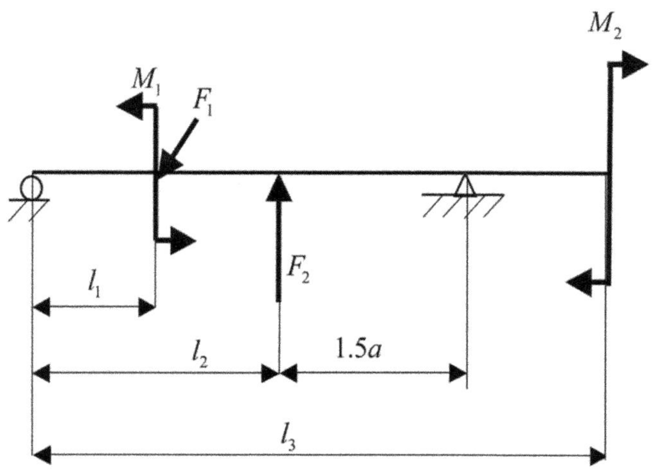

Figure 1p-05

1-06. Determine the torque T_3 and the maximum torque acting in the shaft shown in the Figure 1p-06 if $T_1 = 2T_0$, $T_2 = 4T_0$, $T_4 = 3T_0$, and $T_5 = 6T_0$.

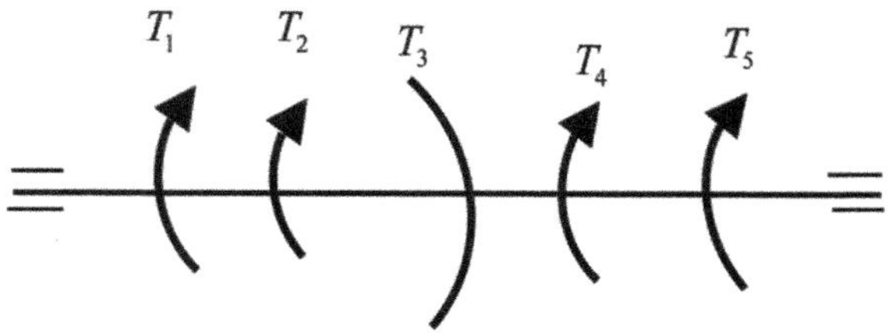

Figure 1p-06

1-07. Based on the third theory of failure, determine the diameter of the shaft shown in Figure 1p-07, considering that the force F_2 is acting in the horizontal plane, while $F_1 = F$, $F_2 = 2F$, $T_1 = T_2 = 2Fa$, $l_1 = a$, $l_2 = 2a$, and $l_3 = 4a$. The shaft is made of steel having $S_y = 60$ ksi, the safety factor $n_\sigma = 2$, $F = 500$ lb, and $a = 2.0$ in.

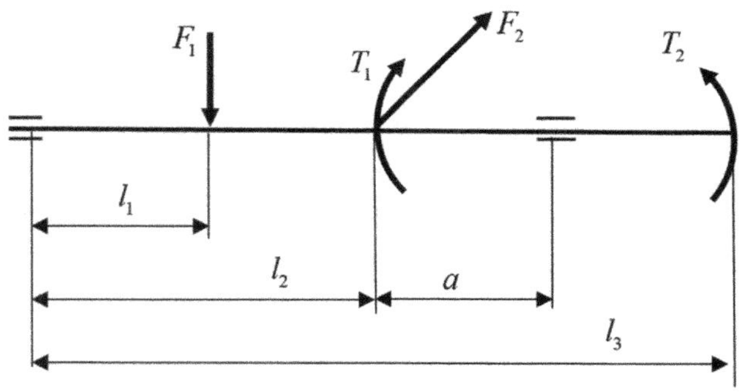

Figure 1p-07

1-08 Using the fourth theory of failure, select the steel for the shaft shown in Figure 1p-08. The shaft is loaded by forces $F_1 = 3F$, $F_2 = 4F$, by a moment $M_1 = 3Fa$, $M_2 = 2Fa$ and by torques $T_1 = Fa$, $T_2 = 2Fa$, and $T_3 = 3Fa$. The diameter of the shaft $d = 22$ mm, the safety factor $n_\sigma = 2$. Take $F = 1000$ N, and $a = 60$ mm.

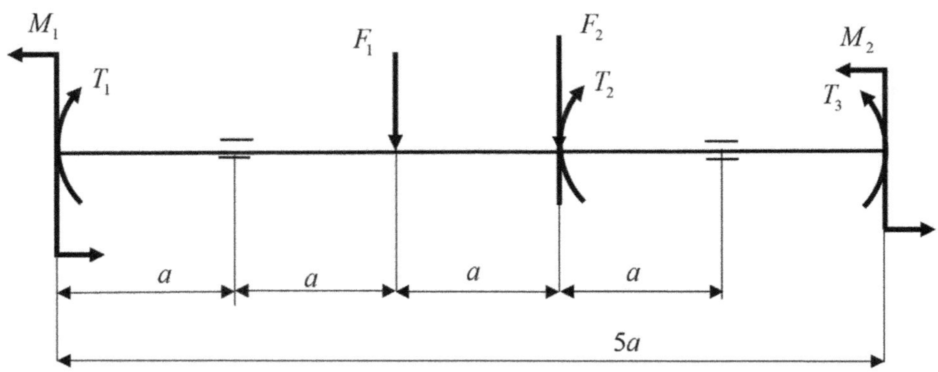

Figure 1p-08

1-09. A shaft shown in Figure 1p-09 is subjected to a repeated bending loading cycle characterized by M_{max} = 2,000 lb-in. The shaft is machined from steel having S_{ut} = 105 ksi and S_{yt} = 62 ksi. Determine the safety factor if the diameter of the shaft is d = 1.0 in.

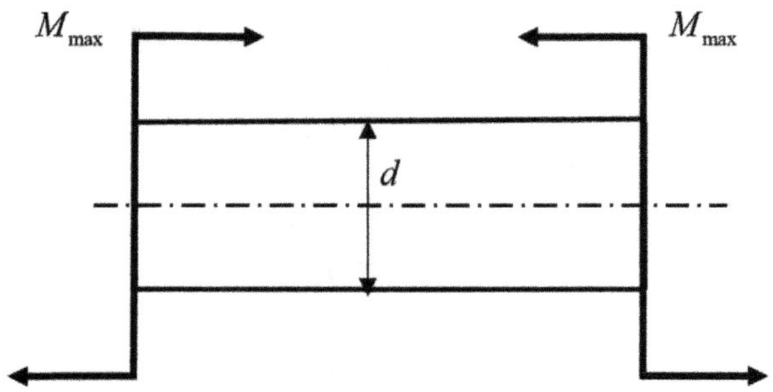

Figure 1p-09

1-10. A shaft is machined from steel having S_{ut} = 118 ksi. The finite life of the shaft is N = 350,000 loading cycles and the diameter of the shaft d = 1.0 in. Determine the fatigue strength for torsional and bending stresses.

CHAPTER 2

Bearings

Bearings serve as supports for rotating machine elements, such as shafts, axles, and others. There are two basic types of bearings: sliding, or journal bearings, and rolling bearings. Of the two, rolling bearings have a significantly broader industrial application. However, in certain mechanical systems, sliding bearings play a very important role and cannot currently be replaced by rolling bearings.

2-1. Sliding Bearings

Sliding bearings are often called journal bearings or sleeve bearings and, sometimes, plain-surface bearings or bushings. They have a variety of applications in mechanical engineering systems.

The applications of sliding bearings in machines and mechanical systems are justified in operational conditions characterized by heavy impact loading and by a combination of heavy loading and high precision rotational velocity. Usually, the operation of these bearings is characterized by very sophisticated lubrication systems. The engineering requirements of the performance of these bearing are extremely high. Here are a few examples of mechanical systems that use these bearings. Internal combustion engines and air compressors use split sliding bearings, consisting of lower and upper shells, that support the main crankshaft journals and crankpin journals. The spindles of high precision metal cutting equipment are supported by sliding bearings.

However, sliding bearings are widely used in assemblies where the operational conditions are associated with relatively light loading, low operational velocities, and limited lubrication.

The operational process of a journal supported by a bearing is characterized by frictional interaction of the contacting surfaces, resulting in generating heat that is dissipated by the bearing enclosure and the shaft. Lubrication of the contact surfaces significantly reduces the friction between the bearing and the journal and assists in evacuation of the heat. In stable operational conditions, the temperature of the assembly containing the bearing and the journal is almost constant. Normally, the operational temperature should correspond to the allowable values for the material of the bearing and the type of lubricant. In cases of overheating, the viscosity of the lubricant drops, resulting in increasing friction between the journal and the bearing, causing seizing of the journal or the damaging of the bearing.

The material of the bearing plays a very important role in the operational process. During initial acceleration, the journal is in direct contact with the bearing, and this results in an intense frictional interaction between the journal and the bearing surface, causing an excessive wear to the bearing. Hence, sliding bearings should be produced from low-friction materials.

In addition, it should be noted that the type of the lubricant should be selected with regard to the operational conditions and the material of the bearing. Sliding bearing are often manufactured from self-lubricating materials.

During the operational processes, sliding bearings could be subjected to the action of radial forces, axial forces, or a combination of both as shown in Figure 2-1.1.

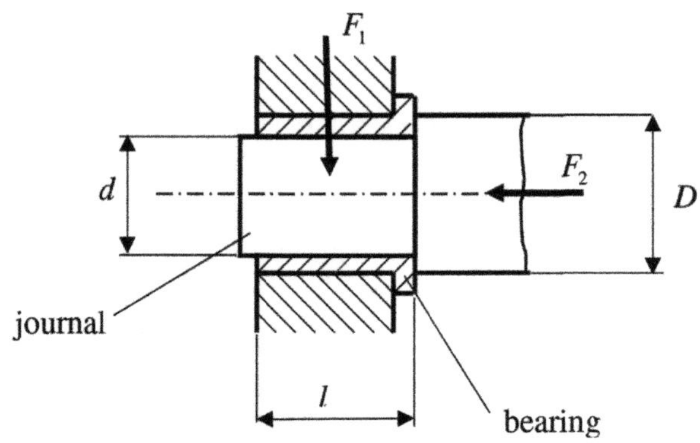

Figure 2-1.1. Radial-thrust sliding bearing.

This figure presents a stepped shaft having diameters d and D, while the shaft is loaded by a radial force F_1 and axial force F_2, and the length of the bearing is l. Based on the analysis of actual structures of sliding bearings, it may be recommended that $l \geq 1.2d$.

Sliding bearings are made of a variety of materials that include Babbitt metal, copper-lead, cast iron, porous metals, plastics, and others. The engineering requirements of these materials are focused on properties such as strength to withstand operational pressure, fatigue strength, resistance to seizing under metal-to-metal contact, thermal conductivity, corrosion resistance, and others.

2-1.1. Materials for Sliding Bearings

The properties and applications of materials for sliding bearings are described below.

2-1.1a. Babbitt or White Metal Alloys

Babbitts represent antifriction alloys comprising, in different proportions, tin, led, antimony, and copper. The tin-base Babbitt is composed of 80 to 90% tin, 3 to 8% copper, and 4 to 14% antimony. Increasing the copper or antimony content increases the hardness and tensile strength, and decreases the ductility. The lead-base white metal alloys comprise from 10 to 15% antimony and up to 10% tin combined with the lead. These Babbitts cause very little wear to the steel journals.

White metal alloys are widely used in the automotive engine industry for the production of split shell bearings. Manufacturing these bearings consists of using the Babbitt as a coating material over flat steel strips. These strips are cut to size, rolled, and machined to form split bearings. Sliding bearings manufactured from Babbitts are machine elements for specific applications.

2-1.1b. Copper-Lead

The composition of this material consists of cooper and 20 to 40% lead. Cooper-lead bearings possess high fatigue strength and load capacity at high operational temperatures. These bearings are used for heavy duty main crankshaft and connecting road journals, for journals of moderate load, speed turbines, and electric motors. Cooper-lead bearings belong to the group of machine elements for general and specific applications.

2-1.1c. Cast Iron

Cast iron possesses good antifriction qualities and is an unexpansive material that is capable of withstanding high load pressures reaching 150 psi (1.03 MPa) and linear velocities up to 130 fpm $\left(0.7\frac{m}{s}\right)$. Sliding bearings made from cast iron have widespread application in industry. However, these bearings require intensive lubrication that limits their use as machine elements for general application.

2.1.1d. Porous Metals

Powder metallurgy allows for sintering porous metals based on plain or leaded bronze, iron, and stainless steel. These sintered metals absorb oil and represent suitable materials for self-lubricating sliding bearings that are used in systems where lubrication during maintenance is restricted. Very often, self-lubricating bearings are sintered to certain sizes and are an off-the-shelf part. In cases when a bearing is fabricated from a porous material, a machining operation that minimizes the blocking of the pores on the internal surfaces of the bearing should be applied. Sliding bearings of porous materials are machine elements for general and specific applications.

2-1.1e. Plastics

The mechanical properties of many plastic materials satisfy the engineering requirements of sliding bearings. These bearings can be molded or machined into a wide variety of shapes and dimensions. They have adequate strength, low coefficients of friction, and require partial or no lubrication. Nylon bearings require no lubrication; however, they have some limitations with regard to applied loads Sliding bearings made of Teflon, which has exceptionally high antifriction properties, are self-lubricating, are highly chemical resistant, and work in wide range of temperatures.

It should be mentioned that plastics absorb moisture resulting in swallowing and, consequently, changing their dimensions. Due to swallowing, plastic bearings could cause seizure of the journal. In order to maintain stable dimensions, bearings should be subjected to appropriate treatments. In some cases this treatment consists in keeping the plastics in boiling water for a certain period of time. Plastic bearings are widely used as machine elements for general and specific applications.

2-1.2. Sliding Friction in Journal Bearings

The operational conditions of a pair of journal-bearing are characterized by the type of sliding friction that is established between the journal and the bearing. There are three basic types of frictional interaction between the surfaces of the journal and the bearing: dry friction, boundary friction, and viscous friction. Depending on the characteristics of certain design and operational factors, it is possible to achieve the desirable type of frictional interaction or frictional regime (regime of friction) between the journal and bearing. The dry friction regime is the interaction between the surfaces of the journal when there is no lubricant. Practically, this definition of dry friction is more an abstraction than a real phenomenon. Even moisture and dust play the role of some kind of lubrication. In addition, it may be considered that there is always a possibility to lubricate the surfaces of the journal and the bearing with grease during the process of assembly. It should be mentioned that in the food industry and in some chemical equipment the presence of oil is not allowed, and the frictional regime in these cases is close to dry friction. The boundary friction is defined as a frictional regime of unstable lubrication of the sliding surfaces by a thin, intermittent film of lubricant. The regime of viscous friction is defined as an operational process in which the surface of the rotating journal is constantly separated from the surface of the bearing by a stable film of liquid lubricant. In this case, the frictional resistance to the rotation is characterized by the internal friction between the layers of the liquid lubricant in their relative motion. The viscosity of the liquid causes the resistance that increases with the increase of the relative velocity between the layers of the lubricant film.

N. Petroff developed the analytical fundamentals associated with viscous friction in sliding bearings in 1883. These fundamentals revealed that under certain operational conditions a pressurized liquid film (wedge) is capable of separating the surface of the rotating journal from the surface of the bearing, preventing contact between these two surfaces and, consequently, eliminating the frictional interaction between these two surfaces. These conditions are characterized by a dimensionless factor of the frictional regime C_r, hence,

$$C_r = \frac{\mu\omega}{2P\psi^2} \tag{2-1.1}$$

where μ is the viscosity of the liquid lubricant, ω is the angular velocity of the journal, P is the characteristic of the unit pressure acting between the surfaces of the journal and the bearing, and ψ is the ratio between the radial clearance and the radius of the bearing. The unit pressure P is determined from the following formula (See Figure 2-1.1.):

$$P = \frac{F_1}{ld} \tag{2-1.2}$$

The graph in Figure 2-1.2 shows the dependence of the friction coefficient f on the regime factor C_r. At the optimal value of the regime factor $C_{r,opt}$ (the point C on the graph), the bearing works in a regime that is characterized by the presence of a complete stable lubricating film separating the surfaces of the journal and the bearing, and the friction coefficient reaches its minimum value $f_{\min}$.

The curve AB corresponds to the low values of the regime factor at which the interaction between the journal and the bearing is characterized by dry friction, while the friction coefficient could be in the range from 0.2 to 0.3. According to formula (2-1.1), the low values of the regime factor correspond to the high values of the unit pressure and the low values of the angular velocity and the viscosity. The curve BC is characterized by the increased regime factor

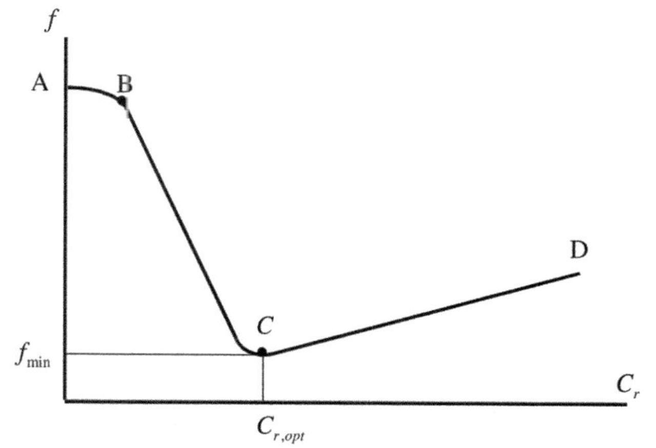

Figure 2-1.2. Graph of the friction coefficient as a function of the regime factor.

and, consequently, by decreased unit pressure and increased angular velocity and viscosity. The values of the regime factor according to the curve BC correspond to the regime of mixed-film and boundary lubrication.

The friction coefficient in the range from 0.05 to 0.1 characterizes these types of lubrication. Usually in these types of lubrication, the surfaces of the journal and bearing are partially rubbing against each other and partially become separated by an unstable thin film of lubricant. The definition of boundary lubrication includes operational conditions of sliding bearing in a range from dry friction (no lubrication) to the presence of a relatively thin film of oil.

The regimes of dry friction and of mixed-film and boundary lubrication are typically applicable to sliding bearings for general applications.

At the point C and its vicinity, the values of the regime factor correspond to the operational conditions of hydrostatic lubrication that are characterized by extremely low friction coefficients ranging from 0.01 to 0.001. Hydrostatic lubrication provides a stable film of lubricant, avoiding direct contact between the journal and the bearing. This type of lubrication requires a special design of the bearings and a continuous supply of pressurized oil. Obviously, a hydrostatic regime of lubrication is associated with high cost. In the automotive industry, hydrostatic lubrication is provided for the movable parts of internal combustion engines. It should be noted that in order to achieve hydrostatic lubrication the bearings should have minimal gaps between movable parts and their enclosures in order to minimize the leakage of the oil through these gaps while keeping the oil pressure at the required level. It is obvious that when the engine (or another system) is not working the journals are contacting the surfaces of the bearings. During the initial starting process, it takes a while for the pressurized oil to reach the bearings and other corresponding parts. During the starting period, the bearings and associated parts are working in the regime of dry friction and are subject to intensive wear. Some authors report that one cold start of a car engine produces wear that is equivalent to about 250 miles of continuous driving. A similar situation occurs during the shutdown of the engine.

The sliding bearings that work in the regime of hydrostatic regime of lubrication belong to the group of machine elements for specific applications and, therefore, this book does not present their design characteristics and methods of calculations.

The further increase of the regime factor that corresponds to the curve CD leads to a hydrodynamic regime of lubrication. This regime of lubrication is characterized by values that may significantly exceed the optimal value of the regime factor; however, in this case these values could also be slightly lower than hydrostatic lubrication. During this type of lubrication, a lubricant film between the journal and the bearing can be created by a rotating journal that pulls in (sucks in) the liquid lubricant (oil) into a wedge-shape space. This space becomes pressurized and the film of oil separates the journal surface from the bearing surface. Unlike hydrostatic lubrication, hydrodynamic lubrication does not require a pump for pressurizing the oil; however, it requires a constant supply of oil from an appropriate reservoir. The oil reservoir should be maintained in a way that the oil supply to the bearing is not interrupted. The sliding bearings of this type of lubrication could be seldom used as machine elements for general application (it would be more justifiable in these cases to use rolling bearings), and, therefore, they are not considered in this book.

Boundary lubrication is the process used for sliding bearings that belong to machine elements for general applications. The design procedure for these bearings is based on the so-called *PV* factor that is discussed below.

2-2. The *PV* Factor for Sliding Bearings

The power required to rotate a journal in a sliding bearing is proportional to the friction force exerted by the bearing and the linear velocity of the journal surface. This power generates heat that is dissipated by the shaft and the enclosure of the bearing. If the generated heat exceeds the dissipated heat, the journal will eventually seize or some kind of other damage will occur. Therefore, the design of sliding bearings intended to operate in the regimes of insufficient lubrication, such as dry friction or mixed-film and boundary lubrication, should be based on methodologies that assure adequate performance of these bearings. The *PV* factor methodology is supported by numerous experimental investigations and provides reliable design solutions. The *PV* factor represents the product of multiplying the unit pressure by the linear velocity of the journal. Thus,

$$P = \frac{F}{dl} \tag{2-2.1}$$

where, as in formula (2-1.2), F is the radial force applied to the journal, d is the diameter of the journal, and l is the length of the bearing. However, formula (1-4.31) has the same structure as formula (2-2.1) and is intended to determine the bearing stress representing contact stress. Therefore, the unit pressure in a sliding bearing is identical to the bearing stress.

The *PV* factor is calculated from the following formula:

$$PV = \frac{FV}{dl} \tag{2-2.2}$$

FV represents power, dl characterizes an area, and $\frac{F}{dl}$ can be defined as a stress. Hence, the PV factor in the physical sense is, actually, associated with power and stress.

The design of sliding bearings is based on the allowable values of the PV factors and on the limits of the maximum values of P and V for the material of the sliding bearing. It should be clear that the values of the parameters P and V represent input data and either PV nor P or V cannot exceed the allowable (or maximum) values for the material. The allowable values of these three parameters are presented in the Appendix A-2-1. It should be stressed that the maximum value of the PV factor is significantly less than the product of multiplying the tabulated maximum values of the pressure and velocity. Therefore, during material selection the pressure or velocity may possess its maximum value but not both. The appropriate information regarding the values of the mentioned parameters is widely available in numerous published sources.

An example of design calculations of a sliding bearing based on the PV methodology is presented below.

Example 2-2.1a: Selecting the material for a sliding bearing.
A journal having a diameter $d = 50$ mm is rotating at a rotational speed $n = 1{,}800$ rpm. The journal is subject to a radial load $F = 15{,}000$ N. We select the material for the sliding bearing.

Solution:
Taking the length of the bearing $l = 1.2d$, we have $l = 1.2 \cdot 50 = 60$ mm. Therefore, $P = \frac{F}{dl} = \frac{15{,}000}{50 \cdot 60} = 5.0$ MPa. The angular velocity of the shaft is $\omega = \frac{\pi n}{30} = \frac{\pi 1{,}800}{30} = 188.4\,\frac{\text{rad}}{\text{s}}$. Hence $V = 0.5\omega d = 0.5 \cdot 188.4 \cdot 50 = 4{,}710\,\frac{\text{mm}}{\text{s}} = 4.71\,\frac{\text{m}}{\text{s}}$. Now we calculate the actual PV factor: $PV = 5 \cdot 4.71 = 23.55\ \text{MPa} \cdot \frac{\text{m}}{\text{s}}$. Referring to Appendix A-2-1, we select the cast bronze having $PV = 31.5\ \text{MPa} \cdot \frac{\text{m}}{\text{s}}$.

Example 2-2.1b: Determining the material for a sliding bearing and the maximum values of the forces applied to the shaft.
The journal of the stepped shaft shown in Figure 2-1.1 has a diameter $d = 1.2$ in., while the larger step has a diameter $D = 1.5$ in. The shaft is rotating at $n = 400$ rpm. We select the material for the sliding bearing and determine the maximum values of the radial force F_1 and axial force F_2 applied to the bearing.

Solution:
The linear velocity of the journal is determined from the following formula: $V = \omega r$, where the radius of journal $r = 0.5d = 0.5 \cdot 1.2 = 0.6$ in. and the angular velocity $\omega = \frac{\pi n}{30} = \frac{\pi 400}{30} = 42\,\frac{\text{rad}}{\text{s}}$. Therefore, $V_1 = 42 \cdot 0.6 = 25.2\,\frac{\text{in.}}{\text{s}} = \frac{25.2 \cdot 60}{12} = 126\ \text{fpm}$. The table in Appendix A -2-1 shows that for most of the tabulated materials the value of the calculated linear velocity $V_1 = 126$ fpm is below the limits. The unit pressure that is similar to the definition of stress equals $P_1 = \frac{F_1}{A_1}$

where $A_1 = ld$. Taking $l = 1.25d = 1.25 \cdot 1.2 = 1.5$ in. we have $P_1 = \dfrac{F_1}{1.5 \cdot 1.2} = 0.555F_1$ psi. Thus, $P_1V_1 = 0.555F_1 126 = 70F_1$. Selecting porous bronze having $PV = 50{,}000$ psi·fpm and the $V_{max} = 1{,}500$ fpm that exceeds 126 fpm, we may write $F_1 = \dfrac{50{,}000}{70} = 714.3$ lb and, therefore, $P_1 = 0.555 \cdot 714.3 = 396.4$ psi. According to the table mentioned above, the limit of the pressure for the selected material is 4,500 psi, and it exceeds the calculated unit pressure P_1.

In the axial direction the contact area represents a ring having the outside diameter D and the inside diameter d. The linear velocity according to the larger diameter exceeds the linear velocity calculated for the diameter of the journal. We have to account for the velocity according to the diameter D. Therefore, $V_2 = 0.5D\omega = 0.5 \cdot 1.5 \cdot 42 = 31.5\ \dfrac{\text{in.}}{s} = 157.5$ fpm < 1,500 fpm (the maximum allowable velocity).

The bearing surface for the axial force is $A_2 = \dfrac{\pi(D^2 - d^2)}{4} = \dfrac{\pi(1.5^2 - 1.2^2)}{4} = 0.636\ \text{in.}^2$, therefore, $P_2 = \dfrac{F_2}{0.636} = 1.572F_2$ psi. Thus, $P_2V_2 = 1.572 \cdot 157.5F_2 = 247.59F_2$ and, consequently, we may write $F_2 = \dfrac{PV}{248} = \dfrac{50{,}000}{248} = 201.6$ lb, and this represents the maximum axial force. The axial unit pressure is $P_2 = 1.572 \cdot 201.6 = 317$ psi < 4,500 psi (the maximum allowable unit pressure).

2-3. Rolling Bearings

Rolling bearings are not machine elements; they represent units or assemblies that comprise several specific elements. These bearings are off-the-shelf products that can be purchased like any bolt or nut without indicating any specifics related to their application. They are sold as completed units that are ready to be installed into the system. In the engineering documentation, the rolling bearing is identified as a part (element), not as an assembly. Actually, from the point of view of the design of mechanical engineering systems, a rolling bearing plays the role of an element for general applications, and for this reason rolling bearings are included in this book and other textbooks. Usually, in courses related to machine elements for general applications corresponding methodologies associated with the design calculations of these elements are presented. However, with regard to rolling bearings, this book is focused on the appropriate selection of the bearings from manufacturers' catalogs. Rolling bearings are produced according to standards and, actually, there are almost no differences between national and international standards. Regardless of the origin of the rolling bearings, their types and dimensions are the same, although they may differ in materials and some manufacturing tolerances. Rolling bearings are an extremely sophisticated mechanical engineering systems. One of the most distinctive characteristics of rolling bearings is their low energy losses when compared with sliding bearings. This is because of rolling friction that significantly reduces the frictional resistance. In addition, rolling bearings are not subjected to the phenomenon of high frictional wear associated with the starting of rotation as when sliding bearings work in the regime of dry friction. Due to the extremely low frictional resistance, rolling bearings are called antifriction bearings.

Rolling bearings are widely used in all kind of mechanical systems, and the role of rolling bearings in industry cannot be overestimated. In many ways, rolling bearings have changed engineering design solutions, making them more sophisticated and efficient.

The main advantages of rolling bearings are extremely low energy losses due to the phenomenon of rolling friction, great performance at moderate loads and velocities, low maintenance, low lubricant consumption, and smaller width compared with sliding bearings. However, rolling bearings have some disadvantages that include limited application at relatively high velocities and loads, particularly, high impact loading, and also in cases requiring high precision of the rotating elements subjected to heavy loading.

The rolling resistance force F can be estimated based on the following empiric formula that resembles the formula for dry friction:

$$F = f_r N \tag{2-3.1}$$

where f_r is a virtual dimensionless rolling friction coefficient, and N is the normal pressure force. In formula (2-3.1), the virtual friction coefficient does not have a physical background as in dry friction. This coefficient for rolling bearings varies from 0.001 to 0.005.

As mentioned above, rolling bearings are available in a variety of dimensions that are internationally standardized. They are manufactured in extremely large quantities in many countries. From the engineering perspective, rolling bearings represent very sophisticated products in their design, performance, accuracy, reliability, maintenance, manufacturing, etc. The improvement of rolling bearings could be expected in the area of increasing of their operational life, but it is more a metallurgical than a design problem. Most likely, engineering graduates will not be needed to develop bearings in their everyday activities, unless they enter into the highly specialized rolling bearing industry. In the vast majority of design projects, engineers have to make the appropriate selection of rolling bearings using the proper methodologies that are discussed below. It should be emphasized that the manufacturing of rolling bearings is based on the most advanced production methods, resulting in a relatively low cost of the bearings. It is justifiable to say that the cost considerations of bearings associated with the development and improvement of mechanical systems almost never represent a business concern since all these expenses will be recovered in a short period.

2-3.1. Types of Rolling Bearings

There are numerous types, groups, and classifications of rolling bearings that are presented in a very comprehensive way in numerous manufacturers' catalogs that include several hundred pages of descriptions, methodologies of selection, and engineering specifications. This textbook provides an introduction to rolling bearings that is intended to address the most general aspects of bearing selection and design considerations associated with their operational specifics in the assemblies.

The general classification of the main types of rolling bearings is based on the characteristics of the forces applied to the bearings. These forces differ by their magnitude and direction. There are three main types of rolling bearings, namely—radial, angular-contact and taper, and thrust bearings. Radial roller bearings predominantly withstand the action of radial forces. Angular-contact and taper rolling bearings are applicable when they are subjected to a combination of radial and axial forces. And, finally, thrust rolling bearings withstand axial forces. Depending on the magnitudes of the forces, these bearings are subdivided into light, medium, and heavy series reflecting their structural design.

The next classification deals with types of rolling bearings based on their design characteristics. Almost all rolling bearings are similar in structure. A bearing comprises an inner ring, an outer ring, and rolling elements. The rolling elements play an extremely important role in the performance of the bearing.

There are two types of rolling elements: balls and rollers, and they determine the two basic types of rolling bearing—namely, ball bearings and roller bearings. The most wide spread and least expensive rolling bearing is the radial ball bearing, see Figure 2-3.1. Figure 2-3.1a shows a partial cutaway view and a cross-section of this bearing, while Figure 2-3.1b presents a partial cross-section, which shows that the radial ball bearing comprises balls 1, an inner ring 2, a separator 3, and an outer ring 4. The rings have corresponding grooves for accommodating the balls that are evenly distributed by the separator that is also called the cage.

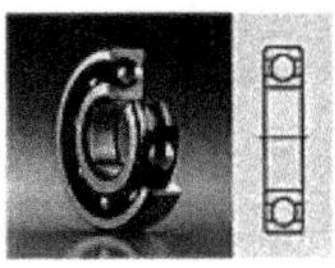

(a)

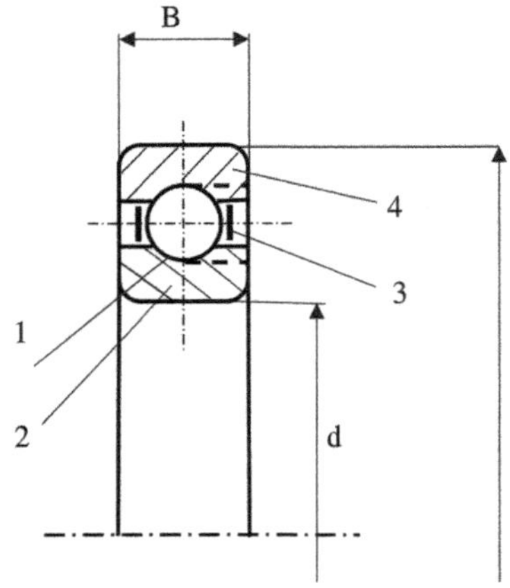

(b)

Figure 2-3.1. Radial ball bearing: (a) partial cutaway view and a cross-section of a radial ball bearing (Courtesy of Emerson Power Transmission Corporation); (b) partial cross-section of a radial ball bearing.

The basic geometric characteristics of the radial ball bearing are the bore diameter *d* of the inner ring, the outside diameter *D* of the outer ring, and the width *B* of the rings that sometimes are called the inner and outer races. Radial rolling bearings are also known as deep-groove ball bearings or the Conrad-type bearings that were introduced at the beginning of the twentieth-century by their inventor. These bearings are capable of withstanding radial loading and some axial loading. Manufacturers' catalogs present the required information mainly for deep-groove ball bearings. Figure 2-3.2 schematically illustrates the way the ball bearings fill the space between the outer and the inner rings placed adjacent to each other.

At the start of the bearing assembly process, the rings touch each other and create a space that is filled by the balls. Then the inner ring is pushed in the direction indicated by the arrow A

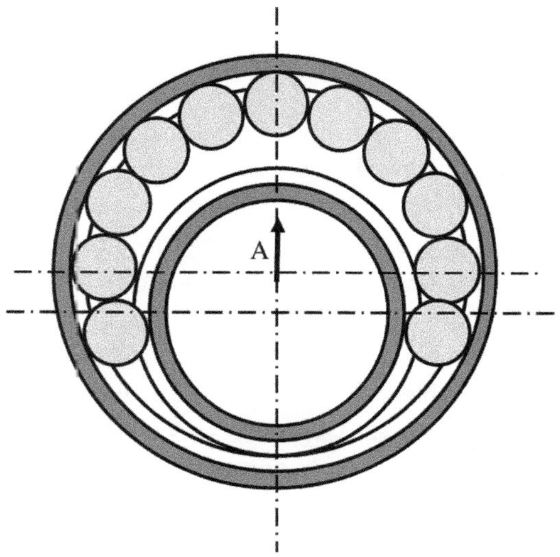

Figure 2-3.2. Positions of the rings and ball bearings, which allow for filling in the balls.

forcing the balls to move around the inner ring. After that, the two halves of the separator cover the balls on both sides, evenly separate the balls, and then these two halves are riveted to each other. The balls can be filled through the special slots in the side of the inner and outer rings. Dashed lines in Figure 2-3.1b show these slots.

Figure 2-3.3 presents a partial cutaway and cross-section of a roller bearing that by analogy with the radial ball bearings is called radial roller bearing. This bearing comprises cylindrical

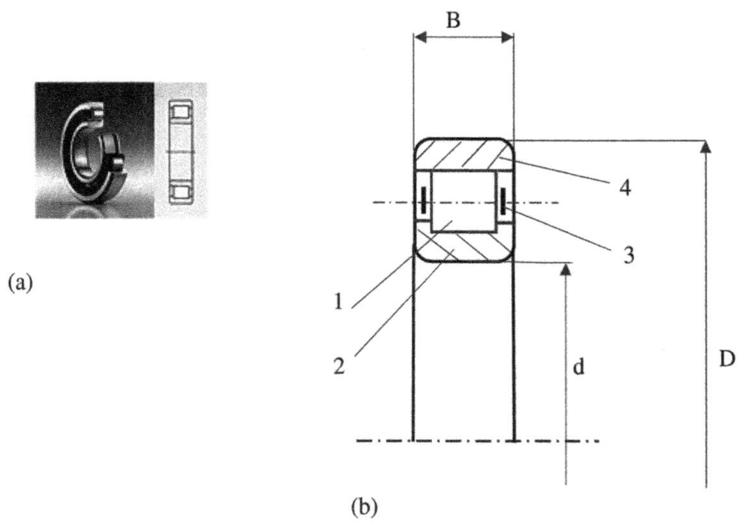

Figure 2-3.3. Radial roller bearing: (a) partial cutaway view and cross-section of a roller bearing (Courtesy of Emerson Power Transmission Corporation); (b) partial cross-section of a roller bearing.

rollers 1, an inner ring 2, a separator 3, and an outer ring 4. The basic geometric parameters are similar to the radial ball bearing and include the bore diameter *d*, the outside diameter *D*, and the width *B*.

When compared to radial ball bearings, roller bearings have an increased load capacity; however, they do not tolerate axial misalignment with the shaft, while the radial ball bearings allow some misalignment. Radial ball bearings are capable of taking a certain combination of radial and axial loading, while radial roller bearings are intended just for radial loading.

The roller elements of roller bearings—the rollers—have different shapes—namely, short, long, needle, tapered, concave, convex, and others.

In accordance with the number of rolling elements, rolling bearings are divided into two basic groups: single-row and double-row bearings (four-row bearings exist also). Single-row bearings were considered above. Figure 2-3.4 provides a sectional view of a double-row ball bearing; while Figure 2-3.5 shows a double-row cylindrical and double-row spherical roller bearings. The structures of these bearings and their basic parameters are similar to single-row bearings. In these figures, the rolling elements 1 are shown in two grooves of the inner and outer rings 2 and 4 having the bore diameters *d* and outside diameters *D*. The separators 3 have more complicated structures in these types of bearings than in single-row bearings.

Some double-row rolling bearings have a spherical raceway in the outer rings that allows for self-alignment of these bearings. The main advantage of double-row bearings is their increased load capacity.

A special type of ball and roller bearings is designed to take a combination of radial and axial loading, and they are called angular-contact ball bearings and tapered roller bearings. Sectional

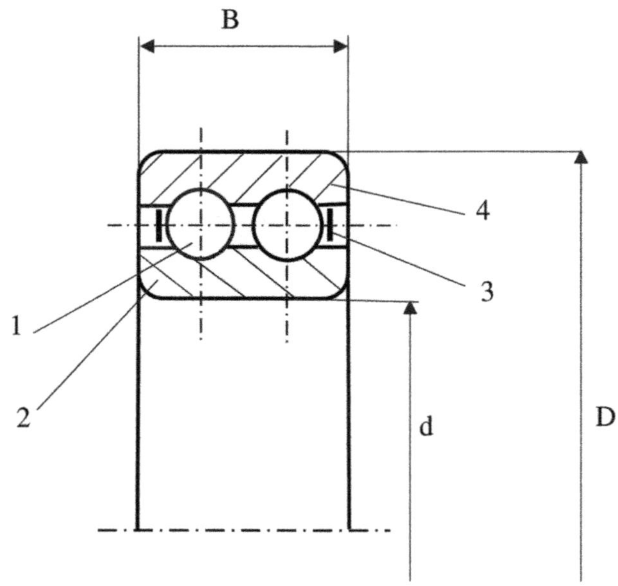

Figure 2-3.4. Double-row ball bearing.

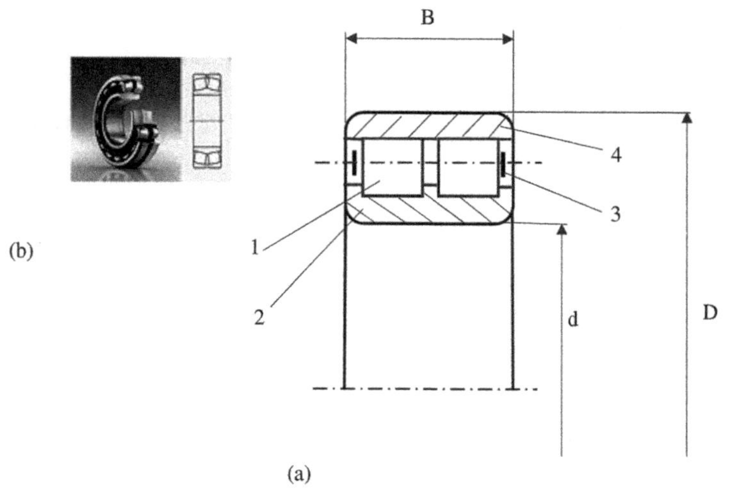

Figure 2-3.5. Double-row roller bearings: (a) partial cross-section of a double-row cylindrical roller bearing; (b) partial cutaway and cross-section of a double-row spherical roller bearing. (Courtesy of Emerson Power Transmission Corporation)

and cutaway views of an angular-contact bearing is presented in Figure 2-3.6 where the ball 1, the inner ring 2, and the separator 3 are similar to the components of a radial ball bearing. The arrow *F* indicates the direction of the axial force that can be applied to the left-hand side of the outer ring, while the right-hand inner side of the ring has a cone shape. In some designs, the outer

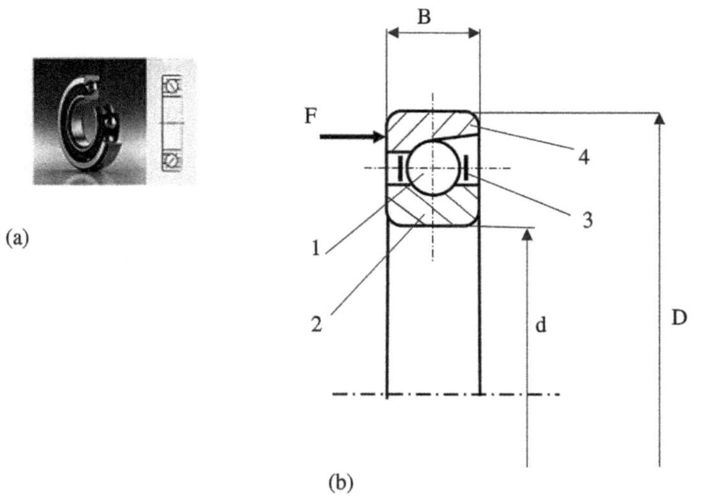

Figure 2-3.6. Angular-contact ball bearing: (a) partial cutaway view and cross section of an angular-contact ball bearing (Courtesy of Emerson Power Transmission Corporation); (b) partial cross-section of an angular-contact ball bearing.

ring can be taken apart. The geometric parameters *d*, *D*, and *B* of angular-contact bearings are the same as for the radial ball bearings; however; in comparison with them the angular-contact ball bearings can withstand significantly higher axial force.

Figure 2-3.7 shows cutaway and cross-section views of a tapered roller bearing, where the inner ring 1 (also called cone) accommodates the tapered rollers 2 (also called cup) that are pushed by the outer ring 3 into the direction indicated by the axial force *F*. The geometric parameters *d*, *D*, and *B* are similar to those in other bearings. When compared to angular-contact bearing, tapered roller bearings can take significantly higher radial and axial loads.

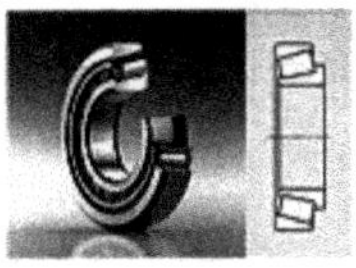

(a)

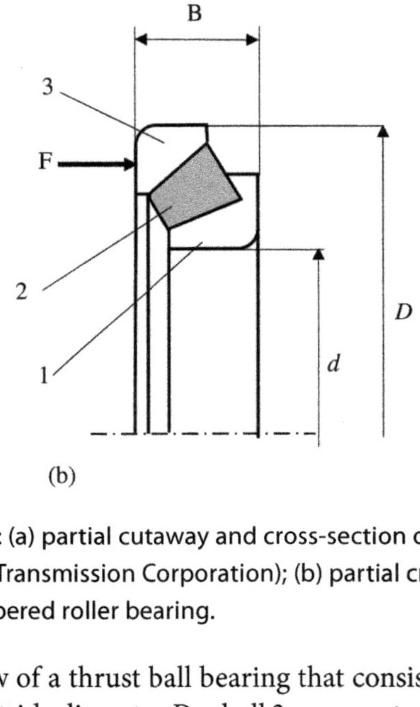

(b)

Figure 2-3.7. Tapered roller bearing: (a) partial cutaway and cross-section of a tapered roller bearing (Courtesy of Emerson Power Transmission Corporation); (b) partial cross-section of a tapered roller bearing.

Figure 2-3.8 shows a sectional view of a thrust ball bearing that consists of an upper ring (plate) 1 having a bore diameter *d* and outside diameter *D*, a ball 2, a separator 3, and a lower ring (plate) 4. The height of the thrust bearing is denoted by *H*.

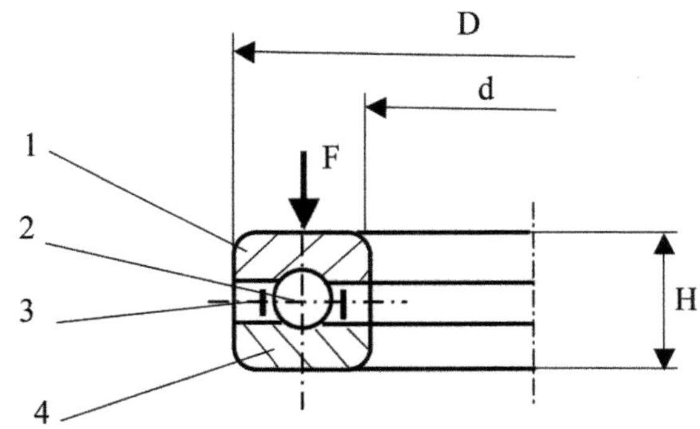

Figure 2-3.8. Thrust ball bearing.

The rolling elements for thrust bearings include balls or rollers. The arrow F in Figure 2-3.8 indicates the direction of the thrust force applied to the bearing.

The rollers in these bearings may have cylindrical, tapered, concave, or other shapes. Roller thrust bearings are characterized by an increased load capacity when compared to ball thrust bearings.

In some situations where the radial dimensions of the shaft assembly are very limited, it is reasonable to use needle bearings, an example is shown in Figure 2-3.9. Very often in these cases, the needles are located between the inner surface of the enclosure belonging to the assembly and the surface of the journal.

Figure 2-3.9. Partial cutaway view of a needle bearing. (Courtesy of Emerson Power Transmission Corporation)

Due to the variety of the geometric parameters in ISO and US Systems, rolling bearings fall into two large groups according to their bore and outside diameters. In these groups, the bearings are subdivided into several series having the same bore diameter or the same outside diameter while other geometric parameters are different. Figure 2-3.10 illustrates the arrangement

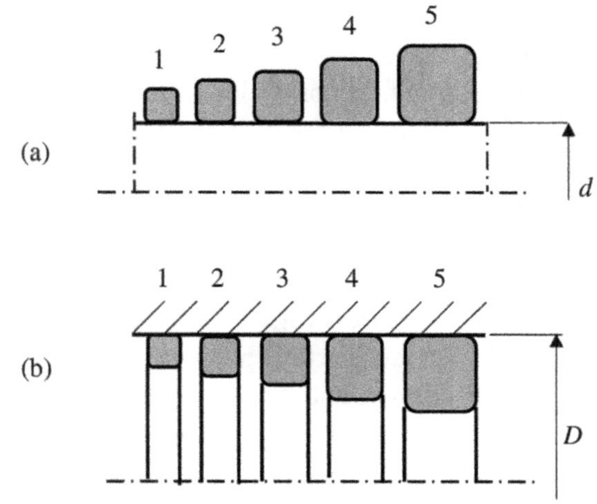

Figure 2-3.10. Rolling bearing series: (a) the group arranged by the bore diameter; (b) the group arranged by the outer diameter.

of bearing series in the group according to the bore diameter (Figure 2-3.10a) and in the group according to the outside diameter (Figure 2-3.10b). For example, the bearings in each group are subdivided into 5 sub-groups that respectively represent the following series: extra-extra light 1, extra light 2, light 3, medium 4, and heavy 5. Some manufacturers have more than five series of rolling bearings. It is clear that the designer has the opportunity to select a rolling bearing that adequately fits to a particular assembly.

2-3.2. Rolling Bearing Materials, Thermal Treatment, and Standardization

The majority of components of rolling bearings are manufactured from high quality steel alloys comprising a high content of carbon and about 1.5% chromium. These components are exposed to heat-treating processes resulting in Rockwell hardness in the range from 61 to 65 HRC.

A variety of stainless steels is used for manufacturing rolling bearings possessing high anticorrosion properties. A special category of bearings that consists of stainless-steel rings and ceramic balls (instead of steel balls) have hardness values that reach about 78 HRC.

Rolling bearings are standardized, and the American Bearing Manufacturing Association (ABMA), as well as the International Standardization Organization (ISO) provide the national and international standards related to these bearings.

More information related to materials of rolling bearings, their dimensions, and characteristics is presented in the manufacturers' catalogs.

2-4. The Role of the Bore Diameter and Service Life in Selecting a Bearing

The stresses in rolling elements, as well in the inner and outer rings, are contact stresses that can be analyzed using the Hertz formulas. The most severe contact stresses in bearings occur between the inner ring and the rolling element because their contact involves the initial touching of two convex surfaces at one point or along a line; the initial contact of rolling elements with the outer ring occurs over curved surfaces. It should be noted that the ball bearings raceway (the groove) has two different radii of curvature—namely, the radius of the race and the radius of the groove. Therefore, it is necessary to determine an equivalent radius of the curvature of the inner ring in order to apply Hertz formulas for calculating the contact stress in the ball bearings.

It is important to remember that stress calculations play a significant role in the design of rolling bearings. However, the actual selection process of rolling bearings for a particular mechanical engineering system is not associated with stress calculations. Rolling bearings are units, not just single elements, and the serviceability of units is associated with many factors, while stress is just one of them.

Many of the important factors associated with the performance of bearings by nature belong to design considerations that can be verified by experimental investigations.

Let us consider the steps of selecting rolling bearings.

Figure 2-3.10a shows that for the same bore diameter the light series bearing has smaller dimensions than the heavier series. The basic difference between these series is that the heavier series of bearings have a longer service life. The service life is defined as the number of revolutions the shaft performs during the failure-free operational time of the mechanical system. The heavier the series of the bearing, the more expensive the bearing will be. Therefore, the essence of proper bearing selection is to find the bearing that for the given bore diameter is characterized by failure-free performance during its service life.

Selection of a rolling bearing is associated with two factors: the bore diameter and the service life. The bore diameter of the bearing depends on the diameter of the journal of the shaft. The strength calculations of the diameter of the journal are not associated with the bearing. However, in order to select a bearing for the calculated diameter of the journal, it is necessary to round up this diameter to the closest larger standard bore diameter of the bearing. It does not make any sense to start the bearing selection process before the corresponding diameter of the shaft is determined. The free of failure service life of a bearing is an engineering requirement that is suggested for each product by the designers based on certain considerations.

The structure of bearing catalogs corresponds to the input parameters of bore diameter and service life. For each series of the same bore diameter, the catalogs provide the information that determines the service life. Usually for each bore diameter, the catalogs start with the information of the light series and then go to the medium and heavy series. Thus, when the bore diameter of the required bearing is already known, the selection process is focused on determining the appropriate value of the bearing service life. The catalogs do not directly provide the values related to bearing service life. However, they do provide the methodologies for determining the service life. These methodologies are based on the results of numerous fundamental experimental investigations and are presented in detail in all manufacturers' catalogs. The methodologies presented in different bearing catalogs are almost identical, or very similar. The main aspects of the methodologies presented in the catalogs are discussed below.

2-4.1 Rolling Bearing Life Ratings

The vast majority of rolling bearing failures are associated with the cyclically acting contact stresses on the surfaces of the rolling elements and raceways of the inner and outer rings causing pitting (spalling) of the contacting surfaces. Pitting represents fatigue failure, and, unfortunately, the materials that are used for manufacturing rolling bearings do not have an endurance limit with regard to contact stresses. Any rolling bearing subjected to cyclic contact stresses could eventually fail. It may happen that during the service life of certain systems some of the bearings should be replaced according to the maintenance schedule. This is why it is very important to provide the possibility to replace the bearings in the assemblies.

The basic operational characteristics of a rolling bearing include the loads and the number of revolutions that a bearing can withstand before its failure. Numerous experimental investigations were carried out with ball and roller bearings that revealed a certain relationship between the load and the number of revolutions the bearing can withstand before failure.

Thus, if in a group of identical bearings, one bearing is subjected to a constant load F_1 and safely completes a number of revolutions L_1, while another bearing under a constant load F_2 safely makes L_2 revolutions, then for all other conditions being equal, we may write the following empiric expression:

$$F_1 L_1^{\frac{1}{a}} = F_2 L_2^{\frac{1}{a}} = const \tag{2-4.1}$$

where $a = 3$ for ball bearings and $a = 3.333$ for cylindrical and tapered roller bearings.

The expression (2-4.1) can be rewritten in the following way:

$$\frac{L_2}{L_1} = \left(\frac{F_1}{F_2}\right)^a \tag{2-4.2}$$

The number of revolutions L represents the bearing service life that can be expressed in hours that a bearing rotates at a certain constant angular velocity before a fatigue failure occurs. It is acceptable to consider fatigue failure of a bearing occurring when an area of 6 mm^2 or 0.01 in.^{2} is affected by pitting (spalling) on a raceway or rolling elements surfaces.

The major rolling bearings manufacturers in accordance with ABMA and ISO define the bearing service life or bearing basic life rating as a certain number of revolutions that 90% of a statistically justifiable group of identical rolling bearings subjected to identical working conditions will complete before failure. The basic life rating or service life L_{10} is defined by ISO to be equal to 1,000,000 revolutions, therefore, $L_{10} = 10^6$ revolutions. The basic life rating accepted by the Timken Company consists of $90 \cdot 10^6$ revolutions. The service life, the basic life rating, the life rating, and the minimum life rating measured in revolutions are synonyms. When the rotational speed (rpm) of the bearing is known, the life rating can be expressed in the number of service hours, and it is denoted by L_{10h}.

2-4.2 Dynamic Load Ratings

The basic dynamic load rating is defined as a constant load by magnitude and direction applied radially to a rotating radial bearing, or a constant axial load centrically applied to a rotating thrust bearing, while for both bearings their service life is in absolute accordance with the basic life rating of the bearing. For some manufacturers, the basic dynamic load rating applies to chromium steel bearings having a minimum hardness of 58 HRC. The above mentioned reliability of 90% associated with the basic life rating (service life) applies to the dynamic load rating. The dynamic load rating is presented in the catalogs for each bearing and is used for calculating the service live of the bearing. During the process of selecting a bearing, the designer should determine the forces applied to the bearing. These forces represent the reactions in the two supports of the shaft that, actually, are the bearings. The methodology for determining the dynamic load rating based on the reactions in the supports is presented below.

As mentioned above, all catalogs are organized in a similar way: the entry parameter for selecting a bearing is the bore diameter of the bearing. It should be clear that neither the service life nor the applied load nor other parameters beside the bore diameter is the initial factor for selecting the required bearing.

In the catalogs the first item in the row (line) describing the bearing is the bore diameter. When the row with the needed bore diameter is found, the engineer should find, in the same row, the dynamic load rating, which in most catalogs is denoted by C. The catalog value of the dynamic load rating should exceed the load applied to the bearing from the shaft. As we will see below, the reactions applied to the bearing are modified by certain empiric load factors in order to calculate the so-called equivalent dynamic load or equivalent load F_e applied to the bearing.

Using equation (2-4.1) we may write

$$F_e L_{10}^{\frac{1}{a}} = C \tag{2-4.3}$$

This equation might be rewritten in the following way:

$$L_{10} = \left(\frac{C}{F_e} \right)^a \tag{2-4.4}$$

According to the majority of bearing manufacturers' catalogs, the equivalent dynamic load F_e is determined by using the following empiric expression:

$$F_e = XVF_r + YF_a \tag{2-4.5}$$

where F_r is the applied radial load, F_a is the applied axial load, V is the rotating ring factor, X is the radial load factor, and Y is the axial load factor.

The applied loads F_r and F_a are actually the radial and axial reactions of the most loaded shaft support that the bearing represents. The rotating factor V depends on which ring is rotating. If the inner ring rotates, as in the majority of cases, then $V = 1$; while $V = 1.2$ if the outer ring rotates. However, for self-aligning bearings $V = 1$ no matter which ring is rotating. The values of the radial load factor X and the axial load factor Y can be found in the manufacturers' catalogs. These factors depend on the type of bearings and on the ratio between the axial and radial loads. Our goal is to select the bearing that can provide the required service life in certain operational conditions.

It is very important to emphasize that for radial roller bearings $X = 1$ and $Y = 0$. Therefore, for these bearings $F_e = F_r$.

Radial ball bearings are the most widely used type of bearings. At the beginning of the selection process based on the bore diameter of the bearing, we can begin by considering bearings from the light series of radial ball bearings. If it happens that the equivalent load F_e exceeds the value of the dynamic load rating C from the light series, we need to consider a bearing in the heavier series of radial ball bearings with the same bore diameter. However, in the vast majority of cases, the equivalent load is less than the dynamic load rating, but the service live is less than required. In these cases, we need to select for the same bore diameter a radial ball bearing having an increased dynamic load rating in the heavier series or continue the search among radial roller bearings. If the search does not deliver the required results, it is necessary to increase the bore diameter to the next higher standard dimension and repeat the search starting with radial ball bearings. Note: in cases when the service life of the lightest series exceeds the required value of the service life, do not decrease the bore diameter of the bearing because the stresses in the shaft could be determined by a high torque that is not producing radial loading on the supports. In these cases, we select the lightest bearing for the required bore diameter.

As mentioned above, the calculation of the equivalent load is associated with accounting for the load factors X and Y that depend on the characteristics of the applied loads and types of bearings.

2-4.3. Static Load Rating

In many industrial applications, rolling bearings rotate very slowly, not exceeding 10 rpm, or do not rotate at all. The definition of static load rating is based on the allowable residual deformation of the contacting elements of a bearing. Therefore, the static load rating represents an equivalent load that results in residual deformation of the rings and roller elements that does not exceed 0.0001 of the diameter of the rolling element. The static load rating C_0 is tabulated in the catalogs next to the dynamic load rating. Depending on the type of bearing, the value of the static load rating may be less or may exceed the dynamic load rating. The equivalent static load F_0 according to bearing catalogs can be determined using the following empiric formula:

$$F_0 = X_0 F_r + Y_0 F_a \tag{2-4.6}$$

where F_r is the radial load, F_a is the axial load, and X_0 and Y_0 are the radial and axial load factors, respectively, and they can be found in the catalogs and handbooks. Based on some catalogs, it is recommended for training purposes to take for radial ball bearings $X_0 = 0.6$ and $Y_0 = 0.5$, while for angular-contact bearings it may be recommended to take $X_0 = 0.5$ and $Y_0 = 0.26$. However, according to other catalogs, the load factor Y_0 depends on the angle of contact, and it ranges from 0.26 to 0.47. For radial roller bearings, the load factor $X = X_0 = 1$ and the load factor $Y = Y_0 = 0$. Therefore, $F_e = F_0 = F_r$. Thus, for cylindrical roller bearings the equivalent dynamic load as well as the equivalent static load are equal. For tapered roller bearings, it is recommended for training purposes to take $X_0 = 0.5$, while for Y_0 it is recommended in this book to take $Y_0 = 0.4$, which according to some catalogs is the lowest value of this load factor for tapered roller bearings.

2-4.4 Calculating the Service Life for Rolling Bearings Based on Dynamic Load Ratings

When the bore diameter is already determined, we select a corresponding radial ball bearing and then have to determine the unknown load factors for the formula (2-4.5) in order to calculate the equivalent radial load F_e. Obviously, these radial and axial loads are the known reactions in the supports that, actually, represent the bearings. Assuming that the inner ring is rotating, we take $V = 1$. In order to calculate the equivalent load, we should first determine the load factors X and Y. The values of these factors can be found in bearing catalogs or other published sources. The values of these factors are also presented in the appropriate tables in the Appendixes A-2-2 and A-2-3. The use of these tables and determining the X and Y factors are shown in the examples below. Determining the load factors X and Y is associated with one more factor, e , that represents the ratio between the axial load F_a and the static load rating C_0:

$$e = \frac{F_a}{C_0} \tag{2-4.7}$$

With introducing the factor e according to formula (2-4.7), it becomes obvious that in order to find the values of the load factors X and Y, we need to find the value of the static load rating C_0 first. Consequently, one of the first steps in the bearing selection process is to determine the load factors X_0 and Y_0. For the bore diameter determined for the project, we find in the manufacturer's catalog the values of X_0 and Y_0 for the light series of the radial ball bearing. We calculate the value of F_0 and compare it with the tabulated value of the static load rating C_0. If $F_0 \le C_0$, we proceed with the evaluation of the applicability of this bearing to the requirements of the project. If $F_0 > C_0$, we go to the next row in the catalog that contains the information of the heavier bearing having the same bore diameter. When the appropriate value of C_0 is found, we then calculate, according to formula (2-4.7), the e factor which should be compared with the ratio $\frac{F_a}{VF_r}$. This comparison leads to two results: (1) $\frac{F_a}{VF_r} \le e = \frac{F_a}{C_0}$ and (2) $\frac{F_a}{VF_r} > e = \frac{F_a}{C_0}$. Depending on these results, we proceed with determining from the catalog the load factors X and Y. In cases when for a single-row radial ball bearing or an angular-contact ball bearing $\frac{F_a}{VF_r} \le e$ then $X = 1$ and $Y = 0$. If for these bearings $\frac{F_a}{VF_r} > e$ then the values of the X and Y should be taken from the appropriate tables in the catalogs. Some of these tables are presented for training purposes in Appendixes A-2-2 and A-2-3. After determining the load factors X and Y, we calculate, according to formula (2-4.5), the value of the equivalent load F_e. Thus, in order to determine the service life of the bearing, we should first calculate F_e, find in the catalog the dynamic load rating C, and use formula (2-4.4). If the calculated service life for a certain bore diameter is insufficient, we go to the next heavier series or even to a higher standard bore diameter.

The relationship between F_e and C can be characterized by the following expression: $F_e = \theta C$, where $0 < \theta \le 1$; then based on formula (2-4.4), we may write $L_{10} = \left(\frac{C}{\theta C}\right)^a = \left(\frac{1}{\theta}\right)^a = L_{10} = \left(\frac{1}{\theta}\right)^a$. Taking, for example, a ball bearing for which $a = 3$, and assuming that $\theta = 1$ then $L_{10} = 1 \cdot 10^6$ revolutions, if $\theta = 0.5$ then $L_{10} = \left(\frac{1}{0.5}\right)^3 = 8 \cdot 10^6$ revolutions, and if

$\theta = 0.15$ then $L_{10} = \left(\frac{1}{0.15}\right)^3 = 296 \cdot 10^6$ revolutions. Usually the service life is measured in hours, and it is denoted by L_{10h}. Thus, taking for the last example a rotational speed of $n = 500$ rpm, we obtain $L_{10h} = \frac{296 \cdot 10^6}{60 \cdot 500} = 9{,}866$ hours. This is a reasonable value for bearing service life; however, for each particular case the bearing service life should be in accordance with the engineering requirements of the project.

In cases when the preliminarily selected bearing from the light series cannot provide the required value of the service life, we select a bearing from a slightly heavier series in order to increase the dynamic load rating C of the bearing; however, this results in the increase of C_0 as well. The increase of C_0 causes a decrease of the factor e. Hence, if the preliminary value of the ratio was $\frac{F_a}{F_r} \le e$, then, it may happen, that due to the increase of C_0 the factor e could decrease to the level that this ratio could become $\frac{F_a}{F_r} > e$. This requires determining the values of the load factors X and Y according to the case when $\frac{F_a}{F_r} > e$ that may change the equivalent dynamic load. Therefore, when at the beginning of the bearing selection process, we started with $\frac{F_a}{F_r} \le e$ and then we went to a heavier bearing series, it is necessary to recalculate the factor e, and if it happens that for the newly selected bearing $\frac{F_a}{F_r} > e$, then we need to determine the appropriate values of X and Y, and, correspondingly, recalculate the equivalent load F_e. However, if the preliminary value of the ratio was $\frac{F_a}{F_r} > e$, then by going to the heavier series of bearings there would be no need to recalculate the factor e and, consequently, the equivalent dynamic load.

It should be stressed that in order to select a bearing the engineering practitioner has to use a bearing catalog. All catalogs contain for each bearing the formulas for F_e and F_0 as well as the corresponding load factors X, Y, X_0, and Y_0.

Just for training purposes, below are presented the rules for determining the load factors X_0 and Y_0 as well as X and Y for rolling radial ball, angular-contact ball, radial roller, and tapered roller single-row bearings.

1) Radial ball bearings
   a) For the equivalent static load:
      $X_0 = 0.6$ and $Y_0 = 0.5$
   b) For the equivalent dynamic load:

      When $\frac{F_a}{F_r} \le e$ then $X = 1$ and $Y = 0$. Therefore, $F_e = F_r$.

      When $\frac{F_a}{F_r} > e$ then use catalog or the table in Appendix A-2-2.

2) Angular-contact bearings
   a) For the equivalent static load:
      $X_0 = 0.5$ and $Y_0 = 0.26$
      When $F_0 < F_r$ then use $F_0 = F_r$
   b) For the equivalent dynamic load:
      When $\frac{F_a}{F_r} \le e$ then $X = 1$ and $Y = 0$. Therefore, $F_e = F_r$.
      For contact angles from $\alpha = 10°$ to $\alpha = 40°$, when $\frac{F_a}{F_r} > e$ then use the catalog or the table in Appendix A-2-3.
      Some catalogs suggest when $\frac{F_a}{F_r} \le 1.14$ then $F_e = F_r + 0.55F_a$, and when $\frac{F_a}{F_r} > 1.14$ then $X = 0.35$ and $Y = 0.57$
3) Radial roller bearings
   Since $F_a = 0$, use $X_0 = X = 1$ and $Y_0 = Y = 0$. Therefore, $F_e = F_r$ and $F_0 = F_r$.
4) Tapered roller bearings
   a) For equivalent static load:
      $X_0 = 0.5$, while the value of the factor Y_0 should be taken from the catalog. In addition, it should be mentioned that if $F_0 < F_r$ then use $F_0 = F_r$.
   b) For equivalent dynamic load :
      When $\frac{F_a}{F_r} \le e$ then $X = 1$ and $Y = 0$. Therefore, for this case $F_e = F_r$.
      However, if for these roller bearings $\frac{F_a}{F_r} > e$ then $X = 0.4$ and Y should be taken from the catalog.

Example 2-4.1a: Selecting a radial ball bearing.
A shaft is rotating at a rotational speed of 180 rpm, and its most loaded support is subjected to a radial force $F_r = 2{,}980$ N and an axial force $F_a = 650$ N. The journal of the shaft has a diameter of 30 mm. We select a ball bearing that will provide a service live of about 22,000 hours.

Solution:
From the SKF bearing catalog, we select a light series radial ball bearing No. 61806 having the bore diameter of 30 mm. For this bearing, $X_0 = 0.6$ and $Y_0 = 0.5$. Hence, according to formula (2-4.6), we determine the equivalent static load $F_0 = 0.6 \cdot 2{,}980 + 0.5 \cdot 650 = 2{,}113$ N. Since the static load rating of this bearing $C_0 = 4.55$ kN exceeds the calculated value of F_0 (4,550 > 2,113), we proceed with this bearing. As the inner ring is rotating, hence $V = 1$. Using the formula (2-4.7), we calculate $e = \frac{F_a}{C_0} = \frac{650}{4{,}550} = 0.142$. Now we determine the ratio $\frac{F_a}{VF_r} = \frac{650}{1 \cdot 2{,}980} = 0.218$. In this case $\frac{F_a}{F_r} = 0.218 > 0.142$, therefore, in order to determine the load

factors X and Y, we have to use the catalog or the table in Appendix A-2-3. According to this table, for $e = 0.109$ the load factors $X = 0.56$ and $Y = 2.3$. Now based on the expression (2-4.5) we calculate the equivalent dynamic load applied to the bearing:
$F_e = XVF_r + YF_a = 0.56 \cdot 1 \cdot 2{,}980 + 2.3 \cdot 650 = 3{,}152.6$ N.
The catalog value of the dynamic load rating for this bearing is C = 7,280 N that exceeds the equivalent dynamic load applied to the bearing that equals F_e = 3,152 N. However, we cannot finalize the decision before we determine the service life of this bearing with regard to the equivalent dynamic load. According to formula (2-4.4), we calculate the service life of the bearing $L_{10} = \left(\frac{c}{F_e}\right)^3 = \left(\frac{7{,}280}{3{,}152.6}\right)^3 = 12.32 \cdot 10^6$ revolutions. Considering that the rotational speed of the shaft is 180 rpm, we convert the calculated service life into hours and we have $L_{10h} = \frac{12.32 \cdot 10^6}{180 \cdot 60} = 1{,}141$ hours. It is now clear that the selected bearing does not satisfy the requirements stated above.

We now select a radial ball bearing with the same bore diameter from the medium series No. 6206 having C = 20.3 kN and C_0 = 11.2 kN. Since the newly selected C_0 exceeds the value of the previous one and, consequently, also exceeds the equivalent static load F_0 and, in addition, it decreases the factor e, therefore there is no need to recalculate the equivalent dynamic load. For this bearing the service life is $L_{10} = \left(\frac{C}{F_e}\right)^3 = \left(\frac{20.3}{3.153}\right)^3 = 266.88 \cdot 10^6$ revolutions or $L_{10h} = \frac{266.88 \cdot 10^6}{180 \cdot 60} = 24{,}711$ hours, exceeding the required service life by about 12%. This is acceptable.

2-5. Design Considerations

It is highly recommended that students become familiar with the bearing manufacturers' catalogs that describe in a very comprehensive way selection, design, assembly, disassembly, lubrication, alignment, maintenance, and other bearing related issues that could not be addresses in this book. These catalogs contain the needed information regarding the housings and enclosures of bearings, the tolerances related to machining of the journals and housings, and many more related engineering considerations.

Some related design considerations are presented below.

The numbering system of bearings contains the information related to the most important characteristics of the bearings. The designation number of a bearing comprises certain numerals and letters. These letters and numerals indicate the dimension of the bore diameter, the type of the rolling elements, the type of structure, the availability of shields, and other specific details. For instance, in the metric bearings starting with the diameter of 20 mm, the last two digits indicate the size of the bore diameter. Therefore, if the two last digits, for example, are 06 then the bore diameter is 6·5 = 30 mm, while if these two digits are 17, then the bore diameter is 17·5 = 85 mm.

In the engineering documentation, usually in the bill of materials the engineer indicates the designation numbers of the bearings and specifies the manufacturer. The majority of manufacturers have similar numbering systems.

It is a good practice to have identical bearings on both sides of the shaft. Avoid using a single bearing installed inside of a rotating element, such as a pulley or a gear.

The diameters of the journals should be machined according to the interference tolerances assuring free rotation of the inner rings together with the shaft, while the inner diameters of the housings or enclosures accommodating the bearing should be machined in accordance with the clearance fits that allow the outer rings to perform a complete revolution in about 24 hours. This will prevent the repeated contacts of the rolling elements at the same points or lines of the raceways of the rings.

The housings usually represent an off-the-shelf product, and in many cases, housings can be selected from the appropriate manufacturers' catalogs. Depending on their design, housings can be divided in two groups: single block and an assembly consisting of lower and upper halves. The latter group is preferred when disassembly is often required to replace a worn out element, such as a drive belt or the bearing itself. The most accurate alignment of the bearing is achieved when the enclosure is a solid body or a frame having a corresponding structure that allows manufacturing the two holes for the bearings in one setup. In this case, the designer should pay special attention to the possibility of disassembling the bearings when necessary. It has happened many times that the bearings were pressed into their places in a certain unit; however, they could not be removed from the assembly for replacement.

It is preferable to rigidly secure the inner rings to the journals of the shaft by retaining rings or by nuts. In cases of using radial ball or roller bearings, it is necessary to provide for the possibility of the axial displacement in both directions of one end of the shaft supported by a bearing due to temperature changes. This resembles the behavior of a beam supported by a movable and a not movable supports. The bearing representing the not movable support should be mounted on the end of the shaft that is carrying a pulley, a gear, or is engaged by a coupling with an assembly, or electrical motor, etc.

In cases of using angular-contact ball bearings or tapered roller bearings, during the assembly process gaps should be provided that will disappear during the thermal expansion of the shaft. The size of these gaps should be determined by consulting the bearing catalogs. In some cases, these bearings could be adjusted by the appropriate preload techniques that are described in detail in bearing catalogs. In case of relatively long shafts subjected to the action of significant radial and axial loads, it is recommended using an arrangement of two angular-contact ball bearings or two tapered roller bearings mounted in tandem back-to-back or face-to-face on one end of the shaft. This pair of the bearings is intended to take the axial loading and part of the radial loading, and to form a not movable support. On the other end of the shaft, a single radial ball or roller bearing playing the role of a movable support should be installed.

Rolling bearings require less lubrication than sliding bearings. Rolling bearings that are shielded on both sides are lubricated at the time of their production and, obviously, do not require lubrication during operation. Both types of lubricants, the liquid and the consistent (grease), are used for rolling bearings. When using liquid lubricants, the level of the oil should not be higher than the center of the lower ball or roller. Lubricant levels higher than recommended can lead to excessive energy losses due to mixing the oil, resulting in raising the oil's temperature and reducing the viscosity of the oil, decreasing the efficiency of the lubrication. In cases of applying consistent lubricants (grease), the bearing housing should be filled to about 60% of its volume.

Chapter 2. Problems

2-01. A journal of a shaft has a diameter d = 60 mm and is rotating at a rotational speed n = 1,200 rpm. Based on the PV factor, select the material for the sliding bearing if the journal is loaded by a force F = 20,000 N.

2-02. A journal of a shaft has a diameter d = 20 mm. Based on the PV factor, select the material for the sliding bearing, and determine the maximum linear velocity of the journal if it is loaded by a force F = 8,000 N.

2-03. A journal of a shaft has a diameter d = 1.5 in. Based on the PV factor, select the material for the sliding bearing, and determine the maximum linear velocity of the journal if it is loaded by a force of F = 1,500 lb.

2-04. A journal of a shaft has a diameter d = 2.0 in. Based on the PV factor, select the material for the sliding bearing, and determine the maximum radial force that could be applied to the journal if its linear velocity V = 450 fpm.

2-05. Select a rolling radial ball bearing based on the following input parameters: the bore diameter d = 35 mm, the radial force F_r = 5,000 N, the axial force F_a = 0, and the rotational speed n = 500 rpm. The bearing should provide a service life of about 10,000 hours. Indicate the manufacturer and the designation number of the selected bearing.

2-06. The NTN angular-contact ball bearing designation No. 7210 has the bore diameter d = 1,9685 in. (50 mm), the dynamic load rating C = 9,300 lb and the static load rating C_0 = 7,050 lb. The bearing is loaded by a radial force F_r = 600 lb and by an axial force F_a = 100 lb, while the rotational speed is n = 3,600 rpm. Determine the expected service life of the bearing.

2-07. The SKF angular-contact ball bearing designation No. 7205BEP has the bore diameter d = 25 mm, the dynamic load rating C = 14,800 N, and the static load rating C_0 = 9,300 N. The bearing is loaded by a radial force F_r = 1,000 N and an axial force F_a = 400 N, while the rotational speed is n = 1,800 rpm. Determine the expected service life of the bearing.

2-08. Select a radial roller bearing based on the following input parameters: the bore diameter d = 25 mm, the radial force F_r = 830 lb, the axial force F_a = 0, and the rotational speed n = 1.600 rpm. The bearing should provide a service life of about 12,000 hours. Indicate the manufacturer and the designation number of the selected bearing.

2-09. The tapered roller bearing SKF No. 32305 J2 has the dynamic load rating C = 60.5 kN and the static load rating C_0 = 63 kN. The bearing is selected for a shaft having the diameters of the journals d = 25 mm. The radial force applied to shaft F_r = 10,000 N, the axial force F_a = 8,000 N, and the rotational speed n = 180 rpm. Determine the expected value of the bearing service life.

2-10. Select a tapered roller bearing and find its expected service life for the following input parameters: the bore diameter d = 90 mm, the radial force F_r = 26,000 N, the axial force F_a = 32,000 N, and the rotational speed n = 800 rpm. Indicate the manufacturer and the designation number of the selected bearing.

CHAPTER 3

Shafts and Related Elements

Shafts represent one of the most widespread machine elements used in mechanical engineering systems. Usually, a shaft carries rotating machine elements that are rigidly secured to it, and the shaft is supported by a pair of bearings. These rotating machine elements transmit to the shaft or from the shaft the power from or to other rotating elements. The distinctive characteristic of shafts is that they transmit torques. Some machine elements possess structures that are similar to shafts; they may rotate, but they do not transmit torques. Shafts that do not transmit torques are called axles. The term shaft is used for shafts as well as for axles, but, technically, an axle is not a shaft.

In some mechanical systems, shafts are supported by more than two supports. As an example, we might consider a camshaft in an internal combustion engine. The camshaft is running in three or more bearings (supports). In this case, the reactions in the supports become statically undetermined, and as it is mentioned earlier, this type of support arrangement is not considered in this course.

Normally, shafts have two identical journals whose diameters are machined in accordance with the inner diameters of sliding bearings or the bore diameters of rolling bearings. Most shafts have a stepped shape, where the diameters of the steps increase beginning at each journal and moving toward the middle of the shaft. If a part of the shaft extends outside the journal, then this part has a diameter smaller than the diameter of the journal. This provides an easy way to assemble the appropriate machine elements on the steps of the shaft and allows high accuracy machining just on the short steps of the shaft where the matching elements should be secured.

Axles, as it is mentioned above, structurally resemble shafts and may or may not rotate; however, unlike shafts they can be mounted in fixed supports.

The machine elements related to shafts include keys, pins, retaining rings, nuts, couplings, and others. Keys and pins are intended to transmit the torque from or to the shaft from or to the rotating element mounted on the shaft. Retaining rings and nuts prevent the axial displacement of the elements mounted on the shaft or axle. The couplings connect the end of the shaft to another shaft that is associated with the source of power, such as an electrical motor, etc.

Shafts are predominantly loaded by torques and bending moments. Therefore, shafts, in the majority of cases, are subjected to the plane stress state, and their stresses can be determined based on the equivalent stresses introduced in the theories of failure. Numerous experimental investigations indicate that for shafts the fourth theory of failure is in closer compliance with the results of actual tests than other failure theories. Therefore, it is reasonable to apply the fourth theory of failure for calculating the stresses in shafts. The rotation of shafts causes alternating of the bending stresses that is associated with the fatigue phenomenon. If the rotation of the shaft is reversible, then the shaft is subjected to the combination of variable bending and torsional stresses. In cases when the shafts are subjected to one-directional torsional stresses, these stresses are static. In general, shafts may experience static and variable stresses, and methodologies for calculating these stresses are presented below.

There are two basic types of shaft calculations: design calculations and verifying calculations. Design calculations are applied in the development of a new assembly when the basic structural layout and the engineering requirements, such as transmitted power and rotational speed, are known. The basic structural design is associated with the type of rotating elements (pulley, gear) and their positioning in the assembly; for example, they could be located between the supports (bearings) or outside the supports. In these cases, when the diameter of the shaft is unknown, the appropriate calculations cannot be performed because the bending moments are also unknown. The bending moments depend, among other factors, on the distance between the supports; however, this distance depends on the width of the bearings and their enclosures and on the width of the rotating elements. The widths of these components depend on the diameter of the shaft. In other words, we need to know the diameter of the shaft in order to begin sketching the assembly. However, we cannot calculate the diameter of the shaft for all the reasons presented above. The solution to this problem comes from the semi-empiric formulas presented below. These formulas allow preliminary calculations of the required diameter of the shaft that represents the bore diameter of the corresponding bearing. Based on this diameter, we may select the bearings and their enclosures, come up with the preliminary design of the steps of the shaft, determine the width of the hubs of the rotating elements, and estimate the distances between the supports, etc. All this allows us to draw the free-body diagram to plot the bending moments and the torque diagrams that are necessary to apply the conventional stress calculations for the shafts.

Verifying calculations are applicable when the structure and the corresponding dimensions are known. In other words, these calculations are intended to verify the strength of an existing shaft. It happens that a machine element of an assembly, for instance a shaft, fails under operational conditions. In this case, the strength verifying calculations are highly justifiable.

These two types of engineering calculations related to strength of the shafts are discussed below.

Obviously, the materials that the machine elements are manufactured from are associated with the strength of these elements. The corresponding characteristics of the materials intended for manufacturing of shafts are presented below.

Generally, shafts and axles are machined from carbon and alloy steels. In most cases, high carbon steel is used for shafts. In corrosive environments, it is advisable to use stainless steel for shafts and axles. If shafts are supported by journal bearings and require elevated hardness, they should be manufactured from high carbon steel and undergo the appropriate thermal treatment.

In cases where impact loading occurs, shafts with hard surfaces and soft cores are required. This can be achieved by making the shaft from low carbon steel with carburization of the surface of its journals and consequent quenching.

3-1. Analysis of the Static Plane Stress State

We analyze a general case related to a shaft subjected to the action of a combination of loads causing axial, shear, bending, and torsional stresses. The goal is to reveal the influence of these stresses on the stress state of the shaft and to clarify the steps of the methodology of shaft strength calculations. The analysis of the stresses is performed in general terms, but can be used for similar calculations of particular cases based on corresponding numeric data. It is desirable to consider an example that reflects an actual interrelationship between the loads, their position on the shaft, and the geometric parameters of the shaft (cross-sectional area and length). As an example, we take a realistic transmission comprising a shaft rotating on two bearings and carrying a machine element exerting three mutually perpendicular forces that are applied to the shaft. This machine element for general applications represents a bevel gear that is discussed in detail in the chapter related to bevel gears and presented later in this book.

In order to simplify the analysis, we replace the stepped shaft with a uniform-diameter shaft having the diameter equal to the diameter of its journals. Therefore, the strength of the actual shaft having corresponding steps is higher than the strength of the uniform-diameter shaft in this example.

Figure 3-1.1a shows a shaft mounted in bearings with a non-movable support A and a movable support B.

The shaft is carrying a bevel gear (the upper part of it is shown in the figure) engaged with another bevel gear (not shown in the figure) that receives power from an electric motor. In the engagement of these gears at point C is acting a system of three mutually perpendicular forces F_1, F_2, and F_3 directed along axes x, y, and z. These forces are interrelated, and their values are determined as follows: $F_1 = 0.3F$, $F_2 = 0.2F$, and $F_3 = F$. The distance between point C and the axis of the shaft, as it is shown in Figure 3-1.1a, equals $2a$. The force F_1 compresses the shaft and creates a moment acting in the vertical plane. The force F_2 also produces a moment in the vertical plane. The force F_3 creates a moment in the horizontal plane and rotates the shaft producing a torque that is transmitted to the far right end of the shaft. This torque is balanced by the torque T_0 shown in Figure 3-1.1a. The value of the diameter of the shaft $d = 0.5a$. (This is estimated based on actual data.)

We need to determine the values of the loading factors in the most loaded cross-sectional areas of the shaft in the vertical and in the horizontal planes. Combining the loads in these two planes, we determine the loads and, consequently, the stresses in the most stressed cross-sectional area.

Figure 3-1.1b shows the free-body diagram in the vertical plane (FBD – xy).

We start by computing the horizontal reaction in the support A. Recalling equation (1-3.1), we may write $R_{Ax} - F_1 = 0$. Substituting here the value of F_1, we obtain

$$R_{Ax} = 0.3F \tag{3-1.1}$$

Based on expression (1-3.3), we compose the equation of the sum of moments about point B $R_{Ay}\,2a - F_2 2a - F_1 2a = 0$. Substituting into this equation the values of F_1 and F_2 and solving it for R_{Ay}, we have

$$R_{Ay} = 0.4F \tag{3-1.2}$$

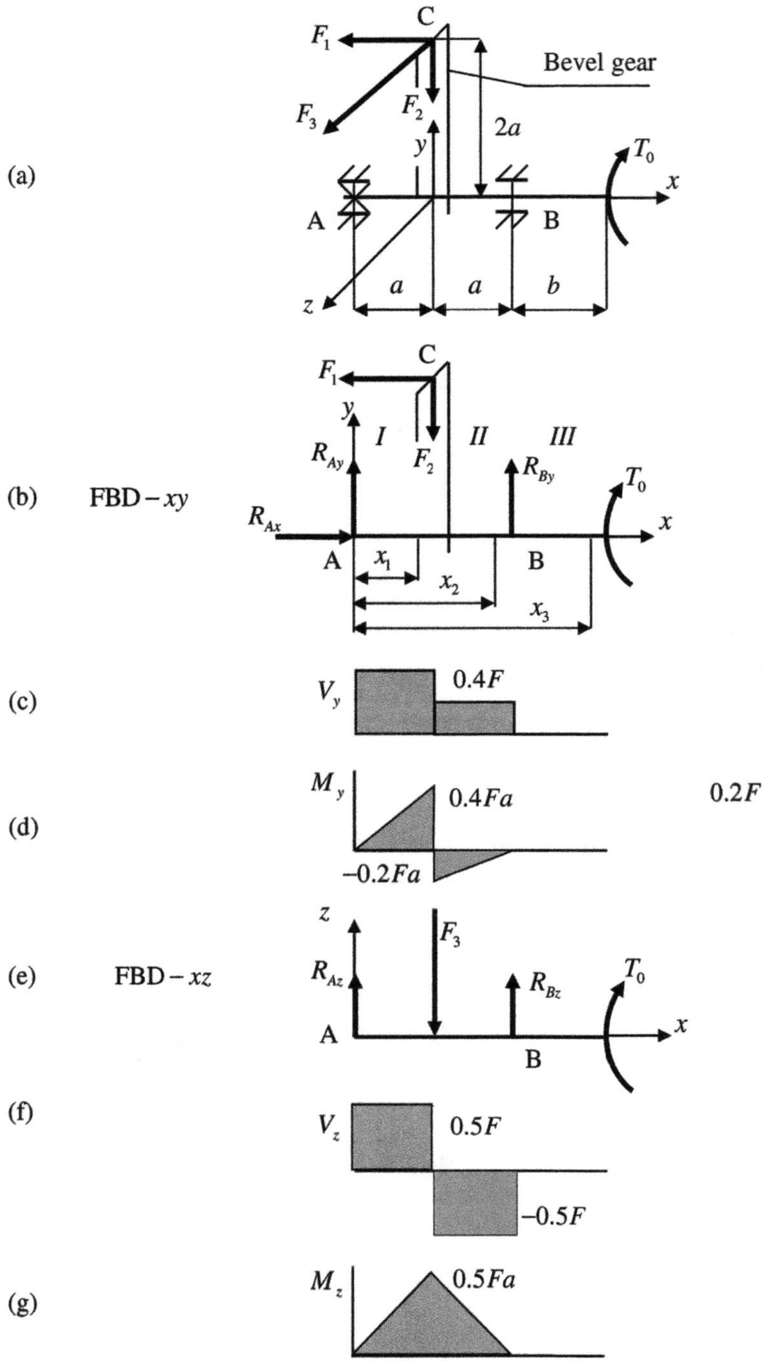

Figure 3-1.1. Shaft subjected to a combined load.

Recalling equation (1-3.2), we write: $R_{Ay} - F_2 + R_{By} = 0$. Substituting here the corresponding values, we obtain

$$R_{By} = -0.2F \tag{3-1.3}$$

Dividing the shaft in the free-body diagram into three intervals, as shown in Figure 3-1.1b, and making cuts indicated by x_1, x_2, and x_3, we determine the shear forces and bending moments acting on these intervals. According to the free-body diagram in Figure 3-1.1b, we determine the limits of the corresponding variables on the first, second, and third intervals. Thus, we may write

$$0 \le x_1 \le a \tag{3-1.4}$$

$$a \le x_2 \le 2a \tag{3-1.5}$$

$$2a \le x_3 \le 2a + b \tag{3-1.6}$$

According to the free-body diagram, we determine the shear force on the first interval:

$$V_{yI} = R_{Ay} = 0.4F \tag{3-1.7}$$

Composing the appropriate equation for the shear forces on the second interval, we may write $V_{yII} = R_{Ay} - F_2$. Substituting into this equation the corresponding values, we obtain

$$V_{yII} = 0.2F \tag{3-1.8}$$

On the third interval we have $V_{yIII} = R_{Ay} - F_2 + R_{By}$. Substituting the corresponding values, we determine that $V_{yIII} = 0$. Equations (3-1.1) and (3-1.2) show that the shear forces are characterized by segments of straight lines parallel to the axis x. Considering that $V_{yIII} = 0$ and applying equations (3-1.7) and (3-1.8), we construct the shear-forces diagram that is presented in Figure 3-1.1c.

Referring to the free-body diagram in Figure 3-1.1b, we compose the equation for the bending moment on the first interval: $M_{yI} = R_{Ay}\,x_1$. Taking into account in this equation the limits of the first interval according to expression (3-1.4), we determine that when $x_1 = 0$ then $M_{yI} = 0$, and when $x_1 = a$ then $M_{yI} = R_{Ay}\,a$. Substituting the value of R_{Ay}, we obtain

$$M_{yI} = 0.4Fa \tag{3-1.9}$$

The equation for the bending moment on the second interval reads $M_{yII} = R_{Ay}x_2 - F_3(x_2 - a) - F_1 2a$. Combining this equation with the expression (3-1.5), we may write that when $x_2 = a$ then $M_{yII} = 0.4Fa - 0.3F2a$, hence, we have

$$M_{yII} = -0.2Fa \tag{3-1.10}$$

In addition, when $x_2 = 2a$ then $M_{yII} = 0.4F2a - 0.3F2a - 0.2Fa = 0$. The bending moment on the third interval can be calculated from the following equation: $M_{yIII} = R_{Ay}x_3 - F_1 2a - F_2(x_3 - a) + R_{By}(x_3 - a)$. Substituting here the corresponding values and accounting that R_{By} is negative, we obtain that $M_{yIII} = 0$. Considering that the bending moments equations on the first and second intervals represent linear functions of the abscissas and taking into consideration equations (3-1.9) and (3-1.10), we plot the bending-moment diagram as it is shown in Figure 3-1.1d.

Now we analyze the free-body diagram for the horizontal plane (FBD – xz) that is presented in Figure 3-1.1e. In order to determine the reaction R_{Az}, we are summing all moments about point B, hence, $R_{Az}\,2a - F_3 a = 0$. Substituting the value of F_3 we calculate the reaction R_{Az} and obtain

$$R_{Az} = 0.5F \tag{3-1.11}$$

In this case it is obvious that $R_{Az} = R_{Bz} = 0.5F$.

The free-body diagram of the shaft for the horizontal plane is also divided in the same three intervals as the free-body diagram for the vertical plane, and, consequently, the limits of the variables on these intervals are described by the same three expressions (3-1.4), (3-1.5), and (3-1.6).

The shear force on the first interval equals

$$V_{zI} = R_{Az} = 0.5F \tag{3-1.12}$$

The shear force on the second interval is determined from the following equation: $V_{zII} = R_{Az} - F_3$. Substituting the corresponding values, we obtain

$$V_{zII} = -0.5F \tag{3-1.13}$$

The shear force on the third interval is determined from the following equation: $V_{zIII} = R_{Az} - F_2 + R_{Bz}$. Substituting the corresponding values, we have $V_{zIII} = 0$. The shear-force diagram is presented in Figure 3-1.1f in accordance with the equations (3-1.12) and (3-1.13).

Considering the bending moments, we find that the equation for the bending moment on the first interval is $M_{zI} = R_{Az}x_1$. Combining this equation with the expression (3-1.4), we have that when $x_1 = 0$ then $M_{zI} = 0$, and when $x_1 = a$ then

$$M_{zI} = 0.5Fa \tag{3-1.14}$$

The equation for the bending moment on the second interval reads $M_{zII} = R_{Az}x_2 - F_3\,(x_2 - a)$. Considering expression (3-1.5) for this equation, we find that when $x_2 = a$ then $M_{zII} = 0.5Fa$, and when $x_2 = 2a$ then

$$M_{zII} = 0 \tag{3-1.15}$$

The equation of the bending moments for the third interval follows:
$M_{zIII} = R_{Az}x_3 - F_3\,(x_3 - a) + R_{Bz}\,(x_3 - 2a)$.
Combining this equation with the expression (3-1.6), we have that when $x_3 = 2a$ then $M_{zIII} = 0$, and when $x_3 = 2a + b$ then $M_{zIII} = 0$ as well. Based on these calculations and on equations (3-1.14) and (3-1.15), we plotted the bending moments diagram that is presented in Figure 3-1.1g.

The continuation of the analysis of this example is presented in Figure 3-1.2 and involves plotting the torque diagram.

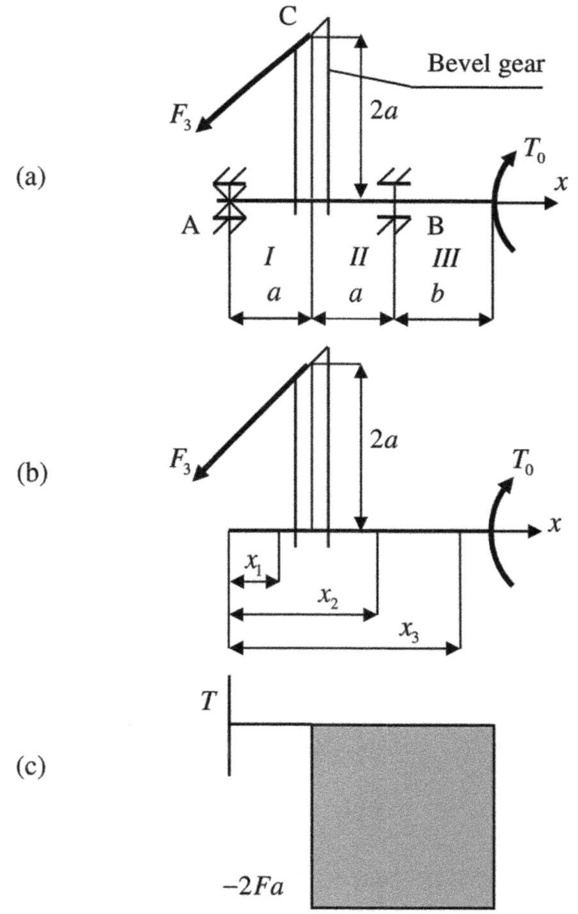

Figure 3-1.2. Shaft, free-body and torque diagrams.

(Continuation of Figure 3-1.1).

Figure 3-1.2a shows the same shaft and the bevel gear as in figure 3-1.1a with the same force *F3* that produces a torque applied to the shaft. Figure 3-1.2a shows that the shaft is divided into the same three intervals that have the same limits.

The free-body diagram of the shaft is presented in Figure 3-1.2b that allows for calculating the torques acting in the shaft. According to the cut having the abscissa x_1, we determine the torque acting on the first interval $T_I = 0$. The equation for the torque acting on the second interval reads

$$T_{II} = -2Fa \tag{3-1.16}$$

The equation for the torque on the third interval is identical to equation (3-1.16), and we have $T_{III} = -2Fa$. Figure 3-1.2c shows the torque diagram on the second and third intervals.

Now we may analyze the loads and the stresses related to the shaft.

The shaft is subjected to the following loads whose maximum absolute values are

1) the axial reaction: $R_{Ax} = |0.3F|$;
2) the shear force in the vertical plane acting on the first interval: $V_{yI} = V_{ver} = |0.4F|$;
3) the bending moment in the vertical plane acting in the cross-sectional area that divides the shaft into the first and second intervals: $M_{yI} = M_{ver} = |0.4Fa|$;
4) the shear force in the horizontal plane acting on the first interval: $V_{zI} = V_{hor} = |0.5F|$;
5) the bending moment in the horizontal plane acting in the cross-sectional area that divides the shaft into the first and second intervals: $M_{zI} = M_{hor} = |0.5Fa|$;
6) the torque acting on the first and second intervals: $T_{II} = T_{III} = T_{\max} = |2Fa|$.

As the diagrams show, the most loaded cross-sectional area is the area that divides the shaft into the first and second intervals.

The loading factors of the same type acting in mutually perpendicular planes and creating mutually perpendicular stresses at the same point may be added to each other on the basis of vector summation. Therefore, we may write

$$V_{\max} = \sqrt{V_{ver}^2 + V_{hor}^2} = \sqrt{(0.4F)^2 + (0.5F)^2} = 0.64F$$

and

$$M_{\max} = \sqrt{M_{ver}^2 + M_{hor}^2} = \sqrt{(0.4Fa)^2 + (0.5Fa)^2} = 0.64Fa.$$

Now we calculate the stresses in order to clarify their influence on the stress state of the shaft. As mentioned above, the diameter of the shaft $d = 0.5a$.

The normal stress σ_x' caused by the reaction R_{Ax} is calculated using the formula (1-4.1), and we obtain $\sigma_x' = 1.53\dfrac{F}{a^2}$. This stress is uniformly distributed over the cross-sectional area and may be added to a normal stress acting on the periphery of the cross-sectional area.

Substituting the value of V_{max} into formula (1-4.19), we calculate the maximum value of the shear stress that is acting in the center of gravity of the cross-sectional area, so $\tau'_{max} = 4.35\frac{F}{a^2}$. Since the stress σ'_x is also acting in the center of gravity of the same cross-sectional area as τ'_{max}, we may use an applicable theory of failure in order to evaluate the equivalent stress in the center of gravity of this area. For this example, both the third and the fourth theories can be applied to determine the equivalent stress, and we will use both of them. Denoting for the third theory of failure the equivalent stress by $\sigma^{III}_{eq.c}$ (subscript "*c*" stays for the "center" of gravity) that acts in the center of gravity of the cross-sectional area and recalling the expression (1-6.7), we may write that $\sigma^{III}_{eq.c} = \sqrt{\left(1.53\frac{F}{a^2}\right)^2 + 4\left(4.35\frac{F}{a^2}\right)^2} = 8.8\frac{F}{a^2}$. Denoting by $\sigma^{IV}_{eq.c}$ the equivalent stress according to the fourth theory of failure and using the expression (1-6.10), we obtain $\sigma^{IV}_{eq.c} = \sqrt{\left(1.53\frac{F}{a^2}\right)^2 + 3\left(4.35\frac{F}{a^2}\right)^2} = 7.7\frac{F}{a^2}$.

Now we calculate the stresses acting on the periphery of the cross-sectional area. The maximum bending stress is determined from the flexure formula (1-4.17). Substituting into this formula the value of the maximum bending moment $M_{max} = 0.64Fa$ and using formula (1-4.18) for the moment of inertia, we obtain $\sigma''_x = 52.2\frac{F}{a^2}$. The maximum torsional stress τ''_{max} can be determined from the torsion formula (1-4.20) by substituting into it the maximum torque $T_{max} = 2Fa$ and the polar moment of inertia calculated by the formula (1-4.21). Therefore we obtain $\tau''_{max} = 81.5\frac{F}{a^2}$.

Using the third theory of failure, we determine the equivalent stress σ^{III}_{eq} according to expression (1-6.7), and we have $\sigma^{III}_{eq} = \sqrt{\left(52.2\frac{F}{a^2}\right)^2 + 4\left(81.5\frac{F}{a^2}\right)^2} = 171\frac{F}{a^2}$. Based on expression (1-6.10), we calculate the equivalent stress σ^{IV}_{eq} according to the fourth theory of failure: $\sigma^{IV}_{eq} = \sqrt{\left(52.2\frac{F}{a^2}\right)^2 + 3\left(81.5\frac{F}{a^2}\right)^2} = 150.5\frac{F}{a^2}$.

First, let us determine the ratio between the equivalent stresses of these two theories of failure, hence $\frac{\sigma^{III}_{eq}}{\sigma^{IV}_{eq}} = \frac{171}{150.5} = 1.136$. Since the two theories of failure are applicable in the engineering practice, we may conclude that a difference in the corresponding calculations of about 13.6% is acceptable. Therefore, about a 13.6% error in calculating the stresses developed in machine elements subjected to the plane stress state may be considered acceptable in engineering design practice. This is justified by using safety factors that compensate for this error with a significant margin of safety. Now, applying the third and fourth theories of failure, we compare the corresponding equivalent stresses describing the plane stress states in the center of gravity of the cross-sectional area of the shaft as well as in the periphery of this area. Calculating the ratios between the corresponding equivalent stresses allows us to compare these stresses and evaluate their influence on the strength of the shaft. These ratios are as follows: $\frac{\sigma_{eq.c}^{III}}{\sigma_{eq}^{III}} = \frac{8.8}{171} = 0.0515$ and $\frac{\sigma_{eq.c}^{IV}}{\sigma_{eq}^{IV}} = \frac{7.7}{150.5} = 0.0512$. These ratios show that the stresses in the center of gravity of the cross-sectional area are about 5.1% of the stresses in the periphery of this area and at the same time they are about 2.667 times less $\left(\frac{13.6}{5.1} = 2.667\right)$ than the acceptable error of the result of the engineering stress calculations. Therefore, it is justifiable to conclude that the stresses caused by the axial and shear forces applied to the majority of shafts and axles that are used in assemblies with other machine elements for general applications are insignificant and can be ignored in the appropriate stress calculations. It becomes obvious that during the process of stress calculations of shafts there is no need to determine the shear forces or to plot their diagrams. It may be noted that adding the stress σ'_x to the stress σ''_x is not necessary because it does not make any noticeable difference in the results of the stress calculations. However, it is required to determine the axial forces applied to the supports because they should be taken into consideration during the selection of rolling bearings or the design calculations of journal bearings.

Based on the analysis presented above, the methodology of stress calculations for shafts in general should comprise the following steps:

1) determining the reactions in the supports in the vertical and horizontal planes
2) plotting the bending moments diagrams in the vertical and horizontal planes
3) plotting the torque diagram
4) determining based on the diagrams the most loaded cross-sectional areas of the shaft
5) applying the appropriate fatigue considerations accounting for the modifying factors, the stress concentration, and the influence of the nonzero mean stress
6) based on the fourth theory of failure, calculating the equivalent stresses in the most loaded cross-sectional areas

Because axles are not subjected to the action of torsional stresses, the methodology of stress calculations for axles is based only on the bending moments that create a unidirectional stress state characterized by normal stresses. All fatigue considerations are applicable to the calculations of axles.

3-2. Engineering Calculations of Shafts

One of the initial steps in design and development of a new assembly or mechanical system comprising shafts subjected to bending and torsional stresses consists in design calculations to determine the preliminary diameter of the shaft. This allows for the preliminary selection of the bearings and their enclosures, estimating the width of these enclosures and of the rotating elements mounted on the shaft, and, finally, sketching the assembly that predetermines the distance between the supports (bearings). When the distances between the supports and other related elements are already determined, it becomes possible to compute the bending moments acting in the shaft. It should be mentioned that the value of the torque applied to the shaft could be calculated based on the power and rotational speed that actually represent the input data related to the design assignment. Considering the bending moment and the torque, it is necessary to perform the preliminary strength calculations of the shaft and, finally, to verify the compliance of the strength of the shaft with the appropriate criteria of strength (allowable stresses). Therefore, we apply two types of calculations of shafts that are discussed above: design calculations to determine the preliminary diameter of the shaft; and verifying calculations that assure the strength of the shaft complies with the required criteria of strength. Usually, the strength requirements are associated with determining the safety factor, while in certain cases the deformation of the shaft is also included in these requirements. Since the shafts could be subject to static or variable stresses, it is necessary to conduct these calculations for both types of stresses.

3-3. Shaft Calculations for Static Stresses

We consider the two types of calculations mentioned above: design calculations to determine the dimensions of a shaft; and verifying calculations, which are used when the dimensions of shaft are known and it is necessary to determine the safety factor.

The operational conditions of shafts are associated with the unidirectional stress state or with the plane stress state. Therefore, we consider design and verifying calculations for both stress states.

3-3.1. Design Calculations for Unidirectional Stress State

Usually a shaft transmits a certain power at a certain rotational speed. If the shaft is subjected to pure torsion, the operational conditions of the shaft are characterized by the unidirectional stress state, and its diameter can be determined based on the torsion formula (1-4.20): $\tau = \frac{Tc}{J}$, where the polar moment of inertia J is computed according to formula (1-4.21): $J = \frac{32Tc}{\pi d^4}$. Combining formulas (1-4.20) and (1-4.21) we may write

$$\tau = \frac{16T}{\pi d^3} \tag{3-3.1}$$

Since the values of the transmitted power P and the angular velocity ω or the rotational speed n are predetermined by the engineering requirements related to the design of the shaft, we can calculate the torque using the following formula:

$$T = \frac{P}{\omega} \tag{3-3.2}$$

while

$$\omega = \frac{\pi n}{30} \tag{3-3.3}$$

Combining equations (3-3.2) and (3-3.3), we determine the torque:

$$T = \frac{30P}{\pi n} \tag{3-3.4}$$

Selecting the corresponding material for the shaft and the appropriate safety factor n_τ, we calculate the allowable torsional stress: $\tau_{all} = \frac{S_{sy}}{n_\tau}$, where S_{sy} is the yield shear stress of the selected material. Substituting this allowable stress τ_{all} into formula (3-3.1) instead of the stress τ and combining this formula with expression (3-3.4), we determine the diameter d of the shaft:

$$d = 3.65\sqrt[3]{\frac{Pn_\tau}{S_{sy}n}} \tag{3-3.5}$$

where the diameter d is in m (meter), the power P is in W (watt), the yield shear stress in Pa (pascal), and the rotational speed n is in rpm.

The design calculations for a non-rotating axle subjected to bending are associated with two cases: pure bending (the shear stresses are absent); and non-uniform bending (the shear stresses are present). In case of pure bending, the bending moment has a constant value and does not depend on the length of axle. In this case the value of the bending moment can be predetermined based on the operational requirements regardless of the length of the axel. Therefore, for pure bending the design calculations could be performed using the following expression that is based on formulas (1-4.17) and (1-4.18):

$$d = 2.166\sqrt[3]{\frac{Mn_\sigma}{S_{yt}}} \tag{3-3.6}$$

where M is the known bending moment.

In cases of non-uniform bending, the bending moment depends on the length of the axle and cannot be determined until the length of the axle is known. However, in a simplified case of bending loading, the designer may have a way to estimate the length of the axle, where using formula (3-3.6) it is possible to perform the design calculation. Without a preliminary estimate of the length of the axle, the design calculations of an axle cannot be performed.

3-3.2. *Design Calculations of a Shaft Subjected to Bending and Torsion*

The operation of a shaft is associated with rotation and transmitting a torque. Usually, if a rotating shaft is subjected to bending, then the bending stresses fluctuate and they cannot be considered as static stresses. Nevertheless, there are mechanical systems in which a so-called torsion bar (or torsion shaft) does not rotate; however, it is loaded by a bending moment and torque that create a plane stress state in the bar (shaft). The values of the torque and bending moment in these cases sometimes represent certain variable loading cycles. If the number of cycles during the service life of these shafts does not exceed 1,000, it is acceptable to apply to these shafts the methodologies of static stress calculation. In addition, it may be considered a case where the shaft rotates in one direction transmitting a torque and is subjected to bending due to a centrifugal force causing static (not alternating) bending stresses in the shaft. Consequently, this case is characterized by a static plane stress state. This case, probably, does not have many practical applications, but it justifies the relevancy of considering the static plane stress state. Figure 3.3.1 shows a shaft 1 with a rigidly attached radial arm 2 carrying a mass m.

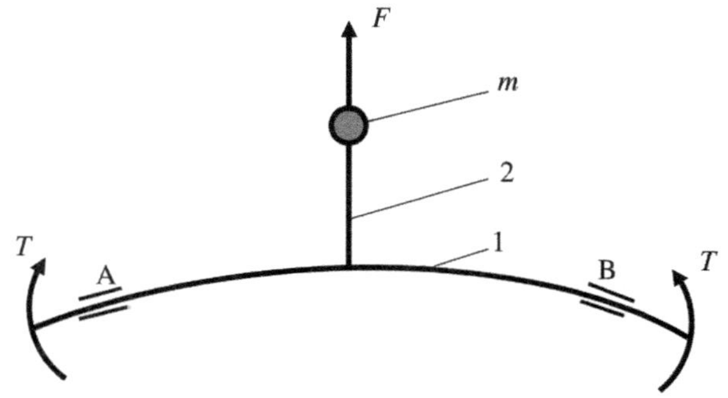

Figure 3-3.1. A rotating shaft subjected to static torsion and bending stresses.

The shaft transmits a torque T and rotates in bearings A and B that are mounted in universal joints as in the fatigue testing machines for pure bending. During the rotation of the shaft, the mass m develops a radial centrifugal force F that bends the shaft in the direction that always coincides with the direction of this force, and, consequently, the stresses in all layers of the shaft remain the same all of the time.

Conventional engineering activities are often associated with the design of an assembly that includes a shaft. The design requirements normally include the information related to the structural composition of the assembly, to the type of a rotational element (pulley, gear, rope drum, etc.) that will be mounted on the shaft, and to the power and the rotational speed associated with the operational process. Therefore, the shaft is subjected to bending and torsional stresses and, consequently, the shaft is in the plane stress state.

As emphasized earlier, in order to come up with the first design sketch of the shaft assembly it is necessary to have the preliminary dimensions of the shaft diameter, bearings, their enclosures, and the width of the hub of the rotating element. Usually, the width of the hub can be estimated based on the diameter of the shaft. The preliminary selection of the bearings and their enclosures can be also accomplished based on the diameter of the shaft. Therefore, the dimension of diameter of the shaft plays a key role in the initial design process of the assembly.

Normally, if the bending moments and the torque diagrams are available and the material is selected, it is possible to calculate the diameter of the shaft. The selection of the material for the shaft does not represent any difficulty; however, the bending moments diagram cannot be constructed if the locations of the bearings and the rotational elements on the shaft are unknown. Fortunately, as mentioned above, there is a semi-empiric methodology for calculating the preliminary diameter of the shaft based on the required torque and on a significantly reduced value of the allowable torsional stress. The value of the reduced allowable torsional stress is justified by extensive experimental investigations.

Using formulas (3-3.4), (1-4.20), and (1-4.21), we determine the torsional stress τ:

$$\tau = \frac{48.7P}{nd^3} \tag{3-3.7}$$

Replacing in this equation τ with the allowable shear stress τ_{all} and solving this equation for the diameter d of the shaft, we obtain

$$d = 3.65\sqrt[3]{\frac{P}{n\tau_{all}}} \tag{3-3.8}$$

where all parameters have the same units as in formula (3-3.5). It is recommended that for steel shafts to select τ_{all} from the range of 11 MPa to 28 MPa.

Formula (3-3.8) shows that the increase of the rotational speed causes a decrease of the diameter of the shaft. So other conditions being equal, the diameter of the shaft rotating at 200 rpm is about two times larger than the diameter of the shaft rotating at 1,200 rpm.

Other sources offer similar semi-empiric formulas for preliminary design calculations of the diameters of the shafts. Hence, since the transmitted power and the required rotational speed of the shaft represent the known input data, we can determine the torque T as

$$T = \frac{63{,}000P}{n} \tag{3-3.9}$$

where the torque T is in lb-in., P is the transmitted power in hp, and n is the rotational speed in rpm. Combining formulas (3-3.7), (3-3.8), and (3-3.9), we obtain an expression for calculating the diameter of the shaft d (in.):

$$d = \sqrt[3]{\frac{5.1T}{S_s}} = \sqrt[3]{\frac{321{,}000P}{nS_s}} \tag{3-3.10}$$

where S_s is the allowable torsional stress in psi. The allowable torsional stresses applicable to this formula range between 4,000 psi and 8,500 psi.

The calculated diameter represents the diameter of the journal, or in other words, this diameter is the bore diameter of the bearing. At the initial stage of the design process, we select a radial rolling bearing based on the bore diameter from the light series of the manufacturer's catalog. According to the outer diameter and width of the selected bearing, we select the corresponding enclosures for the bearings and determine their width. Based on the diameter of the shaft, we determine the width of the rotating element. Now we can draw the preliminary design of the shaft assembly and begin the verifying calculations.

Example 3-3.2a: Design calculations of a shaft subjected to torsion and bending.
A shaft shown in Figure 3-3.2 receives from an electric motor 15 hp of power at 1,800 rpm and transmits this power to a spur gear mounted on the shaft in the middle between bearings A and B.

In this figure the arm CD that is rigidly attached to the shaft represents the gear and the radius of the gear is $r = 1.5$ in. The spur gear is subjected to the action of the tangential force F_t and radial force F_r. The tangential force acts in the horizontal plane, while $T_0 = F_t r$ and $F_r = F_t \tan 20°$. We determine the diameter of the shaft and calculate the maximum values of the torque and bending moment in the most loaded cross-sectional area of the shaft.

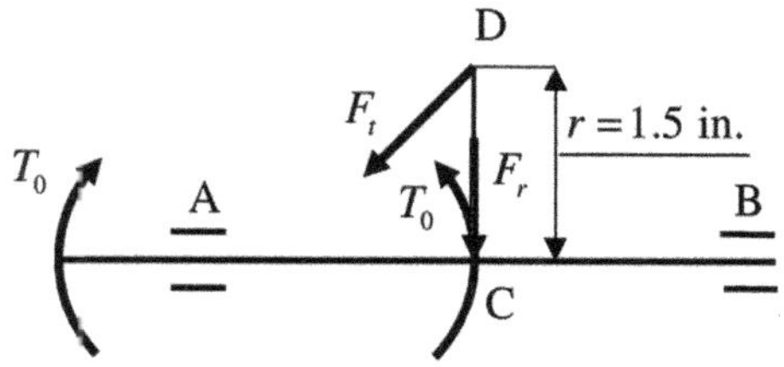

Figure 3-3.2. Shaft subjected to bending and torsion.

Solution:

Obviously, the bending stresses in the shaft are variable; however, we are focusing in this section on the methodologies related to static stresses. This is why it is important to realize that the part of design calculations associated with computing the preliminary value of the diameter of the shaft, selecting the bearings, drawing the sketch of the shaft assembly, drawing the free-body diagram, and plotting the bending moments and torque diagrams are the same for static and variable stresses. Therefore, in this example we present the steps that lead to determining the maximum bending and torsional stresses in the most loaded cross-sectional area of the shaft. The solution for this example is continued below in the section dealing with variable stresses.

Thus, using formula (3-3.10) and selecting $S_s = 5{,}000$ psi, we calculate $d = \sqrt[3]{\dfrac{321{,}000 \cdot 15}{1{,}800 \cdot 5{,}000}} = 0.71$ in. The higher closest standard bore diameter is 0.75 in. We select a radial ball bearing NBC No L58 having the bore diameter of 0.75 in. and the width of 0.5625 in. The estimated width of the enclosure for the bearing should be about 2.0 in. Making a step by increasing the diameter of the shaft to $D = 0.95$ in. to accommodate the gear, we determine the width of the hub of the gear $b = 1.25 \cdot 0.95 = 1.19$ in., we round up to $b = 1.3$ in. Assigning a distance of 1.0 in. between each side of the enclosures, we calculate the distance between the centers of the bearings $l = 2.0 + 2 \cdot 2.0 + 1.3 = 7.3$ in.

Using formula (3-3.9) we calculate the torque $T_0 = \dfrac{63{,}000 \cdot 15}{1{,}800} = 525$ lb-in. Since according to Figure 3-3.2 $T_0 = rF_t$, therefore, $F_t = \dfrac{525}{1.5} = 350$ lb, hence $F_r = 350 \tan 20° = 127.4$ lb. The resultant of these two forces is $F = \sqrt{350^2 + 127.4^2} = 372.5$ lb. Since the force F is applied in the middle between the bearings, the reactions in the supports are equal to $0.5F$ and the maximum bending moment is in the middle of the shaft, and, therefore, $M_{max} = 0.5F\ 0.5l = 0.5 \cdot 372.5 \cdot 0.5 \cdot 7.3 = 679.8$ lb-in. We calculated the values of the maximum torque $T_{max} = T_0 = 525$ lb-in. and the bending moment

M_{max} = 679.8 lb-in acting in the most loaded cross-sectional area of the shaft. In this simple case there is no need to plot the bending moments and torque diagrams. The continuation of the solution is presented in the example 3-5.1a.

3-3.3. Verifying Calculations for Static Stresses

Verifying calculations determine the safety factor of a machine element for which the related geometric parameters and load characteristics are known or could be obtained by direct measuring. The initial steps of verifying calculations involve drawing a free-body diagram, determining the reactions in the supports, and plotting the bending moments and torque diagrams that determine the most loaded cross-sectional area of the shaft. After performing the bending and torsional stresses calculations, we compute the equivalent stress, and upon selecting the material we determine the safety factor. If the safety factor is already assigned, we calculate the diameter of the shaft, and upon comparing with the actual diameter, we finalize the engineering solution regarding the required diameter of the shaft. In cases when the verifying calculations follow the design calculation, the verifying calculations may reveal the need for a change of the preliminary geometric characteristics.

Shafts and axles could be in unidirectional stress state, while shafts can also be in a plane stress state. A shaft subjected to a pure toque, as discussed earlier, is in a unidirectional stress state. The verifying calculations in this case are based on formula (3-3.1) that determines the required diameter of the shaft or the actual safety factor:

$$d = 1.72\sqrt[3]{\frac{T}{\tau_{all}}} \tag{3-3.11}$$

and

$$n_\tau = 0.196\frac{d^3 S_{sy}}{T} \tag{3-3.12}$$

Axles subjected to the action of bending stresses are in a unidirectional stress state and the verifying calculations in this case are based on the flexure formula (1-4.17) combined with the moment of inertia formula (1-4.18). This determines the required diameter of the axle or the actual safety factor:

$$d = 2.166\sqrt[3]{\frac{M}{\sigma_{all}}} \tag{3-3.13}$$

where σ_{all} is the allowable normal stress for the axle and

$$n_\sigma = 0.098\frac{d^3 S_{yt}}{M} \tag{3-3.14}$$

3-4. Calculations of Variable Stresses in a Unidirectional Stress State

Actually, the design and verifying calculations for shafts subjected to variable stresses are based on similar calculations as for static stresses.

As mentioned earlier, a shaft submitted to a pure torque is in a unidirectional stress state. If this shaft reverses the direction of its rotation, the torsional stresses change their directions, and the shaft is considered to be subjected to a completely reversed loading cycle. Therefore, in cases of reversed rotation, all principles of fatigue calculations are applicable to the shaft. We have to calculate according to equation (1-10.4) the actual endurance limit S_e and use it for calculating the diameter of the shaft or the safety factor. Taking an appropriate safety factor n_τ and modifying formulas (3-3.11) and (3-3.12), we determine the required diameter of the shaft or the actual safety factor:

$$d = 1.72\sqrt[3]{\frac{T_{\max} n_\tau}{S_e}} \tag{3-4.1}$$

and

$$n_\tau = 0.196\frac{d^3 S_e}{T_{\max}} \tag{3-4.2}$$

An axle or shaft could be subjected to the action of any of the variable stress cycles. In case of a completely reversed cycle, we determine the actual endurance limit according to equation (1-10.4). Then applying an appropriate safety factor n_σ and modifying formulas (3-3.11) and (3-3.12), we calculate the required diameter of the shaft or the actual safety factor:

$$d = 2.166\sqrt[3]{\frac{M_{\max} n_\sigma}{S_e}} \tag{3-4.3}$$

and

$$n_\sigma = 0.098\frac{d^3 S_e}{M_{\max}} \tag{3-4.4}$$

In cases of a nonzero mean stress cycle, we calculate the diameter or the safety factor based on the following formulas:

$$d = 2.166\sqrt[3]{\frac{M_a n_\sigma}{S_a}} \tag{3-4.5}$$

and

$$n_\sigma = 0.098\frac{d^3 S_a}{M_a} \tag{3-4.6}$$

where the stress amplitude limit S_a is determined using the expression (1-12.8), and the amplitude value of the bending moment is calculated according to formula (1-12.11).

3-5. Calculations of Variable Stresses at a Plane Stress State

These calculations are based on the appropriate static calculations modified by all related fatigue considerations. All design calculations associated with determining the preliminary value of the diameter of the shaft are similar to the static calculations; however, the allowable torsional stress should be closer to the lower limit values. Upon completion of the preliminary calculations of the diameter of the shaft and drawing the sketch of the shaft assembly, it is necessary to draw the free-body diagram and compute the reactions in the supports. After that, plot the bending moment and torque diagrams and determine the values of the bending moment and the torque in the most loaded cross-sectional area of the shaft. These values, depending on the characteristics of the loading cycles, allow us to calculate the values of the normal and torsional stresses. Based on the fatigue considerations, it is possible to determine the equivalent stresses. For these calculations, applying the third or fourth theory of failure is recommended. After selecting the appropriate material for the shaft, we calculate the safety factor, or in cases where the safety factor is given, we determine the diameter of the shaft.

Examples presented below illustrate the use of the described methodology.

3-5.1. Alternating Bending and Steady Torsion

In some cases, an alternating bending and a steady torsion characterize the operational conditions of the shaft. The bending stress is variable and causes the fatigue of the material. Bending generates in a rotating shaft variable stresses characterized by a completely reversed loading cycle. The plane stress state of a machine element can be evaluated by applying the appropriate theory of failure. Using, for instance, the third theory of failure we may write $\sigma_{eq} = \sqrt{\sigma_{max}^2 + 4\tau_{max}^2} \le \frac{S_{yt}}{n}$, where $\sigma_{max} = \frac{M_{max}c}{I}$ and $\tau_{max} = \frac{T_{max}c}{J}$. The values of $M_{max} = |M_{min}|$ and T_{max} are given parameters. Using the formulas (1-4.18) and (1-4.21) for the moment of inertia $I = \frac{\pi d^4}{64}$ and polar moment of inertia $J = \frac{\pi d^4}{32}$, and considering that $c = 0.5d$, we may write

$$\sqrt{\left(\frac{32M_{max}}{\pi d^3}\right)^2 + \left(\frac{32T_{max}}{\pi d^3}\right)^2} = \frac{S_{yt}}{n} \tag{3-5.1}$$

Solving equation (3-5.1) for the diameter d:

$$d = \left\{\frac{32n}{\pi}\left[\left(\frac{M_{max}}{S_{yt}}\right)^2 + \left(\frac{T_{max}}{S_{yt}}\right)^2\right]^{\frac{1}{2}}\right\}^{\frac{1}{3}} \tag{3-5.2}$$

Equations (3-5.1) and (3-5.2) are applicable for static stresses. In case of alternating bending stresses and steady torque, the material of the shaft is subjected to fatigue phenomena. As it is well known, the peripheral layers of the shaft experience the maximum bending and torsional stresses. It is justifiable to assume that the structural fatigue related changes that lead to weakening of the material may reduce the shear strength of the material. Therefore, it would be safe in this case from the point of view of the overall strength of the shaft to relate the equivalent stress to the endurance limit even while the torsion is not alternating. It seems to be reasonable not to apply the stress concentration factor to the torsional stress. In order to apply the third or fourth theory of failure, the endurance limit S_e should be determined, the fatigue stress concentration factor for bending K_f, the amplitude value of the bending moment M_a that in our case equals M_{max}, and the maximum value of the torque T_{max}. Now we modify equation (3-5.2) accounting for the considerations above and obtain the corresponding equation based on the third theory of failure:

$$d=\left\{\frac{32n}{\pi}\left[\left(\frac{K_f M_{max}}{S_e}\right)^2+\left(\frac{T_{max}}{S_e}\right)^2\right]^{\frac{1}{2}}\right\}^{\frac{1}{3}} \tag{3-5.3}$$

Solving equation (3-5.3) for the safety factor, we have

$$n=\frac{\pi d^3}{32\sqrt{\left(\frac{K_f M_{max}}{S_e}\right)^2+\left(\frac{T_{max}}{S_e}\right)^2}} \tag{3-5.4}$$

Similar expressions we obtain applying the fourth theory of failure:

$$d=\left\{\frac{32n}{\pi}\left[\left(\frac{K_f M_{max}}{S_e}\right)^2+\frac{3}{4}\left(\frac{T_{max}}{S_e}\right)^2\right]^{\frac{1}{2}}\right\}^{\frac{1}{3}} \tag{3-5.5}$$

and

$$n=\frac{\pi d^3}{32\sqrt{\left(\frac{K_f M_{max}}{S_e}\right)^2+\frac{3}{4}\left(\frac{T_{max}}{S_e}\right)^2}} \tag{3-5.6}$$

It should be mentioned that some authors suggest in cases of alternating bending and steady torque to present in formulas (3-5.5) and (3-5.6) the ratio between this torque and the yield strength S_{yt}, rather than the ratio between the maximum torque T_{max} and the endurance limit S_e. This suggestion results in a reduced safety margin of the shaft, and in addition contradicts the concept of a single allowable stress regarding the theories of failure.

Example 3-5.1a: Calculations of a shaft subjected to alternating bending and steady torsion (continuation of example 3-3.2a).

We determine the diameter of a shaft transmitting power of 11 kW at 3,600 rpm and subjected to alternating bending and steady torsion.

The results of the calculations for static loading that are presented in example 3-3.2a can be summarized as follows:

1) The dimension of the preliminary diameter of the shaft's journals is $d = 0.75$ in.
2) According to this diameter, we select a radial ball bearing NBC No L58 having the bore diameter of 0.75 in.
3) The diameter of the shaft is increased to $D = 0.95$ in. to accommodate the gear.
4) The maximum values of the torque and bending moment in the most loaded cross-sectional area of the shaft are $T_{max} = 525$ lb-in., and $M_{max} = 679.8$ lb-in.

Solution (continuation):

From Appendix A-1-4, we select an AISI 4150 normalized steel having $S_{ut} = 167{,}250$ psi. Using formulas (1-9.2), (1-9.3), and (1-9.4), we calculate $k_a = 0.705$, $k_b = 0.916$, and $k_c = 1$.

Applying formula (1-9.5) we compute $S_e = 54.\ 05$ psi.

For $\frac{D}{d} = \frac{0.95}{0.75} = 1.267$ and assuming $\frac{r}{d} = 0.1$ we determine from Appendix A-1-5 (Figure 3) that $K_t = 1.65$. Taking $q = 0.6$, we determine from formula (1-10.2) the fatigue stress concentration factor $K_f = 1 + 0.6(1.65 - 1) = 1.39$. Applying the safety factor $n_\sigma = 2$ and using formula (3-5.5), we calculate the diameter of shaft $d = \left\{ \frac{32 \cdot 2}{\pi} \left[\left(\frac{1.39 \cdot 679.8}{54.04 \cdot 10^3} \right)^2 + \frac{3}{4} \left(\frac{525}{54.04 \cdot 10^3} \right)^2 \right]^{\frac{1}{2}} \right\}^{\frac{1}{3}} = 0.7096$ in.

It should be mentioned that in example 3-3.2a the calculations according to formula (3-3.10) resulted in the diameter of 0.71. Actually, we get the same result, and no additional calculations are necessary.

3-5.2. Alternating Bending and Alternating Torsion

As indicated earlier, if the shaft's operational conditions are associated with reversing the direction of its rotation, the torsion stresses should be considered as variable stresses and the stress calculations of the shaft should be based on the principles addressing the fatigue considerations for both type of stresses: bending and torsional stress. For this case, we should determine the amplitude value of the torque T_a and the fatigue stress concentration factor for torsion K_{ft}. using formula (1-12.13). Then, using the third theory of failure, we determine the diameter or the safety factor for this case:

$$d = \left\{ \frac{32n}{\pi} \left[\left(\frac{K_f M_a}{S_e} \right)^2 + \left(\frac{K_{ft} T_a}{S_e} \right)^2 \right]^{\frac{1}{2}} \right\}^{\frac{1}{3}} \tag{3-5.7}$$

and

$$n = \frac{\pi d^3}{32\sqrt{\left(\frac{K_f M_a}{S_e}\right)^2 + \left(\frac{K_{ft} T_a}{S_e}\right)^2}} \quad (3\text{-}5.8)$$

Applying the fourth theory of failure we have

$$d = \left\{ \frac{32n}{\pi} \left[\left(\frac{K_f M_a}{S_e}\right)^2 + \frac{3}{4}\left(\frac{K_{ft} T_a}{S_e}\right)^2 \right]^{\frac{1}{2}} \right\}^{\frac{1}{3}} \quad (3\text{-}5.9)$$

and

$$n = \frac{\pi d^3}{32\sqrt{\left(\frac{K_f M_a}{S_e}\right)^2 + \frac{3}{4}\left(\frac{K_{ft} T_a}{S_e}\right)^2}} \quad (3\text{-}5.10)$$

It is difficult to imagine a shaft that is subjected to a nonzero mean stress bending cycle and a steady torsion, or to alternating bending and a nonzero mean stress torsion cycle. However, it is possible to think about a case when a shaft makes a limited angular displacement less than one revolution in both directions, while the angular displacement in one direction does not equal the angular displacement in the other direction. In this case, the bending and torsional stresses would represent random cycles. However, in the engineering practices associated with the machine elements for general applications, these kinds of situations seldom occur. It does not seem to be reasonable for this book to deal with these cases.

It should be noted that results based on the fourth theory of failure comply with experimental data and, therefore, this theory is recommended for shaft stress calculations.

In addition, in this case, the load factor k_c, contrary to the case of pure torsion, should be taken equal to unity ($k_c = 1$).

Example 3-5.1a: Calculating the diameter of a shaft subjected to variable bending and torsional stresses.

A stepped shaft machined of AISI 1050 steel having $S_{ut} = 105$ ksi and $S_{yt} = 60$ ksi is subjected to the action of a bending moment $M_{\max} = 400$ lb-ft and a torque $T_{\max} = 500$ lb-ft. The ratio between the diameters is $\frac{D}{d} = 1.15$ and the ratio of the fillet radius to the smaller diameter is $\frac{r}{d} = 0.125$. Using a safety factor $n_\sigma = 2.5$, we determine the diameter of the shaft if its rotation is reversible.

Solution:

The calculations should be carried out for a completely reversed cycle with regard to bending and torsional stresses. First, we determine the endurance limit modifying factors k_a and k_c based on formulas (1-9.2) and (1-9.4): $k_a = 0.793$ and $k_c = 1$. Since the diameter is unknown, we assume that $k_b = 0.75$, keeping in mind that upon determining the diameter this factor might be corrected. Now we determine the value of the endurance limit using formula (1-9.5): $S_e = 0.793 \cdot 0.75 \cdot 1 \cdot 0.5 \cdot 105 = 31.23$ ksi. Based on the input data $\frac{D}{d} = 1.15$ and $\frac{r}{d} = 0.125$, we find in Appendix A-1-5 (Figures 2 and 3) that the corresponding theoretical stress concentration factors for the torsion and bending stresses are $K_{tt} = 1.28$ and $K_t = 1.52$, respectively. In the next step, we calculate the fatigue stress concentration factors using formula (1-10.2) for torsional and bending stresses accounting that the notch sensitivity factor $q = 0.6$. Hence, $K_{ft} = 1 + 0.6(1.28 - 1) = 1.168$ and $K_f = 1 + 0.6(1.52 - 1) = 1.312$. Now we calculate the amplitude values of the bending moment and torque according to formulas (1-12.11) and (1-12.13): $M_a = \frac{3{,}200 - (-3{,}200)}{2} = 3{,}200$ lb-ft and $T_a = \frac{3{,}880 - (-3{,}880)}{2} = 3{,}880$ lb-ft. Using formula (3-5.9) we determine the required diameter of the shaft: $d = \left\{ \frac{32 \cdot 2.5}{\pi} \left[\left(\frac{1.312 \cdot 400 \cdot 12}{31.23 \cdot 1{,}000} \right)^2 + \frac{3}{4} \left(\frac{1.168 \cdot 500 \cdot 12}{31.23 \cdot 1{,}000} \right)^2 \right]^{\frac{1}{2}} \right\}^{\frac{1}{3}} = 1.92$ in. Using this value for the diameter, we recalculate the size factor using expression (1-9.3). Hence, for the diameter 1.78 in. we obtain $k_b = 0.828$. Now we recalculate the value of the endurance limit using the corrected value of the size factor, and we have $S_e = 34.46$ ksi. We have to recalculate the diameter based on the adjusted value of the endurance limit and upon the corresponding calculations; we obtain $d = 1.82$ in. It should be mentioned that it was not necessary to recalculate the diameter since we should always take the closest larger diameter from the available standard bore diameters of the bearings. Hence, converting both diameters into the metric system in order to select a metric bearing, we obtain 48.77 mm and 46.23 mm. The closest standard bore diameter of an appropriate bearing is 50 mm.

3-6. Deflection and Angle of Rotation of Shafts

The deflection often occurs due to static loading that is present even before the shaft starts its rotation. When the shaft starts its rotation, this deflection causes the development of centrifugal forces that promote further deflection of the shaft. These centrifugal forces are associated with rotating machine elements, such as pulleys, gears, sprockets, etc. The deflection of the shaft results in a variety of negative effects that disturb the operational process of the machine elements and, consequently, of the assembly. Specifically, the deflection of the shaft imposes additional forces on the bearings, gears, and other elements. Thus, the bearings become misaligned and more stressed and the contact stresses in the teeth of the gears increase. The main concern regarding deflections of shafts is associated with gears and bearings, and, therefore, the shaft should possess the required stiffness. Some machine elements such as pulleys, sprockets, and rope drums that are associated with flexible links like belts, chains, and ropes are not sensitive to the deflection of shafts.

An exaggerated schematic of shaft deflection is shown in Figure 3-5.1, where shaft 1 is supported by bearings A and B and is carrying a rotating element 2, while the deflection of the shaft is h and the angle of rotation of the shaft cross-sectional area is θ. It should be emphasized that in this book we present the methods for calculating the deflection as well as the angle of rotation caused by static loading factors.

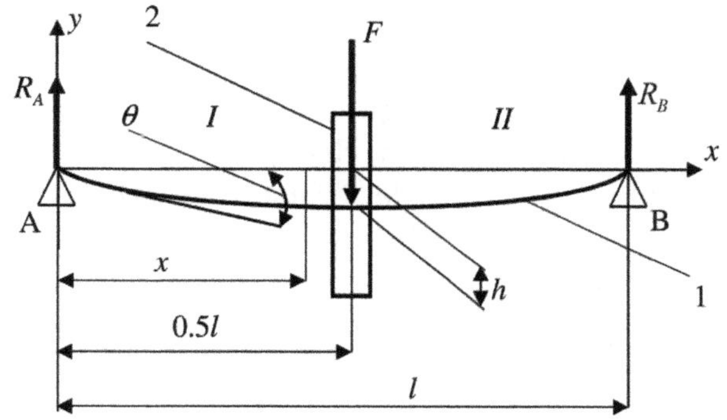

Figure 3-5.1. Shaft deflection and angle of rotation.

Some authors recommend that shaft deflection between its supports should not exceed $0.0003l$, where l is the distance between the supports. It is also recommended that the deflection of the shafts in places that carry cylindrical toothed gears should not exceed $\frac{0.03}{P}$ (in.) where P is the diametral pitch or $0.03m$ (mm), where m is the module; while for bevel gears the deflection of the shaft should not exceed $\frac{0.007}{P}$ (in.) or $0.007m$ (mm).

For the angle of rotation, it is recommended that it should not exceed 0.001 radian for a journal bearing and 0.01 radian for a radial ball bearings and 0.05 radian for roller bearings with spherical roller elements.

The deflection and the angle of rotation are associated with the action of the loads exerted by the rotating element 2 and applied to the shaft 1. The calculations of the deflection and the angle of rotation are based on the differential equation of the deflection curve that is discussed in the course of Strength of Materials (Mechanics of Materials). The differential equation of the deflection curve reads

$$EI\frac{d^2y}{dx^2} = M(x) \tag{3-6.1}$$

where E is the modulus of elasticity, I is the moment of inertia of the shaft cross-sectional area, and $M(x)$ is the bending moment as a function of x, and y is the displacement of the rotating mass as a function of x. Or in other words, y characterizes the deflection of any abscissa x. The angle of rotation is calculated using the expression

$$\theta = \frac{dy}{dx} \tag{3-6.2}$$

Obviously, R_A and R_B are the reactions in the supports A and B, and $R_A = R_B = 0.5F$.

The shaft is divided into two intervals I and II as shown in Figure 3-5.1. The bending moment on the first interval represents $M(x) = R_A x = 0.5Fx$, while the limits of the first interval are $0 \le x \le 0.5l$. In order to solve equation (3-6.1), we have to determine the initial conditions related to the deflection curve. It is clear that in the supports the deflection equals zero and in the middle of the shaft the angle of rotation also equals zero. The angle of rotation has its maximum value in the supports and decreases to zero in the middle of the shaft. Accounting for these considerations, we begin solving differential equation (3-6.1). For the first interval this equation reads $EI\frac{d^2y}{dx^2} = 0.5Fx$. Dividing this equation by EI, we obtain

$$\frac{d^2y}{dx^2} = \frac{0.5Fx}{EI} \tag{3-6.3}$$

The initial conditions are

for $$x = 0 \quad h_0 = 0; \quad \frac{dy}{dx} = \theta \tag{3-6.4}$$

where h_0 and θ are the initial deflection and angle of rotation, respectively. The solution of differential equation (3-6.3) at initial conditions (3-6.4) reads

$$y = \frac{Fx^3}{12EI} + \theta x \tag{3-6.5}$$

Taking the first derivative from equation (3-6.5) we may write

$$\frac{dy}{dx} = \frac{Fx^2}{4EI} + \theta \tag{3-6.6}$$

In this equation, when $x = 0.5l$ the angle of rotation equals zero. Therefore, we equate to zero the left-hand side of equation (3-6.6) and obtain $0 = \frac{Fl^2}{4 \cdot 4EI} + \theta$. We determine the value of the initial angle of rotation θ that represents the angle of rotation at the support A: $\theta = -\frac{Fl^2}{16EI}$. Substituting the value of θ into equation (3-6.6) we have

$$y = \frac{Fx}{4EI}\left(\frac{x^2}{3} - \frac{l^2}{4}\right) \tag{3-6.7}$$

This equation describes the relationship between y and x just on the first interval. The maximum deflection in our case occurs in the middle of the shaft, therefore, substituting into equation (3-6.7) $x = 0.5l$ we determine that the deflection h in the middle of the shaft is characterized by the following equation:

$$h = -\frac{Fl^3}{48EI} \tag{3-6.8}$$

The first derivative from equation (3-6.7) represents the angle of rotation:

$$\frac{dy}{dx} = \frac{F}{4EI}\left(x^2 - \frac{l^2}{4}\right) \tag{3-6.9}$$

Because the maximum angle of rotation occurs in the supports A and B, it should be clear that equations (3-6.7) and (3-6.9) are valid just on the first interval for which $0 \le x \le 0.5l$. Considering this, we take $x = 0$, and we calculate the maximum angle of rotation:

$$\theta = -\frac{Fl^2}{16EI} \tag{3-6.10}$$

Actually, this value was already obtained above.

The methodology of calculating the deflection and angle of rotation discussed above is applicable to any structural design of a shaft assembly; however, for each case the appropriate bending moment should be determined. Very often, the shafts have a stepwise shape that complicates the calculations of the deflection and the angle of rotation. In order to simplify these calculations, some sources recommend calculating the equivalent uniform diameter of the shaft d_e using the following formula:

$$d_e = \frac{1}{\sqrt[4]{\sum_{1}^{k} \frac{l_j}{l d_j^4}}} \tag{3-6.11}$$

where l_j and d_j are the length and diameter of a certain step of the shaft, l is the total length of the shaft, j is the orderly number of the step, and k is the number of the steps. According to this equation, the error in determining the equivalent diameter is in range of about 20%.

Formulas to calculate deflections for typical axles and shaft arrangements are presented in Figure 3-6.1.

$$h = -\frac{Fl^3}{3EI} \quad (3\text{-}6.12) \qquad \theta_B = -\frac{Fl^2}{2EI} \quad (3\text{-}6.14)$$

$$h = -\frac{Ml^2}{2EI} \quad (3\text{-}6.13) \qquad \theta_B = -\frac{Ml}{EI} \quad (3\text{-}6.15)$$

(a)

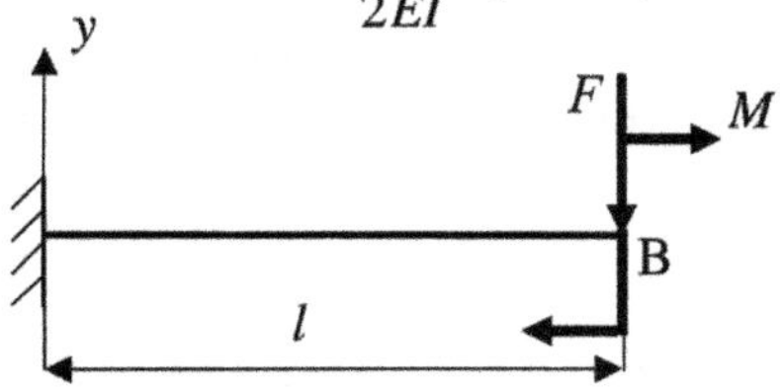

$$h = -\frac{Fa^2(l-a)^2}{3EIl} \quad (3\text{-}6.16)$$

$$h = -\frac{Ma(2l^2 - 6al + 4a^2)}{6EIl} \quad (3\text{-}6.17)$$

$$\theta_A = -\frac{Fa(l-a)(2l-a)}{6EIl} \quad (3\text{-}6.18)$$

$$\theta_A = -\frac{M(2l^2 - 6al + 3a^2)}{6EIl} \quad (3\text{-}6.19)$$

(b)

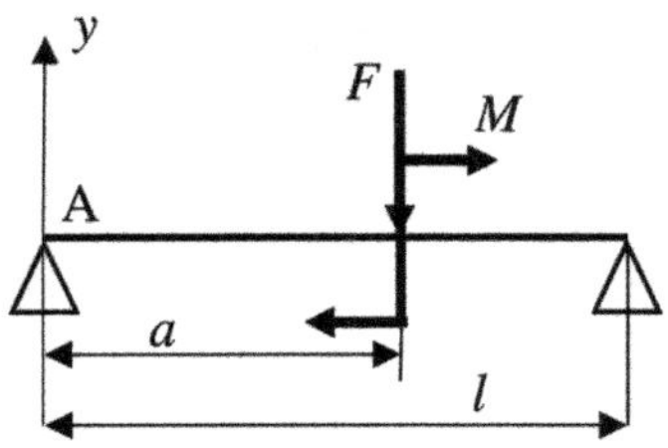

(c)

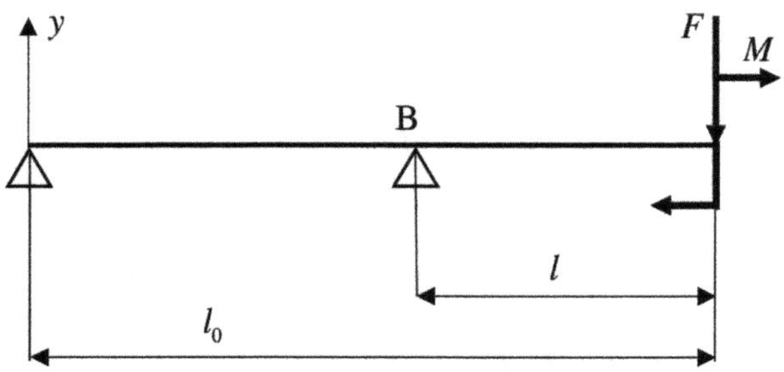

$$h = -\frac{Fl^2(l + l_0)}{3EI} \quad (3\text{-}6.20) \qquad \theta_B = -\frac{Fll_0}{3EI} \quad (3\text{-}6.22)$$

$$h = -\frac{Ml(3l - 2l_0)}{6EI} \quad (3\text{-}6.21) \qquad \theta_B = -\frac{Ml_0}{3EI} \quad (3\text{-}6.23)$$

Figure 3-6.1. Deflections and angles of rotation for typical axles and shafts.

For a uniform round bar of length l mounted in a fixed support as shown in Figure 3-6.1a and loaded at the free end B by a force F and a moment M the deflection and angle of rotation in the point B can be calculated from the following formulas:

$$h = -\frac{Fl^3}{3EI} \qquad (3\text{-}6.12)$$

$$h = \frac{Ml^2}{2EI} \tag{3-6.13}$$

and

$$\theta_B = -\frac{Fl^2}{2EI} \tag{3-6.14}$$

$$\theta_B = \frac{ML}{EI} \tag{3-6.15}$$

In Figure 3-6.1b a uniform round shaft of length l is supported by two bearings at its ends, and it is loaded by a force F and a moment M at the distance a from the support A. The deflection and angle of rotation can be determined from the following formulas:

$$h = -\frac{Fa^2(l-a)^2}{3EIl} \tag{3-6.16}$$

$$h = \frac{Ma(2l^2 - 6al + 4a^2)}{6EIl} \tag{3-6.17}$$

and

$$\theta_A = -\frac{Fa(l-a)(2l-a)}{6EIl} \tag{3-6.18}$$

$$\theta_A = \frac{M(2l^2 - 6al + 3a^2)}{6EIl} \tag{3-6.19}$$

A uniform cantilever shaft is shown in Figure 3-6.1c. The shaft is supported by two bearings located at the distance $l_0 - l$, where l_0 is the length of the shaft and l is the length of the cantilever. The cantilever is loaded on its free end by a force F and a moment M. The formulas for computing the deflection of the cantilever under the force and the angle of rotation in point B follows:

$$h = -\frac{Fl^2(l + l_0)}{3EI} \tag{3-6.20}$$

$$h = \frac{Ml(3l - 2l_0)}{6EIl} \tag{3-6.21}$$

and

$$\theta_B = -\frac{Fl(3l + 2l_0)}{6EI} \tag{3-6.22}$$

$$\theta_B = \frac{Ml_0}{3EI} \tag{3-6.23}$$

It should be noted that in cases when the force and the moment are producing the deflections and the angles of rotation in the same plane, the total deflection and angle of rotation represents a superposition of the corresponding values. In cases when the deflections and the angles of rotation occur in two mutually perpendicular planes, the total deflection and the angle of rotation should be determined based on the vector sum of the corresponding values.

3-7. Critical Speed of Shafts

Consider a shaft rotating on two bearings that is loaded by a radial force exerted by the rotational element that is mounted between the supports. This radial force is related to the operational process of the shaft assembly. Under the action of this force, the axis of the shaft deflects, creating an eccentricity that further promotes the deflection due to the centrifugal forces associated with the masses of the rotating element and the shaft. Actually, the deflection of the shaft can occur due to the centrifugal forces of an imbalanced rotational element without any additional radial forces.

Figure 3-7.1 illustrates the combined deflection of a shaft produced by the radial force F_0 and centrifugal force F_c that results in displacing the center of gravity of the rotating element 2 to the point C located on the distance h from the geometric axis of the shaft and the rotating element.

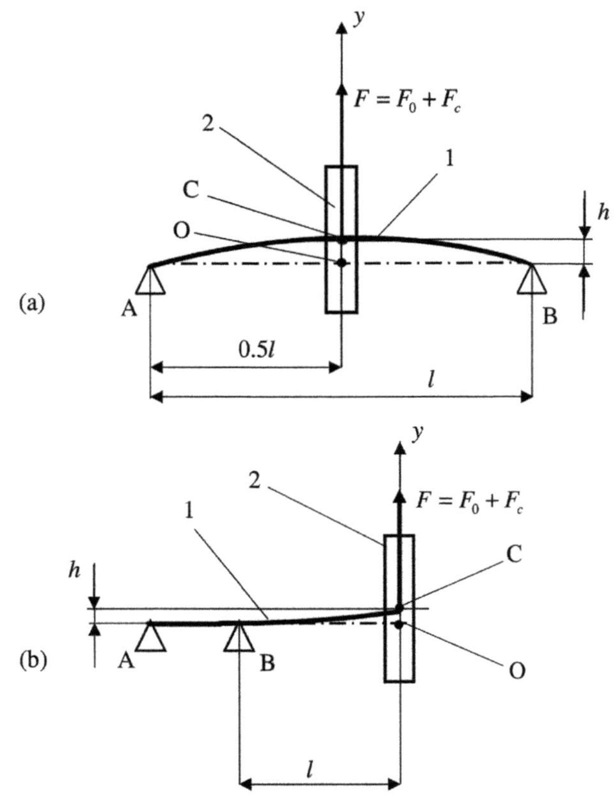

Figure 3-7.1. Deflection of shafts and centrifugal forces.

This distance represents the eccentricity that actually leads to the occurrence of centrifugal forces applied to the shaft. With the increase of the rotational speed, the centrifugal force may increase to a level where the deflection could cause stresses that may become dangerously close to the strength limit of the shaft. In Figure 3-7.1a the rotating element 2 is placed between the supports (bearings) A and B, while in Figure 3-7.1b the rotating element 2 is placed on the cantilever outside of the bearings. The centrifugal force F_c applied to the shaft can be determined from the following formula:

$$F_c = m_0\omega^2 h \tag{3-7.1}$$

where m_0 represents the mass of the shaft combined with the mass of the rotating element concentrated on one point, C, and ω is the angular velocity of the shaft. Due to the radial force and the centrifugal force, F_c, the shaft deflects in the y direction and the trajectory of point O creates a circle. If the centrifugal force increases, the eccentricity increases and the radius of this circle increases as well. The radius of this circle represents the deflection. It should be emphasized that only the centrifugal force increases with the increase of the angular velocity of the shaft and, consequently, just the centrifugal force causes the increase of the deflection that in its turn results in the increase of the bending stresses in the shaft. Therefore, it is necessary to analyze the role of the centrifugal force in the increase of the stresses in the shaft. This analysis is based on determining the characteristics of motion of the center of gravity of the rotating element. Thus, the center of gravity is in plane motion that can be determined by two similar equations of motion describing the rectilinear displacement of this center in two mutually perpendicular directions, while the amplitude of this center in each direction has the same value. Obviously, the deflection of the shaft represents this amplitude. In order to analyze the amplitude, we may analyze the motion in any direction. For convenience we analyze motion along axis y on which we indicate the deflection of the shaft. Since the center of gravity moves along a circle, then the projection of the centrifugal force on the axis y will change according to a harmonic function such as sine or cosine. Accounting for all these considerations, we can compose the differential equation of motion of the rotating element into the direction of y axis:

$$m_0\frac{d^2y}{dt^2} + Ky = F_c \sin\omega t \tag{3-7.2}$$

where K is the stiffness coefficient of the shaft in the direction of the axis y, and t is the running time. Substituting into equation (3-7.2) the value of the centrifugal force according to formula (3-7.1), we obtain $m_0\frac{d^2y}{dt^2} + Ky = m_0\omega^2 h \sin\omega t$. Dividing this equation by m_0, we may write

$$\frac{d^2y}{dt^2} + \omega_0^2 y = \omega^2 h \sin\omega t \tag{3-7.3}$$

where

$$\omega_0^2 = \frac{K}{m_0} \tag{3-7.4}$$

while ω_0 is the natural frequency of the shaft.

The initial conditions of motion for the point O of the shaft in the direction of y are

$$\text{for } t = 0 \quad y = 0; \ \frac{dy}{dt} = 0 \tag{3-7.5}$$

The solution of the differential equation (3-7.3) with the initial conditions of motion (3-7.5) reads

$$y = \frac{\omega^2 h(\omega \sin \omega_0 t - \omega_0 \sin \omega t)}{\omega_0(\omega^2 - \omega_0^2)} \tag{3-7.6}$$

The denominator of equation (3-7.6) contains the term $(\omega^2 - \omega^2{}_0)$ that explains the reason why the system may have a critical velocity that is dangerous to the system. Indeed, if

$$\omega^2 - \omega_0^2 = 0 \tag{3-7.7}$$

then the deflection y tends to infinity (∞). This represents the conditions of resonance of the shaft assembly and these conditions could be dangerous for the shaft, causing damage to it due to increased stresses that may approach the limit of its strength. Analyzing equations (3-7.6) and (3-7.7), we conclude that the condition of resonance is not associated with the applied radial forces to the shaft, to the eccentricity, or to the value of the deflection. But it is related to a specific case where the angular velocity of the shaft equals its natural frequency of the shaft assembly. Hence, based on these conclusions, it becomes clear that if $\omega_0 \neq \omega$ then there is no resonance. Therefore, formula (3-7.4) shows that in order to avoid the resonance it is necessary that the value $\sqrt{\frac{K}{m_0}} \neq \omega$. Hence, the critical angular velocity ω_{cr} of the shaft assembly equals its natural frequency of this assembly and can be calculated from the following formula:

$$\omega_{cr} = \sqrt{\frac{K}{m_0}} \tag{3-7.8}$$

It is more convenient to operate with the rotational speed of the shaft in rpm, hence

$$n_{cr} = \frac{30}{\pi}\sqrt{\frac{K}{m_0}} \tag{3-7.9}$$

The stiffness coefficient K depends on the structures of the corresponding systems and can be measured or determined based on the fundamentals of the course Strength of Materials (Mechanics of Materials) or obtained from numerous published sources. In general, $K = \frac{F}{h}$. For the case shown in Figure 3-7.1a where the rotating mass is located in the middle between the bearings, the stiffness coefficient K_1 based on formula (3-6.8) can be determined from the following equation:

$$K_1 = \frac{F}{h} = \frac{48EI}{l^3} \tag{3-7.10}$$

where E is the modulus of elasticity and I is the moment of inertia of the shaft cross-sectional area. Therefore, the critical speed of the shaft is associated with its deflection caused by static loading. Combining formulas (3-7.9) and (3-7.10), we obtain for the case mentioned above the critical speed n_{1cr} equals

$$n_{1cr} = \frac{30}{\pi}\sqrt{\frac{48EI}{l^3 m_0}} \tag{3-7.11}$$

For the case where the rotating element is located on a cantilever outside of the bearings (Figures 3-6.1c and 3-7.1b), the stiffness coefficient is calculated according to the following formula:

$$K_2 = \frac{3EI}{l^2(l + l_0)} \tag{3-7.12}$$

Combining formulas (3-7.9) and (3-7.12), we obtain

$$n_{2cr} = \frac{30}{\pi}\sqrt{\frac{3EI}{m_0 l^2 (l + l_0)}} \tag{3-7.13}$$

As an example, we consider a steel shaft according to Figure 3-7.1 with the rotating element located between the bearings. The critical rotational speed for this shaft can be determined from formula (3-7.11) that is modified to the following:

$$n_{1cr} = 15\sqrt{\frac{3Ed^4}{\pi l^3 m_0}} \tag{3-7.14}$$

Thus, the shaft has a diameter d = 1.5 in. (38.1 mm), l = 6.0 (152 mm), and the mass of the rotating element $m_0 = 10\ \frac{\text{lb-s}^2}{\text{ft}}$ (slug) or (146 kg). Using formula (3-7.14), we substitute E in psi, d and l in inches, and m_0 in $\frac{\text{lb-s}^2}{12\text{ft}}$. Hence, we calculate $n_{cr} = 15\sqrt{\frac{3 \cdot 30 \cdot 10^6 \cdot 1.5^4 \cdot 12}{\pi 6.0^3 \cdot 10}} = 13{,}468$ rpm.

For the ISO units we install into formula (3-7.14) E in $\frac{\text{N}}{\text{m}^2}$, d and l in m, and m_0 in kg.

Therefore, $n_{cr} = 15\sqrt{\frac{3 \cdot 207 \cdot 10^9 \cdot 0.0381^4}{\pi 0.152^3 \cdot 146}} = 13{,}523$ rpm.

According to formula (3-7.13), the calculation of the critical speed is associated with the moment of inertia and mass of the shaft assembly. In most practical cases, these characteristics are unknown, and due to many structural issues of the shaft assembly, it is too complicated to compute the accurate values of these characteristics. In order to overcome the difficulties, some authors have suggested analytical methods for calculating approximate values of the critical speed based on simplified formulas. It should be noted that near the critical speed the shaft assembly experiences high vibrations followed by a load noise that signals a possible failure due to resonance. It is recommended that shaft assemblies avoid the range of rotational speeds between $0.7n_{cr}$ and $1.3n_{cr}$. Therefore, the exact value of the rotational speed is not very important, and the approximate results of these calculations mentioned above are acceptable for practical needs. Thus, according to the Rayleigh-Ritz methodology, the critical speed of shafts shown in Figure 3-7.1a and 3-7.1b can be calculated using the following formula:

$$n_{cr} = \frac{30}{\pi}\sqrt{\frac{g}{h}} \tag{3-7.15}$$

where g is the acceleration of gravity, and h is the deflection of the shaft due to static radial forces and the weight of the assembly. The deflection due to the weight is, in most cases, negligible; while actually, formula (3-6.8) determines the deflection caused by the radial static force F only. Therefore, in order to use formula (3-7.15), the deflection should be already determined.

It is important to introduce the approach that is used in the Rayleigh-Ritz methodology of calculating the critical speed of shafts with multiple rotating elements exerting forces that cause corresponding deflections of the shaft. According to this methodology, the critical speed of a shaft with multiple rotating masses can be calculated from the following formula:

$$n_{cr} = \frac{30}{\pi}\sqrt{\frac{g\sum_1^j W_i h_i}{\sum_1^j W_i h_i^2}} \tag{3-7-16}$$

where W_i is the equivalent weight of the corresponding rotating mass combined with the exerted force, h_i is the corresponding deflection, i is the orderly number of the mass, and j is the number of masses.

The methodology of determining the critical rotational speed is associated with the first or lowest critical speed that represents the most practical interest. The steps presented above for calculating the critical speed are applicable to any structural design of a shaft assembly.

3-8. Keys and Couplings

Keys are intended to transmit torque (or power) from a shaft to a rotating element mounted on the shaft or to transmit torque from a rotating element to the shaft. A coupling transmits torque (or power) from one shaft to another shaft or to an appropriate component.

3-8.1. Keys

A key is a machine element that allows transmitting toque from the shaft to a rotating element or from the element to the shaft. Most of the keys have a prismatic shape of a square or rectangular cross-sectional area and are usually called parallel keys. They are placed in corresponding keyseats machined in the shafts, while hubs have appropriate keyways accommodating the keys.

The cross-sectional dimensions of the keys are standardized, and the selection of a key is based on the diameter of the shaft. Appropriate tables for selecting keys are available in numerous published sources. Standard dimensions of the width and height in inch and metric series keys can be found in the sources presented in the Appendix A-3-1 and A-3-2, respectively.

Usually the designer selects the width and height of the key according to the diameter of the shaft and calculates the length L of the key with regard to its strength. Figures 3-8.1a and 3-8.1b show a drawing of an assembly comprising the hub 1 of the rotating element, the shaft 2, and the parallel key 3. Figure 3-8.1b represents a sectional view AA of the shaft assembly.

Figure 3-8.1c shows the top view of a keyseat with the inserted key having the length L. The keyseat shown in this figure is machined using an end mill tool having a diameter equal to the width b of the key. Note that between the key and the upper surface of the keyway there should be a gap δ that prevents the misalignment between the geometric axes of the rotating element

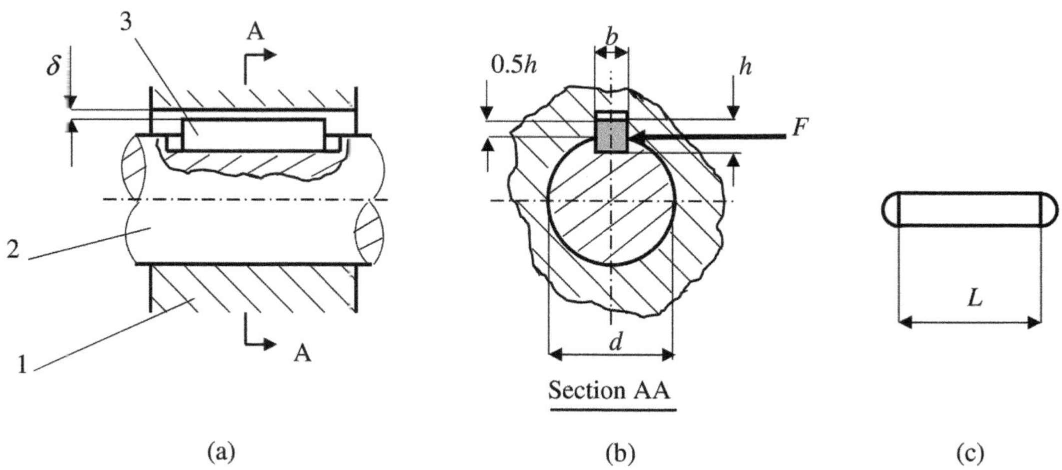

Figure 3-8.1. A shaft assembly including a key and a rotating element.

and the shaft in cases of a clearance fit between the shaft and the hub. In cases of assembling the shaft and hub by applying an interference fit, a plain gib key or a gib head key, as shown in Figures 3-8.2a and 3-8.2b, respectively, might be used.

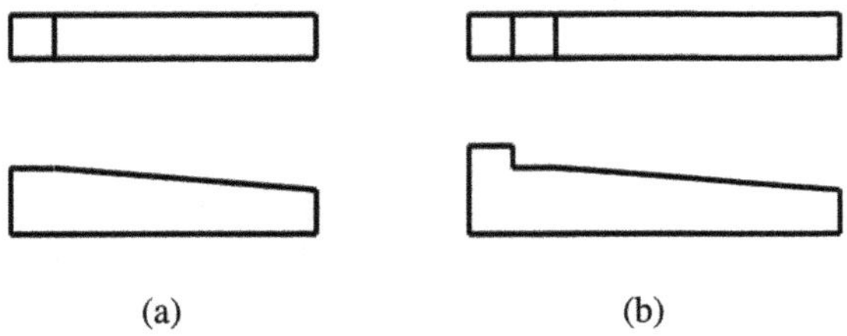

Figure 3-8.2. Plain and gib head keys.

These keys, sometimes called tapered keys, secure the hub from moving in the axial direction. The gib head key shown in Figure 3-8.2b can be removed by forcing a wedge between the head of the key and the hub.

For relatively small diameter shafts, it is advisable to use Woodruff keys, a drawing of which is presented in Figure 8-3.3, where $b \times l$ is the nominal key size, h is the height of the key, L is the actual length, and ε is the distance below the center of the shaft.

Usually keys are made from low carbon steels in order to prevent damage to the keyseat or the keyway.

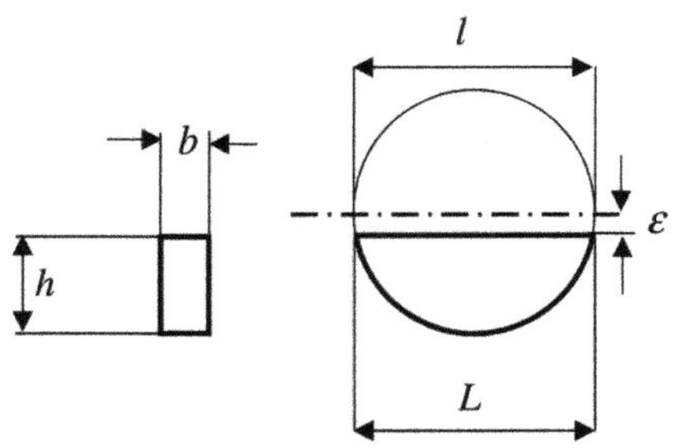

Figure 3-8.3. Woodruff key.

3-8.2. *Calculations of Keys*

It is acceptable to use the same cross-sectional area of the key for a certain, relatively small, range of diameters. The tables for the selection of keys indicate the range of diameters for each standard key. However, for preliminary calculations it is acceptable to consider a square key having the size of its side equal to 25% of the diameter of the shaft. In reality the key is subjected to the action of a distributed force; however, it is justifiable to replace it with a concentrated force F as shown in Figure 3-8.1b. In order to calculate the stresses in the key, it is necessary to determine the value of this force. Since the diameter of the shaft and the transmitted torque are known, we can determine the force applied to the key using the following formula:

$$F = \frac{2T}{d} \tag{3-8.1}$$

In general, the key is subjected to shear and bearing stresses. According to the Figure 3-8.1, the cross-sectional area A_s related to shear stress equals $A_s = bL$, while the cross-sectional area A_b related to the bearing stress equals $A_b = 0.5hL$. The shear stress acting in the key can be calculated using the following formula: $\tau = \frac{F}{A_s} = \frac{2T}{dbL}$. To assure the strength of the key, it should be $\tau \le \frac{S_{sy}}{n_\tau}$ where S_{sy} and n_τ are the shear stress, yield strength, and the safety factor for the key, respectively. Combining the expressions presented above, we obtain the formula for calculating the required length of the key L_τ with respect to shear stresses:

$$L_\tau = \frac{2Tn_\tau}{dbS_{sy}} \tag{3-8.2}$$

The bearing stress is computed from the formula $\sigma = \frac{F}{A_b} = \frac{2T}{0.5dhL} = \frac{S_y}{n_\sigma}$, where S_y is the tensile yield stress limit and n_c is the safety factor for the key with respect to bearing stresses. Rearranging this expression, we determine the length of the key with respect to bearing stress:

$$L_\sigma = \frac{2Tn_\sigma}{0.5dhS_y} \tag{3-8.3}$$

Obviously, we select the longer of these two keys. These calculations are applicable for the preliminary calculations of the key. In cases where the final calculations of the diameter of the shaft are carried out, the cross-sectional dimensions of the key should be selected from the appropriate published sources.

The stock material for parallel keys according to the standard cross-sectional areas is sold in standard length, and each key is cut to its appropriate length.

The keyseats in the shafts cause stress concentration that should be accounted for in the shaft calculations related to variable stresses. The keyseats are commonly machined by circular sled runner mills or by end mills as shown in Figures 3-8.4a and 3-8.4b, respectively.

It is reasonable to use the fatigue stress concentration factors indicated in this figure for shaft calculations; however, it should be noted that some authors report different values for these factors. It is advisable to consult appropriate published sources for additional related information.

It should be noted that due to machining and assembly considerations, normally, the shafts obtain stepped shape, while the keyways usually are located on enlarged diameters that actually increase the strength and stiffness of the shafts.

3-8.3 Couplings

A coupling is a machine element that transmits torque, or power, from one shaft to another. There are two basic types of couplings: rigid and flexible. A sleeve mounted between two adjacent ends of shafts and transmitting torque by pins or keys is an example of a rigid coupling, also known as a box coupling. This type of couplings requires high accuracy in the alignment of the

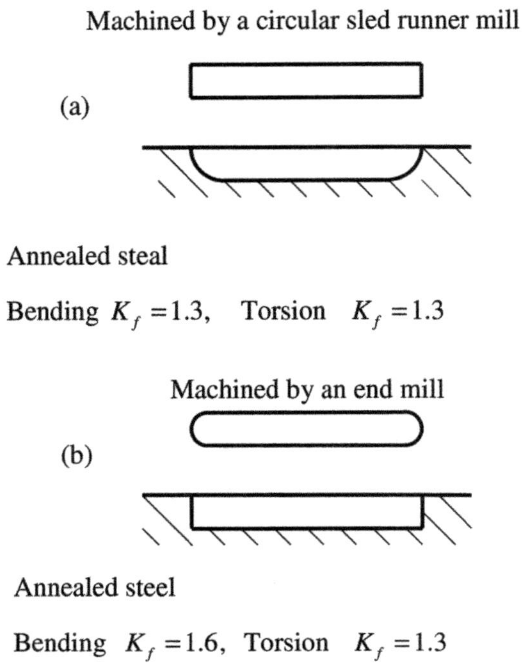

Figure 3-8.4. Types of keyseats.

shafts. A simple rigid coupling has the shape of a bushing or a sleeve that is secured to the shafts by a pair of identical pins or keys. A coupling with pins may play the role of a safety device that prevents damage to important elements of the assembly in case of an overload. In this case, the pin is sheared by a stress that exceeds the predesigned strength of the pin.

Flexible couplings allow misalignment between the shafts and are capable of softening some random irregularities of the transmitted torques. These couplings may have engaging and disengaging devices. However, these couplings are for assemblies that have specific applications, and, therefore, these couplings are not included in this book. It is recommended that you consult numerous catalogs that present the variety of off-the-shelf rigid and flexible couplings.

The stress calculations for box couplings that have pins determine the shear and bearing stresses. It should be realized that the shear and bearing stresses occur in two places simultaneously. Therefore, the shear and bearing stresses should be calculated using the following formulas: $\tau = \dfrac{F}{2A_s}$ and $\sigma_{br} = \dfrac{F}{2A_b}$, where $A_s = \dfrac{\pi d^2}{4}$ is the cross-sectional area of the pin (d is the pin diameter) and $A_b = td$ is the bearing surface (t is the wall thickness of the sleeve).

3-9 Design Project

Engineers often design shaft assemblies based on the input data comprising the values of the transmitted power and rotational speed. The basic steps of the design process of a common shaft assembly are presented below.

It is necessary to design a shaft assembly that transmits power of $P = 15$ kW at a rotational speed $n = 1{,}200$ rpm to a spur gear mounted in the middle between the shaft supports. The gear has a diameter (pitch diameter) $D_g = 65$ mm. The rotation is reversible, the service life of the bearings is 15,000 hours, and the safety factor $n_\sigma = n_\tau = 2$.

The design process comprises several steps that involve engineering calculations and drawing. Each project has its own objectives and conditions, but the steps presented below are applicable for common engineering tasks.

1. Design calculation of the shaft diameter.

Using formula (3-3.8) and accepting $\tau_{all} = 23$ MPa, we calculate the preliminary value of the diameter of the shaft $d = 3.65\sqrt[3]{\dfrac{P}{n\tau_{all}}} = 3.65\sqrt[3]{\dfrac{15\cdot 10^3}{1{,}200\cdot 23\cdot 10^6}} = 0.02985\text{ m}$. Normally, in order to select the bearing we round off the diameter to the value of the closest higher number corresponding to availability of standard radial ball bearings. In our case, we take $d = 30$ mm, and we select from the SKF catalog a radial ball bearing No. 6206-2Z having a bore diameter of 30 mm, an outer diameter of $D = 62$ mm, the width of B = 16 mm, the dynamic load rating $C = 20.3$ kN, and the static load rating $C_0 = 11.2$ kN. The bearing is double shielded.

2. Sketching the shaft assembly.

Having the width, the bore diameter, and outer diameter of the bearing, we estimate the dimensions of the width of the bearing enclosures, the width of the hub of the rotating element, and the distances between the hub and the enclosures. Based on all these design considerations, we draw a sketch of the shaft assembly that is presented in Figure 3-9.1.

This is a preliminary drawing, and it is not made to the scale. This drawing shows shaft 1 carrying bearing 4 mounted on the shaft by an interference fit. Housing 5 accommodates the

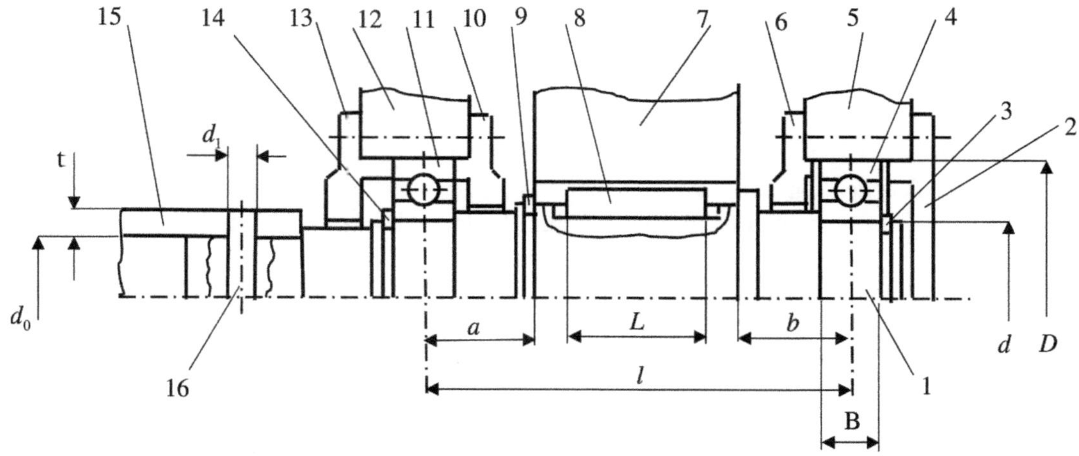

Figure 3-9.1. Design of a shaft assembly.

bearing by its outer diameter with a clearance fit. This bearing is placed between the shaft shoulder and a retaining ring 3 and is restricted from the axial displacement relative to the shaft. The retaining ring is selected from a manufacturer catalog. The covers 2 and 6 of the bearing housing do not touch the outer ring of the bearing, providing gaps on both sides of the outer ring that allow expansion or contraction of the shaft due to temperature change. Housing 12 accommodates bearing 11 that is located between the shaft shoulder and a retaining ring 14. The bearing is rigidly secured on both sides of the outer ring by covers 10 and 13, preventing the shaft assembly from axial displacement in order to avoid any axial interference of its left-hand side with the adjacent assembly. This bearing represents the non-movable support, while the right-hand bearing is the movable support.

As shown in the drawing the shaft increases its diameter by certain increments from the bearings toward the rotating element 7. Usually, the increments are about from 2 to 5 mm of the diameter; thus, the diameter of the shaft carrying the rotating element is increased to 35 mm. The steps on the shaft are needed for a variety of design considerations: the bearings should be positioned at certain places on the shaft; and if the seals are mounted in the inner covers of the bearings, then the surfaces of the shaft should be ground or polished. For these two reasons, we need to make two increased steps of the shaft on both sides of the rotating element. It is extremely important to provide an easy and a non-restricted way to place the rotating element on the shaft. This also requires the additional step of a shoulder on the shaft. In addition, different intervals of the shaft require different degrees of accuracy and different surface finish. All these design considerations lead to corresponding stepped shape of the shaft. In our case, the right-hand side of the rotating element is lining to the shoulder, while a retaining ring 9 locks the left-hand side of this element. The parallel key 8 is located in a keyseat that is machined by an end mill. The key transmits the torque from the shaft to the rotating element. The width of the hub of the rotating element is taken $1.2d = 1.2 \cdot 35 = 42$ mm. Accounting for the width of the bearing and the width of the housings and the covers, and taking 10 mm as the distance between the cover and the hub on each of its sides, we come up with the distances a and b from the center of the bearings to the sides of the hub to be 32 mm on each side. Therefore, the distance between the bearings $l = 2 \cdot 32 + 42 = 106$ mm.

Figure 3-9.1 shows that the box coupling 15 and pin 16 connect the shaft assembly to the source of power.

3. Calculating the reactions and plotting the bending and torque diagrams.

In order to calculate the reactions in the supports and to plot the bending diagrams, we need to determine the forces applied to the shaft. Using the formula (3-3.4), we determine the torque applied to the shaft $T = \dfrac{30 \cdot 15 \cdot 10^3}{\pi 1{,}200 \cdot} = 119.43\ \text{N} \cdot \text{m}$. Since the rotating element is a spur gear having the pitch diameter $D_g = 65$ mm, we can determine the tangential force F_t applied to the gear, hence, $F_t = \dfrac{T}{0.5D_g} = \dfrac{119.43 \cdot 10^3}{0.5 \cdot 65} = 3{,}675$ N. As it is discussed later in this book, the spur gear creates a radial force F_r that is applied to the shaft, and it is equal $F_r = F_t \tan 20° = 3{,}675 \tan 20° = 1{,}338$ N. These two forces are perpendicular to each other and could be replaced by a resultant force F that is equal to their vector summation $F=\sqrt{3{,}675^2+1{,}338^2}=3{,}911$ N. The free-body diagram of the shaft assembly is shown in Figure 3-9.2a.

(a)

(b)

(c)

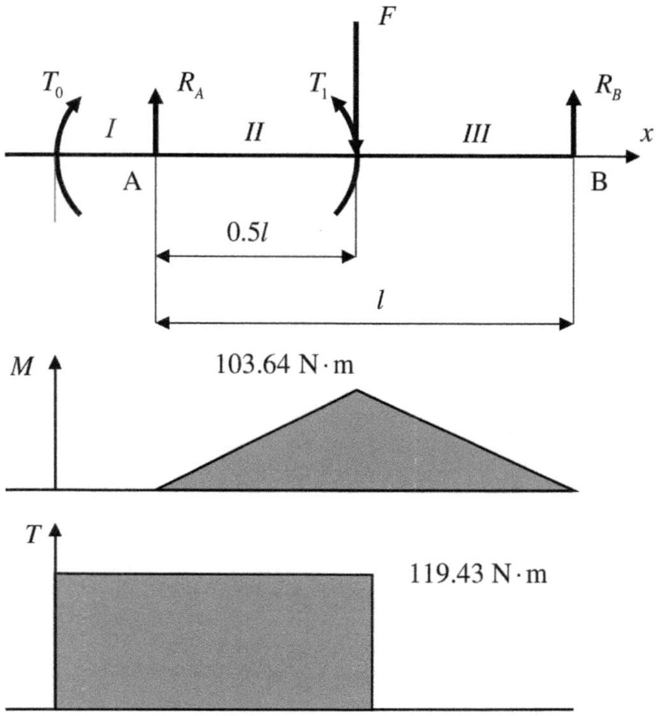

Figure 3-9.2. Reactions, bending moments and torque diagrams.

Since the force F is applied in the middle between the supports A and B, the reactions R_A and R_B are equal, therefore, $R_A = R_B = 0.5F = 0.5 \cdot 3.911 = 1{,}955.4$ N. The torques applied to the shaft are T_0 and T_1, while $T_0 = T_1 = T$. The shaft is divided into three intervals, while on the first interval the bending moment equals zero. At the end of the second interval, the bending moment equals $M_{II} = R_A 0.5l = 1{,}955.4 \cdot 0.5 \cdot 106 = 103{,}638$ N·mm. At the end of the third interval, the bending moment equals zero. Therefore, the maximum bending moment is $M_{max} = 103.64$ N·m. Figure 3-9.2b shows the diagram of the bending moments. The torques are equal on the first and second intervals, while on the third interval the torque equals zero. The torque diagram is presented in Figure 3-9.2c, while the maximum tongue $T_{max} = T = 119.43$ N·m. The most loaded cross-sectional area of the shaft is characterized by M_{max} and T_{max}, and it is located in the middle of hub (or in the middle between the bearings). The diameter of the shaft in the most loaded cross-sectional area is preliminarily taken as 35 mm.

4. Verifying calculations of the shaft diameter.

The shaft is subjected to completely reversed bending and torsion cycles. The shaft will be machined from an AISI 1022 annealed steel having $S_{ut} = 450$ MPa and $S_{yt} = 317.4$ MPa. We need to determine the endurance limit modifying factors. Using expression (1-9.2), we calculate the surface finish factor $k_a = 0.917$. Based on the expression (1-9.3), we calculate the size factor $k_b = 0.862$. From expression (1-9.4) we determine the load factor $k_c = 1$. Using the formula (1-9.5), we calculate the endurance limit $S_e = 0.917 \cdot 0.862 \cdot 1 \cdot 0.5 \cdot 450 = 177.85$ MPa. Since the keyseat in the shaft is machined using an end mill tool, we take the fatigue stress concentration factor for bending $K_{fb} = 1.6$ and for torsion $K_{ft} = 1.3$. In addition, we should determine the stress concentration factors due to the step in the shaft that is characterized by $\frac{D}{d} = 1.15$ and $\frac{r}{d} = 0.15$. According to Appendix A-1-5 (Figures 2 and 3), we determine the theoretical stress concentrations factors for torsion and bending, and they are $K_{tt} = 1.2$ and $K_{tb} = 1.5$. Hence, taking $q = 0.6$ and using formula (1-10.2) we calculate $K_{f1t} = 1.12$ and $K_{f1b} = 1.3$. We should account for each type of loading the larger of the two fatigue stress concentration factors. Therefore, since $K_{fb} > K_{f1b}$ (1.6>1.3) and $K_{ft} > K_{f1t}$ (1.3>1.12), we determine $K_{fb} = 1.6$ and $K_{ft} = 1.3$. Based on formulas (1-12.10) and (1-12.12), we determine the amplitude values of the bending moment and torque: $M_a = \frac{103.64 - (-103.64)}{2} = 103.64\ \text{N} \cdot \text{m}$ and $T_a = \frac{119.43 - (-119.43)}{2} = 119.43\ \text{N} \cdot \text{m}$.

Using formula (3-5.9), we perform verifying calculations of the diameter of the step of the shaft that carries the gear $d = \left\{ \frac{32 \cdot 2}{\pi} \left[\left(\frac{1.6 \cdot 103.64 \cdot 10^3}{177.85} \right)^2 + 0.75 \left(\frac{1.3 \cdot 119.43 \cdot 10^3}{177.85} \right)^2 \right]^{\frac{1}{2}} \right\}^{\frac{1}{3}} = 28.932$ mm.

Rounding up this result to 30 mm, we could select a bearing having the bore diameter of 25 mm instead of the preliminary selected bore diameter of 30 mm. However, at this step of the calculations we cannot finalize the decision, because we do not yet know the dimension of the diameter of the shaft that carries the coupling. Obviously, this diameter should be smaller than the bore diameter of the bearing. Therefore, the next step is calculating the diameter that carries the coupling.

5. Calculating the diameter of the shaft carrying the coupling.

Now we have to verify the diameter d_0 on the left-hand side of the shaft that carries the box coupling 15. We have to calculate the endurance limit for this part of the shaft that is subjected to a completely reversed torsion cycle. This part of the shaft has a step on its right-hand side and a hole for the pin representing two raisers of stress concentration. The fatigue stress concentration factor associated with the step is estimated to be $K'_{fo} = 1.3$ and with the hole $K''_{fo} = 1.9$. The larger of these two fatigue stress concentration factors is $K''_{fo} = 1.9$ that is accounted for in the corresponding calculations. The surface finish factor was calculated as $k_a = 0.917$. Accounting for a two-step reduction of the diameter, we assume that $d_0 = 25$ mm. Based on expression (1-9.3), we calculate the size factor $k_b = 0.876$, and according to expression (1-9.4), we determine the load factor $k_c = 0.577$. Based on formulas (1-10.3) and (1-10.4), we calculate the endurance limit $S_e = 0.917 \cdot 0.876 \cdot 0.577 \cdot \dfrac{1}{1.95} \cdot 0.5 \cdot 450 = 54 \cdot 9$ MPa. Combining formulas (1-4.20) and (1-4.21), we calculate $d_0 = \sqrt[3]{\dfrac{16 \cdot 119.43 \cdot 10^3 \cdot 2}{\pi 54.9}} = 27.998$ mm. We take the diameter $d_0 = 28$ mm.

6. Bearing calculations.

Now it is clear that the preliminary selected bearing with the bore diameter of 30 mm could represent the proper selection if its service life is in the compliance with the requirements of the project. With regard to the diameter of the step of the shaft having 35 mm, it becomes obvious that the applied design considerations are justifiable. However, if a bearing with the bore diameter of 30 mm is not capable of providing the required service life, we should go to the bore diameter of 35 mm, and the diameter of the step should be increased to 40 mm.

In order to determine the service life, we first need to calculate, according to formula (2-4.6), the equivalent static load F_o. In our case $F_r = 1{,}955.4$ N and $F_a = 0$, while the load factor $X_0 = 0.6$. Therefore, $F_0 = 0.6 \cdot 1{,}955.4 = 1{,}173$ N. Since for the selected bearing $C_0 = 11{,}200$ N and $F_0 < C_0$, we may proceed with the selected bearing. Since $F_a = 0$ based on formula (2-4.7), we compute $e = 0$. Therefore, the load factor $X = 1$, and the load factor $Y = 0$. In our case $V = 1$ since the inner ring is rotating. Using formula (2-4.5), we determine the equivalent dynamic load $F_e = 1{,}955.4$ N. Applying formula (2-4.4) and recalling that for the selected bearing $C = 20{,}300$ N, we calculate the service life of the bearing: $L_{10} = \left(\dfrac{C}{F_e}\right)^3 = \left(\dfrac{20{,}300}{1{,}955.4}\right)^3 = 1{,}119 \cdot 10^6$ revolutions or $L_{10h} = \dfrac{1{,}119 \cdot 10^6}{60 \cdot 1{,}200} = 15{,}540$ hours. Hence, the calculated service life slightly exceeds the required service life, and the selected bearing is acceptable for this project.

7. Calculating the pin diameter and wall thickness of the coupling.

We calculate the diameter d_1 of pin 16 based on shear stresses. The torque T creates a couple applied to the diameter d_0 developing in the pin shear and bearing stresses. So, we may write $0.5F_1 d_0 = \mathrm{T}$ and $F_1 = \dfrac{T}{0.5d_0} = \dfrac{119.43 \cdot 10^3}{0.5 \cdot 28} = 8{,}531$ N. Because the shear and bearing stresses occur on both ends of the pin, we base the stress calculations on the force $0.5F_1$. The pin and the coupling are made of the same material as the shaft. Accounting that $S_{sy} = 0.577 S_{yt}$ we have $S_{sy} = 0.577 \cdot 317.4 = 183.14$ MPa. The pin diameter d_1 can be determined using the formula $d_1 = \sqrt{\dfrac{2F_1 n_\tau}{\pi S_{sy}}} = \sqrt{\dfrac{2 \cdot 8{,}853.1 \cdot 2}{\pi 183.14}} = 7.847$ mm. We take the closest standard dimension of the pin $d_1 = 8$ mm. The wall thickness of the coupling is calculated based on the following formula: $t = \dfrac{0.5F_1 n_\sigma}{d_1 S_{yt}} = \dfrac{0.5 \cdot 8{,}853.1 \cdot 2}{8 \cdot 317.4} = 3.487$ mm. We take $t = 5$ mm.

8. Key calculations.

We select the AISI 1015 steel having S_{yt} = 285 MPa as the material for the key. For the shaft having a diameter d_2 = 35 mm, we select according to Appendix A-3-2 the following dimensions of the parallel key: b = 10 mm and h = 8 mm. The required length of the key with respect to shear stress is calculated according to formula (3-8.2) that reads $L_\tau = \frac{2Tn_\tau}{dbS_{sy}}$, where $d = d_2$ = 35 mm, and $S_{sy} = 0.577S_y$. Hence, $L_\tau = \frac{2\cdot 119.43\cdot 10^3\cdot 2}{35\cdot 10\cdot 0.5\cdot 285} = 14.95$ mm. Based on formula (3-8.3), we determine the length of the key with respect to bearing stress: $L_\sigma = \frac{2Tn_\sigma}{0.5\cdot dhS_y} = \frac{2\cdot 119.43\cdot 10^3\cdot 2}{0.5\cdot 35\cdot 8\cdot 285} = 11.97$ mm. Hence, $L_\tau > L_\sigma$, and we take L = 15 mm.

9. Calculations of the deflection and the angle of rotation.

Since the shaft has a stepped structure, it is recommended we replace it with a uniform shape having an equivalent diameter according to formula (3-6.11). The diameters of the steps of the shaft on both sides of the hub of the rotating element are equal 32 mm.

$$d_e = \frac{1}{\sqrt[4]{\sum_1^k \frac{l_j}{ld_j^4}}} = \frac{1}{\sqrt[4]{\frac{8}{106\cdot 30^4} + \frac{24}{106\cdot 32^4} + \frac{42}{106\cdot 35^4} + \frac{24}{106\cdot 32^4} + \frac{8}{106\cdot 30^4}}} = 32.63 \text{ mm}$$

We need to determine the moment of inertia of the shaft having the equivalent diameter, hence, $I = \frac{\pi d_e^4}{64} = \frac{\pi 32.63^4}{64} = 55{,}618 \text{ mm}^4$. Using formula (3-6.8), we calculate the deflection h of the shaft between the bearings $h = -\frac{Fl^3}{48EI} = \frac{3{,}675\cdot 106^3}{48\cdot 207\cdot 10^3\cdot 55{,}618} = -0.00792$ mm.

It is recommended that the deflection of the shaft carrying a gear should not exceed $0.0003l$. In our case $0.0003l = 0.0003\cdot 106 = 0.0318$ mm that exceeds the calculated deflection 0.00792 mm.

The angle of rotation is calculated using formula (3-6.10): $\theta = -\frac{Fl^2}{16EI} = -\frac{3{,}675\cdot 106^2}{16\cdot 207\cdot 10^3\cdot 55{,}618} = -0.000224$ rad, that is less than the allowable angle of rotation for our case 0.01 rad.

10. Calculations of critical rotational speed.

Based on formula (3-7.15), we calculate the critical rotational speed for our case: $n_{cr} = \frac{30}{\pi}\sqrt{\frac{g}{h}} = \frac{30}{\pi}\sqrt{\frac{9.8\cdot 10^3}{0.00792}} = 10{,}627$ rpm.

11. Assembly and disassembly.

a) Assembly

First, we install key 8 into the keyseat of shaft 1, put in place rotating element 7, and secure it with retaining ring 9. In our case, the bearings are lubricated for life and have shields on both sides.

However, if the bearings would require lubrication then covers 6, 10, and 13 should be able to accommodate appropriate seals that are installed preliminarily. We put in place covers 6 and 10 assuming that there is sufficient room between the covers and the hub to install the screws that secure the covers to housings 5 and 12. Then by pushing these housings by hand, we place them on the bearings. Now we press bearings 4 and 11 onto the shaft and secure them with retaining rings 3 and 14. Then using bolts, we rigidly secure the housings to the frame (not shown in the sketch). After that, we tighten the screws securing covers 6 and 10 to housings 5 and 12, respectively. Now we secure cover 2 to housing 5. Then we need to perform an adjustment procedure during the securing of cover 13. Normally, this adjustment is associated with selecting the appropriate thickness of shims that should be installed between the flange of the cover and the housing or between the cover and the outer ring of the bearing in order to prevent any axial displacement of the bearing. After that, we secure the cover to the housing using appropriate screws. Now we install coupling 15 and press in pin 16.

The shaft assembly is ready to be connected by coupling 15 to the source of power.

b) Disassembly

First, we have to disconnect the source of power and remove pin 16 and coupling 15. After that we unscrew the screws securing all four covers to the housings and unscrew the bolts securing the housings to the frame. Now we remove the housings with the covers 2 and 13 and using appropriate pliers remove retaining rings 3 and 14. After that, using appropriate pullers, we take off the bearings from the shaft. The pullers should be engaged with the inner rings of the bearings. Now we remove retaining ring 9 and press off the shaft from the rotating element. After that, we remove the key.

This concludes the sequence of steps of taking apart the shaft assembly.

Chapter 3. Problems

3-01. Calculate the bending stress in the axle shown in Figure 3p-01. The diameter of the axle d = 30 mm. The axle is loaded by a force F = 600 N and moments M_1 = 250 N·m and M_2 = 300 N·m, while l = 300 mm.

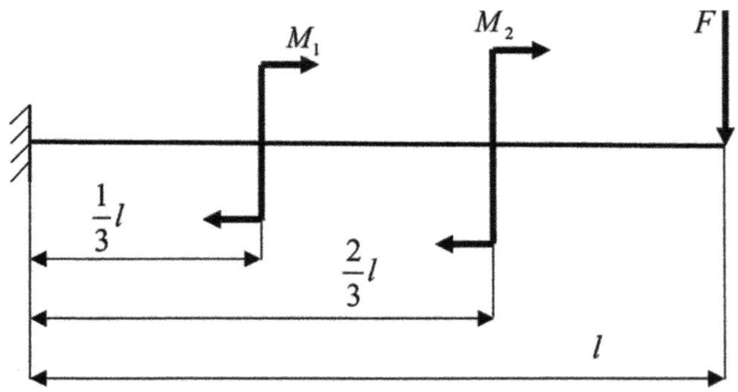

Figure 3p-01

3-02. Determine the equivalent stress according to the fourth theory of failure for the axle shown in Figure 3p-02 The axle has a diameter d = 35 mm and it is loaded by forces F_1 = 400 N and F_2 = 350 N, a moment M = 280·N·m, and a torque T_0 = 400 N·m, while l = 300 mm.

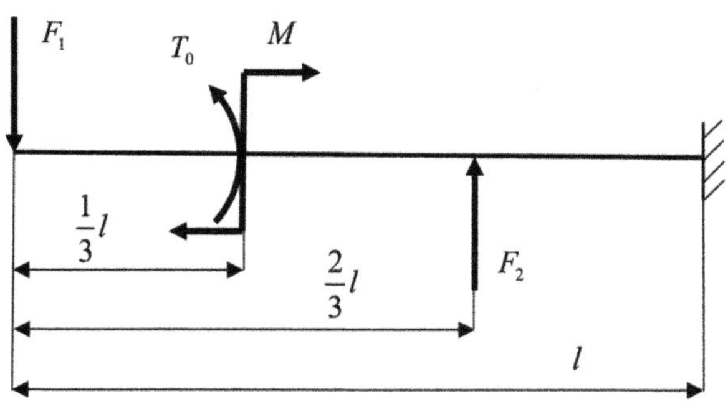

Figure 3p-02

3-03. Determine the equivalent stress according to the fourth theory of failure for the shaft shown in Figure 3p-03. The shaft has a diameter d = 1.5 in. and it is loaded by forces $F_1 = F_2$ = 500 lb, moments $M_1 = M_2$ = 6.000 lb-in., and torques $T_1 = T_2$ = 8,000 lb-in., while l = 15.0 in.

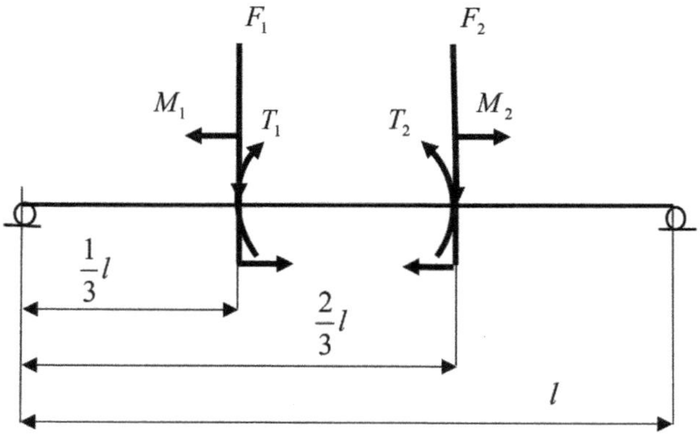

Figure 3p-03

3-04. Determine the equivalent stress according to the fourth theory of failure for the shaft shown in Figure 3p-04. The shaft has a diameter d = 70 mm and it is loaded by forces $F_1 = 2{,}500$ N, $F_2 = 1{,}500$ N, and $F_3 = 2{,}000$ N, moments $M_1 = 1{,}500$ N · m, $M_2 = 2{,}000$ N · m, and $M_3 = 1{,}000$ N · m, and torques $T_1 = 2{,}000$ N · m, $T_2 = 1{,}500$ N · m, and $T_3 = 3{,}500$ N · m, while l = 400 mm.

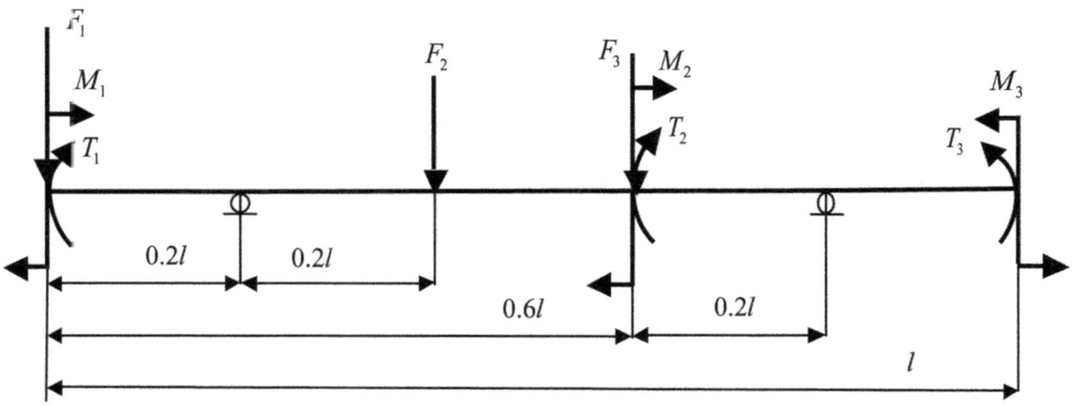

Figure 3p-04

3-05. Applying the design calculation methodology, determine the preliminary diameter of the shaft. The shaft transmits power of 15 kW at the following rotational speeds – 3,600 rpm, 1,800 rpm, and 1,200 rpm. Determine the diameter for each of the speeds using the following allowable stress τ_{all} = 20 MPa.

3-06. Figure 3p-06 shows a bar that is machined from an AISI 1050 annealed steel having $S_{ut} = 98{,}750$ psi . The bar is loaded by a force $F = 500$ lb, a moment $M = 2{,}000$ lb-in, and a torque $T = 2{,}500$ lb-in. The loading develops a completely reversed cycle. Using the fourth theory of failure, determine the diameter of the bar taking the safety factor $n_\sigma = 2$ and assuming the size factor $k_b = 0.85$, while $l = 12.0$ in.

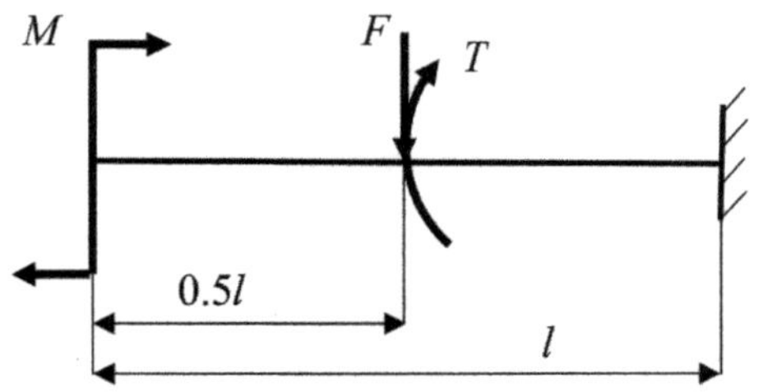

Figure 3p-06

3-07. The shaft shown in Figure 3p-07 is machined from an AISI 1060 annealed steel having $S_{ut} = 626$ MPa. The diameter of the shaft $d = 45$ mm and its length $l = 300$ mm. The shaft is rotating in one direction being supported by two bearings. The shaft is loaded by a force $F = 7{,}500$ N and torques $T_1 = T_2 = 600$ N · m. It should be accounted that due to the keyseat the fatigue stress concentration factor for bending $K_f = 1.6$. Based on the fourth theory of failure, determine the safety factor of the shaft.

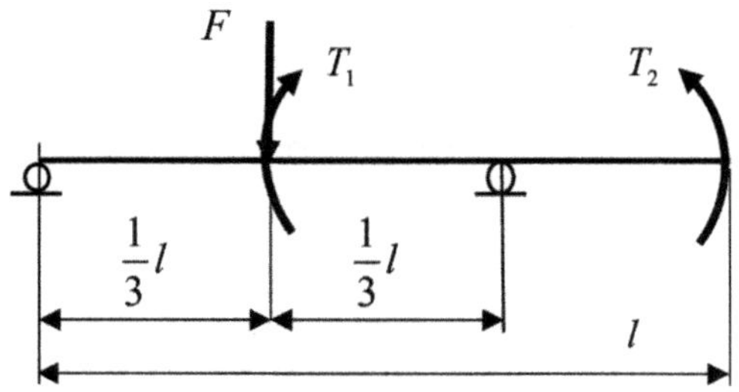

Figure 3p-07

3-08. The shaft shown in Figure 3p-08 is machined from a normalized AISI 1050 steel having S_{ut}= 748.7 MPa and S_{yt} = 428 MPa. The diameter of the shaft d = 45 mm and its length l = 360 mm. The rotation of the shaft is supported by two bearings and the shaft is loaded by a force F = 3,500 N, a moment M = 450 N · m and torques $T_1 = T_2 = 550$ N · m. Based on the fourth theory of failure, determine the safety factor of the shaft for the case of a completely reversed cycle and for a repeated cycle.

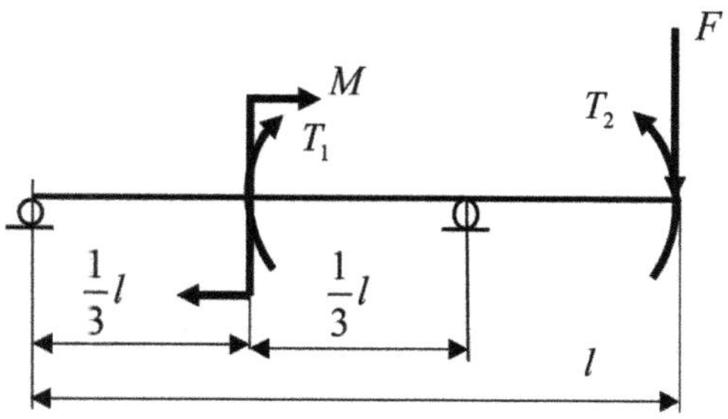

Figure 3p-08

3-09. The shaft shown in Figure 3p-09 is machined from an AISI 1060 normalized steel having S_{ut} = 112,500 psi and S_{yt} = 61,000 psi. The diameter of the shaft d = 2.0 in. and its length l = 18 in. The shaft is supported by two bearings, and it is loaded by moments $M_1 = M_2$ = 6,000 lb-in and torques $T_1 = T_2$ = 7,000 lb-in. It should be accounted that due to the keyseat the fatigue stress concentration factor for bending stresses K_f = 1.6 and for torsion stresses K_{ft} = 1.3. Based on the fourth theory of failure, determine the safety factor of the shaft for the case of a completely reversed cycle and of a repeated cycle.

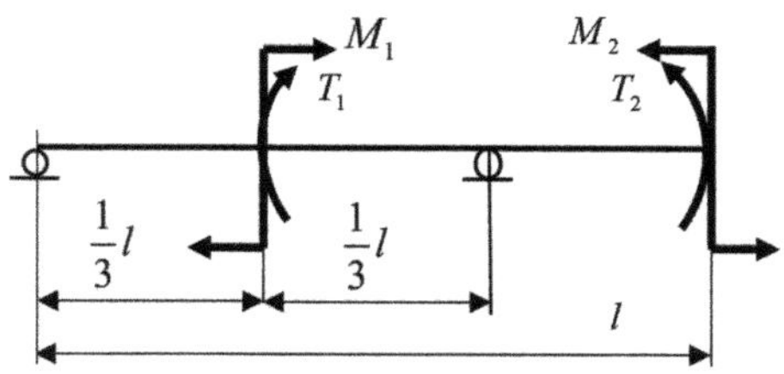

Figure 3p-09

3-10. The steel shaft shown in Figure 3p-10 has a diameter d = 45 mm and its length l_0 = 180 mm, while the length of the cantilever l = 60 mm . The shaft is loaded by a force F = 5,000 N. Determine the deflection at point A, the angle of rotation at point B of the shaft, and compute the critical rotational speed of the shaft.

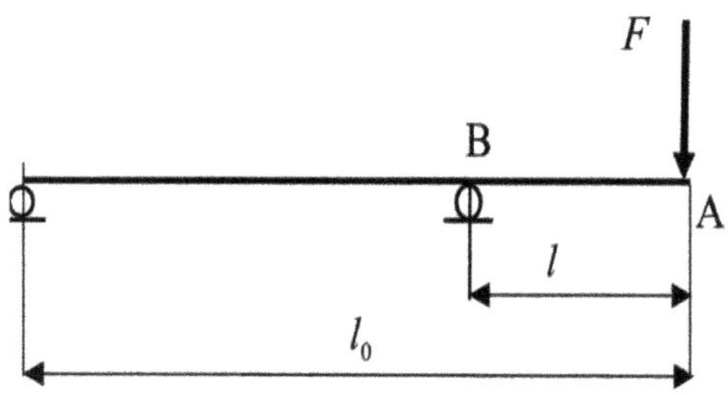

Figure 3p-10

CHAPTER 4

Belt Transmissions

Transmissions represent a very significant group of mechanical engineering systems that sometimes are referred to as drives. Drives transmit power at a specific rotational speed to an appropriate system comprising a rotating mechanical link that performs the desired function. The gearbox in a car is a transmission that transmits the power from the engine to the wheels. The question may arise: why do we need an intermediate mechanical system between the source of power and the mechanical system that performs the required rotational function? Why is the source of power not connected directly by a coupling to the system that performs the desired operation?

Here is the answer. The existing sources of power, or energy, that are used in mechanical rotational systems are internal combustion engines and electric motors. The basic characteristics of these two sources of energy are their mechanical stability in terms of power and torque that requires relatively high rotational speeds that cannot be controlled (electrical motors) or allow very limited control of their operational speeds (engines). Obviously, that the endless variety of operational processes require an endless variety of operational speeds that in the vast majority of cases are lower than the nominal speeds of the motors and engines. This explains the need in transmissions that are able to deliver the power from the energy sources to the mechanical systems at predominantly reduced speeds. The operational speeds of the majority of working processes are lower than the nominal speeds of the energy sources. Actually, in the vast majority of mechanical systems transmissions represent an integrated part of the systems. The electrical motors of alternating current are manufactured in a certain range of nominal speeds and do not allow any adjustment of its nominal speeds.

In many countries, there are entire industries devoted to production of all types of transmissions. It should be noted that electric motors of direct current allow a wide control of the rotational speed, but their application is limited due to cost considerations.

Internal combustion engines are used predominantly in transportation, construction, and similar industries. In the vast majority of industrial facilities, commercial and residential buildings electric motors of alternating current represent the main sources of energy.

Thus, the belt transmission is one of the most widely used drives for mechanical engineering systems.

In this book we consider two types of widespread transmissions: friction based transmissions and engagement based transmissions. In this chapter, we discuss belt drives that are friction transmissions, while in later chapters we discuss toothed gear drives that are engagement transmissions.

As it is mentioned above, the power sources for the majority of mechanical systems are alternating current electric motors. The number of built in poles defines the structural design of these motors. Depending on the number of poles, the electric motor develops a certain constant nominal rotational speed. Obviously, the minimum number of poles in a motor is two. For the United States the standard current frequency of 60 Hz the nominal rotational speed of a motor having two poles is 3,600 rpm. For motors having 4, 6, and 8 poles, the nominal speeds are 1,800 rpm, 1,200, and 900 rpm, respectively. With the increase of the number of poles, the rotational speed of the motor decreases; however, its cost, weight, and overall dimensions increase noticeably. The power of electric motors is available in a wide standard row starting from 0.5 hp and continuing through increments to very high numbers. Usually, each standard power from this row is available in the above mentioned four most applicable nominal speeds indicated below. Considering the cost, weight, overall dimensions, and availability, it is recommended for general application to use motors having speeds of 3,600, 1,800, 1,200 rpm, and 900 rpm.

4-1. Basic Parameters of Belt Transmissions

A belt transmission comprises a driving pulley (sheave), a driven pulley (sheave), and a flexible element, a belt that is wound around the pulleys. Certain tension forces are applied to both strands of the belt in order to develop frictional interaction between the belt and pulleys. The friction forces applied from the pulleys to the belt enable transmitting the power from the driving pulley to the driven pulley. The driving pulley receives the power straight from the source of power and, therefore, it has the same rotational speed as the source of power. The diameters of driving and driven pulleys are usually different and, consequently, their rotational speeds are different. Power (ignoring some frictional loses) is transmitted by the belt to the driven pulley that usually has a reduced rotational speed. However, there are some systems, such as fatigue-testing machines, where the driven rotating element has a higher rotational speed than the driving element. Figures 4-1.1a and 4-1.1b show schematics of belt transmissions consisting of a driving pulley 1 and a driven pulley 3 that are in contact with belt 2.

The belt transmission in Figure 4-1.1a is a horizontal open drive, and both pulleys of this drive are rotating in the same direction. The transmission in Figure 4-1.1b is a closed-belt drive, where the pulleys of this drive are rotating in opposite directions. Closed drives have a limited application.

It is acceptable to denote all parameters of the driving rotational element with the subscript 1 and the driven rotational element with the subscript 2. Therefore, the driving pulley has a diameter d_1 and the contact angle (wrap angle) φ_1, while the diameter and contact angle of the driven pulley are d_2 and φ_2, respectively. The center distance between the pulleys is C. The driving pulley is rotating at the angular velocity ω_1, and the driven pulley is rotating at the angular velocity ω_2, while in the majority of cases $\omega_1 > \omega_2$. The linear velocity of the belt V is, obviously, the same in each point of the periphery of both pulleys, therefore, denoting by V_1 and V_2 the corresponding linear velocities of the driving and driven pulley, respectively, we may write $V_1 = V_2 = V = \omega_1 \frac{d_1}{2} = \omega_2 \frac{d_2}{2}$. Actually, due to the inherent slip between the belt and the pulley, the

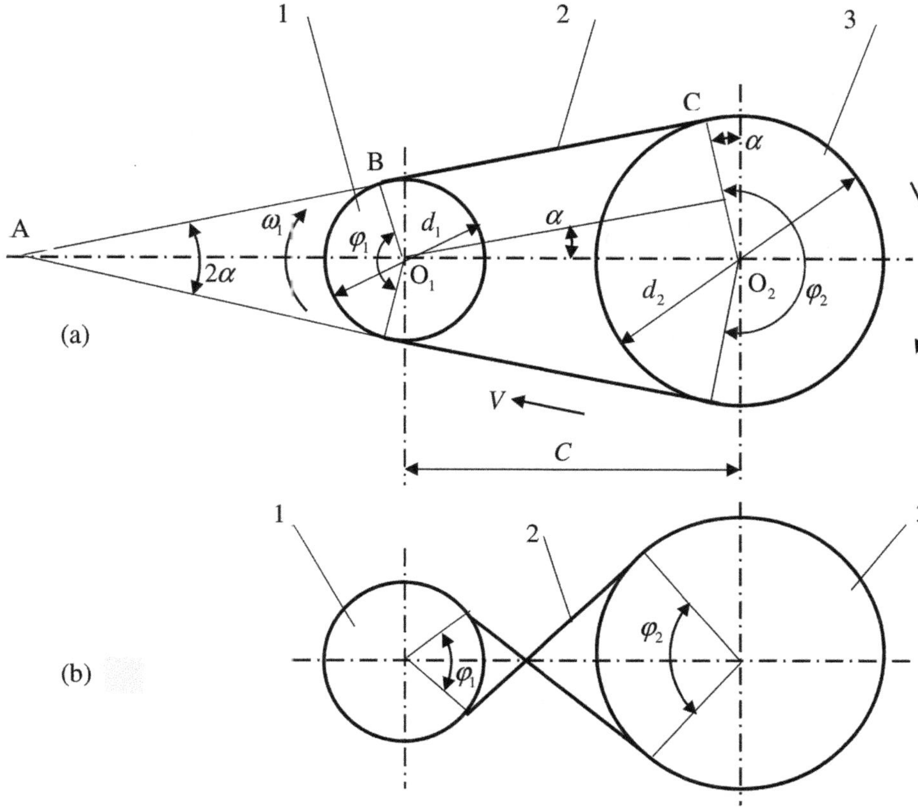

Figure 4-1.1. Belt transmissions.

linear velocity of the driven pulley is slightly less than the velocity of the driving pulley, and this is reflected in the expression $V_2 = V_1(1\text{-}\xi)$, where ξ is the slip coefficient that depends on the type of belt material and varies from 0.01 to 0.02.

The ratio between the angular velocity of the driving pulley and driven pulley is the velocity ratio:

$$i = \frac{\omega_1}{\omega_2} \tag{4-1.1}$$

while $i > 1$. It should be emphasized that due to the belt slip, the velocity ratio is not a constant value, but the instability of this ratio is negligible. The considerations presented above allow us to calculate the velocity ratio using the following formula:

$$i = \frac{d_2}{d_1} \tag{4-1.2}$$

Since in general, the torque $T = \frac{P}{\omega}$, where P is the power, therefore, $T_1 = \frac{P}{\omega_1}$ and $T_2 = \frac{P}{\omega_2}$. Obviously, $T_1 < T_2$ and the ratio between the torques equals the velocity ratio:

$$\frac{T_2}{T_1} = i \tag{4-1.3}$$

As shown below, the angle of contact φ_1 plays an important role in the operation of belt transmissions; therefore, it is desirable to determine the relationship between this angle and other parameters of the transmission. Based on the similarity of the triangles AO_1B and AO_2C (Figure 4-1.1a) we may write $\frac{d_1}{2\overline{AO_1}} = \frac{d_1}{2(C+\overline{AO_1})}$, therefore $\overline{AO_1} = \frac{d_1 C}{d_2 - d_1}$. From the triangle AO_1B we have $\sin\alpha = \frac{d_1}{2AO_1} = \frac{d_1(d_2-d_1)}{2d_1C} = \frac{d_2-d_1}{2C}$. Since $\alpha = \frac{\pi - \varphi_1}{2}$ we may write that $\sin\frac{\pi-\varphi_1}{2} = \frac{d_2-d_1}{2C}$ and after corresponding procedures, we obtain

$$\varphi_1 = 2\cos^{-1}\frac{d_2-d_1}{2C} \tag{4-1.4}$$

Based on Figure 4-1.1a, we calculate the length of the belt L for the open belt drive $L = 2C\cos\alpha + \frac{d_1}{2}(\pi - 2\alpha) + \frac{d_2}{2}(\pi + 2\alpha)$. Replacing α with the value presented above and performing corresponding transformations, we have

$$L = \sqrt{4C^2 - (d_2-d_1)^2} + \pi d_2 - \frac{\varphi_1}{2}(d_2 - d_1) \tag{4-1.5}$$

This formula is applicable for preliminary calculations of open flat drives and V-belt drives where $d_1 \le d_2$.

It is important to note that it is always preferable for open-belt transmissions to have the tight strand on the bottom and the slack strand on the top because this arrangement increases the belt contact angle on the driving pulley.

Example 4-1.1a: Calculating basic parameters.
An open horizontal, flat-belt drive transmits power of 7.5 hp $= 7.5 \cdot 550\ \frac{\text{ft-lb}}{s}$ at the rotational speed of 1,200 rpm to the driving pulley that has a 9 in. diameter. The driven pulley has a diameter of 30 in. and it is located 98 in. from the driving pulley. Determine the velocity ratio of the transmission, the rotational speed of the driven pulley, the linear velocity of the belt, the tangential force applied to the pulleys, the torques on both pulleys, the contact angle on the driving pulley, and the length of the belt.

Solution:
Using formula (4-2.2), we determine the velocity ratio $i = \frac{d_2}{d_1} = \frac{30}{9} = 3.333$, therefore, $n_2 = \frac{n_1}{i} = \frac{1{,}200}{3.333} = 360.04$ rpm. The linear velocity of the belt $V = 0.5d_1\omega_1$, thus, substituting the

corresponding values, we have $V = \frac{\pi 0.5 \cdot 9 \cdot 1{,}200}{30 \cdot 12} = 47.1 \ \frac{\text{ft}}{s}$, or $V = 47.1 \cdot 60 = 2{,}826 \ \frac{\text{ft}}{\text{min}}$. The tangential force applied to the pulleys can be determined using the following formula: $F_t = \frac{P}{V} = \frac{7.5 \cdot 550}{47.1} = 87.58 \text{ lb}$. Therefore, $T_1 = 87.58 \cdot 0.5 \cdot 9 = 394.1$ lb-in., and based on the formula (4-1.3), we obtain $T_2 = 394.1 \cdot 3.333 = 1{,}313.56$ lb-in., Using formula (4-1.4), we determine the contact angle on the driving pulley: $\varphi_1 = 2\cos^{-1}\frac{d_2 - d_1}{2C} = 2\cos^{-1}\frac{30-9}{2 \cdot 98} = 2\cos^{-1} 0.107$ or $\cos\frac{\varphi}{2} = 0.107$. Therefore, $\varphi_1 = 2 \cdot 83.86° = 167.72°$. The length of the belt is calculated using formula (4-1.5), and it is $L = \sqrt{4 \cdot 98^2 - (30-9)^2} + \pi 30 - \frac{167.72}{2 \cdot 57.3}(30-9) = 260.59$ in.

4-1.1. *Characteristics of Belt Transmissions*

The common industrial applications of belt drives involve transmitting power between two parallel shafts that are rotating in the same directions. These transmissions provide a number of advantages: relatively low cost, simplicity of design, low operational noise, low maintenance, easy installation, dampening of shock loads, ability to operate at high speeds, ability to transmit power between relatively long distances, preventing overloading of the system, no need for lubrication, and others.

One disadvantage is the inherent elastic slip of the belt over the pulleys, causing the velocity ratio to deviate from a constant number, which prevents the use of belt transmissions in certain systems where the constant velocity ratio is a necessity (for instance, in computing systems). The elastic slip between the belt and pulleys cannot be eliminated by increasing the tension in the belt. Another disadvantage of belt transmissions is the generation of electrical sparks between the belt and the pulleys that might ignite certain gases that could leak during some chemical processes in closed industrial facilities. The relatively short service time of the belts and the relatively large dimensions of the pulleys are also among the disadvantages of belt transmissions. In addition, it should be mentioned, that the belts impose relatively large radial forces on the shafts.

Usually, the transmitted power of belt drives does not exceed 50 kW; however, there are belt transmissions that transmit 100 kW and more. The velocity ratio of belt drives in most cases is limited to $i = 6$, but in some cases this ratio reaches $i = 10$.

4-2. Types of Belts

Depending on the structure of the cross-sectional areas, we consider three types of belts: flat belts, wedge-type belts, and round belts. Wedge-type belts are more commonly known as V-belts. Figures 4-2.1a, 4-2.1b, and 4-2.1c show the cross-sectional dimensions of flat belts, V-belts, and round belts, respectively. This book considers most widespread belt drives to be flat belts and V-belts.

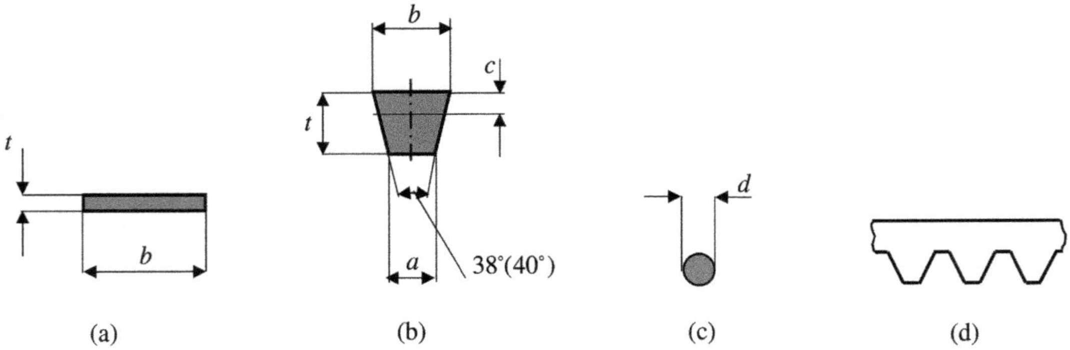

Figure 4-2.1. Types of belts.

Figure 4-2.1d shows a cogged V-belt that is more flexible than a regular V-belt and is characterized by a reduced bending stress when being wound around the sheave.

In addition to these types of belts, there is a flat belt that is called the synchronous belt, which has a toothed profile. The teeth of this belt mate with a pair of toothed pulleys having the same pitch. These belts with their pulleys work similarly to chains with their sprockets and are used in internal combustion engines to transmit from the crankshaft to the camshaft the rotational motion according to the required velocity ratio. These belts are also known as timing belts. Actually, these belts belong to the group of machine elements for specific applications and are not included in this book.

4-2.1. Flat Belts

Flat belts are usually manufactured from natural or synthetic materials that are often reinforced by high-strength fabric, providing high resistance to stretching. The primary belt materials are leather, cotton, rubber, and some synthetic materials, such as urethane, polyurethane, polyamide, polyester, elastomer, and others. Lately, flat belts are made from synthetic material such as polyurethane reinforced by steel cords or steel fabric. Covering the belts with certain coatings increases the friction coefficient between the belt and pulley, improving the efficiency of the transmission. Due to the relatively low specific weight of flat belts, they generate smaller centrifugal forces than V-belts. Since flat belts are characterized by relatively small thickness, the bending stresses generate a relatively low level of heat, allowing the belts to run at relatively high linear velocity that is in range from 3,000 to 10,000 $\frac{ft}{\text{min}}$ $\left(\text{from 15 to 50 } \frac{m}{s}\right)$. It should be noted that certain surface coatings contain materials that prevent the buildup of static electricity and, consequently, sparking.

Pulleys for flat belts are made of cast iron, steel, aluminum, and other materials. It is accepted that the rim of the pulley is about 10% wider than the belt. Usually, the surface of the rim is crowned in the range of 0.012 to 0.1 inch (from 0.3 to 2.5 mm). Belts running over convex crowned pulleys have the tendency to align themselves with the center of the pulley. The crowned surface causes a centrifugal force that keeps the belt in the center of the pulley. As shown in Figure 4-2.2, when the belt 1 is moved away from the center to the left-hand side of pulley 2, for example, the centrifugal force F_c that acts constantly becomes resolved in a normal component F_{cn} and a tangential component F_{ct} that pulls the belt to the center of the pulley.

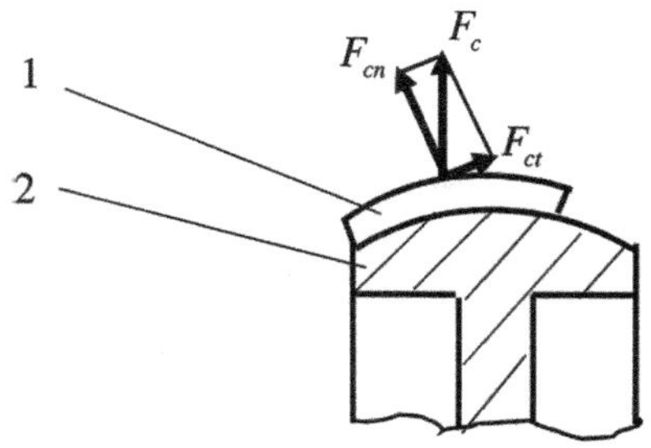

Figure 4-2.2. Centrifugal force due to a crowned pulley.

It should be mentioned that the efficiency of flat belt transmissions is relatively high, above 98%.

The design process of a flat-belt drive involves determining the diameter of the smaller pulley. The smaller the diameter of the pulley, the higher the bending stresses are in the part of the belt that is wound around the pulley. It is important to note that assigning the minimum diameter of the driving (smaller) pulley should be based on recommended ratios between the diameter and the thickness of the belt $\frac{d_1}{t}$. This prevents excessive bending stresses that occur in the belt during its wrapping around the pulley. The bending stresses acting in the belt represent a repeated cycle. Based on numerous experimental data, it is recommended to use $\frac{d_1}{t} \geq 25$ for belt velocities $V \leq 10{,}000\ \frac{\text{ft}}{\text{min}}\left(50\frac{m}{s}\right)$, while increasing this ratio to 40 for velocities $15{,}000\frac{\text{ft}}{\text{min}}\left(80\ \frac{\text{m}}{s}\right)$.

4-2.2. V-Belts

Figure 4-2.1 shows that the wedge angle of the V-belt is 38° that is in accordance with the U.S. standard. It is 40° according to the ISO standard. As shown below, the decrease of this angle allows the same diameter of the pulley to transmit increased power. However, when the wedge angle is smaller than 38°, it becomes more and more difficult to pool out the belt from the wedge, which disturbs the functionality of the belt drive. Numerous experimental investigations show that the optimal value of the wedging angle is in the order of 38° or 40°.

V-belts possess significant advantages when compared with flat belts, and V-belt transmissions became the most widely used transmissions among other belt drives. The most significant advantage of V-belt drives, as mentioned above, consists in the ability to transmit much more power than a flat-belt drive having the same overall dimensions. The interaction between a

V-belt and a pulley having a corresponding V-type grove is characterized by a so called wedging effect of friction between the belt and the pulley (sheave). The analysis of the wedging effect of friction explains the reason V-belts transmit significantly increased power for the same pulley diameters of flat-belt drives. Figure 4-2.3 shows the interaction between a wedge and a prism that resembles the operational process of a V-belt.

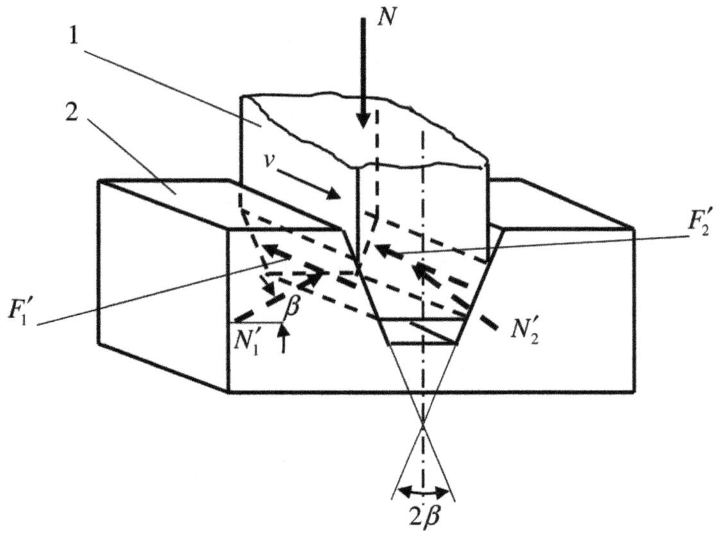

Figure 4-2.3. Wedging effect of friction.

In this figure, wedge 1 imitates the V-belt sliding in motionless prism 2 that imitates the sheave. The wedge is moving in the direction indicated by the arrow v and is loaded by a normal pressure force N that causes the reactions N_1' and N_2' applied perpendicularly to the inclined surfaces of the wedge from the prism. These reactions exert on each of the inclined surfaces of the wedge friction forces N_1' and N_2' directed oppositely the arrow v. Since the wedge and the prism are symmetrical, then $N_1' = N_2'$ and $F_1' = F_2'$, while $F_1' = fN_1'$, where f is the friction coefficient between the wedge and the prism. Considering the equilibrium of the wedge in the vertical direction, we may use the equation of statics projecting on the vertical axis the forces applied to the wedge, hence, $-N + N_1' \sin\beta + N_2' \sin\beta = 0$, therefore, $N = N_1' \sin\beta + N_2' \sin\beta$. From here we have $N_1' + N_2' = \dfrac{N}{\sin\beta}$. Because $F_1' + F_2' = F$, where F is the total friction force between the wedge and the prism, and consequently, between the V-belt and the sheave, we may write that $F = \dfrac{fN}{\sin\beta}$. Denoting by f' the effective friction coefficient, we have

$$f' = \frac{f}{\sin\beta} \tag{4-2.1}$$

where $\beta = 0.5 \cdot 38° = 19°$. Since, $\sin 19° = 0.325$, then $f' = 3.071f \approx 3f$. Formula (4-2.1) shows that decreasing the angle β results in the increase of the effective coefficient of friction f'.

For instance, in cases when $\beta = 5°$ the effective coefficient of friction $f' = 11.473\,f$, but at such small angles, the V-belt becomes strongly wedged in the groove of the sheave, and it takes a significant effort to drive it out from the groove.

In comparison with flat belts, V-belts have a larger thickness and, consequently, are subjected to higher bending stresses generating higher levels of heat that destroys the material, resulting in lower service time. V-belts are subjected to the action of relatively higher centrifugal forces that increase the stresses in the belt. Because of all of this, usually the operational linear velocity of V-belts is reduced and is in the range from 1,500 to 6,500 $\frac{ft}{\text{min}}\left(\text{from } 7.5 \text{ to } 32.5\frac{m}{s}\right)$. It should be noted that the preferable linear speed for V-belts is around 4,000 $\frac{\text{ft}}{\text{min}}\left(20\frac{\text{m}}{s}\right)$.

V-belts are made of natural and synthetic materials such as technical fabric, urethane, polyurethane, polyamide, polyester, elastomer, and others and are reinforced by synthetic and steel fabric.

V-belts are standardized by their cross-sectional dimensions and closed length. They can be subdivided into four groups based on their cross-sectional dimensions: classical, narrow, light duty, and banded. The classical V-belts are mostly intended for general applications and, therefore, are considered in this book. The notations of standard cross sections of classical V-belts are A, B, C, D, and E, while E is the largest cross-section. The dimensions of the cross section of classical profiles of V-belts for the U.S. System (inches) and ISO System (metric) are presented in Appendix A-4-1. The identification of classical V-belts is standardized and is based on a combination of a letter with numerals. Hence, a V-belt identified as A70 means that the belt has a size A cross section, while 70 indicates the standard length designation. The data related to the standard length designation is presented in Appendix A-4-2. It should be noted that this Appendix uses a definition of a belt standard datum length. This definition is introduced by ISO just for classical V-belts, and it replaces the old definition of the pitch length. A similar change is introduced for the pitch diameter of the sheaves. The datum length is associated with the datum diameters. Figure 4-2.4 shows the difference between the pitch and datum diameters. This figure also shows the appropriate position of the V-belt in the grove of the sheave, where the belt 1 is immersed into the groove of pulley 2 below the upper edge of the grove, creating a gap between its inner surface and the bottom of the groove.

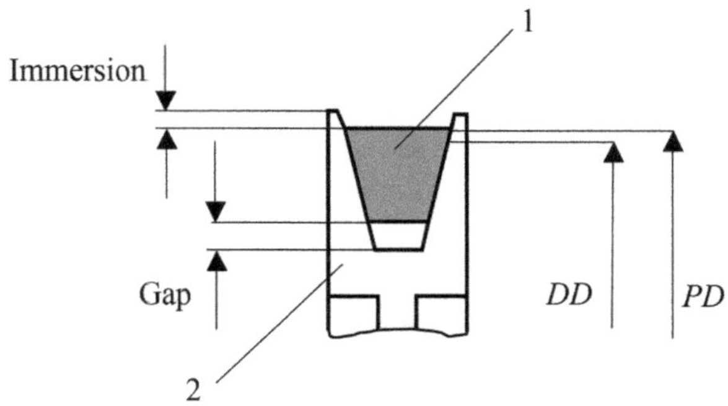

Figure 4-2.4. The correct position of the V-belt in the groove, and the pitch (PD) and datum (DD) diameters.

This gap ensures deeper positioning of the belt in the groove due to the creep deformation of the belt that occurs during the operational time. If the belt protrudes above the groove, the belt develops shoulders that do not allow the belt to immerse deeper when it is needed during a readjustment of the tension of the belt. The datum length and datum diameter are associated with the parameters of the cross section of the belt. According to Figure 4-2.1b the cross-sectional area A_V of V-belts is calculated from the following formula:

$$A_V = \frac{t(a+b)}{2} \tag{4-2.2}$$

and the formula for determining the distance c from the neutral layer to the top of the belt reads

$$c = \frac{t(2a+b)}{3(a+b)} \tag{4-2.3}$$

Minimum datum diameters of sheaves for classical V-belts are presented in Appendix A-4-3.

It should be noted that the center distance between pulleys for flat belts can be in a wide range, while for V-belts the distance is limited by the availability of standard datum (pitch) lengths, and it is recommended that

$$C < 3(d_1 + d_2) \tag{4-2.4}$$

Classical V-belts of A and B sizes are often used individually or in an arrangement of several parallel belts (a multi-belt drive). It is advisable to limit the number of belts in a multi-belt drive to 4; however, there are V-belt transmissions employing 8 belts in parallel. It is preferable to use the larger belt sizes C, D, and E for multi-belt drives.

4-2.3. Tensioners

From time to time during the operation of a belt transmission, the tension of the belt should be readjusted due to the elongation of the belts because of creep deformation. There are numerous designs of tensioners for adjustment of belt tension and for increasing the contact angle between the belt and the pulley. Sometimes the readjustment can be achieved by loosening the bolts of a special sledge carrying the supports of one pulley and tightening them after the adjustments. There are spring and weight tensioners that automatically maintain the required tension. The appropriate information related to the selection of tensioners is available in the catalogs of belt manufacturers.

4-3. Forces Acting on Belt Transmissions

In order to clarify the peculiarities of the operational process of belt transmissions, we consider the frictional interaction of a flexible link moving around a round surface. Figure 4-3.1 helps to explain some of these peculiarities. In this figure, the round non-rotating element 1 is rigidly secured to a non-movable support. A non-deformable weightless flexible link 2 is making contact over the arc AB with non-rotating element 1 and passes over a rotating roller 3, while a load 4 is secured to the flexible link 2. A load 5 is attached to the other end of the flexible link 2.

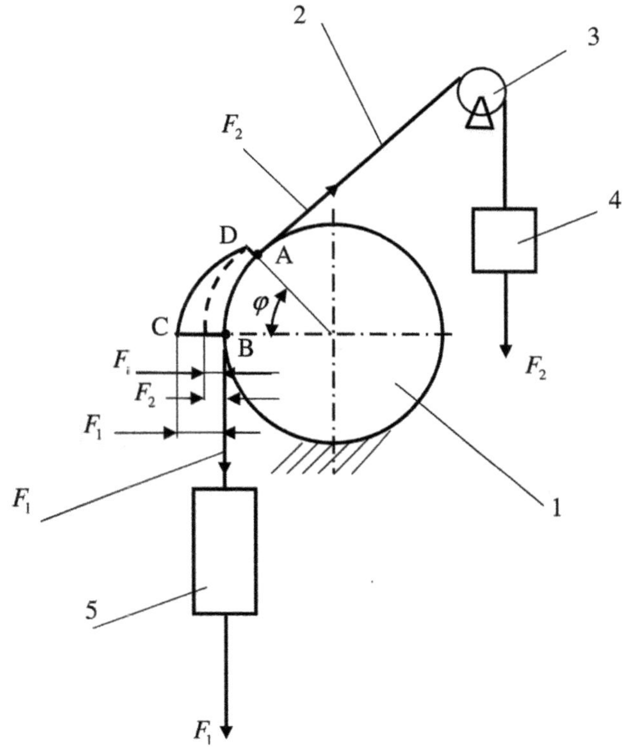

Figure 4-3.1. Friction of a flexible link.

The force of gravity F_1 developed by the load 5 exceeds the force of F_2 that represents the weight of load 4; however, the system is in equilibrium. The area ABCD represents the diagram of the friction forces applied from the non-rotating element to the flexible link. At the same time, this diagram shows how the tension force applied to the link increases along the arc AB from F_2 to F_1 (the forces on the diagram are shown in a decreased scale compared to the forces shown on the link). The increase of the tension in the link can be explained by the increase of friction on the arc AB between the round non-rotating element 1 and link 2 due to the increase of the normal pressure force coming from the flexible link 2 . The normal force applied from the link to the round element in point B exceeds the normal force applied to point A. The diagram shows that $F_1 = F_2 + F_t$, where F_t is, actually, the tangential force that rotates the pulleys in belt transmissions.

The friction between the belt and a pair of pulleys creates a tangential friction force that is applied by the driving pulley to the belt and from the belt to the driven pulley, causing the rotation of the driven pulley. The product of multiplying this tangential force by the linear velocity of the belt represents the power transmitted by the belt drive. The tangential force depends on the normal pressure forces applied by the belt to the pulleys and on the friction coefficient on the interface between the belt and pulleys. The value of the friction coefficient is always known. The normal pressure forces applied from the belt to the pulleys are developed by preloading the strands of the belt, applying a certain tension to the motionless pulleys and creating in each strand the same tension.

Figure 4-3.2 presents a schematic of a driving pulley having the diameter d_1 and a contact angle φ_1 with a belt where each strand is subjected to tension forces F_0. It should be clear that the deformations of the belt imposed by the preloading forces F_0 are elastic and this is necessary to create friction between the belt and the pulleys.

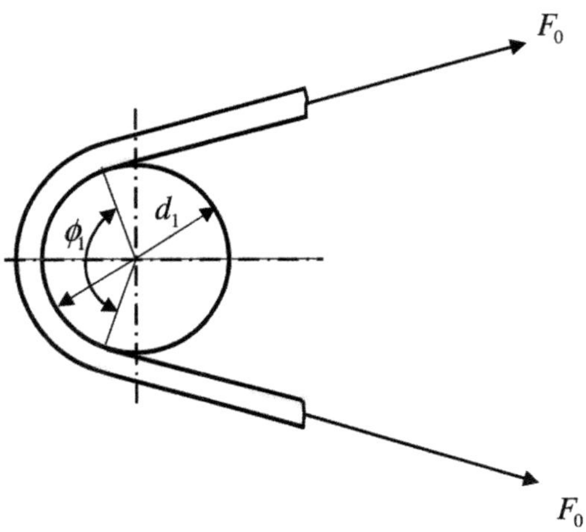

Figure 4-3.2. Preloading the belt.

When the torque is applied to the driving pulley in the direction indicated by the angular velocity ω_1, the bottom strand of the belt should overcome the working resistance of the driven pulley. This results in additional elastic deformation (elongation) of the bottom strand, while the upper strand contracts to the same measure of deformation as the bottom strand is stretched. The tension forces in the strands of the belt are changed proportionally to these deformations. Therefore, in the bottom strand (tight strand) the tension force is increased by a certain force increment ΔF, while in the upper strand the force is decreased by the same increment. Sometimes the upper strand is referred to as a slack strand. Actually, the upper strand remains tight being subjected to the preliminary elastic deformation. However, in cases where the center distance is relatively large, it could be seen a slack in both strands. This is explained by the well-known fact that it is practically impossible to eliminate the slack of a horizontal flexible link caused even by its own weight since the tension forces in the horizontal direction should tend to infinity. When the transmission is operating, the tension force in the bottom strand increases to the value of F_1, while in the upper strand the tension force drops to F_2. The difference between these forces represents the tangential force F_t that creates the torques on the pulleys. This tangential force is equal:

$$F_t = F_1 - F_2 \tag{4-3.1}$$

Based on this equation, it becomes apparent that the increment of the force $\Delta F = 0.5\ F_t$, and, therefore, we may write

$$F_1 = F_0 + 0.5F_t \tag{4-3.2}$$

and

$$F_2 = F_0 - 0.5F_t \tag{4-3.3}$$

During the operation of the transmission, the parts of the belt that are wound around the pulleys develop centrifugal forces. These forces cause an increase of the tension in the strands of the belt, while at the same time decreasing the pressure force applied to the pulleys that results in a decrease of the pressure force between the pulleys and the belt, and, consequently, in the decrease of the transmitted power. In some extreme cases, the belt begins to skid, and the transmission stops working. It should be mentioned that centrifugal force plays a noticeable role in decreasing the frictional interaction between the pulley and the belt at linear velocities of the belt exceeding 3,000 $\frac{ft}{\text{min}}\left(15\frac{\text{m}}{s}\right)$, however, at velocities in the range of 5,000 $\frac{\text{ft}}{\text{min}}\left(25\frac{\text{m}}{s}\right)$ the efficiency of the drive decreases noticeably.

The interaction between the forces applied from the belt to the pulley can be understood from Figure 4-3.3, which represents a schematic view of the driving pulley 1 in contact with the belt 2.

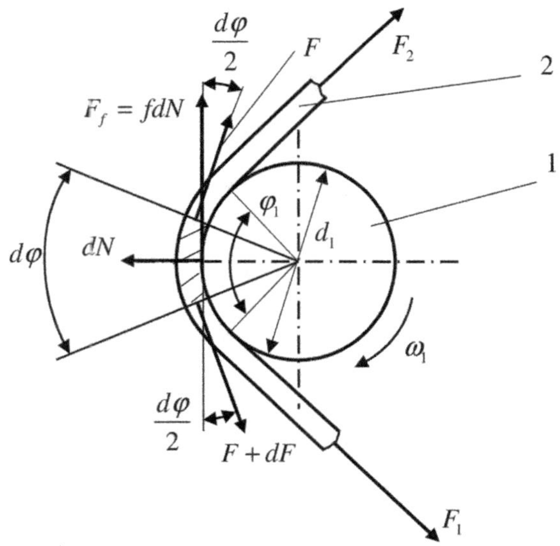

Figure 4-3.3. Forces acting in belt transmissions.

In order to reveal the relationship between the forces applied to the belt, we analyze the equilibrium of an infinitesimal element $d\varphi$ that is cut off from the belt as shown in Figure 4-3.3 by dashed lines. This element is subjected to a force differential dN that represents the resultant reaction of the pulley to the pressure forces caused by the tension forces related to the strands. This force differential dN acts in the radial direction in relation to the pulley and creates a friction force differential fdN in the tangential direction. A tension force F that is applied perpendicularly to the upper plane of the cut creates an angle $\frac{1}{2}d\phi$ with the line tangential to the pulley. On the opposite side of this element acts a similar tension force $F + dF$ that due to the friction between this element and the pulley is increased by a force differential dF. The tension forces F and $F + dF$ applied to both sides of this infinitesimal element develop a normal pressure force from the element to the pulley that reacts in the opposite direction of this element by the reaction force differential dN applied to the belt element. The friction force F_f that is applied to the pulley is perpendicular to the normal pressure force and equals to $F_f = fdN$, where f is the friction coefficient between the belt element and the pulley. Since the belt element is in equilibrium, we apply the equations of statics to determine the relationship between the applied forces. The sum of projections of the forces on the direction of the force dN equals zero, hence $-(F+dF)\sin\frac{d\varphi}{2} + dN - F\sin\frac{d\varphi}{2} = 0$, or $-2F\sin\frac{d\varphi}{2} - dF\sin\frac{d\varphi}{2} + dN = 0$. Ignoring the higher order of the product of two infinitesimals in this equation and considering that for very small values of angles $\sin\frac{d\varphi}{2} \approx \frac{d\varphi}{2}$ we may write

$$dN = Fd\varphi \tag{4-3.4}$$

Equation (4-3.4) has two unknown parameters; therefore, we compose the equation of the sum of moments about the center of the pulley and we have $-(F+dF)\frac{d_1}{2} + f\frac{d_1}{2}dN + F\frac{d_1}{2} = 0$, or

$$-dF + f\,dN = 0 \tag{4-3.5}$$

Combining equations (4-3.4) and (4-3.5), we obtain

$$\frac{dF}{F} = fd\varphi \tag{4-3.6}$$

Taking integrals from both sides of the equation (4-3.6), we have $\int_{F_2}^{F_1}\frac{dF}{F} = \int_0^{\varphi_1} fd\varphi$, this gives us that $\ln\frac{F_1}{F_2} = f\varphi_1$. Transforming this equation further, we have

$$F_1 = F_2 e^{f\varphi_1} \tag{4-3.7}$$

L. Euler in 1775 first obtained formula (4-3.7) considering the interaction of a non-deformable weightless flexible link with a motionless round element. The influence of the contact angle φ_1 on the transmitted power can be demonstrated by the following example. Assume the angle of contact is 120°. Converting this angle into radians, we obtain $\frac{120^\circ}{57.3} = 2.094$ rad, hence, $e^{2.094} = 8.119$. Increasing the contact angle to 170°, we have $e^{2.967} = 19.43$; this translates to an increase of the transmitted power by $\frac{19.43}{8.119} = 2.39$ times. The calculations of the forces F_1 and F_2 could be associated with the tangential force $F_t = \frac{P}{\omega}$ that is applied to the pulleys. Combining equations (4-3.1) and (4-3.7), we may write $F_t = F_2(e^{f\varphi_1} - 1)$. Solving this equation for F_2 we have

$$F_2 = \frac{F_t}{e^{f\varphi_1} - 1} \tag{4-3.8}$$

Combining equations (4-3.7) and (4-3.8), we determine

$$F_1 = \frac{F_t e^{f\varphi_1}}{e^{f\varphi_1} - 1} \tag{4-3.9}$$

Because $F_1 + F_2 = 2F_0$ and taking into account equations (4-3.8) and (4-3.9), we may write

$$F_0 = \frac{F_t(e^{f\varphi_1} + 1)}{2(e^{f\varphi_1} - 1)} \tag{4-3.10}$$

In addition, the belt is subjected to the action of centrifugal forces that increase the stresses in the cross-sectional area of the belt. Based on the schematic presented in Figure 4-3.4, we determine the centrifugal forces acting on the flat belt. We consider the equilibrium of the infinitesimal element (it is dashed in the figure) having the length $\frac{d_1}{2}d\varphi$ that is cut off from the belt. This element is subjected to the action of the centrifugal force differential dF_c that is balanced by two centrifugal forces F_c applied to the planes of the cut off element.

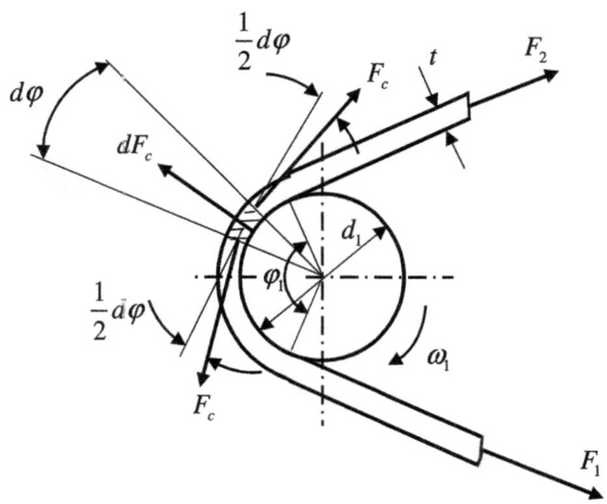

Figure 4-3.4. Centrifugal forces.

Summing the projections of the forces to the direction of dF_c, we may write $dF_c - 2F_c \sin\frac{d\varphi}{2} = 0$. Considering that in our case, $\sin\frac{d\varphi}{2} \approx \frac{d\varphi}{2}$ we have

$$dF_c = F_c d\varphi \tag{4-3.11}$$

The differential of a centrifugal force dF_c can be determined from the following expression:

$$dF_c = \frac{V^2 dm}{0.5d_1} \tag{4-3.12}$$

where dm is the mass differential of the belt element and $V = 0.5\omega_1 d_1$. The length of the arc of the belt element is $0.5d_1 d\varphi$, and its cross-sectional area is tb, where t is the belt thickness and b is its width. The volume dv of the mass element equals $dv = 0.5d_1 tbd\varphi$. Hence, the mass differential $dm = \frac{\gamma}{g}dv$, where γ is the specific weight (density) of the belt material and g is the acceleration of gravity. Therefore, we may write $dm = 0.5\frac{\gamma}{g}d_1 tbd\varphi$. Substituting this equation into equation (4-3.12), we have $dF_c = 0.5d_1\frac{\gamma}{g}tb\frac{V^2}{0.5d_1}d\varphi = \frac{\gamma}{g}tbV^2 d\varphi$. Combining this equation with equation (4-3.11), we determine the centrifugal force acting on the flat belt $F_c = \frac{\gamma}{g}tbV^2$, where the product of tb represents the cross-sectional area of a flat belt (that is different from the area of a V-belt or round belt). Therefore, replacing this product by A, we may write

$$F_c = \frac{\gamma}{g}AV^2 \tag{4-3.13}$$

where A is the cross-sectional area of an actual belt.

This equation does not contain the diameters of the pulleys nor the length and contact angles of the belt. It is similar to the centrifugal force that occurs in a hollow cylinder having t as the thickness of its wall and b as its width (length).

One more significant aspect related to the design of belt transmissions is associated with the calculations of the shafts carrying the pulleys and bearings supporting the shafts. In order to perform these calculations, we need to determine the forces exerted by the pulleys and applied to the shafts. As shown in Figure 4-3.5, the forces F_1 and F_2 applied to the belt are transferred to the point O, while R is the resultant of these two forces applied to the shaft.

The vector summation of these two forces is presented in the following equation:

$$R = \sqrt{F_1^2 + F_2^2 - 2F_1F_2\cos\varphi_1} \tag{4.3.14}$$

or $R = \sqrt{F_1^2 + F_2^2 + 2F_1F_2\cos 2\alpha}$, where $2\alpha = \pi - \varphi_1$.

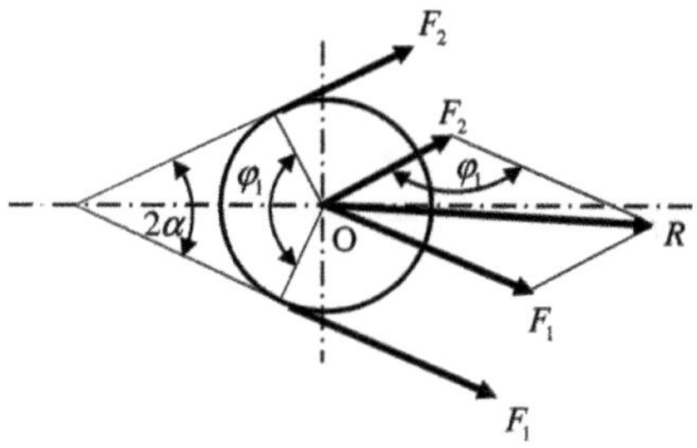

Figure 4-3.5. Resultant force applied to the shaft.

The analysis of the forces acting in belt transmissions are based predominately on Euler's equation that is obtained on assumptions that do not completely reflect the peculiarities of the operational process of belt transmissions. However, numerous experimental data support the applicability of this equation for calculations of belt transmissions.

4-4. Stresses in Belts

In general, during the operation of a transmission the belt is subjected to tensile stresses because of tension in the belt strands, to centrifugal forces, and to bending stress as a result of wrapping around the pulley. The most loaded cross-sectional area of the belt is indicated by point B in Figure 4-4.1, which represents a schematic view of the driving pulley with the flat belt subjected to the tension forces F_1 and F_2.

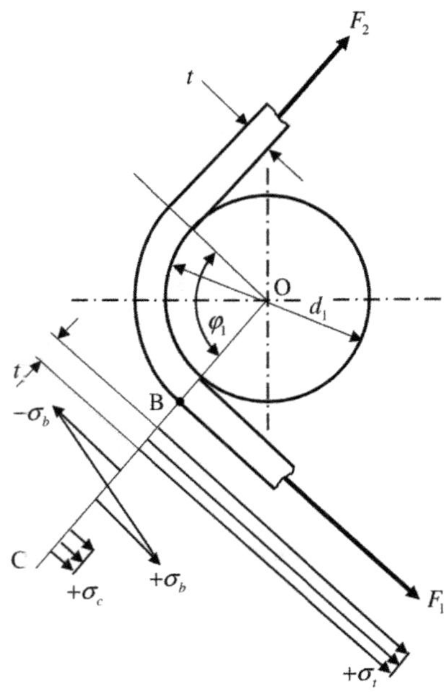

Figure 4-4.1. Stresses in a flat belt.

Because the radius of the driving pulley is smaller than the driven pulley, the bending stress in the part of the belt that is in contact with the driving pulley is higher than the part of the belt in contact with the driven pulley. The cross-sectional area for a flat belt is $A_F = tb$. The tension stress σ_t is calculated using the following formula:

$$\sigma_t = \frac{F_1}{A} \tag{4-4.1}$$

The diagram of this stress is plotted on the plane that passes through point B and crosses the belt along line OC. This diagram shows the distribution of the normal stress over the thickness t of the belt. Next to this diagram, the diagram of the bending stresses is plotted. The negative bending stress $-\sigma_b$ represents the compression stress in the inner layer of the belt that is in touch with the pulley, while the outer layer is stretched by the stress $+\sigma_b$. At pure bending, the strain ε of the outer and inner layer of a flat belt, as well as a V-belt, can be determined by the formula $\varepsilon = \dfrac{y}{0.5d_1 + t - y}$, where y is the distance between the neutral layer and the top of the belt, while t is the thickness of the belt. Taking into account that $\sigma = \varepsilon E$, where E is the modulus of elasticity, we can calculate the normal stress σ_b in the outer layer as

$$\sigma_b = \frac{yE}{0.5d_1 + t - y} \tag{4-4.2}$$

For flat belts $y = 0.5t$, while for V-belts $y = c$, where c is determined from formula (4-2.3). Taking $y = 0.5t$ we determine the bending stress in the flat belt $\sigma_{Fb} = \dfrac{tE}{d_1 + t}$. However, since t is significantly smaller than d_1 it is accepted that

$$\sigma_{Fb} \approx \frac{tE}{d_1} \tag{4-4.3}$$

The normal stress σ_c due to the centrifugal force F_c that is determined from formula (4-3.13) can be calculated from the following formula:

$$\sigma_c = \frac{F_c}{A} = \frac{\gamma}{g} v^2 \tag{4-4.4}$$

The diagram of the centrifugal stress is also shown in Figure 4-4.1.

Considering all of these stresses, we calculate the maximum stress acting in the belt as

$$\sigma_{\max} = \sigma_t + \sigma_b + \sigma_c = \frac{F_1}{A} + \frac{yE}{0.5d_1 + t - y} + \frac{F_c}{A} \tag{4-4.5}$$

Example 4-4.1a: Calculating the forces applied to the belt and the shaft, and the maximum stress acting in the belt.

A horizontal A size V-belt drive transmits power $P = 2.0 \text{ hp} = 2 \cdot 550 \dfrac{\text{ft-lb}}{s}$ at $n_1 = 900$ rpm to the driving pulley having $d_1 = 3.0$ in., the driven pulley $d_2 = 9$ in., the center distance $C = 25$ in., and the friction coefficient between the belt and sheave $f = 0.35$. We determine the contact angle φ_1 of the driving pulley, the tangential force F_t, the preloading force F_0, the tension forces in the strands F_1 and F_2, the force applied to the shaft, and the stresses acting in the belt.

Solution:

a) Forces

Using formula (4-1.4), we determine the contact angle $\varphi_1 = 2\cos^{-1}\left(\frac{9-3}{2\cdot 25}\right) = 2\cdot 83.1° = 166.2°$.

The linear velocity of the belt $V = \frac{\pi 0.5 d_1 n_1}{30\cdot 12} = \frac{\pi 0.5\cdot 3\cdot 900}{30\cdot 12} = 11.675\ \frac{\text{ft}}{s} = 700\ \frac{\text{ft}}{\text{min}}\left(3.55\frac{\text{m}}{\text{s}}\right)$.

Now, we calculate the tangential force: $F_t = \frac{P}{V} = \frac{2\cdot 550}{11.65} = 94.22$ lb. According to formula (4-2.1), the effective friction coefficient $f' = \frac{0.35}{\sin 19^{\circ}} = 1.075$, we calculate $e^{f'\varphi_1} = e^{\frac{1.075\cdot 166.2}{57.3}} = 22.6$. Based on formulas (4-3.8), (4-3.9), and (4-3.10), we determine the forces F_2, F_1, and F_0: $F_2 = \frac{84.22}{22.6-1} = 4.36$ lb, $F_1 = \frac{94.22\cdot 22.6}{21.6} = 98.58$ lb, and $F_0 = \frac{94.22(22.6+1)}{2(22.6-1)} = 51.48$ lb. Applying formula (4-3.13), we calculate the centrifugal force $F_c = \frac{\gamma}{g} tbV^2$, where the product of tb represents the cross-sectional area of a flat belt, while we are dealing with a V-belt. Using formula (4-2.2), we calculate the cross-sectional area A_V of the A size V-belt and obtain $A_V = \frac{0.3125(0.2182+0.5)}{2} = 0.122\ \text{in.}^2$, or $A_V = 8{,}48\cdot 10^{-4}\ \text{ft}^2$. Taking the specific weight (density) of the material of the V-belt $\gamma = 153.8\ \frac{\text{lb}}{\text{ft}^3}$ and taking into account the acceleration of gravity $g = 32.17\ \frac{\text{ft}}{\text{s}^2}$, using formula (4-3.13) we calculate the centrifugal force $F_c = \frac{153.8}{32.17}\cdot 8.48\cdot 10^{-4}\cdot 11.675^2 = 0.553$ lb. Now based on formula (4-3.14), we determine the force applied to the shaft: $R = \sqrt{98.48^2 + 4.36^2 - 2\cdot 98.48\cdot 4.36\cos 166.2°} = 102.72$ lb.

b) Stresses

In formula (4-4.1) we replace A with A_V and calculate the stress $\sigma_t = \frac{F_1}{A_V} = \frac{98.58}{0.122} = 808.03$ psi. According to formula (4-2.3), we calculate the distance c from the neutral layer to the top of the A-size belt: $c = \frac{0.3125(2\cdot 0.2182+0.5)}{3(0.2182+0.5)} = 0.1358$ in. Replacing in formula (4-2.16) y with c and taking for the material of the belt $E = 11{,}000$ psi, we calculate the bending stresses $\sigma_b = \frac{cE}{0.5d_1 + t - c} = \frac{0.1358\cdot 11{,}000}{0.5\cdot 3.0 + 0.3125 - 0.1358} = 891$ psi. Replacing in formula (4-3.3) A with A_V, we calculate the normal stresses due to the centrifugal force $\sigma_c = \frac{F_c}{A_V} = \frac{0.553}{0.122} = 4.53$ psi. Now using formula (4-4.4), we determine the maximum stress:

$\sigma_{max} = \sigma_{1t} + \sigma_b + \sigma_c = 808.03 + 891 + 4.53 = 1{,}703.6$ psi.

It may by noted that the centrifugal stress in this case $\left(V = 700\,\frac{\text{ft}}{\text{min}}\right)$ is about 0.266% of the total stress and is negligible, while the bending stress is about 52% and typically plays a significant role in the V-belt drive.

4-5. Design of Flat-Belt Transmissions

The design process for a flat-belt transmission starts with the proper selection of the belt's material. As mentioned above, flat belts are made from natural and synthetic materials that in some cases are reinforced by cords or steel fabric. Depending on the type of belt materials there are certain differences in the design procedures of flat-belt drives. Corresponding design examples are presented below.

4-5.1. Design of Flat-Belt Transmissions Made From Common Materials

Common materials for flat belts include leather, cotton, rubber, and some non-reinforced synthetic polymers.

The standard thicknesses and widths of the belts can be found in published sources. These characteristics for some common rubber, leather, and cotton flat belts are presented in Appendix A-4-9. Numerous experimental investigations of belt drives show that for each of these materials a certain allowable stress characterized by the ratio r_b between the minimum dimension of the pulley diameter and the thickness of the belt $r_b = \frac{d_1}{t}$ can be determined. The corresponding data is presented in Appendix A-4-10. Thus, in cases where the preload stress in the strands of the belt is in a practical order of 0.26 ksi (1.8 MPa) and the ratio $\frac{d_1}{t} = 30$, then the allowable stress for a leather belt σ_{all} = 0.275 ksi (1.9 MPa), while the average value of the allowable stress for a leather belt is σ_{all} = 0.326 ksi (σ_{all} = 2.25MPa). Comparison of the allowable stress of a leather belt with the average value of the ultimate strength S_{ut} = 4.06 ksi (S_{ut} = 28 MPa) of the leather material shows that the allowable stress is reduced more than 12 times compared to the ultimate strength. This means that the strength safety factor incorporated in the belts is high enough to ensure the reliable operation of the drive for a long time (according to some sources, for about 15 years). Therefore, based on operational conditions and other related design considerations, we choose the material of the belt and its minimum standard thickness. Accounting for design considerations associated with the available space for the belt drive, its estimated weight, and other factors, we take an appropriate value of the pulley-thickness ratio and determine the minimum diameter of the driving (smaller) pulley. The assignment related to the design of the belt drive comprises the information about the transmitted power, the rotational speed of the driving pulley, the velocity ratio, the center distance, and the friction coefficient between the belt and the pulley. These data are sufficient to calculate the tension force in the tight strand of the belt. Increasing the calculated tension force by a corresponding safety factor (usually, about 2) and decreasing the allowable stress by applying an appropriate loading factor $C_b = 0.6$, we calculate the width of the belt. If the result exceeds the standard width of the belt for the chosen thickness, we increase the thickness and perform a similar iteration.

An example demonstrating the design steps involving numeric calculations is presented below.

Example 4-5.1a: Determining the parameters of a leather flat-belt drive.
It is necessary to design a flat-belt transmission for the following input data: the transmitted power $P = 2.2$ kW, the rotational speed of the driving pulley $n_1 = 3{,}600$ rpm, the velocity ratio $i = 3.5$, the center distance $C = 900$ mm, and the friction coefficient $f = 0.3$.

Solution:
Trying to design a compact drive, we select from Appendix A-4-9 the thickness and the width of a leather belt and from Appendix A-4-10 the ratio $\frac{d_1}{t} = 30$. For this ratio, the allowable stress for the leather belt $\sigma_{all} = 1.9$ MPa and taking the standard minimum thickness of the belt $t = 5$ mm, we have $d_1 = 30 \cdot 5 = 150$ mm.

Thus, $d_2 = 3.5 \cdot 150 = 437.5$ mm, and we round up to $d_2 = 440$ mm. Now we calculate the tangential force: $F_t = \frac{2 \cdot 30P}{\pi n_1 d_1} = \frac{2 \cdot 30 \cdot 2.2 \cdot 1{,}000}{\pi 3{,}600 \cdot 0.150} = 77.85$ N. Using formula (4-1.4), we can calculate the contact angle of driving pulley, hence, $\varphi_1 = 2\cos^{-1}\left(\frac{440-150}{2 \cdot 900}\right) = 2\cos^{-1} 0.161 = 2 \cdot 80.72° = 161.45°$.

Based on formula (4-3.9), we determine the force acting in the tight strand of the belt $F_1 = \frac{77.85e^{\frac{0.3 \cdot 161.45}{57.3}}}{e^{\frac{0.3 \cdot 161.45}{57.3}} - 1} = 136.3$ N. Taking a safety factor $n_\sigma = 2$ and reducing the allowable stress applying the loading factor $C_b = 0.6$, we determine the required width of the belt: $b = \frac{F_1 n_\sigma}{t C_b \sigma_{all}} = \frac{136.3 \cdot 2}{0.005 \cdot 0.6 \cdot 1.9 \cdot 10^6} = 0.0478$ m, we take $b = 48$ mm. The maximum standard width of this belt material is 63 mm, therefore, the selection of the belt is acceptable.

4-5.2. Design of Flat-Belt Transmissions Using Synthetic Materials Reinforced by Steel Cords

Flat belts manufactured from polymer materials reinforced by steel fabric are widely used in flat-belt transmissions. The Gates® Corporation offers numerous structural constructions of flat belts made of polyurethane synthetic material reinforced by steel cords. The catalog of Gates®

Corporation presents information on minimum pulley diameters with the corresponding thicknesses and widths of the flat belts made of polyurethane reinforced with steel cords. The catalog offers a variety of combinations of the minimum diameters of the pulleys for the same belt thickness with different width of the belts. The catalog provides the required corresponding data that allows selecting the appropriate set of the pulley (its diameter) with the related thickness and width of the belt. Depending on the minimum diameter of the pulley and the thickness and width of the belt, the catalog provides the information regarding the ultimate tensile strength and the allowable force F_{1all} acting in the tight strand of the belt for open and endless welded belts. This information allows for the selection of the appropriate flat belt for the transmission. An example of selecting a flat belt made from polyurethane reinforced by steel cords is presented below.

Example 4-5.2a: Design of an open flat-belt drive using a polyurethane belt reinforced by steel cords.

It is necessary to design a flat-belt transmission that transmits power $P = 7.5$ kW at the rotational speed of the driving pulley $n_1 = 3{,}600$ rpm. The velocity ratio between the driving and driven pulley is $i = 4.25$, the center distance between the pulleys is 450 mm. The safety factor $n_\sigma = 2$.

Solution:

Referring to the appropriate Gates® Corporation catalog, we determined that the welded endless belt Gates® BFL20 having the minimum diameter of the pulley $d_1 = 48$ mm, the belt thickness $t = 2$ mm, and the belt width $b = 50$ mm has the allowable tension force in the tight strand $F_{1all} = 4{,}832$ N.

The torque applied to the driving pulley $T_1 = \dfrac{P}{\omega_1} = \dfrac{7.5 \cdot 1{,}000 \cdot 30}{\pi 3{,}600} = 19.9\ \text{N} \cdot \text{m}$. Having the diameter of the driving pulley $d_1 = 48$ mm, we determine the tangential force for our belt transmission. Hence, $F_t = \dfrac{2T_1}{d_1} = \dfrac{2 \cdot 19.9}{0.048} = 829.2$ N. The diameter of the driven pulley equals $d_2 = id_1 = 4.25 \cdot 48 = 204$ mm. Now applying formula (4-1.4), we determine the contact angle of the smaller pulley, hence, $\varphi_1 = 2\cos^{-1}\dfrac{d_2 - d_1}{2C} = 2\cos^{-1}\left(\dfrac{204 - 48}{2 \cdot 450}\right) = 2\cos^{-1} 0.173 = 2 \cdot 80° = 160°$.

We convert the contact angle into radians $\varphi_1 = \dfrac{160}{57.3} = 2.792$ rad and taking for our case $f = 0.6$, we calculate using formula (4-3.9) the force acting in the tight strand of the belt. Thus, $F_1 = \dfrac{829.2e^{0.6 \cdot 2.792}}{e^{0.6 \cdot 2.792} - 1} = 1{,}020.2$ N.

Accounting for the above mentioned loading factor $C_b = 0.6$, we determine the actual safety factor n of the belt.

Hence, $n = \dfrac{F_{1all}C_b}{F_1} = \dfrac{4{,}832 \cdot 0.6}{1{,}020.2} = 2.84$, while the required safety factor $n_\sigma = 2$. Obviously, the selected belt satisfies the requirements of the current task.

The force acting in the slack strand of the belt equals $F_2 = F_1 - F_t = 1{,}020.2 - 829.2 = 191$ N. Now, according to formula (4-3.14), we calculate the force applied to the shafts of the transmission:

$R = \sqrt{1{,}020.2^2 + 191^2 - 2 \cdot 1{,}020.2 \cdot 191 \cos 160°} = 1{,}201.46$ N.

Hence, all required parameters for the designed transmission are determined.

4-6. Selecting V-belts for Belt Transmissions

The input data for a design project of a V-belt transmission includes the information regarding the transmitting power at a certain rotational speed of a sheave (usually the driving sheave), the velocity ratio between the sheaves, and the approximate distance between the sheaves. Let us demonstrate the basic steps of selecting a classical V-belt for a belt drive. Thus, the transmitted power $P = 1.0$ hp, the rotational speed of the driving sheave $n_1 = 1{,}200$ rpm, the velocity ratio $i = 2.6$, and the center distance $C = 21.0$ in.

1) Selecting the diameter for the driving sheave.

Since the A-size V-belt is the least expensive and most compact, we begin the design process with the A-size belt in order to verify its applicability to the current project. From Appendix A-4-3 we select the minimum datum diameter of the driving sheave, that is $d_1 = 3.0$ in. for an A-size V-belt.

2) Determining the diameter of the driven sheave, the contact angle of the driving sheave, and the length of the belt.

Accounting for the velocity ratio, we calculate the diameter of the driven pulley $d_2 = 2.6 \cdot 3 = 7.8$ in., and we take the closest value of this diameter using the table in Appendix A-4-5. We select $d_2 = 8.0$ in., and compute the actual velocity ratio: $i = \frac{8.0}{3.0} = 2.667$. Now according to formula (4-1.4), we compute the contact angle: $\varphi_1 = 2\cos^{-1}\left(\frac{8-3}{2 \cdot 21}\right) = 2 \cdot 76.23° = 152.5°$. According to formula (4-1.5), we calculate the approximate length of the belt: $L = \sqrt{4 \cdot 21^2 - (8.0-3.0)^2} + \pi 8.0 - \frac{152.5}{2 \cdot 57.3}(8.0-3.0) = 60.188$ in. Now, from the table in Appendix A-4-2 we select the closest standard length of the V-belt. Hence, we select the A-size V-belt with the standard designation A60 that corresponds to the standard datum length of 61.3 in. Therefore, the center distance should be recalculated based on the selected standard datum length.

3) Calculating the power rating of V-belts.

Computing the power rating of classical V-belts is based on empiric formulas that are available in numerous published sources. The power rating represents the maximum allowable value of power for a single belt subjected to the operational conditions associated with the size of the driving sheave, its rotational speed, and the velocity ratio of the transmission. In Appendix A-4-8 are empiric formulas for calculating the power rating for different sizes of V-belts. These formulas are adapted from the *Machinery's Handbook* (Edition 30). The following formula allows us to calculate the power rating for A-size V-belts:

$$P_A = d_1 n_1 10^{-3}\left[1.004 - \frac{1.652}{d_1} - 1.547\cdot 10^{-4}\,(d_1 n_1 10^{-3})^2 - 0.2120\cdot \log(d_1 n_1 10^{-3})\right] + 1.652 n_1 10^{-3}\left(1 - \frac{1}{K_{SR}}\right) \quad (4\text{-}6.1)$$

where P_A is the power rating of the A-size V-belt and K_{SR} is the speed ratio correction factor taken from the corresponding table in Appendix A-4-4. It should be noted that if as in our case $i > 1.64$ then $K_{SR} = 1{,}1106$ Substituting the corresponding data into formula (4-6.1), we may write

$$P_A = 3.0\cdot 1{,}200\cdot 10^{-3}\left[1.004 - \frac{1.652}{3.0} - 1.547\cdot 10^{-4}\,(3.0\cdot 1{,}200\cdot 10^{-3})^2 - 0.2120\cdot \log(3.0\cdot 1{,}200\cdot 10^{-3})\right] + 1.652\cdot 1{,}200\cdot 10^{-3}\left(1 - \frac{1}{1.1106}\right) = 1.4034 \text{ hp.}$$

Now, depending on the standard length destination and if $\varphi_1 \neq 180°$ we have to multiply the result obtained from formula (4-6.1) by the length correction factor K_L and the arc of contact correction factor K_φ the values of which are presented in the corresponding tables in Appendix A-4-6 and Appendix A-4-7, respectively. Hence, in our case $K_L = 0.97$ and $K_\varphi = 0.93$. Therefore, the corrected value of the power rating is $P_A = 0.97\cdot 0.93\cdot 1.4034 = 1.266$ hp. The number of the required belts is calculated from the following formula:

$$N_b = \frac{P}{P_A} \quad (4\text{-}6.2)$$

Hence, $N_b = \dfrac{1.0}{1.266} = 0.79$. Thus, for our case it is sufficient to use one A-size belt.

In cases when the required number of parallel A-size belts exceeds four, it is necessary to verify the applicability of B-size belts, etc.

4-7. Design Project

We need to design a V-belt drive that the driving sheave receives power of 3 hp at 1,800 rpm and transmits to the driven sheave that should have a rotational speed of 500 rpm. The approximate center distance is 25 in. The safety factor of the shafts is $n_\sigma = 2.5$. The drive is reversible.

The following engineering calculations and drawing represent a significant part of the activities of many engineers.

1. Selection of the V-belt size and the minimum diameter of the sheave.

We start with verifying the applicability of the A-size V-belt for our project. From the table in Appendix A-4-3 we determine that the standard datum diameter for the driving sheave is $d_1 = 3.0$ in.

2. Velocity ratio, diameter of the driven sheave, center distance, contact angle, and the length of the belt.

The velocity ratio $i=\frac{n_1}{n_2}=\frac{1{,}800}{500}=3.6$. Using formula (4-1.2), we determine the diameter of the driven sheave: $d_2 = id_1 = 3.6\cdot 3 = 10.8$ in. From the table in Appendix A-4-5, we select the corresponding standard datum diameter of the driven sheave $d_2 = 10.6$ in. We assume that the preliminary center distance equals $C = 25.0$ in., which should be $25.0 < 3(3.0 + 10.6) = 40.8$. Based on formula (4-1.4), we calculate the contact angle on the driving sheave:

$\varphi_1 = 2\cos^{-1}\left(\frac{10.6-3.0}{2\cdot 25}\right) = 2\cos^{-1} 0.152$, therefore, $\varphi_1 = 2\cdot 81.257° = 162.514°$.

Applying formula (4-1.5), we determine the length of the belt:

$$L=\sqrt{4\cdot 25^2-(10.6-3.0)^2}+\pi 10.6-\frac{164.34}{2\cdot 57.3}(10.6-3.0)=71.8 \text{ in.}$$

According to the calculated length of the belt, we select from the table in Appendix A-4-2 the closest standard length designation number of the belt, and it is A68 that corresponds to the standard length of the belt of 69.3 in. This requires us to recalculate the center distance and the contact angle φ_1, however, we will do this later if the selected V-belt size A will be chosen.

3. Horsepower rating for the A-size belt.

Using the empiric formula (4-6.1) for calculating the horsepower rating, we first need to determine the speed ratio correction factor that depends on the velocity ratio of the drive. From the table in Appendix A-4-4, we determine that $K_{SR} = 1.1106$. Now substituting the corresponding data into formula (4-6.1), we may write

$$P_A = 3.0\cdot 1{,}800\cdot 10^{-3}\left[1.004-\frac{1.652}{3.0}-1.547\cdot 10^{-4}\,(3.0\cdot 1{,}800\cdot 10^{-3})^2-0.2120\cdot\log(3.0\cdot 1{,}800\cdot 10^{-3})\right]$$

$$+1.652\cdot 1{,}800\cdot 10^{-3}\left(1-\frac{1}{1.1106}\right)=1.881 \text{ hp.}$$

According to the standard length designation of the belt and since in our case $\varphi_1 \neq 180°$, we have to multiply the result obtained from formula (4-6.1) by the length correction factor K_L and the arc of contact correction factor K_φ. The values of these factors are presented in the tables in Appendix A-4-6 and Appendix A-4-7, respectively. Thus, in our case $K_L = 1.0$ and $K_\varphi = 0.96$ Applying these factors, we obtain the corrected value of the power rating that is $P_A = 1.0\cdot 0.96\cdot 1.881 = 1.806$ hp.

Now we calculate the number of needed belts: $N_b=\frac{Pn}{P_A}=\frac{3.0}{1.806}=1.661$.

Therefore, it is required to take $N_b = 2$.

4. Correcting the center distance and the contact angle.

Incorporating the standard datum length of the belt that is equal to 69.3 in., we determine the corrected value of the center distance using the following formula:

$$C=0.25\{[L_d-0.5\pi(d_2-d_1)]+\sqrt{[L_d-0.5\pi(d_2-d_1)]^2+2(d_2-d_1)^2}\} \quad (4\text{-}7.1)$$

where L_d is the datum length of the belt.

Substituting the appropriate data, we write

$$C = 0.25\ \{[69.3 - 0.5\pi\ (10.6 + 3)] + \sqrt{[69.3 - 0.5\pi(10.6+3)]^2 + 2(10.6-3)^2}\} = 24.56 \text{ in.}$$

According to formula (4-1.4), we recalculate the contact angle of the driving sheave:

$$\varphi_1 = 2\cos^{-1}\frac{10.6-3}{2\cdot 24.56} = 2\cos^{-1}0.155 = 2\cdot 81.083°, \text{ therefore, } \varphi_1 = 162{,}166°.$$

5. Sketching the drive assembly.

It is required that the transmission assembly consist of an appropriate electrical motor generating 3.0 hp at 1,800 rpm and carrying the driving sheave that transmits the power to the driven sheave, which is supported by a pair of bearings accommodated by bearing enclosures. We begin with the preliminary calculations of the diameter of the shaft for the driven pulley. This diameter represents the bore diameter of the bearings and allows us to estimate the distance between the bearings and the distance between the bearing's enclosure and the pulley. We need to know these distances in order to calculate the reactions in the supports and determine the bending moments acting in the shaft. Using formula (3-3.10) and taking the allowable stress S_s = 4,000 psi, we calculate the diameter of the shaft for the driven pulley: $d = \sqrt[3]{\frac{321{,}000\cdot 3}{500\cdot 4{,}000}} = 0.784$ in. (19.91 mm). We preliminarily select a metric radial ball bearing SKF No. 6304 with the bore diameter of 20 mm, outer diameter of 52 mm, and the width of 15 mm. This allows us to perform the preliminary sketch of the assembly. Figure 4-7.1 shows the preliminary sketch of the V-belt transmission, where the electric motor 1 is carrying the driving sheave 4 secured to the shaft 6 of the motor by a flat gib key 5. The power is transmitted to the driven sheave 7 by a pair of size A V-belts 2 and 3.

The driven shaft 12 has a cantilever 9 that carries the sheave 7 secured by a flat gib key 8. This design allows replacement of the belts without disassembly of the transmission. The flat gib keys prevent the sheaves from displacement in the axial direction. The estimated width of the sheaves is about 1.5 in. Now we can verify the diameter of the shaft that carries the driven sheave.

6. Calculating the radial force and the torque applied to the driven shaft.

Since the driving shaft that carries the driving sheave represents the shaft of the electric motor, there is no need to perform engineering calculations for this shaft. The shafts of electric motors are designed with a sufficient margin of safety that provide extremely high reliability. We have to perform the calculations just for the driven shaft that carries the driven sheave.

The driven shaft is subjected to the action of a radial force and a corresponding torque.

However, we first have to determine the tangential force F_t that allows us to calculate the forces F_1 and F_2 applied to the strands of the belts. Thus, $F_t = \frac{P}{V}$, where $P = 3.0\cdot 550 = 1{,}650$

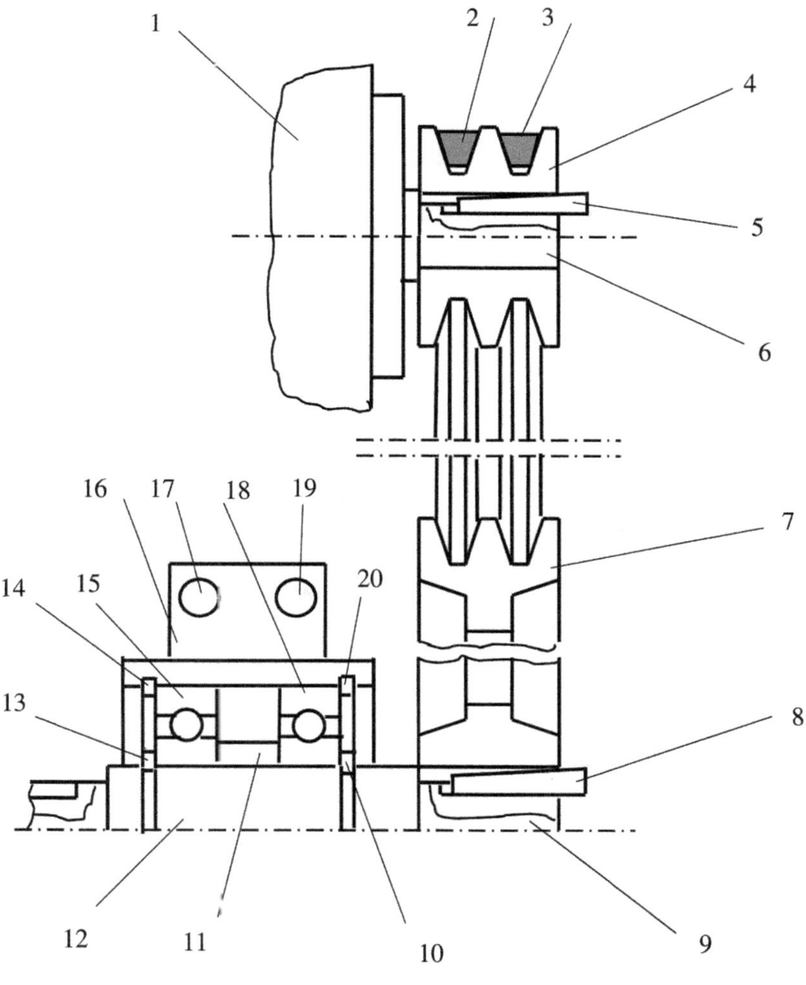

Figure 4-7.1. Sketch of the V-belt assembly.

$\frac{\text{ft-lb}}{s}$ and $V = \frac{0.5 \cdot \pi 3.0 \cdot 1{,}800}{30 \cdot 12} = 23.55\ \frac{\text{ft}}{s}$, therefore, $F_t = \frac{1{,}650}{23.55} = 70.06$ lb. Taking for a belt made from polyurethane the friction coefficient $f = 0.6$ and accounting for the wedging effect of friction results in an effective friction coefficient $f' \approx 3.0\,f$, we calculate $e^{f'\varphi_1} = e^{0.6 \cdot 3.0 \cdot \frac{162}{57.3}} = 162.22$. Using formulas (4-3.9) and (4-3.8), we determine the forces F_1 and F_2: $F_1 = \frac{70.06 \cdot 162.22}{162.22 - 1} = 76.53$ lb and $F_2 = \frac{70.06}{162.22 - 1} = 0.435$ lb. Hence, the radial force R applied to the shaft force according to formula (4-3.14) is $R = \sqrt{76.53^2 + 0.435^2 - 2 \cdot 76.53 \cdot 0.435 \cos 162^\circ} = 76.94$ lb. Now we calculate the torque applied to the driven shaft: $T_2 = 0.5\,F_t d_2 = 0.5 \cdot 70.06 \cdot 10.6 = 371.32$ lb-in.

7. Reactions. Bending, and torque diagrams.

Based on the sketch in Figure 4-7.1 we draw the free-body diagram that allows us to determine the reactions, the bending moments and the torque applied to the driven shaft. The free-body diagram is presented in Figure 4-7.2a, where the estimated distance between the bearings is $l_1 = 2.0$ in. and the estimated distance from the right-hand bearing to the center of the sheave is $l_2 = 2.25$ in.

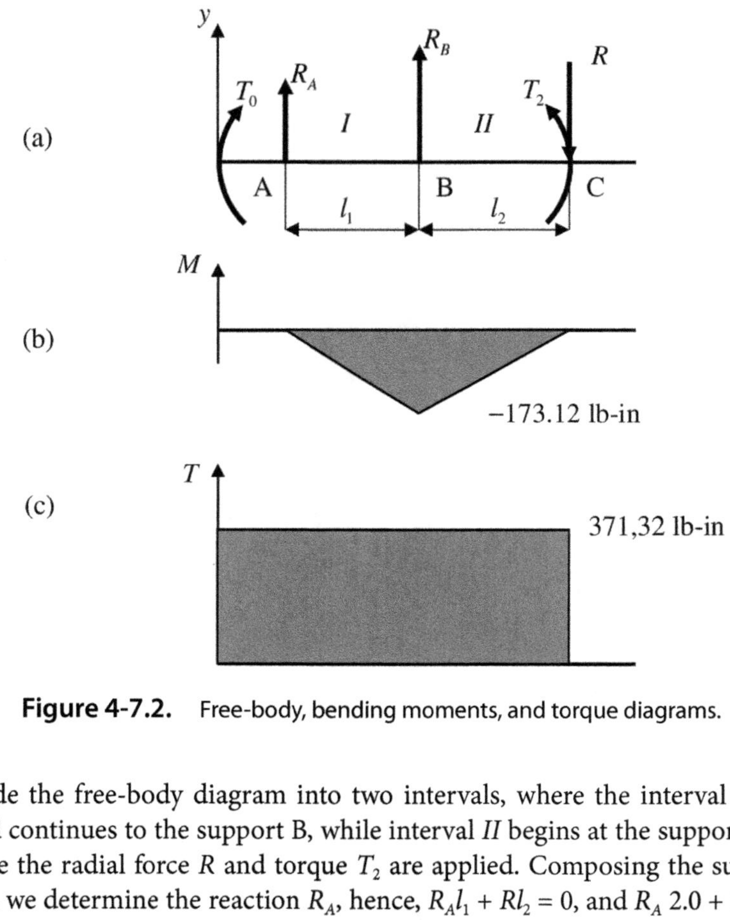

Figure 4-7.2. Free-body, bending moments, and torque diagrams.

We divide the free-body diagram into two intervals, where the interval *I* begins at the support A and continues to the support B, while interval *II* begins at the support B and ends at point C, where the radial force *R* and torque T_2 are applied. Composing the sum of moments about point B, we determine the reaction R_A, hence, $R_A l_1 + R l_2 = 0$, and $R_A\ 2.0 + 76.94 \cdot 2.25 = 0$. Therefore, $R_A = -86.56$ lb. Projecting the forces on the *y* axis, we calculate the reaction R_B, hence, $-86.56 + R_B - 76.94 = 0$, therefore, $R_B = 163.5$ lb. The bending moment at the end of the first interval equals $M_I = R_A l_1 = -86.56 \cdot 2.0 = -173.12$ lb-in. Calculating the bending moment at the end of the interval *II*, we obtain $M_{II} = R_A\ (l_1 + l_2) + R_B l_2 = -86.56\ (2.0 + 2.25) + 163.5 \cdot 2.25 = 0$. The bending moments diagram is shown in Figure 4-7.2b. The torque T_2 applied to the driven shaft is balanced by the torque T_0 and, therefore, $T = T_2 = T_0 = 371.32$ lb-in. The torque diagram is presented in Figure 4-7.2c. The bending-moments diagram and the torque diagram show that in cases of a uniform shaft the most loaded cross-sectional area is in point B, where $M_{max} = M = |173.12|$ lb-in and $T_{max} = T_2 = 371.32$ lb-in. However, the shaft has a stepped shape, and the bore diameter of the bearing exceeds the diameter of cantilever where the sheave is positioned. We need to calculate the diameter of the cantilever and the diameter of the shaft where the bearing is mounted.

8. Calculations of the driven shaft.

a) Cantilever

Figure 4-7.1 shows that the cantilever, where the driven sheave is mounted, has a keyseat that is machined by an end mill. Therefore, we have a combination of two stress concentration factors (step and keyseat) affecting the strength of the cantilever having a smaller diameter. This shaft is subjected to completely reversed loading cycles of bending stresses and torsion stresses because the drive is reversible. The distance from the middle of the sheave to the cross-section of the step on the shaft is 1.5 in.; therefore, the bending moment in this section equals $M_0 = 76.94 \cdot 1.5 = 115.41$ lb-in. In order to determine the fatigue stress concentration factors, we take $\frac{D}{d} = 1.1$ and $\frac{r}{d} = 0.05$, and determine from Appendix A-1-5 (Figures 2 and 3) the theoretical stress concentration factors for torsion and bending stresses: $K_{tt} = 1.55$ and $K_{tb} = 1.85$, respectively. Applying formula (1-10.2) and taking into account that $q = 0.6$, we calculate the fatigue stress concentration factors for torsion: $K_{ft} = 1 + 0.6(1.55-1) = 1.33$; and for bending: $K_{fb} = 1 + 0.6(1.85-1) = 1.51$. The fatigue stress concentration factors due to the keyseat for bending $K_{fbk} = 1.6$ and for torsion $K_{ftk} = 1.3$. Comparing the values of these factors with previously calculated factors we have to select the larger values. Therefore, we take for the torsion stress $K_{ft} = 1.33$ and for the bending stress $K_{fb} = 1.6$. Adjusting the bending moment according to the stress concentration factor we obtain $M_a = K_{fb}M_0 = 1.6 \cdot 115.41 = 184.66$ lb-in. Accordingly correcting the torque, we have $T_a = K_{ft}T_{max} = 1.33 \cdot 371.32 = 493.85$ lb-in.

Selecting from the table in Appendix A-1-4 the AISI 1015 annealed steel having $S_{ut} = 56$ ksi, we determine the endurance limit using formula (1-9.5) $S_e = k_a k_b k_c S'_e$. Hence, based on formula (1-9.2) we calculate $k_a = 0.983$. In order to avoid confusion we denote the diameter of the cantilever by d_{2c} and assuming that $d_{2c} = 1.2$ in., we determine from formula (1-9.3) that $k_b = 0.861$. According to formula (1-9.4), $k_c = 1$, therefore, $S_e = 0.983 \cdot 0.861 \cdot 1 \cdot 0.5 \cdot 56 = 23.7$ ksi.

Applying formula (3-5.9) based on the fourth theory of failure and using the safety factor $n_\sigma = 2.5$, we calculate

$$d_{2s} = \left\{ \frac{32 \cdot 2.5}{\pi} \left[\left(\frac{184.66}{23.7 \cdot 10^3} \right)^2 + 0.75 \left(\frac{493.85}{23.7 \cdot 10^3} \right)^2 \right]^{\frac{1}{2}} \right\}^{\frac{1}{3}} = 0.77 \text{ in. } (19.6 \text{ mm}).$$

If we round up the diameter of the cantilever to 20 mm, it coincides with the preliminary selected bore diameter of the bearing. This is not acceptable from the design point of view. We need to recalculate the diameter of the shaft journals in order to obtain the proper bore diameter of the bearings. After that, we will finalize the decision regarding the diameter of cantilever that should be at least 20 mm.

9. Determining the diameter of the shaft to select the proper bearing.
Because we determined that the most loaded cross-sectional area of the shaft is located in point B (in the middle of the right-hand bearing), we can determine the proper dimension of the shaft diameter for the bearing. The bending moment and the torque acting in the most loaded cross-sectional areas are $M_{max} = M_a = 173.12$ lb-in. and $T_{max} = T_a = 376.32$ lb-in, respectively. It should be mentioned that to the right of the bearing (support B) there is a groove in the shaft that causes stress concentration. In this case, we should determine the diameter d of the groove and consider that D is the diameter for the bore of the bearing. Taking $\frac{D}{d}=1.1$ and $\frac{r}{d}=0.1$, we determine from the Appendix A-1-5 (Figures 5 and 6) the theoretical stress concentration factors for bending and torsion: $K_{tb} = 2.2$ and $K_{tt} = 1.75$, respectively. Therefore, we calculate $K_{fb} = 1 + 0, 6(2.2-1) = 1.72$ and $K_{ft} = 1 + 0.6(1.75-1) = 1.45$. The endurance limit is the same as in the calculation of the cantilever. Now we determine $M_a = 1.72 \cdot 173.12 = 297.6$ lb-in. and $T_a = 1.45 \cdot 371.32 = 538.4$ lb-in. Hence, using formula (3-5.9), we calculate the diameter of groove d_2', so $d_2' = \left\{ \frac{32 \cdot 2.5}{\pi} \left[\left(\frac{297.6}{23.7 \cdot 10^3} \right)^2 + 0.75 \left(\frac{538.4}{23.7 \cdot 10^3} \right)^2 \right]^{\frac{1}{2}} \right\}^{\frac{1}{3}} = 0.84$ in. (21.34 mm).

Therefore, the diameter of the driven shaft $d_2 = 1.1d'_2 = 1.1 \cdot 0.84 = 0.924$ in. (23.46 mm). Now we select the SKF No. 61905 double shielded bearing having the bore diameter of 25 mm. Therefore, the diameter of the cantilever is 20 mm (0.785 in.) according to the calculations presented above.

10. Finalizing the selection of the bearing.
The newly selected bearing No. 61905 has the outside diameter $D = 42$ mm, the width is 9 mm, the dynamic load rating $C = 7.02$ kN, and the static load rating is $C_0 = 4.3$ kN. The radial load $F_r = R_B = 163.5$ lb or $F_r = 726.7$ N. The axial load $F_a = 0$. In this case $F_0 = 0.6 \cdot 726, 7 = 456$ N. Since $F_0 < C_0$ (456<4,300), we may proceed.

According to formula (2-4.5), we determine the equivalent dynamic load considering that in our case V = 1 because the inner ring is rotating. In this case $X = 1$ and $Y = 0$, and, consequently, $F_e = F_r = 726.7$ N. Applying formula (2-4.4) and taking $a = 3$, we determine the bearing service life: $L_{10} = \left(\frac{C}{F_e} \right)^3 = \left(\frac{7{,}020}{726.7} \right)^3 = 901 \cdot 10^6$ revolutions or $L_{10h} = \frac{901 \cdot 10^6}{60 \cdot 500} = 30{,}033$ hours. We may try to select a lighter bearing having a shorter service life.

In the SKF catalog, the only lighter bearing having the bore diameter of 25 mm is the bearing No. 61805 that has the dynamic load rating $C = 4.36$ kN. Hence, $L_{10} = \left(\frac{4360}{726.7} \right)^3 = 215.97 \cdot 10^6$ revolutions, therefore, $L_{10h} = \frac{215.97 \cdot 10^6}{60 \cdot 500} = 7{,}199$ hours, that seems to be insufficient. Thus, we accept the previously selected bearing.

11. Calculations of the key.

According to Appendix A-3-1, we select the parallel square key having the width $b = 0.25$ in. and the height $h = 0.25$ in. The material for the key we select from the table in Appendix A-1-4. We take the AISI 1015 annealed steel having $S_{yt} = 41.25$ ksi.

The required length of the key based on the shear stress can be calculated using formula (3-8.2): $L_\tau = \dfrac{2Tn_\tau}{dbS_{sy}}$, where $n_t = 2.5$ and $S_{sy} = 0.5S_{yt}$, therefore, $L_\tau = \dfrac{2 \cdot 371.32 \cdot 2.5}{0.785 \cdot 0.25 \cdot 0.5 \cdot 41{,}250} = 0.458$ in.

The required length of the key with respect to bearing stress is determined using formula (3-8.3): $L_\sigma = \dfrac{2Tn_\sigma}{0.5dhS_{yt}}$, where $n_\sigma = 2.5$. Therefore, $L_\sigma = \dfrac{2 \cdot 371.32 \cdot 2.5}{0.5 \cdot 0.785 \cdot 0.25 \cdot 41{,}250} = 0.458$ in. We take the length of the key as 0.50 in.

12. Calculation of the angle of rotation of the cantilever.

The deflection of the shaft for belt drives does not play a noticeable role, and it is not required to be calculated.

Since the shaft has a stepped structure, it is recommended to replace it with a uniform shape having an equivalent diameter. As mentioned above, the bore diameter is 25 mm (0.984 in.), and the distance between the bearings is 2.0 in., while the distance between the bearing and the center of the sheave is 2.25 in., and the diameter of cantilever is 0.785 in. Using formula (3-6.11), we calculate the equivalent diameter of the uniform shaft: $d_e = \dfrac{1}{\sqrt[4]{\dfrac{2}{4.25 \cdot 0.984^4} + \dfrac{2.25}{4.25 \cdot 0.785^4}}} = 0.852$ in.

Taking the modulus of elasticity for steel $E = 30 \cdot 10^6$, we calculate the angle of rotation using formula (3-6.22), where $l = l_2 = 2.25$ in., $l_0 = l_1 + l_2 = 2.0 + 2.25 = 4.25$ in., and $F = R = 76.94$ lb.

Hence, $\theta = \dfrac{Fl_2[l_2 + 2(l_1 + l_2)]}{6EI} = \dfrac{76.94 \cdot 2.25[2.25 + 2(2 + 2.25)]64}{6 \cdot 30 \cdot 10^6 \cdot \pi 0.852^4} = 0.0004$ rad, that is less than the allowable angle of rotation for our case 0.01 rad.

13. Calculation of the critical rotational speed.

Combining formulas (3-7.9) and (3-7.14), we obtain an expression for calculating the critical rotational speed for our case: $n_{cr} = \dfrac{30}{\pi}\sqrt{\dfrac{3EI}{ml^3}}$, where m is the equivalent mass associated with the force R causing the deflection of the shaft. Since $R = 76.94$ lb, then $m = \dfrac{76.94}{32.2} = 2.39\ \dfrac{\text{lb-s}^2}{\text{ft}}$ (slug). It should be noted that the diameter of the cantilever 0.785 in. is used in the formula of the moment of inertia and the length of the cantilever $l_2 = 2.25$ in. is used in the formula of the critical speed. Calculating the moment of inertia of the shaft, we have $I = \dfrac{\pi 0.785^4}{64} = 0.0186\ \text{in}^4$.

Hence, $n_{cr} = \dfrac{30}{\pi}\sqrt{\dfrac{3 \cdot 30 \cdot 10^6 \cdot 0.0186 \cdot 12}{2.39 \cdot 2.25^3}} = 8{,}207$ rpm.

14. Assembly and disassembly.

a) Assembly

We begin with mounting sheave 4 onto the shaft 6 of the electric motor 1 and secure the sheave to the shaft by the key 5. Now we assemble the electric motor with the frame (not shown) by means of bolts and nuts (not shown) but not tightening the threaded connections. After that, we mount by an interference fit bearing 15 onto the left-hand side of the shaft 12 and secure the bearing with the retaining ring 13. Then we place distance-bushing 11, mount bearing 18, and install retaining ring 10. After that, we install by a clearance fit the bearing-shaft assembly into bearing housing 16 and secure this assembly by retaining rings 14 and 20. It should be mentioned that the gap remaining between the bearings and the retaining rings is sufficient to handle the temperature expansion of the shaft. Now we rigidly secure the bearing housing to the frame (not shown) by installing the bolts into the holes 17 and 19. Then we mount sheave 7 onto the cantilever 9 and secure the sheave with key 8. After that, we install V-belts 2 and 3 and adjust the tension in the belts by displacing the electric motor along the longitudinal slots in the frame. Now we tighten the bolts that rigidly secure the electric motor to the frame.

b) Disassembly

First we loosen the bolts securing the electric motor to the frame and remove the belts. Pulling out key 5 we take off sheave 4. Now pulling out key 8 we remove sheave 7. After that, removing retaining ring 20 we take apart the bearing-shaft assembly from the bearing housing. After removing retaining rings 10 and 13, we pull out bearing 15 using an appropriate puller and take off bushing 11. Then we pull out bearing 18 using a corresponding puller. After that, we remove retaining ring 14 and unscrew the bolts securing the bearing housing to the frame and taking apart the housing.

Chapter 4. Problems

4-01. The diameter of the driving pulley of an open horizontal flat-belt drive $d_1 = 150$ mm, the center distance between the pulleys $C = 650$ mm, and the velocity ratio $i = 3.5$. Determine the contact angle of the driving pulley and the length of the belt.

4-02. Calculate the tension force in the tight strand (F_1) of the open horizontal flat-belt and V-belt transmissions for the contact angles of $\varphi_1 = 90°$ and $\varphi_1 = 180°$ if the tension force in the slack strand (F_2) is 100 lb, the wedge angle $2\beta = 38°$, and the friction coefficient $f = 0.3$. Compare the ratios between the corresponding forces of the flat and V-belt drives.

4-03. Determine the bending stress in the belt of a flat-belt transmission. The diameter of the driving pulley is 150 mm and the thickness of the belt is 5 mm. The modulus of elasticity of the belt is 75 MPa.

4-04. An open horizontal flat-belt drive transmits power of 7.5 kW to the driving pulley at 1,800 rpm. The diameter of the driving pulley is 90 mm, the velocity ratio is 4.0, the center distance is 600 mm, and the friction coefficient between the pulley and the belt is 0.3. Determine the radial force applied to the shafts.

4-05. Determine the stresses in the belt of a flat-belt transmission due to the centrifugal force and due to the bending. The diameter of the driving pulley is 7.5 in., and its rotational speed is 1,200 rpm. The thickness of the belt is 0.25 in., and the modulus of elasticity of the belt is 11,000 psi, while the weight density of the belt is 154 $\frac{\text{lb}}{\text{ft}^3}$.

4-06. Design an open horizontal flat-belt drive transmitting power of 1.5 hp to the driving pulley at 1,800 rpm, while the velocity ratio is 4.2, and the center distance is 40 in. The belt is made from leather. The friction coefficient is 0.35. Determine the radial force applied to the shafts. Take safety factor $n_\sigma = 2$. (Refer to Example 4-5.1a)

4-07. Design an open horizontal flat-belt drive transmitting power of 2.2 kW to the driving pulley at 1,800 rpm, while the velocity ratio is 4.0 and the center distance is 550 mm. The belt is made from polyurethane reinforced by steel cords. Determine the radial force applied to the shafts. Take the safety factor $n_\sigma = 2$. (Refer to Example 4-5.2a)

4-08. Determine the power rating for the B-size V-belt drive based on the following input data: the minimum diameter of the driving sheave $d_1 = 5.4$ in, the rotational speed of the driving sheave $n_1 = 1{,}800$ rpm, the velocity ratio $i = 4$, and the center distance $C = 68$ in.

4-09. Determine the power rating for the C-size V-belt drive based on the following input data: the minimum diameter of the driving sheave $d_1 = 9.0$ in, the rotational speed of the driving sheave $n_1 = 1{,}800$ rpm, the velocity ratio $i = 4$, and the center distance $C = 110$ in.

4-10. Determine the power rating for the D size V-belt drive based on the following input data: the minimum diameter of the driving sheave $d_1 = 13.0$ in, the rotational speed of the driving sheave $n_1 = 1{,}800$ rpm, the velocity ratio $i = 4$, and the center distance $C = 160$ in.

CHAPTER 5

Basics of Gearing and Cylindrical Gears

The use of toothed power transmissions is extremely widespread. Toothed power transmissions, or power drives, are often called reducers because they deliver a reduced rotational speed from an energy source to the operational link performing a desired task. Toothed power drives are off-the-shelf products, and information contained in manufacturers' catalogs aids in their selection. However, the selection process is based on engineering calculations that engineers should be familiar with.

Toothed transmissions can transmit power up to tens of thousands kW. The operational processes of these drives are characterized by an extremely wide range of velocities, starting with very low velocities and reaching up to 3,000 $\frac{\text{ft}}{\text{min}}$ $\left(150\frac{\text{m}}{s}\right)$ and more. These transmissions allow velocity ratios in the range of thousands and even more. Toothed drives are used in a wide variety of operational conditions characterized by extremely heavy loading, as in hot rolling mills and rotating furnaces, and in extremely precise mechanisms, such as watches and dial indicators. Some very important features of toothed transmissions are the constancy of their velocity ratios, their high efficiency (up to 0.99), and their significant service time.

A toothed transmission comprises at least a pair of meshed toothed gears. The manufacturing of toothed gears relies on sophisticated and costly machining equipment.

During the operation of a pair of meshed gears, the tooth of the driving gear enters the tooth space of the driven gear and exerts pressure to the adjacent tooth forcing the gear to rotate. The tooth drive can transmit power between parallel, perpendicular, and crossing axes. There are toothed drives with external and internal engagement of the teeth; however, the most commonly used drives have an external engagement. As mentioned above, the majority of gear transmissions are used to reduce the rotational speed; however, they can be used for increasing the rotational speed as well.

The profiles of the teeth of a meshing pair of gears should satisfy specific requirements. The most important of these requirements include the constancy of the instantaneous values of the velocity ratio; the serviceability of the pair of gears regardless of the number of teeth; the allowance of certain inaccuracies of the distance between the centers of the gears; and the simplicity of fabrication of the tooth profile.

There are several types of gears that are briefly described below:

1) External spur gears are cylindrical gears having straight teeth oriented parallel to their axes of rotation and transmit rotation between parallel axes. In an external engagement, the meshed gears rotate in opposite directions. The teeth of these gears exert tangential and radial forces.

2) In an internal spur gear engagement, the larger gear has internal teeth, while the smaller gear has external teeth (as in an external pair of gears), and both gears rotate in the same direction.
3) External helical cylindrical gears have teeth oriented at an angle to the axes of rotation and transmit rotation between parallel axes, while the meshed gears rotate in the opposite direction. Helical gears exert tangential, radial, and axial forces. A special group of helical gears is called herringbone gears. Their axial forces act in opposite directions and cancel out each other.
4) Crossed helical gears are cylindrical gears that transmit rotation between non-parallel axes. They exert tangential, radial, and axial forces.
5) Straight bevel gears have a conical shape, and their teeth are oriented in the radial direction to the center of rotation so that the thickness and the height of the teeth decreases toward the center of rotation of the gear. Straight bevel gears transmit rotation between intersecting axes and exert tangential, radial, and axial forces.
6) Spiral bevel gears also transmit rotation between intersecting axes; however, their curved teeth in comparison with straight bevel gears provide smoother engagements between the teeth. These gears also exert tangential, radial, and axial forces.
7) Worm gears transmit rotation between axes at right angles and are characterized by relatively high velocity ratios and relatively high energy losses. The energy losses are associated with the extensive friction between the worm and the worm gear teeth. Worm gears exert tangential, radial, and axial forces.

There are some more specific types of gears; however, they have limited applications.

We begin the study of gearing with spur gears, which are straight cylindrical gears whose structure is less complicated than other gears. This allows for a simplified analysis of the basic principles that are applicable to other types of gears. Figure 5-1.1 shows a pair of engaged meshing spur gears.

Figure 5-1.1. These gears are off-the-shelf gears whose inner diameters can be machined to the required dimensions. (Courtesy of KHK-USA, Inc.)

5-1 Basics of Teeth Engagement

The fundamental condition of the serviceability of a pair of teeth profiles is that they are conjugate. Figure 5-1.2 shows a mechanical system comprising two elements having profiles 1 and 2.

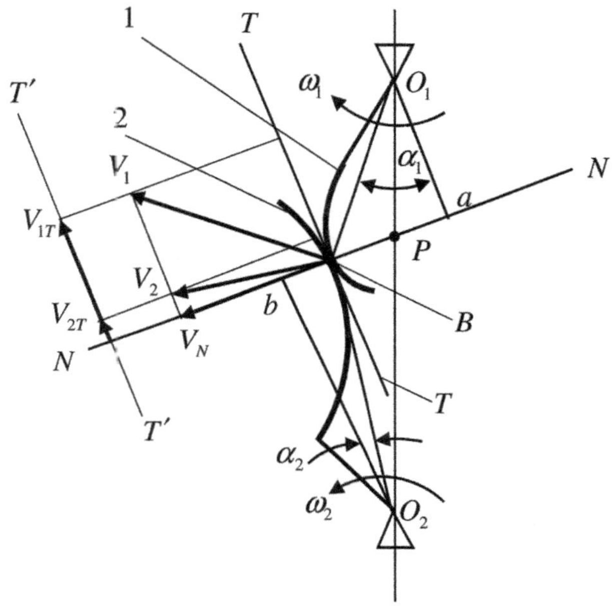

Figure 5-1.2. Conjugate profiles.

Element 1 is rotating with an instantaneous angular velocity ω_1 around its center O_1 and is forcing element 2 to rotate with an instantaneous angular velocity ω_2 around its center O_2. The common normal *NN* to profiles 1 and 2 passing through the point of their contact *B* intersects with the centerline $\overline{O_1O_2}$ in point *P* that represents the instantaneous center of the relative rotation for both elements. Hence, the profiles are conjugate if the common normal to both profiles intersects with the centerline in a point that represents the instantaneous center of relative rotation for both elements.

The analysis of the instantaneous linear velocities of each profile reveals some important peculiarities that describe the interaction of the conjugate profiles. Connecting the centers of rotation with the point of contact of the profiles by straight lines, we determine the instantaneous values of the corresponding radii $\overline{O_1B}$ and $\overline{O_2B}$. The vectors V_1 and V_2 that are perpendicular to the corresponding radii represent the instantaneous velocities of the mutual point *B* of the profiles 1 and 2. The perpendiculars from the centers O_1 and O_2 onto the normal *NN* in the points *a* and *b* create with the radii $\overline{O_1B}$ and $\overline{O_2B}$ two corresponding angles α_1 and α_2. It can be seen that the angle between the vector V_1 and the common normal *NN* equals α_1, while the angle between the vector V_2 and this normal equals α_2. Projecting these two vectors onto the common normal *NN* and accepting an assumption that these two projections of the velocities are equal, we may write that $V_1 \cos \alpha_1 = V_2 \cos \alpha_2 = V_N$. However, this assumption should be justified. Thus, in cases where $V_1 \cos \alpha_1 > V_2 \cos \alpha_2$ then the profiles would tend to deform each other, while in cases where $V_1 \cos \alpha_1 < V_2 \cos \alpha_2$ then the contact between these profiles would be lost. These two cases contradict the serviceability of a pair of meshed teeth and, therefore, they are irrelevant. The accepted assumption does not have contradictions, and for the conjugate profiles the projections of the velocity vectors must be equal to each other. Hence, the serviceability of these profiles is possible just in cases when these two projections of the velocities are equal and, therefore, we may write that $\overline{O_1B}\omega_1 \cos\alpha_1 = \overline{O_2B}\omega_2 \cos\alpha_2$. This equation yields the following expression:

$$\frac{\omega_1}{\omega_2} = \frac{\overline{O_2B}\cos\alpha_2}{\overline{O_1B}\cos\alpha_1} = i \tag{5-1.1}$$

From the triangles O_2bP and O_1aP, we determine $\overline{O_2B}\cos\alpha_2 = \overline{O_2b}$ and $\overline{O_1B}\cos\alpha_1 = \overline{O_1a}$. Substituting these values into expression (5-1.1), we obtain

$$i = \frac{\omega_1}{\omega_2} = \frac{\overline{O_2b}}{\overline{O_1a}} \tag{5-1.2}$$

From the similarity of triangles BO_1a and BO_2b we have that $\frac{\overline{O_2b}}{\overline{O_1a}} = \frac{\overline{O_2P}}{\overline{O_1P}}$. Combining this proportion with expression (5-1.2), we write

$$i = \frac{\omega_1}{\omega_2} = \frac{\overline{O_2P}}{\overline{O_1P}} \tag{5-1.3}$$

Because $\overline{O_1P}$ and $\overline{O_2P}$ may represent the radii of corresponding imaginary circles of a pair of meshing gears, and these circles are revolving on each other without slipping, then the ratio (5-1.3) may be the velocity ratio of this pair of gears.

Therefore, based on the considerations presented above, it could be stated that the principal rule of toothed gear engagement is that the common normal to the interacting conjugate profiles passing through their contact point intersects the gear's centerline at a point that divides this centerline into line segments whose lengths are in inverse proportion to the instantaneous velocity ratio.

In Figure 5-1.2 the line *TT* is tangent to the profiles 1 and 2 in point *B* and, obviously, this line is perpendicular to common normal *NN*. For the clarity of the schematic, the line *TT* is moved parallel to itself into the position *T′T′*. The projections of the instantaneous velocity vectors V_1 and V_2 onto the tangent line *T′T′* are the vectors V_{1T} and V_{2T}, respectively, while

$$V_{1T} = \overline{PB}\omega_1 \tag{5-1.4}$$

and

$$V_{2T} = \overline{PB}\omega_2 \tag{5-1.5}$$

The difference ΔV between these vectors according to formulas (5-1.4) and (5-1.5) equals

$$\Delta V = \overline{PB}\omega_1 - \overline{PB}\omega_2 = \overline{PB}(\omega_1 - \omega_2) \tag{5-1.6}$$

Actually, ΔV according to expression (5-1.6) represents the instantaneous relative velocity between profile 1 and profile 2 at the point B. It characterizes the effectiveness of the operational process of gear transmissions, the wear of the teeth, and energy loses. According to expression (5-1.6), in a pair of gears when $\Delta V = 0$ then $\omega_1 = \omega_2$ and, therefore, the velocity ratio in this case is $i = 1$.

It is important to emphasize that in cases when point P moves along the centerline during the rotation of the profiles the velocity ratio of those profiles is a variable value, while in cases when this point remains motionless on the centerline the velocity ratio is constant. Depending on the geometric parameters of the conjugate profiles, this point could remain motionless or move along the centerline. Because the velocity ratio for gear transmissions should remain constant, we need to select an appropriate curve for the profiles of the teeth that will satisfy this requirement. The ultimate goal in selecting or synthesizing tooth profile curves is to obtain the maximum beneficial features of the profile. Actually it is difficult to obtain a single profile that provides all desirable features. The profile of teeth for the vast majority of power gear transmissions is discussed below.

5-2. Structure and Parameters of Meshing Gears

Several curves satisfy the principal requirements of the engagement of conjugate profiles and, in addition, could provide a constant velocity ratio between these profiles. According to the suggestion of L. Euler, the involute was adopted as the curve for profiles of toothed power transmissions. Usually, a toothed transmission comprises a pinion and a gear, while the pinion is the driving gear, and it is smaller in diameter. It is acceptable that all geometric parameters related to the pinion have a subscript 1, while the gear's parameters have a subscript 2. Figure 5-2.1 shows the basic parameters related to the engagement of the pinion and the gear based on involute profiles of their teeth.

The pitch circles of the pinion and the gear are imaginary circles that contact in an imaginary pitch point P that is located on the centerline O_1O_2, and these circles revolve on each other without slipping. One of the distinctive features of the involute is that the pitch point remains in the same position and divides the centerline into two line segments O_1P and O_2P, having their lengths in inverse proportion to the velocity ratio that also remains constant. The pitch circles of the pinion and the gear rotate with the angular velocities ω_1 and ω_2, while the velocity ratio is $i = \frac{\omega_1}{\omega_2} = \frac{d_2}{d_1} = \frac{\overline{O_2P}}{\overline{O_1P}}$, where $\overline{O_1P}$ and $\overline{O_2P}$ are the radii of the contacting pitch circles.

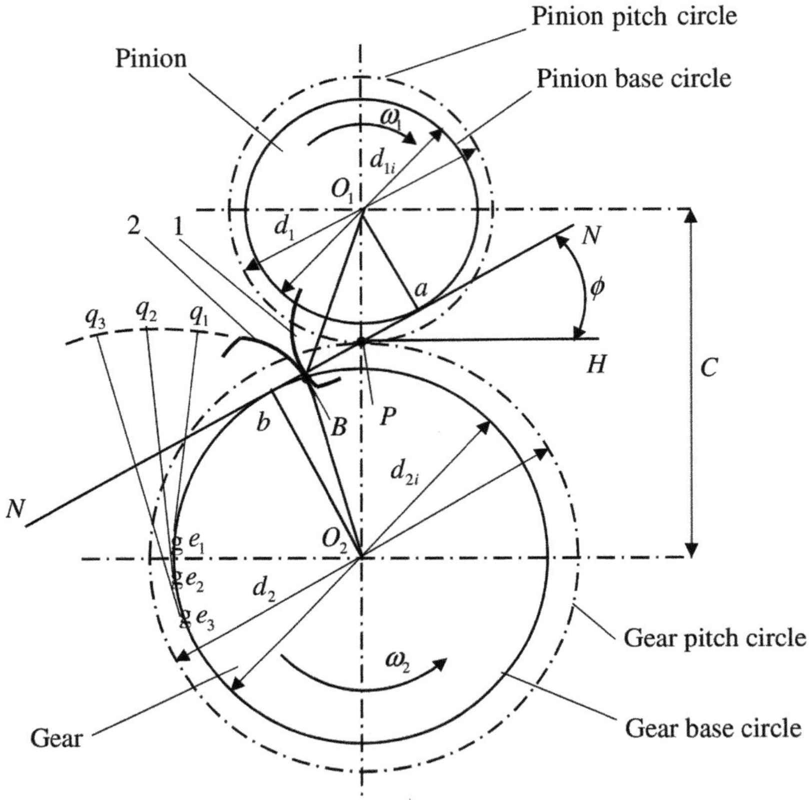

Figure 5-2.1. Involute engagement.

It should be clear that in a real pinion and in a real gear there are no contacting pitch circles, and their diameters d_1 and d_2 cannot be measured in a direct way using conventional measuring instruments. These diameters can be calculated based on certain geometric relationships that are discussed below. They can be measured indirectly by applying special tooling attachments.

The pinion transmits the rotation to the gear due to the engagement of involute teeth 1 and 2 that belong to the pinion and gear, respectively. The common normal *NN* to the profiles of the teeth passes through the point of contact *B* of these involute profiles and through the pitch point *P*. The common normal *NN* does not change its position during the operation of the gears. The line *PH* represents a tangent line to the pitch circles in the pitch point and is, obviously, perpendicular to the centerline. The lines *NN* and *PH* create an angle ϕ that is called the pressure angle, while the common normal *NN* is called the pressure line. Line segments O_1a and O_2b are perpendicular to the pressure line and represent the radii of the pinion and the gear base circles, while d_{1i} and d_{2i} are the diameters of these base circles. These diameters are called base diameters.

The line segment *ab* is called the line of action or the generating line. The pressure angle can be determined using the following expression:

$$\phi = \cos^{-1}\frac{d_{1i}}{d_1} = \cos^{-1}\frac{d_{2i}}{d_2} \quad (5\text{-}2.1)$$

During the gears rotation, the contact point *B* moves along the line of action that results in the constancy of the velocity ratio.

The next distinctive feature of the involute is that the velocity ratio remains the same when the centerline changes its length in a certain range (due to the adequate manufacturing tolerances associated with the machining of the center holes). Due to the allowable tolerances, the center distance *C* of the mashing gears can become slightly shorter or longer, resulting in an increase or decrease of the pitch circles of the pinion and the gear. However, in these cases the base diameters d_{1i} and d_{2i} remain unchanged; therefore, according to expression (5-2.1), the change in the pitch circles causes a corresponding change in the pressure angle, while the velocity ratio remains the same.

The generation (or drawing) of the involute profile of the tooth belonging to the gear, as shown in Figure 5-2.1, can be achieved by rolling a tangent line around the base circle without slippage. This can be explained in the following way: imagine one end of a cord is secured to the base circle and wound around this circle in a clockwise direction, while the other end is connected to a pencil and its tip is located at point *B*. By unwinding the cord starting at the point *B*, the tip of the pencil draws an involute that begins at point, *B* creating the profile 2 and continues along the dashed curve and passes through the points q_1, q_2, q_3, and so on. The line segments e_1q_1, e_2q_2, and e_3q_3 show the successive positions of the cord and represent the instantaneous radii of curvature of the involute, while these radii are tangent to the gear base circle at the points e_1, e_2, and e_3.

One of the important features of the involute profile involves the straightening of the curved profile as the diameter of the base circle increases, and when this diameter tends to infinity then the profile becomes a straight line. This feature plays an essential role in creating a teeth manufacturing tool, called a rack, that has straight-line tooth profiles. More details related to the rack as a manufacturing tool are discussed later.

5-2.1. Basic Geometric Parameters of Spur Gears

An image of an off-the-shelf spur gear and the nomenclature for the basic geometric parameters of spur gears are shown in Figure 5-2.2, where *d* is the diameter of the pitch circle that represents a theoretical geometric parameter that defines the structural characteristics of the gear.

The circular pitch is the distance on the pitch circle between the point of one tooth to the identical point of the adjacent tooth. The tooth thickness and tooth space width are measured on the pitch circle and the sum of them equals the circumference of the pitch circle.

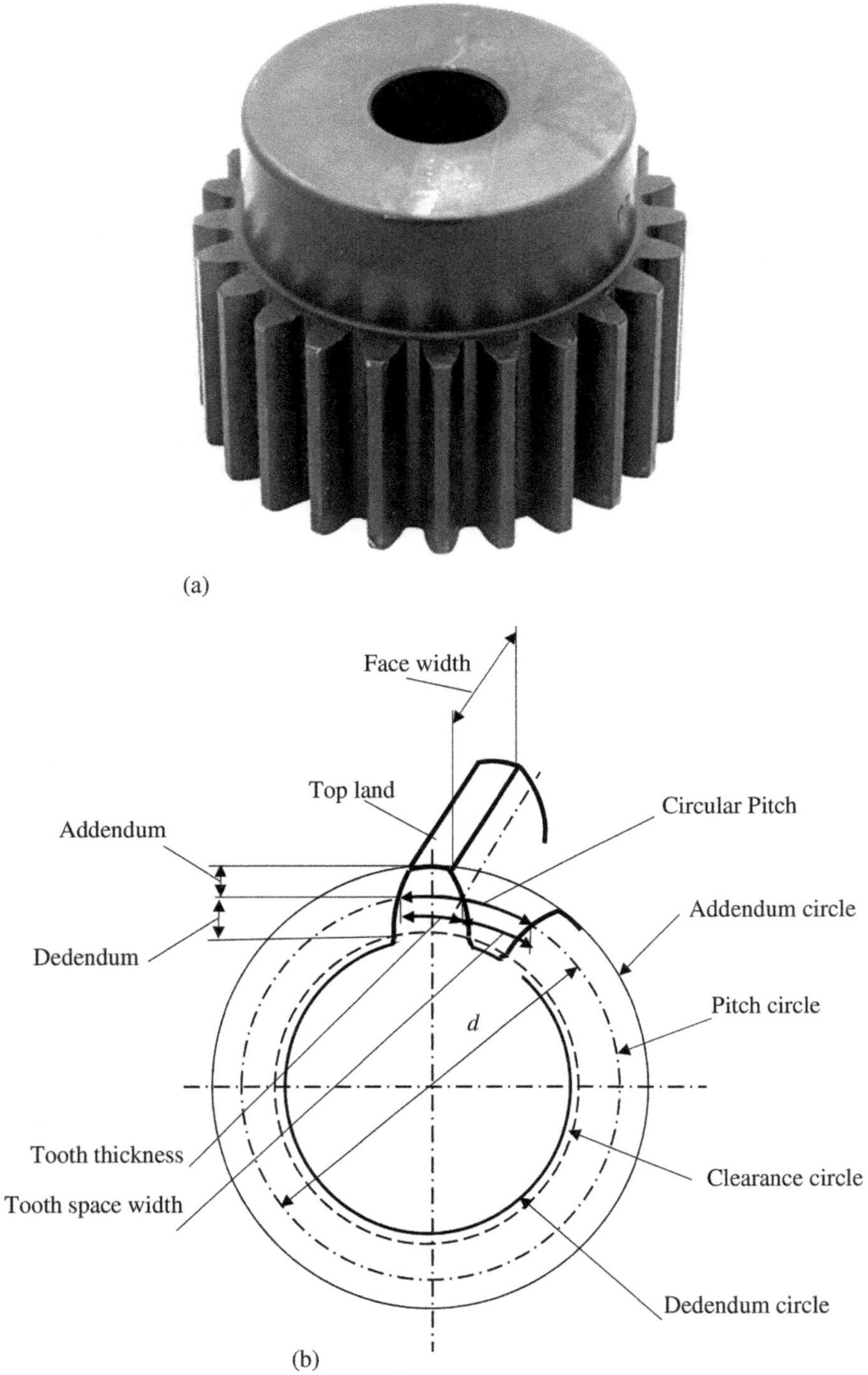

Figure 5-2.2. Spur gear and its parameters: (a) spur gear (Courtesy of Martin Sprocket & Gear, Inc.); (b) spur gear parameters.

The addendum is the part of the length of the tooth from the pitch circle to the top land of the tooth. The dedendum is the part of the length of the tooth from the pitch circle to the bottom of the tooth. The addendum circle has the diameter of the blank cylinder for machining the teeth. The dedendum circle has the diameter of the roots of the teeth, and it is used for manufacturing the gear. The addendum and dedendum diameters are sometimes referred as the outer and inner diameters of the gear. Conventional measuring tools cannot directly measure these diameters if the gears have an odd number of teeth. The tooth face width represents the width of the blank cylinder intended for machining the teeth. The clearance between the top land and the dedendum circle is predetermined by the clearance circle that is tangent to the addendum circle of the mating gear.

Denoting the circular pitch by p and the number of teeth by N, we may write

$$\pi d = pN \tag{5-2.2}$$

Solving this equation for the diameter d, we have

$$d = \frac{pN}{\pi} \tag{5-2.3}$$

Denoting

$$P = \frac{\pi}{p} \tag{5-2.4}$$

We introduce a very important parameter of gearing: the diametral pitch P (or just the pitch), which is accepted as the basic gear parameter in the United States and other countries that use the inch as a measure of length. The units of P are reciprocal of inches. The diametral pitch defines the number of teeth per inch. It should be kept in mind that the larger the number of the pitch the smaller are the geometric parameters of the teeth and gears, and consequently, a larger number of teeth could be placed in an inch. From equation (5-2.4), we obtain the formula for calculating the pitch diameter:

$$d = \frac{N}{P} \tag{5-2.5}$$

Equation (5-2.4) shows that $p = \frac{\pi}{P}$, hence, we may write

$$pP = \pi \tag{5-2.6}$$

Pitches are standardized, and Appendix A-5-1 presents the values of the standardized coarse and fine pitches. It is important to notice that the units of a pitch are $\frac{1}{\text{in.}}$, for instance, $P = 12\ \frac{1}{\text{in.}}$ means 12 teeth per inch.

In countries that use metric units, a different approach in defining the basic parameter of gears is accepted. According to formula (5-2.4), the pitch P is defined as $\frac{\pi}{p}$, while in the metric system the ratio $\frac{p}{\pi}$ is defined as module m, and it is the basic parameter of gears. Therefore, we have

$$m = \frac{p}{\pi} \tag{5-2.7}$$

In the metric system, the larger the module the larger are the geometric parameters of the teeth and gear. The modules are expressed in millimeters, for instance, m = 12 mm. It should be clear that neither the pitch nor the module can be directly measured by measuring tools. Combining formulas (5-2.3) and (5-2.7), we obtain the formula for calculating the pitch diameter of the gear:

$$d = mN \tag{5-2.8}$$

The pitch diameters of gears play an important role in all engineering calculations associated with gearing. The relationship between the diametral pitch and the module is Pm = 25.4.

The modules are also standardized, and Appendix A-5-2 lists the standard values of modules.

The pressure angle ϕ is standardized and has the same value for pitch and module system gears. Currently the accepted values of the pressure angles are 20° an 25°, while the angle of 20° is used more often. In the past, a pressure angle of 14.5° was used. In this book the value of the pressure angle is considered to be equal to 20°. The geometric parameters of the teeth are standardized for the pitch and module-based systems. Thus, the addendum $a = \frac{1}{P}$ for coarse and fine pitches. Therefore, the diameter of the addendum circle is

$$d_a = \frac{N}{P} + \frac{2}{P} = \frac{N+2}{P} \tag{5-2.9}$$

The dedendum for coarse pitches (P < 20) is $b = \frac{1.25}{P}$ and for fine pitches (P > 20) is $b' = \frac{1.2}{P} + 0.002$ in. Hence, the diameters of the dedendum circles for coarse and fine pitches are

$$d_d = \frac{N}{P} - \frac{2 \cdot 1.25}{P} = \frac{N - 2.5}{P} \tag{5-2.10}$$

and

$$d'_d = \frac{N}{P} - \frac{2 \cdot 1.2}{P} - 0.002 = \frac{N - 2.4}{P} - 0.002 \text{ in.} \tag{5-2.11}$$

The addendum in the module system is $a = m$, therefore, the addendum diameter is

$$d_a = m(N+2) \tag{5-2.12}$$

The dedendum is $b = 1.25\,m$, hence, the diameter of the dedendum circle is

$$d_d = m(N - 2.5) \tag{5-2.13}$$

The relationship between the base circle diameter and pitch diameter is determined by the following expression:

$$d_b = d\cos\phi \tag{5-2.14}$$

Therefore, the base pitch equals

$$p_b = p\cos\phi \tag{5-2.15}$$

It should be mentioned that the base circles, like the pitch circles, cannot be measured directly by conventional tools.

Gears of the same pitch and the same pressure angle regardless of their number of teeth can be meshed, creating a toothed transmission. The same is true of the module-based gears.

However, pitch gears are not interchangeable with module gears.

5-3. Contact Ratio, Interference, and Minimum Number of Teeth

In order to achieve the continuity of engagement of the meshing teeth, it is necessary to satisfy the following condition: before a pair of meshing teeth (the first pair) exits from the engagement, the next pair of teeth (the second pair) must be already engaged. As shown in Figure 5-2.1, the contact point B of two meshing profiles is moving along the line of action. It is possible to indicate on this line the length of the line segment that characterizes the motion of the contact point from the beginning (entry point) of the engagement of a pair of teeth to the end (exit point) of the engagement of this pair of teeth. Denoting the length of this line segment as the length of the line of action L_a, we determine its value using the following formula:

$$L_a = \frac{\sqrt{(N_1+2)^2 - N_1^2\cos^2\phi} + \sqrt{(N_2+2) - N_2^2\cos^2\phi} - (N_1+N_2)\sin\phi}{2P} \tag{5-3.1}$$

This formula is derived for the standard value of the addendum.

For the normal operation of a pair of meshing gears, an overlap of the engagements of two pairs of adjacent teeth is required. It is acceptable to evaluate the degree of this overlap by a parameter that is called the contact ratio c_r. This parameter can be determined as the ratio between the length of the line of action according to the expression (5-3.1) and the base pitch from the formula (5-2.15):

$$c_r = \frac{\sqrt{(N_1+2)^2 - N_1^{\prime 2}\cos^2\phi} + \sqrt{(N_2+2)^2 - N_2^2\cos^2\phi} - (N_1+N_2)\sin\phi}{2\pi\cos\phi} \tag{5-3.2}$$

The expression (5-3.2) comprises the gear parameters representing the numbers of teeth and the pressure angle. Therefore, the contact ratio depends on these parameters and does not depend on the diametral pitch or module, or other gear parameters. The minimum value of the contact ratio is considered to be 1.1; however, in order to provide a smooth engagement, it is recommended that the contact ratio be in excess of 1.4

The tendency to compactness of the assemblies in general and to the meshing pairs of gears in particular leads to the decrease of the numbers of teeth in a pair of gears. As a result, it may happen that in the gear having a fewer number of teeth occurs a restraining of the addendum tip of the mating (larger) gear from a free disengagement with the smaller gear. The addendum tip of the larger gear tends to undercut the dedendum root of the smaller gear in order to create a space for disengagement. There are relatively small gears that have teeth with undercuts; however, these gears are used in applications where the strength of the teeth is not essential. This kind of interaction between the teeth that is characterized by the tendency of the addendum tip to penetrate into the dedendum root of the mating gear represents teeth interference.

According to the analytical approach to teeth interference, it can be mentioned that as in the case of the contact ratio, the interference also depends on the pressure angle and number of teeth of the mating pairs. In this book, we do not consider the analytical approach to the problem of interference because practicing engineers do not need to deal with this issue. This will become apparent from the following discussions.

The pressure angles represent standard values that are incorporated in the tooling for manufacturing gears. The only parameter that the engineers can control is the number of teeth in a meshing pair. Considering that the velocity ratio is a given factor, it is desirable to have the least amount of teeth on the pinion, while the number of teeth of the gear is determined by the following formula: $N_2 = iN_1$. In order to avoid interference in a pair of mashing gears having a standard configuration of teeth, we can use the following formula for calculating the minimum number of teeth of the smaller gear (pinion):

$$N_{1\min} \geq \frac{4i}{(2i+1)\sin^2\phi} \tag{5-3.3}$$

Therefore, in order to avoid interference of the teeth during their engagement, the minimum number of teeth on the pinion should be assigned based on formula (5-3.3).

For gears having a pressure angle of 20° the expression (5-3.3) becomes the following:

$$N_{1\min} \geq \frac{34.2i}{2i+1} \tag{5-3.4}$$

In cases where the number of teeth of the larger gear tends to infinity, $N_2 \to \infty$, as in a rack, and consequently, the velocity ratio also tends to infinity, $i \to \infty$, the minimum number of teeth on the pinion according to the expression (5-3.3) can be determined from the following expression:

$$N_{1\min} \geq \frac{2}{\sin^2\phi} \tag{5-3.5}$$

This case represents the engagement of a pinion with a rack. It occurs when the diameter of the base circle tends to infinity, the radius of curvature of the involute also tends to infinity, and consequently, the involute converts into a straight line.

The conversion of the involute profile into a straight-line profile provides a significant advantage that allows for the creation of a relatively simple cutting tool for manufacturing gears.

5-4. Gear Manufacturing Methods

Gear manufacturing is a wide and relatively complicated field of study. We can address in a very limited way the most general principles of forming the teeth of gears. Teeth can be machined on universal milling machines equipped with indexing heads as well as on specialized metal working machines intended just for cutting all kinds of teeth. Gears are produced using machining, casting, forging, cold rolling, injection molding, and other manufacturing processes. The most widespread manufacturing methods of gearing are single-tooth milling and production by specialized metal-cutting machines.

5-4.1. Single-Tooth Milling

In this method, the metal-cutting tool has the shape of the tooth space and is secured to the spindle of the milling machine. End mills and gear cutters are the tooling for gear manufacturing. The gear blank is mounted in an indexing head that is hand-fed to the rotating cutting tool that gradually removes metal to the required depth. When the first tooth space is machined, the work-piece is moved back to its initial position, and it is rotated by the indexing head on an angle that corresponds to the circular pitch. The cutting process repeats itself until the gear is completed.

It should be emphasized that theoretically the configuration of the tooth depends on the number of teeth in the gear. However, as a practical matter, geometric differences among the teeth in a close range of numbers of teeth is not so critical. This allows for the use of the same tool for cutting teeth in a specified range of number of teeth in the gears. Actually, sets comprising from 8 to 25 tools are used for cutting gears in a wide practical range of number of teeth. A designated set of tools is necessary for each pitch or module of the gears. The main disadvantages of this method include the decreased accuracy and increased noise of the engagement, low productivity, and the need for a significant number of tools. This method of gear manufacturing is used for small-scale production, repair, and restoration.

In addition, there is a special gear manufacturing process based on broaching. During this machining process, all teeth of a gear are fabricated at one setup. This process is predominantly used for manufacturing gears with internal engagement; however, there are special arrangements that allow fabrication of gears with external engagement.

5-4.2. Teeth Generating Method

The tremendous need for power gear transmissions stimulated the development of sophisticated and specialized metal-cutting equipment. The machines specialize in the production of certain types of gears. Thus, certain machines produce cylindrical gears, while another group of machines produces bevel gears, and so on. All of these machines generate teeth on the workpiece by cutting tooling that interacts with the gear blank by imitating the engagement process of a pair of rotating gears. The machining process is completely automated, and the teeth geometry complies with the theoretical profile for each number of teeth in the gear. The working regime of these machines allows adjustments for the required diametrical pitch or module and number of teeth.

This method of generating gear teeth became possible, in part, due to the above mentioned feature of the involute curved profile converting into an involute straight line profile when the diameter of the base circle tends to infinity. Figure 5-4.1 shows a rack and a schematic explaining some details of the machining process of teeth using the method of generation with a rack as a cutting tool.

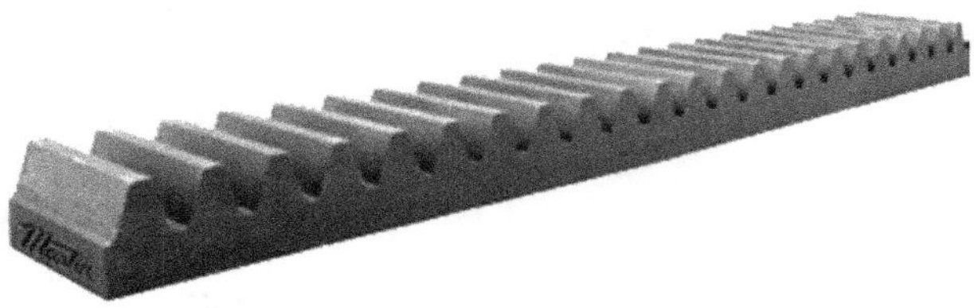

(a)

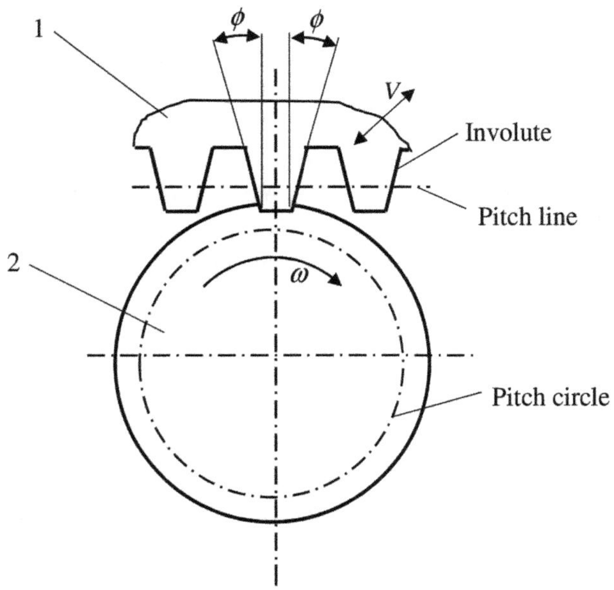

(b)

Figure 5-4.1. (a) A rack (Courtesy of Martin Sprocket & Gear, Inc.); (b) generating teeth using a rack.

Rack 1 has involute teeth with a pressure angle ϕ. The rack is reciprocating in the vertical direction indicated by the arrow V, and at the same time is slowly rolling around the gear blank 2. The gear blank is rotating about the vertical axis as shown by the arrow ω. The interaction of the rack with the gear blank resembles the actual engagement of a pair of gears. The reciprocating movement of the rack consists of a cutting stroke and a return stroke. At the end of the cutting stroke, the rack withdraws radially from the gear blank to avoid rubbing the cutting edges. At the end of the return stroke, the rack is incrementally fed to the gear blank and repeats the cutting cycle. The machine stops automatically when the full depth of tooth space is achieved. This process is called shaping. There are round shapers with involute profiles that work similarly to the rack. The shaping process is also completely automated.

Another process for generating teeth is called hobbing that is based on using a hob as a cutting tool whose structure resembles a threaded component similar to a screw. Figure 5-4.2 shows the structure of the hob and introduces the basics related to the hobbing process. Hob 1 in its axial section resembles the rack having the corresponding straight line involute profile cutting elements that are oriented according to the appropriate pressure angle.

The hob rotates about the horizontal axis as indicated by the arrow Ω. The threads of the hob have longitudinal slots for forming appropriate cutting teeth. Gear blank 2 is rotating about a vertical axis as shown by the arrow ω. The hob and gear blank are engaged, and the cutting

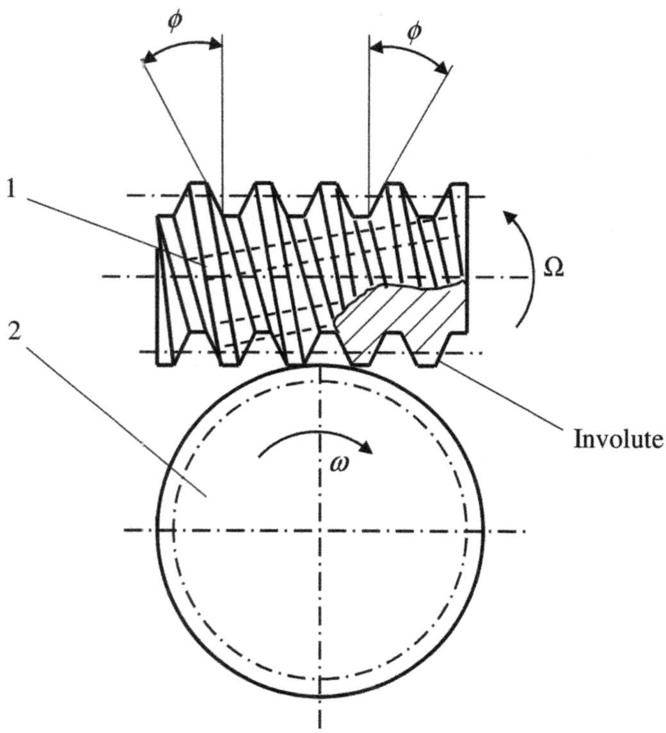

Figure 5-4.2. Generating teeth using a hob.

process is not interrupted until the process is finished. The hob is fed incrementally during each revolution of the gear blank until the whole depth of the teeth is achieved. Hobbing is completely automated and is much more productive than shaping because it operates continuously compared to the shaping process that consists of cutting and return strokes.

There are other tooth manufacturing processes based on the method of tooth generating.

In many mechanical engineering applications, the surfaces of the teeth are subjected to thermal treatment in order to obtain the desired surface hardness. Thermal treatment of metal components causes warping of the material that results in the geometric distortion of the components. Usually after thermal treatment, the machine elements should undergo finishing operations. The finishing operations comprise shaving, honing, lapping, burnishing, grinding, and others. The application of finishing processes depends, among other factors, on the surface hardness achieved during the thermal treatment. Finishing processes are based on the method used for generating the teeth.

5-5. Gear Trains

Very often, design considerations impose limits on the applicability of a set of two gears solving an actual problem. For example, transmitting power between two remote axes requires a pair of gears having large diameters that need a relatively large space (not to mention the high cost of manufacturing these gears). Another example is the necessity of a relatively high velocity ratio that would be very problematic for a pair of gears to achieve. The solution for these problems is the use of certain structures of gear transmissions.

Mechanical engineering systems comprise a variety of structural combinations of toothed drives for transmitting power. These structural combinations of gear drives are called gear trains, and a pair of meshed gears is the simplest gear train. The calculations of a gear train determine the velocity ratio of the train or, in other words, the velocity ratio between the input and output shafts. This velocity ratio of a gear train is called the train value or the train ratio. The variety of gear train combinations can be analyzed by considering the typical gear arrangements that compose these structural combinations. The typical arrangements are classified as simple gear trains, compound gear trains, and epicyclic, or planetary, gear trains.

It should be mentioned that transmitting power by a gear train is associated with power losses that are caused by the frictional interaction between the gears and between the components of the bearing. In enclosed transmissions, there are additional power losses associated with the mixing of the oil. Power loses can vary from 1% to 4% of the input power. The loses in one pair of gears, accounting for loses in the bearings and oil mixing, normally does not exceed 5%.

5-5.1. Simple Gear Train

Figure 5-5.1 shows an example of a simple gear train that is characterized by an arrangement of several gears mounted on their shafts. The intermediate gears can be engaged with two neighboring gears, and the input and output gears are engaged with one neighboring gear.

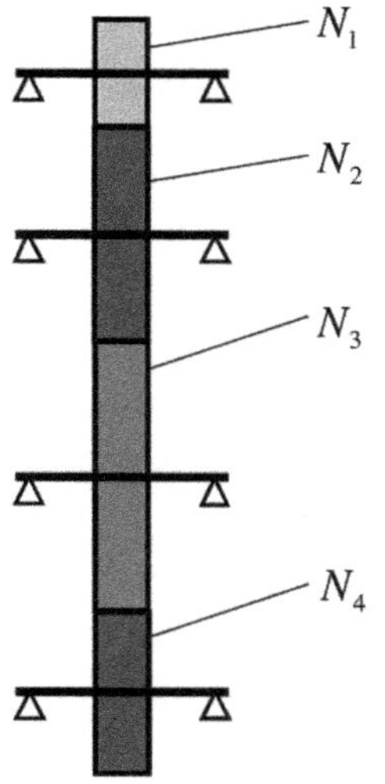

Figure 5-5.1. Simple gear train.

The velocity ratio between the first and second gears is $i_{12} = \frac{n_1}{n_2} = \frac{N_2}{N_1}$. Hence, the rotational speed of the second gear is $n_2 = \frac{n_1 N_1}{N_2}$. The velocity ratio between the second and third gears is $i_{23} = \frac{n_2}{n_3} = \frac{N_3}{N_2}$; therefore, the rotational speed of the third gear is $n_3 = \frac{n_2 N_2}{N_3}$. Replacing n_2 with its value, we have $n_3 = \frac{n_1 N_1}{N_2} \cdot \frac{N_2}{N_3} = \frac{n_1}{i_{12} i_{23}}$ that shows that the velocity ratio between the first and third gears is $i_{13} = \frac{n_1 i_{12} i_{23}}{n_1} = i_{12} i_{23}$. Therefore, the velocity ratio of this gear train can be calculated in the following way: $i_{14} = i_{12} i_{23} i_{34} = \frac{N_2 N_3 N_4}{N_1 N_2 N_3} = \frac{N_4}{N_1}$. Since the velocity ratio i_{14} represents the train value i_{tv} of this gear train, we may write that $i_{tv} = \frac{N_4}{N_1}$. Hence, the train value of a simple train is the ratio of the number of teeth of output and input gears.

If the direction of the rotation of the input and output gears is the same, than the train value is positive, otherwise it is negative. The simple gear train presented in Figure 5-5.1 contains four gears that create three engagements. Because the number of engagements is relatively small, we can see that input and output gears are rotating in the opposite directions; therefore, the train value is negative. Multiplying the train value calculated above by a coefficient $(-1)^3 = -1$, we determine the sign of this train value. Based on the above considerations, if a simple train consists of j gears, the train value is

$$i_{tv} = \frac{N_j}{N_1}(-1)^{j-1} \tag{5-5.1}$$

The intermediate gears in a simple train are called idlers. The use of simple trains is limited to cases of transmitting rotation to a distant shaft where a belt drive is not applicable or for some other cases.

5-5.2. Compound Gear Trains

If one of the shafts of a gear train carries two gears with a different number of teeth, this gear train is a compound train. Figure 5-5.2 shows a compound gear train comprising input and output gears and three shafts that each carries two gears with a different number of teeth.

The train value for this compound train is the ratio between the rotational speeds of the input and output gears. Assuming that the rotational speeds of the input and output shafts are known, we may write that $i_{tv} = i_{18} = \frac{n_1}{n_8}$. Unlike simple trains where the intermediate gears do not influence the value of the velocity ratio, in compound trains all the gears should be accounted for in order to calculate the train value. The velocity ratio i_{18} equals the product of multiplying the velocity ratios of each pair of the series of gears in the train. Hence, we write that $i_{18} = i_{12} i_{34} i_{56} i_{78} = \frac{N_2 N_4 N_6 N_8}{N_1 N_3 N_5 N_7}$. In this expression, the numerator represents a product of multiplying the number of teeth of all driven gears, while the denominator equals the product of

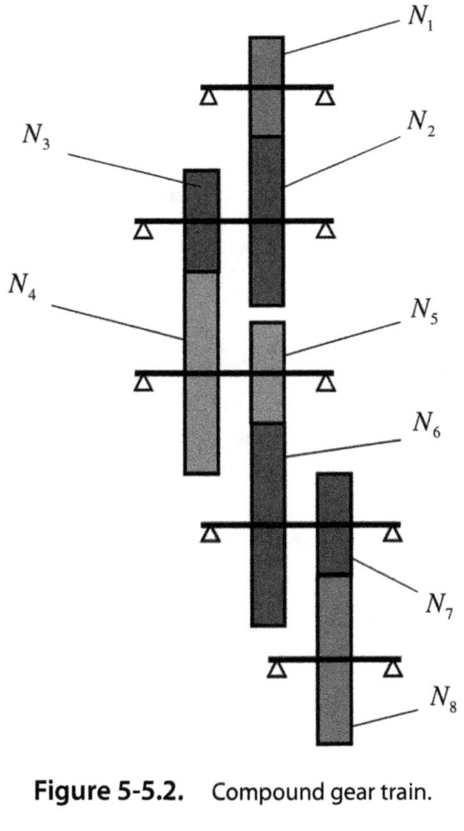

Figure 5-5.2. Compound gear train.

multiplying the numbers of teeth of all driving gears. Following the accepted order of gear indication, we see that the even numbers of gears are the driven gears, while the odd numbers are the driving gears. Based on the considerations above, we may present the general formula for calculating the train value for a compound train having j gears (j being an even number):

$$i_{tv} = \frac{N_2 N_4 \cdots N_j}{N_1 N_3 \cdots N_{j-1}} (-1)^e \tag{5-5.2}$$

where e is the number of external engagements that in many cases $e = 0.5\,j$.

In multi-step gearboxes, it is desirable that the input and output shafts have the same geometric axes. This type of gear train is called reverted. Figure 5-5.3 shows a reverted compound gear train where the input gear 1 and output gear 4 have the same geometric axis.

The reason behind this structure of the gearbox is the possibility of machining the hole for the bearings in one setup.

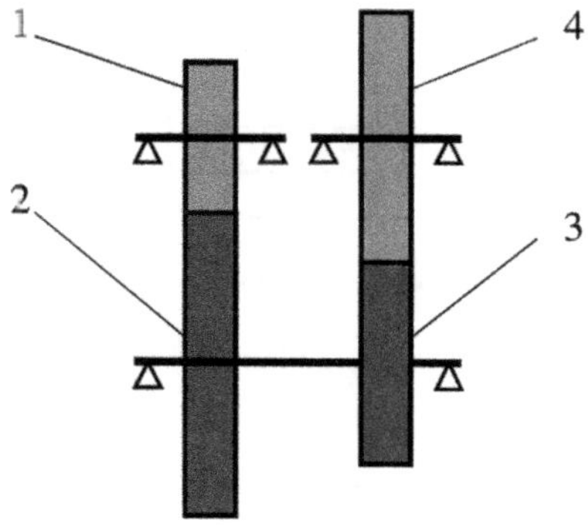

Figure 5-5.3. Reverted compound gear train.

5-5.3. Planetary, or Epicyclic, Gear Train

Gear trains having one or more gears whose geometric axes rotate about other axes are planetary, or epicyclic, gear trains. These gear trains have many advantages compared to compound gear trains. Among these advantages is their ability to deliver extremely high velocity ratios, reaching 10,000 and more. Planetary gear trains allow control of the output of the rotational speed; they are relatively compact and lightweight. The operation of these trains is smooth and quiet. In addition, they are capable of transmitting relatively high power. However, they require highly accurate manufacturing and are more complicated to maintain. There are numerous structures of planetary trains, and they differ in complexity and performance. One of a relatively simple example of a planetary gear train is shown in Figure 5-5.4.

This planetary gear train comprises the following links: gear 1, the sun gear; gear 2, the planet gear; gear 3, the ring gear having an internal engagement; and the arm or carrier 4. The gears 1, 2, and 3 have the numbers of teeth N_1, N_2, and N_3 and rotate at angular velocities ω_1, ω_2, and ω_3, respectively. Arm 4 carries gear 2 and rotates at angular velocity ω_4 as the arrow

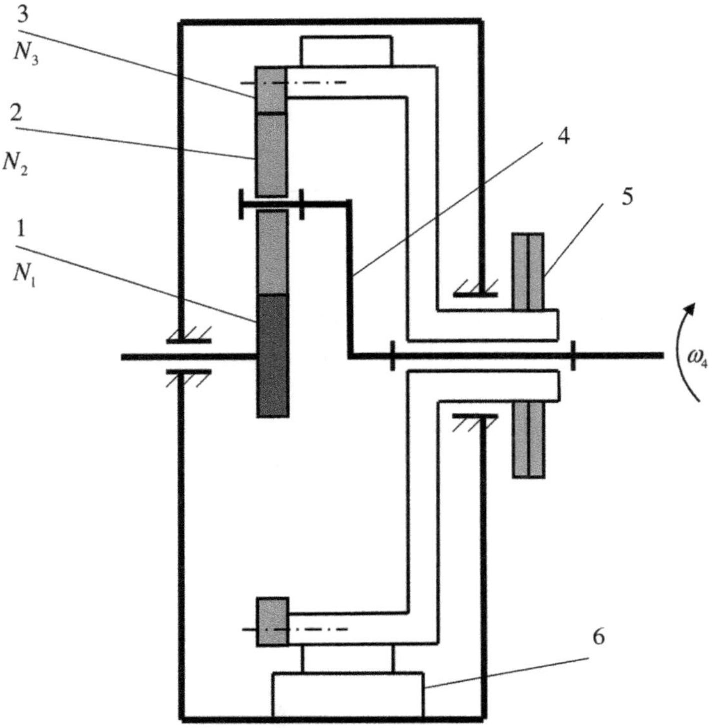

Figure 5-5.4. Planetary gear train.

in figure 5-5.4 indicates. The motion of the planet gear (the arm very often carries three or more planet gears) around the sun gear resembles the motion of the planets around the sun. This explains the reason behind the name of this type of gear train. Gear 3 carries a clutch 5, and, obviously, they both rotate with the same angular velocity ω_3. The clutch could be used for braking the gear 3 or to deliver to this gear a certain rotational speed. In some trains, a band brake 6 performs the braking of gear 3. Very often, the ring (gear 3) is rigidly connected to the enclosure. In this case, the planetary gear train represents a one-degree-of-freedom mechanical system. Gear 1 is usually the input link; however, when gear 3 is not braked (is free to rotate) the planetary gear train has two independent output links: the ring gear 3 and the arm 4. In this case the train represents a two-degree-of-freedom mechanical system. The rotational movement of these two output links could be altered independently according to certain requirements.

Usually the shaft of the arm indicated by the arrow in Figure 5-5.4 plays the role of the output link. In order to calculate the angular velocity of this output shaft or the velocity ratio between the input and output shafts, we have to impose a certain desirable condition on the other output link—to the ring gear. It should be clear that we have to know the characteristics of motion of one of these two output links, then we can determine the characteristics of motion of the other link. This is because we can compose just one equation that describes the relationships between the links of this train and, therefore, we can determine just one unknown. In order to compose this equation, we use the method of R. Willis that is based on analyzing the relative angular velocities between gears 1, 2, and 3 relative to the arm. Thus, the relative angular velocity of any of the rotational links (1, 2, or 3) relative to the arm 4 equals the difference between the absolute value of the angular velocity of a certain link and the absolute value of the angular velocity of link 4. Hence, the angular velocities of the links 1, 2, and 3 relative to link 4 can be presented as $\omega_1 - \omega_4$, $\omega_2 - \omega_4$, and $\omega_3 - \omega_4$, respectively. Therefore, the velocity ratio between gear 1 and gear 3 reads $i_{13} = \frac{\omega_1 - \omega_4}{\omega_2 - \omega_4} \cdot \frac{\omega_2 - \omega_4}{\omega_3 - \omega_4} = \frac{\omega_1 - \omega_4}{\omega_3 - \omega_4}$. This expression yields

$$\omega_1 = i_{13}\omega_3 + \omega_4(1 - i_{13}) \tag{5-5.3}$$

Based on the number of teeth in gears 1, 2, and 3, we determine $i_{13} = \frac{N_2 N_3}{N_1 N_2} = \frac{N_3}{N_1}$. Combining this expression with formula (5-5.3), we obtain

$$\omega_1 = \frac{N_3}{N_1}\omega_3 + \omega_4\left(1 - \frac{N_3}{N_1}\right) \tag{5-5.4}$$

It should be emphasized that equation (5-5.4) contains two unknowns: w_3 and w_4, while the solution of this equation is possible if we assign the value of one of the unknowns, which is usually related to the value of the angular velocity of the ring (very often $\omega_3 = 0$).

Solving equation (5-5.4) for the angular velocity of the arm, we may write

$$\omega_4 = \frac{\omega_1 - \omega_3 \frac{N_3}{N_1}}{1 - \frac{N_3}{N_1}}(-1) \tag{5-5.5}$$

The multiplier (−1) relates to the number of external engagements, which in this case is 1. From the formula (5-5.5), we can determine the velocity ratio i_{14} of the planetary gear train:

$$i_{14} = \frac{\omega_1}{\omega_4} = \frac{\omega_1\left(1 - \frac{N_3}{N_1}\right)}{\omega_1 - \omega_3 \frac{N_3}{N_1}} \tag{5-5.6}$$

As it became apparent, the number of teeth of the planet gear 2 is not present in formula (5-5.6), therefore, the planet gear does not influence the train value of the planetary gear train but plays the role of an idler.

Example 5-5.3a: Planetary gear train calculations.
a) For the planetary gear train shown in Figure 5-5.4 we take the rotational speed $n_1 = 3{,}600$ rpm, $N_1 = 20$, $N_3 = 240$, and $n_3 = 280$ rpm. We need to determine the rotational speed of the arm and the velocity ratio of the planetary gear train.

Solution:

Using formula (5-5.5) we determine $n_4 = \dfrac{3{,}600 - 280\dfrac{240}{20}}{1-\dfrac{240}{20}}(-1) = 21.82$ rpm. Applying formula (5-5.6), we obtain $i_{14} = \dfrac{3{,}600}{21.82} = 165$.

b) For the same planetary gear train we take $n_3 = 0$ and determine the same parameters as in the previous example.

Solution:
Using formulas (5-5.5) and (5-5.6) we determine

$$n_4 = \frac{3{,}600}{1-\dfrac{240}{20}}(-1) = 327.3 \text{ rpm and } i_{14} = \frac{3{,}600}{327.3} = 10.999.$$

5-6. Gear Materials

Gears are predominantly made of steels; however, cast iron, bronze, and polymers are also used in gear manufacturing. More often, gears are produced from alloyed steels that contain nickel, manganese, silicon, chromium, molybdenum, tungsten, vanadium, and other alloying elements.

The presence of nickel increases the strength and hardness without compromising the ductility of the steel. Chromium, in comparison with nickel, results in higher strength, hardness, and toughness; however, it decreases the ductility. Chromium steels are characterized by high wear resistance and by good machinability. Adding manganese and vanadium to steel alloys results in greater strength and, at the same time, provides higher toughness. Alloying with molybdenum increases the strength, the toughness, and hardness without affecting the ductility. The alloying elements are present in steels in certain combinations providing desirable features to the steel. Chrome-molybdenum steels (AISI 4130) have increased strength and are easily heat treated and machined. The combination of chromium, molybdenum, and manganese in steel alloys (AISI 4140) results in high fatigue and torsional strength and high impact resistance. Steels containing molybdenum, nickel, and chromium (AISI 9310) are characterized by high levels of hardness, strength, and toughness and have good machinability.

5-7. Engineering Calculations of Gears

The objectives of engineering calculations of gears are associated with creating new gear drives (a new product) and verifying the strength of the newly developed gears or verifying the strength of existing gears. Sometimes, an existing machine element fails under normal operational conditions due to insufficient strength. In such cases, it is necessary to carry out verifying calculations that may lead to replacement of the material of the element or to a change of some geometric parameters of the element. In order to perform verifying calculations, we must know the characteristics of the forces applied to the element and its angular velocity. These characteristics could be obtained by analyzing the power and the rotational speed related to this element.

More often, engineers are involved in development of new mechanical systems. In these situations, the designers obtain the necessary initial technical information to perform the corresponding engineering calculations of the new machine elements. With regard to a meshing pair of gears as new elements, this information comprises the values of the output power and the rotational speed of the driven gear. The rotational speed of the pinion (the driving gear) and the velocity ratio between the pinion and the gear are associated with the selection of energy sources (electrical motor, engine, etc.). The engineering requirements for developing a new gear transmission normally include the information regarding the structure and desired dimensions of the assembly. For example, the overall dimensions of the transmission should not exceed certain limits, the center distance between the gears should be in a certain range, the relative orientation of the gears, and other requirements that relate to the assembly. The methodologies of design calculations for gear drives, presented in this book, allow us to calculate the appropriate parameters of the meshing pinion and gear. These methodologies are based on strength calculations of the pair—pinion and gear. The geometric parameters of the pinion and gear obtained as a result of the design calculations have a preliminary character, and they should undergo verifying calculations whose methodologies are also presented in this book.

It should be noted that the required accuracy of the geometric parameters of the meshing pair of gears with regard to their strength largely depends on the production scale of the newly developed transmission. Thus, in cases where it is required to manufacture just a few transmissions, it is justifiable to incorporate low-cost materials and apply during the design calculations an adequately elevated strength safety factor (at least 2 or higher). Based on these calculations, it is justifiable to manufacture the required limited number of drives. It is also justifiable to avoid special thermal treatments and costly machining processes, such as grinding, shaving, burnishing, etc. These drives may be slightly overweight and have slightly increased overall dimensions, but this should not present any problems. However, in cases of mass production, it is justifiable to use high-quality materials, appropriate thermal treatments, and corresponding surface finish. The results of the design calculations should undergo verifying calculations based on the most precise available methods. Normally, before mass production it is a common practice to manufacture a pilot group of gear drives and carry out extensive testing of these drives. After successful testing, mass production begins.

We considered just two possible scenarios associated with the design of a gear transmission, while the decisive factor for these scenarios is associated with the required quantity of the transmissions. In real-life, different situations could take place regarding the design of gear transmissions, and different decisive factors should be accounted for before accepting the design and development strategy for these transmissions. However, it seems that the approach discussed above to apply certain design and development considerations with regard to the gear transmission could represent some practical interest.

With regard to the most precise methods of verifying calculations, it should be underlined that in many countries national standards for calculations of gear strength are accepted. For example, the International Standardization Organization (ISO) offers its standards; in the United States the standards of gear strength calculations are developed by the American Gear Manufacturers Association (AGMA); in Germany by the German Institute for Standardization (DIN); in Japan by the Japanese Gear Manufacturers Association (JGMA); in the United Kingdom by the United Kingdom Gear Standards, and others. It is advisable for the corresponding cases, specifically in cases of mass production, to apply the methodologies of verifying calculations based on AGMA standards or other national standards that are widely published.

A significant role in the efficiency of the operational processes of gears involves the level of manufacturing accuracy of the gears. Thus, the level of accuracy of gear manufacturing is associated with altering certain stresses in the teeth and the smoothness of the engagement. Several countries have accepted gear quality standards that reflect the accuracy of gear manufacturing. For example, AGMA has standards for gear quality, while ISO has standards for gear accuracy. AGMA developed standards for quality grades from grade Q3 through grade Q15—the higher the quality grade, the higher the accuracy of manufacturing. According to ISO standards, the accuracy level decreases from class 0 to class 13. Some empiric coefficients that are used for stress calculations of the teeth according to AGMA, ISO, and other methodologies reflect the accuracy level of the gears. The values of these coefficients could be determined from the widely published standards mentioned above. It is obvious that the accuracy of gear production is proportional to the cost of gear manufacturing. In addition, it is clear that different applications of gear transmissions require different levels of accuracy. This text does not deal with the accuracy of gear drives.

It should be emphasized that in order to compensate for the inaccuracies of the values of the empiric coefficient associated with the stress calculations of the teeth, applying elevated values of strength related safety factors, the minimum value of which should be not less than 2, is recommended.

National and international standards are available in numerous publications. The engineering practitioner could choose the corresponding standards for verifying calculations of the strength of the teeth of a gear transmission.

This book presents a general approach to the methodologies of design calculations, as well as verifying strength calculations of gears based on conventional theoretical and experimental studies related to the strength calculations of teeth subjected to variable loading and experiencing bending and surface (contact) stresses.

In general, the teeth of a gear are subjected to fluctuating bending and contact stresses. If the rotation of a gear is one directional, the bending stresses represent a repeated cycle; while if the rotation is reversible, the bending stresses are acting according to the completely reversed cycle. Contact stresses are always characterized by a repeated cycle.

Steel gears have an endurance limit with regard to bending stresses and, therefore, could be calculated for infinite life. However, steels do not have an endurance limit for the contact stresses and, consequently, each gear may eventually fail if subjected to certain number of loading cycles. The failure due to contact stresses is called pitting and represents the distortion of the contacting surfaces that become covered by black points. These black points are solidified oil impregnations that filled very small recesses of missing material that resembles, in some way, very small scales. The pitting process rapidly destroys the surface. Appropriate strength calculations determine the duration of the normal service life of a component subjected to contact stresses.

There are two types of gear transmissions: open and enclosed. In an open transmission, the gears do not have an adequate lubrication and are exposed to the environment containing abrasive dust. The surfaces of teeth in open drives are subjected to an intensive wear, and often the surfaces are worn out before the pitting occurs. Usually, these transmissions fail due to the fracture of the head of the tooth.

Enclosed transmissions are gearboxes in which the gears receive adequate oil lubrication and are protected from abrasive dust. The teeth surface wear in these transmissions is minimal, and the tooth failure occurs due to contact stresses.

5-7.1. Forces in Spur Gear Engagement

In order to perform stress calculations, we first define the loading factors applied to the teeth. Figure 5-7.1 shows an engagement of pinion 1 with gear 2 having the pressure angle ϕ. The pinion rotates at the angular velocity ω_1 transmitting power to the gear that is forced to rotate at the angular velocity ω_2.

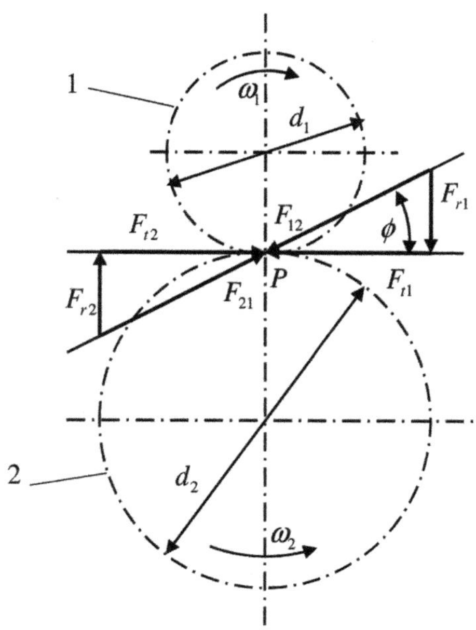

Figure 5-7.1. Forces in spur gear engagement.

The pinion exerts a force F_{12} applied to the gear in the pitch point P and is directed along the pressure line. The gear develops a reaction F_{21} that opposes the force F_{12}. These forces are resolved into tangential forces F_{t1} and F_{t2}, and radial forces F_{r1} and F_{r2} that are equal. Hence, $F_{12} = F_{21} = F$, $F_{t1} = F_{t2} = F_t$, and $F_{r1} = F_{r2} = F_r$. The tangential force F_t develops the torques $T_1 = 0.5F_td_1$ and $T_2 = 0.5F_td_2$. The relationships between these forces can be expressed by the following equations: $F_t = F\cos\phi$ and $F_r = F\sin\phi$. Since the information related to the power, rotational speed of the pinion, and the velocity ratio between the pinion and gear should be known at the beginning of the strength calculations, we can use formula (5-3.4) to calculate the minimum number of teeth of the pinion N_1, and consequently, determine the number of teeth in the gear: $N_2 = iN_1$. Calculating the pitch diameters for the pitch system or module system gears according to the formulas $d = \dfrac{N}{P}$ or $d = mN$, we can determine the tangential force applied to the gears using the following formula:

$$F_t = \frac{2W}{\omega d} \tag{5-7.1}$$

where W is the power. In the preceding chapters, we use P for denoting power, but it is widely accepted in gearing to use this letter for denoting the pitch. Therefore, in order to avoid some confusion we use W instead of P to designate the power. It is obvious that we should substitute $\omega_1 d_1$ or $\omega_2 d_2$ into formula (5-7.1), while the result will be the same in both cases. With regard to the radial force we may write

$$F_r = F_t \tan\phi \tag{5-7.2}$$

We calculate the pressure force acting along the pressure line using the following formula:

$$F = \frac{F_t}{\cos\phi} \tag{5-7.3}$$

Manufacturing of gears and performing their adjustments during assembly in a gear drive are associated with inevitable inaccuracies that cause misalignment between the meshing teeth and results in additional stresses in the teeth. The process of engaging teeth at relatively high rotational speeds creates some dynamic loading applied to the teeth. Some random fluctuating of the nominal loading causes additional stresses in the teeth. It should be noted that there are additional irregularities in the operation of a pair of gears that represent certain problems for the strength of the teeth. It is very difficult to develop an analytical methodology to account for all of these problems. Usually, the influence of these kinds of problems are accounted for in the engineering calculations by applying empiric coefficients and factors whose approximate values are determined based on actual data. Sometimes these coefficients and factors relate to the loads applied to the teeth, resulting in the increase of the loading, while sometimes these coefficients and factors are associated with the characteristics of the strength resulting in a decrease of the allowable strength of the material.

Since the teeth are subjected to variable loading, it is necessary to account for the stress concentration that occurs near the root of the tooth. The theoretical value of the coefficient related to the stress concentration in many cases can be determined analytically based on the theory of elasticity; however, the fatigue stress concentration factor is determined using an empiric formula.

Numerous experimental data show that the load (force) applied to the teeth should be increased by multiplying it by several empiric factors. It is acceptable to account for the random fluctuating of the nominal load by the random loading factor k_r, the value of which according to numerous published sources might be taken in the range of 1.2 to 1.4. The effect of the continually changing load applied to the teeth during their engagement should be accounted for by the dynamic loading factor k_d that according to available empiric data for average operational velocities is in the range of 1.3 to 2.0. The misalignment of the teeth in engagement can be accounted for by a misalignment factor k_m that according to different published sources averages from 1.1 to 1.3. In order to simplify the calculations that are intended to provide experience to the reader, it is acceptable for this book to take for spur gears the following values of the factors mentioned above: $k_r = 1.25$, $k_d = 1.6$, and $k_m = 1.15$.

The values of these coefficients are recommended just for preliminary engineering calculations that allow estimating the approximate values of the geometric parameters of the gears and of the teeth, and consequently, also of the characteristics of their strength. For more accurate results, it is recommended to perform verifying calculations based on the corresponding American Gear Manufacturers Association (AGMA) standards or other national standards related to gearing. It may be also recommended to apply where appropriate the computer software programs to refine the obtained results.

5-7.2. *Bending Stresses in Spur Gears Subjected to Static Loading*

The analytical approach to determine the bending stresses in gear teeth was carried out by W. Lewis, who in 1892 published his formula that still represents the basis for engineering calculations of teeth.

In operational conditions, the forces acting in a pair of teeth are applied to the point of contact of the teeth, and this point moves along the pressure line, while at the stage of the end of the engagement of the pair of teeth the next pair of teeth is already engaging and shares the working load. This presents a considerable challenge to develop a methodology of stress calculation that reflects the process in all its stages. The advantage of Lewis' analytical approach consists in consideration of a hypothetical condition where the pressure force is applied in the vicinity of the tip of the tooth, creating the maximum stress in the cross-sectional area of the bottom of the tooth. Figure 5-7.2a presents a tooth loaded in a point that is close to its tip by the force F that is parallel to the force acting along the pressure line.

This force is resolved into two components representing the tangential F_t and radial F_r forces. The most loaded cross-sectional area of the tooth is shown by a dashed line and is located on the distance l from the tangential force. The width of the tooth in this area equals h. The tangential force develops bending stresses in this cross-sectional area, the diagram of which is shown in Figure 5-7.2b. The radial force compresses the tooth imposing compression stresses σ_c

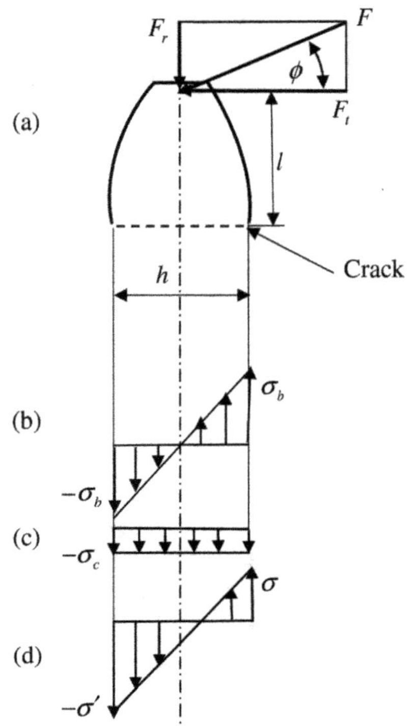

Figure 5-7.2. Normal stresses in a tooth.

as shown on the diagram in Figure 5-7.2c. The superposition of the bending and compression stresses is presented in the diagram in Figure 5-7.2d. This diagram shows that the compression stress σ' has the maximum absolute value. However, in reality the teeth fail due to cracks on the sides of the teeth that are subjected to tension, as shown by the arrow in Figure 5-7.2a. This could be explained by the ability of materials to have a higher strength on compression than on tension, and on the influence of stress concentration that is significantly more destructive at tension than at compression.

Therefore, the bending stress calculation is performed for the stress σ. Accounting for the load correction coefficients, we may write $\sigma = \sigma_b - \sigma_c = k_r k_d k_m \left(\frac{12F_t l 0.5h}{bh^3} - \frac{F_r}{hb} \right)$, where b is the tooth face width (Figure 5-2.2). Combining this equation with formula (5-7.2) and performing conventional transformations, we have

$$\sigma = k_r k_d k_m \frac{F_t}{b} \left(\frac{6l - h \tan \phi}{h^2} \right) \tag{5-7.4}$$

The expression (5-7.4) can be presented in the following way:

$$\sigma = k_r k_d k_m \frac{\pi F_t}{pb} \left[\frac{p(6l - h \tan \phi)}{\pi h^2} \right] \tag{5-7.5}$$

where p is the circular pitch. The expression in the brackets reflects the relationship between the geometric parameters of the teeth and does not depend on the diametral pitch or module; however, it depends on the number of teeth of the gear. Denoting the expression in the brackets by $\frac{1}{y}$, where y is the tooth form factor, we may write

$$\frac{1}{y}=\frac{p(6l-h\tan\phi)}{\pi h^2} \tag{5-7.6}$$

For given values of the pair of parameters l and h, formula (5-7.6) allows plotting of graphs to determine the tooth form factor depending on the number of teeth of the gear. However, determining the values of these pairs is associated with certain difficulties. For spur gears, the tooth form factor can be determined with adequate accuracy using the following simplified formula that is presented in published sources:

$$y=0.52\left(1-\frac{5.6}{N}\right) \tag{5-7.7}$$

The minimum number of teeth that prevents the interference phenomenon of spur gears can be calculated, according to some authors, using formula (5-3.5); hence, in cases when $\phi = 20°$, we have $N_{min} = 17$. It is recommended to take $N_{min} \geq 18$. According to formula (5-7.7), for a 17-tooth gear $y = 0.349$. Increasing the number of teeth results in the increase of the tooth form factor and a corresponding decrease in the stress of the tooth. The maximum value of the tooth form factor according to formula (5-7.7) cannot exceed 0.52. This indicates that the increase of the number of teeth of the pinion is more important for the pinion than for the gear since the gear always has a significantly larger number of teeth than the pinion.

Combining expressions (5-7.1), (5-7.5), and (5-7.6) we obtain

$$\sigma=\frac{2\pi Wk_rk_dk_m}{\omega dpby} \tag{5-7.8}$$

Substituting into formula (5-7.8) and into the similar formulas below ω_1, d_1, and y_1 or ω_2, d_2, and y_2 we calculate the stress σ_p in the pinion or in the gear σ_g. For pitch system expression (5-7.8) for the pinion reads

$$\sigma_p=\frac{2WPk_rk_dk_m}{\omega_1 d_1 by_1} \tag{5-7.9}$$

while for module system we obtain

$$\sigma_p=\frac{2WPk_rk_dk_m}{\omega_1 md_1 by_1} \tag{5-7.10}$$

The relationship between the bending stresses in the gear and in the pinion is expressed by the following equation:

$$\sigma_g = \sigma_p \frac{y_1}{y_2} \tag{5-7.11}$$

Equations (5-7.9) and (5-7.10) are derived for static loading. However, as mentioned above, the teeth are subjected to variable stresses according to a repeated loading cycle if the gears are rotating in one direction only or to the completely reversed loading cycle if the operation of the gears is reversible. In both cases, the stress calculations should be performed accounting for the fatigue considerations. It should be emphasized that the methodologies of stress calculations for a repeated cycle and a completely reversed cycle are different and each cycle is discussed separately.

5-8. Verifying Calculations for Bending Stresses in Spur Gears Subjected to Variable Loading

The formulas (5-7.9) and (5-7.10) would be acceptable if the service life of the pinion is limited to 1000 cycles. However, during the long service time that may be characterized by the number of cycles exceeding 10^3, the teeth of the gears are subjected to variable stresses and the stress calculations should include fatigue considerations. The stress concentration that occurs at the root of the teeth plays a significant role in cases of variable stresses and should be accounted for in formulas (5-7.9) and (5-7.10) by applying the fatigue stress concentration factor K_f. The theoretical stress concentration factor that depends on the ratio between the fillet radius r_f and the thickness of the tooth h can be determined from Figure 5-8.1a, where the structure of the sample subjected to stress concentration resembles the structure of the tooth in Figure 5-8.1b.

The tooth thickness at the pitch diameter can be determined for a pitch system using the formula $s = \frac{0.5\pi}{P}$, while for a module system $s = 0.5\pi m$. However, the thickness of the teeth in

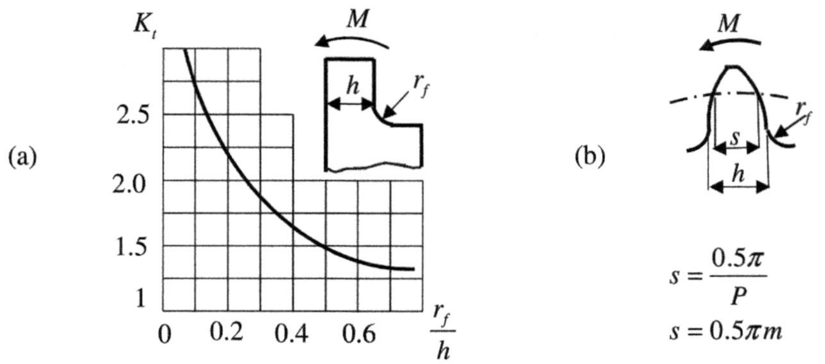

Figure 5-8.1. Theoretical stress concentration factor.

the section passing through the root of the tooth may exceed the thickness of the tooth at the pitch circle. The fillet radius varies from $\frac{0.157}{P}$ to $\frac{0.3}{P}$ or from 0.157m to 0.3m, more often the fillet radius $r_f = \frac{0.3}{P}$ or $r_f = 0.3m$. For approximate estimation of the ratio $\frac{r_f}{h}$, we assume that it may vary from 0.1 to 0.2, while in some cases to 0.3. This corresponds to the values of the theoretical stress concentration factor K_t from 1.75 to 2.75. Taking the notch sensitivity factor $q = 0.6$, we determine using formula (1-10.2) that the fatigue stress concentration factor K_f ranges from 1.45 to 2.05.

It should be emphasized that in cases when the gears are rotating just in one direction the teeth are subjected to a repeated loading cycle, while in cases of a reversible drive the teeth are subjected to a completely reversed loading cycle. The verifying calculations for these cycles differ in the applied load and the fatigue related allowable stresses. All these differences are addressed in the following subsections.

5-8.1. Verifying Calculations of Bending Stresses for Repeated Cycles

In this cycle, the mean stress and the stress amplitude are equal, hence $\sigma_m = \sigma_a$ because $\sigma_{\min} = 0$ and $M_{\min} = 0$ (see Chapter 1 for details). The bending stresses for the pinion in this case are calculated based on the amplitude value of the bending moment M_a that is equal

$$M_a = \frac{M_{\max}}{2} \tag{5-8.1}$$

and the stress amplitude limit S_a is determined from formula (1-12.14) and repeated here:

$$S_a = \frac{S_e S_{ut}}{S_{ut} + S_e} \tag{5-8.2}$$

Based on these considerations, the formulas (5-7.9) and (5-7.10) for calculating bending stresses in the pinion σ_{rc} for the repeated cycle should be modified to the following:

$$\sigma_{rc} = \frac{WPk_r k_d k_m K_f}{\omega_1 d_1 b y_1} \leq \frac{S_a}{n_b} \tag{5-8.3}$$

and

$$\sigma_{rc} = \frac{Wk_r k_d k_m K_f}{\omega_1 m d_1 b y_1} \leq \frac{S_a}{n_b} \tag{5-8.4}$$

where n_b is the safety factor with regard to bending stresses, while the formulas (5-8.3) and (5-8.4) are applicable for pitch and module system gears, respectively.

It should be underlined that before applying the approach based on the Modified Goodman line methodology discussed in Chapter 1 to the nonzero mean stress cycles, it is necessary to compare the values of tanα and tanβ that are calculated according to formulas (1-12.4) and

(1-12.5), respectively. If in cases when tan α < tan β, the approach mentioned above is applicable. However, for a repeated cycle $\tan\beta = \frac{\sigma_a}{\sigma_m} = 1$. Analyzing formula (1-12.4) shows that the fatigue test sample has the maximum value of tanα, since S_e = 0.5 S_{ut}, hence, $\tan\alpha_{\max} = \frac{0.5(S_{ut} - S_{yt})}{S_{yt} - 0.5S_{ut}}$. Taking a realistic relationship between S_{ut} and S_{yt}, it becomes clear that tan $\alpha_{\max}$ < 1. Therefore, for repeated cycles it is justifiable to apply the above mentioned methodology without calculating the actual value of tanα.

In order to avoid possible confusion when performing calculations for pitch system gears and module system gears, it is highly recommended to use the corresponding formulas for the U.S. Customary units and the SI units, respectively.

Example 5-8.1a: Verifying calculations for a repeated cycle of bending stresses for a pitch system.

A 22-tooth pinion having a diametral pitch $P = 16\ \frac{1}{\text{in.}}$ transmits power of 10 hp at a rotational speed of n_1 = 1,800 rpm. The tooth face width b = 0.8 in. We need to calculate the bending stresses in the teeth of the pinion. The rotation is one directional.

Solution:

We have to determine the angular velocity, the pitch diameter, and the tooth form factor for the pinion. Therefore, $\omega_1 = \frac{\pi 1{,}800}{30} = 188.4\ \frac{\text{rad}}{\text{s}}$, $d_1 = \frac{22}{16} = 1.375$ in., and using formula (5-7.7), we have $y_1 = 0.52\left(1 - \frac{5.6}{22}\right) = 0.388$. Based on Figure 5-8.1, we estimate that K_t = 1.75, therefore, using formula (1-10.2), we calculate K_f = 1.45. Taking k_r = 1.25, k_d = 1.6, and k_m = 1.15 and using formula (5-8.3), we calculate $\sigma_{rc} = \frac{10 \cdot 550 \cdot 12 \cdot 16 \cdot 1.25 \cdot 1.6 \cdot 1.15 \cdot 1.45}{188.4 \cdot 1.375 \cdot 0.8 \cdot 0.388} = 43{,}799$ psi.

Example 5-8.1b: Verifying calculations for a repeated cycle of bending stresses in a module system.

A 22-tooth pinion having a module m = 1.75 mm is transmitting power of 3.0 kW at 1,800 rpm and is rotating in one direction. The tooth face width b = 20 mm. The pinion is made from AISI 4150 normalized steel having S_{ut} = 800 MPa. We must determine the safety factor of the pinion.

Solution:

We calculate $\omega_1 = 188.4\ \frac{\text{rad}}{\text{s}}$ and y_1 = 0.388. As it is suggested above, we take k_r = 1.25, k_d = 1.6, k_m = 1.15, and K_f = 1.45. Now we compute $d_1 = 1.75 \cdot 22 = 38.5$ mm. Using expression (1-9.2), we determine k_a = 0.777. Considering that the height of the rectangular cross-sectional area of the root of the tooth does not exceed 8 mm, we take k_b = 1. According to expression (1-9.4), we found that $k_c = 1$. We determine the endurance limit $S_e = 0.777 \cdot 1 \cdot 1 \cdot 0.5 \cdot 800 = 310.8$ MPa. Based on formula (5-8.2), we calculate $S_a = \dfrac{310.8 \cdot 800}{800 + 310.8} = 223.84$ MPa. Using expression (5-8.4), we determine

$\sigma_{rc} = \dfrac{4.0 \cdot 10^3 \cdot 1.25 \cdot 1.6 \cdot 1.15 \cdot 1.45}{188.4 \cdot 0.00175 \cdot 0.0385 \cdot 0.020 \cdot 0.388} = 101.57 \cdot 10^6$ MPa. Hence, $101.57 \cdot 10^6 = \dfrac{223.84 \cdot 10^6}{n_b}$,

therefore, $n_b = \dfrac{223.84}{101.57} = 2.2$. The safety factor for the pinion is acceptable.

5-8.2. Verifying Calculations of Bending Stresses for a Completely Reversed Cycle

For this cycle according to formula (1-12.11), we have $M_a = \dfrac{M_{\max} - (-M_{\min})}{2}$, however, since $|M_{\max}| = |M_{\min}|$, therefore, $M_a = M_{\max}$. For the completely reversed cycle, the endurance limit represents the stress amplitude limit. Hence, for this cycle the expressions for the pitch system and module system (5-7.9) and (5-7.10) can be presented, respectively, as follows:

$$\sigma_{crc} = \frac{2WPk_r k_d k_m K_f}{\omega_1 d_1 b y_1} \le \frac{S_e}{n_b} \qquad (5\text{-}8.5)$$

and

$$\sigma_{crc} = \frac{2Wk_r k_d k_m K_f}{\omega_1 m d_1 b y_1} \le \frac{S_e}{n_b} \qquad (5\text{-}8.6)$$

where σ_{crc} is the bending stress for the completely reversed cycle.

Example 5-8.2a: Verifying bending stress calculations for a completely reversed cycle of a pitch system.

A spur gear set with the diametral pitch $P = 10\dfrac{1}{\text{in.}}$ consists of a pinion and gear having $N_1 = 18$ teeth and $N_2 = 51$ teeth and transmits power of 4.0 hp. The pinion rotates at $n_1 = 1{,}800$ rpm. The teeth face width $b = 1$ in. The gears are manufactured from AISI 1340 annealed steel having $S_{ut} = 102{,}000$ psi. The gear drive is reversible. We need to determine the safety factors for the teeth of the pinion and the gear.

Solution:

First, we have to determine the endurance limit based on formula (1-9.5). Using formula (1-9.2), we calculate $k_a = 0.793$. Since the height of the rectangular cross-sectional area of the root of the tooth does not exceed 0.3 in., we take $k_b = 1$. According to formula (1-9.3), $k_c = 1$. Now we calculate the endurance limit $S_e = 0.793 \cdot 1 \cdot 1 \cdot 0.5 \cdot 102{,}000 = 40{,}443$ psi. Taking from Figure 5-8.1 since $\dfrac{r}{h} = 0.225$, we find that $K_t = 2.0$. Using formula (1-10.2) and taking $q = 0.6$, we calculate

$K_f = 1 + 0.6(2.0\text{-}1) = 1.6$. Based on the suggestion above, $k_r = 1.25$, $K_d = 1.6$, and $K_m = 1.15$. Applying formula (5-7.7), we calculate $y_1 = 0.52\left(1 - \frac{5.6}{18}\right) = 0.358$. The diameter of the pinion and its angular velocity are $d_1 = \frac{18}{10} = 1.8$ in. and $\omega_1 = \frac{\pi 1{,}800}{30} = 188.4\ \frac{\text{rad}}{\text{s}}$. The transmitted power is $W = 4.0 \cdot 550 \cdot 12 = 26{,}400\ \frac{\text{in.-lb}}{s}$. Now we calculate the stresses in the teeth of the pinion using formula (5-8.5): $\sigma_{crc} = \frac{2 \cdot 10 \cdot 26{,}400 \cdot 1.25 \cdot 1.6 \cdot 1.15 \cdot 1.6}{188.4 \cdot 1.8 \cdot 1.0 \cdot 0.358} = 16{,}004$ psi. Hence, $16{,}004 = \frac{s_e}{n_{1b}}$, therefore, the safety factor for the pinion is $n_{1b} = \frac{S_e}{16{,}004} = \frac{40{,}443}{16{,}004} = 2.526$. The safety factor for the gear can be calculated using the ratio between the tooth form factors of the gear and pinion. The tooth form factor for the gear equals $y_2 = 0.52\left(1 - \frac{5.6}{51}\right) = 0.463$. Therefore, using formula (5-7.11), we calculate $n_{2b} = \frac{y_1 n_{1b}}{y_2} = \frac{0.463 \cdot 2{,}526}{0.358} = 3.268$. The safety factors for the gears are acceptable.

Example 5-8.2b: Verifying bending stress calculations for a completely reversed cycle for a module system.

A 25-tooth pinion having a module $m = 3.5$ mm is transmitting power of 5.5 kW at 3,600 rpm. The tooth face width $b = 22$ mm. We need to determine the bending stress acting in the pinion. The drive is reversible.

Solution:

For the pinion, we calculate its angular velocity, pitch diameter, and the tooth form factor: $\omega_1 = \frac{\pi 3{,}600}{30} = 376.8\ \frac{\text{rad}}{\text{s}}$, $d_1 = 3.5 \cdot 25 = 87.5$ mm, and $y_1 = 0.52\left(1 - \frac{5.6}{19}\right) = 0.404$, respectively.

From the previous example, we take the load correction factors and the fatigue stress concentration factor k_r, k_d, k_m, and K_f. Now using formula (5-8.6), we calculate the stress

$$\sigma_{crc} = \frac{2 \cdot 5.5 \cdot 10^3 \cdot 1.25 \cdot 1.6 \cdot 1.15 \cdot 1.6}{376.8 \cdot 0.0035 \cdot 0.0875 \cdot 0.022 \cdot 0.404} = 39{,}684 \text{ MPa.}$$

5-9. Design Calculations for Bending Stresses

Engineering practice is very often associated with the design and development of new gear transmissions. Analyzing formulas (5-7.9) and (5-7.10) that are for verifying calculations, it can be seen that if the value of the tooth face width b is given, then these formulas could be modified for design calculations.

Several published sources based on numerous experimental investigations introduce an empiric coefficient that characterizes the relationship between the circular pitch and the tooth face width. For pitch-based systems, the tooth face width coefficient η_b is defined as follows:

$$\eta_b = \frac{\pi}{p} b = Pb \quad (5\text{-}9.1)$$

Similarly, we obtain for module-based systems

$$\eta_b = \frac{\pi}{p} b = \frac{b}{m} \quad (5\text{-}9.2)$$

Therefore, we may write

$$b = \frac{\eta_b}{P} = \eta_b m \quad (5\text{-}9.3)$$

For spur gears, according to factual data presented in numerous publications, the coefficient η_b ranges from 5 to 15. Obviously, the design calculations based on this coefficient should be verified in the later steps of the design.

As the first step of the design calculations, it is required to select the type of material and estimate its endurance limit. The surface factor k_a depends on the ultimate strength S_{ut} and can be calculated upon selection of the material. However, the size factor k_b depends on dimensions that become known after the results of the design calculations are obtained. It is justifiable to consider that for gears for general applications the tooth thickness at the root rarely exceeds 0.6 in. (15 mm); therefore, we recommend for the preliminary estimation of the endurance limit to take $k_b = 0.95$. In addition, since for bending $k_c = 1$, we may take for the gear design calculations that $S_e = k_a\, 0.95 \cdot 1 \cdot 0.5\, S_{ut} = 0.475\, S_{ut}$.

The considerations related to the estimation of the fatigue stress concentration factor K_f presented for the verifying calculations are applicable for design calculations.

Taking into account that $d = \frac{N}{P}$ and $b = \frac{\eta_b}{P}$, while $d = mN$ and $b = m\eta_b$, we modify the expressions (5-8.3) and (5-8.4) for bending stress calculations with the regard to the repeated cycle to the following shapes:

$\sigma_{rc} = \dfrac{PWk_r k_d k_m K_f}{\omega \dfrac{N}{P} \dfrac{\eta_b}{P} y} \le \dfrac{S_a}{n_b}$ and $\sigma_{rc} = \dfrac{Wk_r k_d k_m K_f}{\omega mmNm\eta_b y} \le \dfrac{S_a}{n_b}$. Therefore, we obtain

$$\sigma_{rc} = \frac{WP^3 k_r k_d k_m K_f}{\omega N \eta_b y} \le \frac{S_a}{n_b} \quad (5\text{-}9.4)$$

and

$$\sigma_{rc} = \frac{Wk_r k_d k_m K_f}{\omega m^3 N \eta_b y} \le \frac{S_a}{n_b} \quad (5\text{-}9.5)$$

Expressions (5-9.4) and (5-9.5) allow us to derive design calculation formulas for the repeated and completely reversed cycles with regard to bending stresses. Ones again, in all these formulas we substitute ω_1, d_1, and y_1 or ω_2, d_2, and y_2 for the pinion and the gear, respectively. These formulas are presented below.

5-9.1. Design Calculations for Bending Stresses for Repeated Cycles

Solving expressions (5-9.4) and (5-9.5) for the diametral pitch and module, respectively, we obtain the equations for design calculations for pitch and module system spur gears subjected to a repeated bending loading cycle:

$$P=\sqrt[3]{\frac{\omega_1 N_1 \eta_b y_1 S_a}{W k_r k_d k_m K_f n_b}} \tag{5-9.6}$$

and

$$m=\sqrt[3]{\frac{W k_r k_d k_m K_f n_b}{\omega_1 N_1 \eta_b y_1 S_a}} \tag{5-9.7}$$

Here we repeat the formula (5-8.2) $S_a=\dfrac{S_e S_{ut}}{S_e+S_{ut}}$ and recall that η_b should be in the range of 5 to 15.

Example 5-9.1a: Design calculation of pitch system gears for bending stresses for a repeated cycle.

We need to design an open spur gear set having a 20-tooth pinion rotating at 1,200 rpm and transmitting power of 4.0 hp to the gear that is rotating at 400 rpm. The safety factor is n_b = 2.

The rotation is one directional.

Solution:

The angular velocity of the pinion is $\omega_1=\dfrac{\pi 1{,}200}{30}=125.6\ \dfrac{\text{rad}}{\text{s}}$ and the tooth form factor according to formula (5-7.7) equals $y_1=0.52\left(1-\dfrac{5.6}{20}\right)=0.374$. We select for the pinion the AISI 4140 normalized steel having S_{ut} = 148,000 psi. According to expression (1-9.2), we calculate the factor k_a = 0.738. We take k_b = 0.95 and considering that k_c = 1, we determine

$S_e = 0.738\cdot 0.95\cdot 1\cdot 0.5\cdot 148{,}000 = 43{,}690$ psi.

Based on formula (5-8.2), we compute $S_a=\dfrac{43{,}690\cdot 148{,}000}{43{,}690+148{,}000}=33{,}732$ psi. We take

η_b = 10, k_r = 1.25, k_d = 1.6, k_m = 1.15, and K_f = 1.75.

Using formula (5.9.6), we determine

$$P=\sqrt[3]{\frac{125.6\cdot 20\cdot 10\cdot 0.374\cdot 33{,}732}{4\cdot 550\cdot 12\cdot 1.25\cdot 1.6\cdot 1.15\cdot 1.75\cdot 2}}=11.39\frac{1}{\text{in.}}$$

We take the smaller standard diametral pitch $P=11\frac{1}{\text{in.}}$. The pitch diameter of the pinion is $d_1=\frac{20}{11}=1.818$ in. The velocity ratio equals $i=\frac{1{,}200}{400}=3$, therefore, $d_2 = 3\cdot 1.818 = 5.454$ in. The width of the gears is $b=\frac{\eta_b}{P}=\frac{10}{11}=0.909$ in. Modifying formula (5-7.11) allows us to determine the ultimate strength for the gear material $S_{utg}=S_{utp}\frac{y_1}{y_2}$. Hence, $y_2=0.52\left(1-\frac{0.56}{20\cdot 3}\right)=0.515$. Therefore, $S_{utg}=\frac{0.374}{0.515}148{,}000=107{,}480$ psi. We select for the gear the AISI annealed steel having S_{ut} = 108,000 psi.

Example 5-9.1b: Design calculations of module system gears for bending stresses for a repeated cycle.

An open gear drive transmits power of 4 kW at the rotational speed of the pinion of 1,800 rpm to the gear. The number of teeth of the pinion N_1 = 18, the number of teeth of the gear is N_2 = 54, and the safety factor is n_b = 2. We need to determine he module of the gear set and select the material for the gear. The rotation is one directional.

Solution:

We select for the pinion the AISI 1340 annealed steel having S_{ut} = 703.8 MPa.

We take the values of the load factors and the fatigue stress concentration factor from the previous example: k_r = 1.25, k_d = 1.6, k_m = 1.15, and K_f = 1.75. The angular velocity of the pinion $\omega_1=\frac{\pi 1{,}800}{30}=188.4\ \frac{\text{rad}}{\text{s}}$, and the tooth form factor $y_1=0.52\left(1-\frac{5.6}{18}\right)=0.358$. Now according to expression (1-9.2), we calculate k_a = 0.793. Taking k_b = 0.95 and k_c = 1, we determine the endurance limit $S_e = 0.793\cdot 0.95\cdot 1\cdot 0.5\cdot 703.8 = 265.1$ MPa. Using formula (5-8.2), we determine $S_a=\frac{265.1\cdot 703.8}{265.1+703.8}=192.565$ MPa. Taking $\eta_b = 6$ we calculate the module using equation (5-9.7):

$$m=\sqrt[3]{\frac{Wk_rk_dk_mK_fn_b}{\omega_1N_1\eta_by_1S_a}}=\sqrt[3]{\frac{4\cdot 10^3\cdot 1.25\cdot 1.6\cdot 1.15\cdot 1.75\cdot 2}{188.4\cdot 18\cdot 6\cdot 0.358\cdot 192.565\cdot 10^6}}=0.00298\ \text{m}.$$

We have to take the closest higher standard value of the module m = 3 mm. Now we determine the tooth form factor for the gear $y_2=0.52\left(1-\frac{5.6}{54}\right)=0.466$. Modifying formula (5-7.11), we determine the required ultimate strength for the gear material. Hence,

$$S_{utg}=S_{utp}\frac{y_1}{y_2}=703.8\frac{0.358}{0.446}=564.93\ \text{MPa}.$$

We select for the gear the AISI 4620 normalized steel having S_{ut} = 574 MPa.

5-9.2. Design Calculations for Bending Stresses for Completely Reversed Cycles

Similarly to the repeated cycle, we derive design calculations formulas for gears subjected to completely reversed bending cycles:

$$P = \sqrt[3]{\frac{\omega_1 N_1 \eta_b y_1 S_e}{2Wk_r k_d k_m K_f n_b}} \tag{5-9.8}$$

and

$$m = \sqrt[3]{\frac{2Wk_r k_d k_m K_f n_b}{\omega_1 N_1 \eta_b y_1 S_e}} \tag{5-9.9}$$

The formulas for design calculations of spur gears with regard to completely reversed cycles of bending stresses in comparison with the repeated cycles contain the terms $2W$ and S_e, while η_b is in the range of 5 to 15.

Example 5-9.2a: Design calculations of pitch system gears for bending stresses for a completely reversed cycle.

For comparison, this example is based on the input data of Example 5-9.1a that deals with a repeated cycle and shows the differences between the parameters of the gears depending on the type of the cycle.

Hence, it is required to design an open gear drive having a 20-tooth pinion rotating at 1,200 rpm and transmitting power of 4.0 hp to the gear that is rotating at 400 rpm. The safety factor is $n_b = 2$. The transmission is reversible.

Solution:

The angular velocity of the pinion is $\omega_1 = \frac{\pi 1{,}200}{30} = 125.6\ \frac{\text{rad}}{\text{s}}$ and the tooth form factor according to formula (5-7.7) equals $y_1 = 0.52\left(1 - \frac{5.6}{20}\right) = 0.374$. We select for the pinion the AISI 4140 normalized steel having S_{ut} = 148,000 psi. According to expression (1-9.2), we calculate the factor $k_a = 0.738$ and we take $k_b = 0.95$ and $k_c = 1$. Hence, we compute the endurance limit: $S_e = 0.738 \cdot 0.95 \cdot 1 \cdot 0.5 \cdot 148{,}000 = 43{,}690$ psi.

We take $\eta_b = 10$, $k_r = 1.25$, $k_d = 1.6$, $k_m = 1.15$, and $K_f = 1.75$. Using formula (5.9.8), we calculate $P = \sqrt[3]{\frac{125.6 \cdot 20 \cdot 10 \cdot 0.374 \cdot 33{,}732}{2 \cdot 4 \cdot 550 \cdot 12 \cdot 1.25 \cdot 1.6 \cdot 1.15 \cdot 1.75 \cdot 2}} = 9.857\frac{1}{\text{in.}}$. In Example 5-9.1a we have $P = 11.39\frac{1}{\text{in.}}$, therefore, $\frac{11.39}{9.857} = 1.155$, which means that in this case for a completely reversed cycle the diametral pitch should be about 15% smaller than for a repeated cycle.

We take the smaller standard diametral pitch $P = 9\ \frac{1}{\text{in.}}$. The pitch diameter of the pinion is $d_1 = \frac{20}{9} = 2.222$ in. The velocity ratio equals $i = \frac{1{,}200}{400} = 3$, therefore, $d_2 = 3 \cdot 2.222 = 6.666$ in. The face width of the gears is $b=\frac{\eta_b}{P}=\frac{10}{9}=1.111$ in. Modifying formula (5-7.11), we determine the required ultimate strength for the gear material: $S_{utg}=S_{utp}\frac{y_1}{y_2}$. Now we calculate the gear tooth form factor: $y_2=0.52\left(1-\frac{0.56}{20\cdot 3}\right)=0.515$. Hence, $S_{utg}=\frac{0.374}{0.515}148{,}000=107{,}480$ psi. We select for the gear the AISI 4340 annealed steel having S_{ut} = 108.000 psi.

Example 5-9.2b: Design calculations of module system gears for bending stresses for a completely reversed cycle.

This example allows us to compare the modules of the repeated and completely reversed cycles because we are using the same input data as in example 5-9.1b, which deals with a repeated cycle.

It is needed to design an open gear drive transmitting power of 4 kW with the rotational speed of the pinion of 1,800 rpm to the gear. The number of teeth of the pinion N_1 = 18, and the safety factor is 2. The transmission is reversible.

Solution:

We select for the pinion the AISI 1340 annealed steel having S_{ut} = 703.8 MPa.

We take the values of the load factors and the fatigue stress concentration factor from the previous example: k_r = 1.25, k_d = 1.6, k_m = 1.15, and K_f = 1.75. The angular velocity of the pinion $\omega_1=\frac{\pi 1{,}800}{30}=188.4\ \frac{\text{rad}}{\text{s}}$, and the tooth form factor $y_1=0.52\left(1-\frac{5.6}{18}\right)=0.358$. Now according to expression (1-9.2), we calculate k_a = 0.793 and we take k_b = 0.95 and k_c = 1, and determine the endurance limit S_e = 0.793·0.95·1·0.5·703.8 = 265.1 MPa. Taking η_b = 6 we calculate the module using equation (5-9.9): $m=\sqrt[3]{\frac{2Wk_rk_dk_mK_fn_b}{\omega_1N_1\eta_by_1S_e}}=\sqrt[3]{\frac{2\cdot 4\cdot 10^3\cdot 1.25\cdot 1.6\cdot 1.15\cdot 1.75\cdot 2}{188.4\cdot 18\cdot 6\cdot 0.358\cdot 265.1\cdot 10^6}}=0.00323$ m.

In the example 5-9.1b, we obtained that m = 0.00298 m. Therefore, $\frac{0.00323}{0.00298}=1.108$. Thus, the module in this case of a completely reversed cycle should be about 8% larger than in case of a repeated cycle.

We have to take the closest higher standard value of the module: in this case m = 3 mm.

5-10. Surface Durability Calculations for Spur Gears

As mentioned above, the teeth of gears are subjected to variable bending and contact stresses. Bending stresses are critical for open transmissions, while contact stresses are the predominate reason for the failure of enclosed gear transmissions.

Contact stresses, also called Hertzman stresses, cause surface pitting that gradually destroys the surface of the teeth. There is no endurance limit for contact stresses; however, there is a sufficient accumulation of empiric data that supports engineering calculations for predicting loading conditions that prevent pitting during a practical and acceptable period of time. These calculations are based on the Hertz analytical investigations of the surface stresses occurring between two rolling cylinders. Some of the related equations are presented and discussed in Chapter 1. The engagement of involute teeth is an interaction of curved surfaces that are simultaneously rolling and slipping for most of the engagement. Pure rolling occurs in the pitch point of the engagement and also when the velocity ratio $i = 1$. However, it is observed that the distortion of the gear surface begins near the pitch point. This justifies the application of the Hertz theory for the analysis of the contact stresses in the teeth.

The allowable surface durability contact stresses S_H for preliminary calculations, as several authors recommend, can be taken in the range from $0.333H_B$ in ksi to $0.377H_B$ in ksi or from $2.3H_B$ in MPa to $2.6H_B$ in MPa, while $H_B \leq 350$ and the service life is up to 10^8 cycles. It should be emphasized that the contact stresses in the teeth of the pinion and the gear are the same. However, because the teeth of the pinion compared with the teeth of the gear experience an exceeding number of loading cycles, it is highly recommended that the hardness of the pinion material exceeds the mating gear material by at least 50 units of H_B.

5-10.1. Verifying Calculations of Surface Durability

Formula (1-4.30) is for two contacting cylinders having constant radii; however, the teeth contact in points that have instantaneous radii of curvature of the involute. In Figure 5-2.1, the line segments aP and bP represent the radii of curvature ρ_1 and ρ_2 in the contact point of the tooth profiles of the pinion and the gear. These radii equal $\rho_1 = \frac{d_1}{2}\sin\phi$ and $\rho_2 = \frac{d_2}{2}\sin\phi$. The pressure force in the teeth engagement is located on the common normal NN to the teeth profiles and is determined from formula (5-7.3). Accounting for all these considerations, we modify formula (1-4.30) to the following shape:

$$\sigma_H = 0.418\sqrt{\frac{4WE(i \pm 1)}{l\omega_1 d_1^2 i \cos\phi \sin\phi}} \qquad (5\text{-}10.1)$$

where E is the modulus of elasticity.

Replacing in this formula l by b and accounting that $\sin\phi\cos\phi = 0.5\sin 2\phi$, and also substituting the loading factors, we obtain

$$\sigma_H = 1.182\sqrt{\frac{WEk_r k_d k_m (i \pm 1)}{\omega_1 d_1^2 bi \sin 2\phi}} \leq \frac{S_H}{n_H} \qquad (5\text{-}10.2)$$

where S_H is the allowable contact stress and n_H is the safety factor with regard to contact stresses. If the safety factor is given, the verifying calculations allow us to select the material for the gears.

Example 5-10.1a: Verifying calculations of contact stresses for pitch system gears.
A gear drive transmits power of 5.5 hp at the rotational speed of the pinion $n_1 = 3{,}600$ rpm having the diametral pitch $P = 10\ \frac{1}{\text{in.}}$, $N_1 = 24$, $N_2 = 72$, and the tooth face width 1.5 in. Using the safety factor $n_H = 2$, we determine the material for the pinion and for the gear.

Solution:

Since the gears are made from steel, the modulus of elasticity $E = 30 \cdot 10^6$ psi. The angular velocity of the pinion is $\omega_1 = \frac{\pi 3{,}600}{30} = 376.8\ \frac{\text{rad}}{s}$. The diameter of the pinion and the velocity ratio are $d_1 = \frac{24}{10} = 2.4$ in. and $i = \frac{72}{24} = 3$. Taking the factors $k_r = 1.25$, $k_d = 1.6$, and $k_m = 1.15$, we may now use formula (5-10.2) to calculate the surface stress:

$$\sigma_H = 1.182\sqrt{\frac{5.5 \cdot 550 \cdot 12 \cdot 30 \cdot 10^6 \cdot 1.25 \cdot 1.6 \cdot 1.15(3+1)}{376.8 \cdot 2.4^2 \cdot 1.5 \cdot 3\sin 2 \cdot 20°}} = 47{,}219 \text{ psi.}$$

Therefore, $\frac{S_H}{n_H} = 47.219$ ksi, or $S_H = 2 \cdot 47.219 = 94.438$ psi. We take $S_H = 0.35\ H_B$ expressed in ksi, therefore, $H_B = \frac{S_H}{0.35} = \frac{94.438}{0.35} = 269.82$. We select for the gear AISI 5160 normalized steel having $H_B = 269$, while for the pinion the hardness should be at least 50 units higher; therefore, we take the AISI 4150 normalized steel having $H_B = 321$.

Example 5-10.1b: Verifying calculations of contact stresses for module system gears. The input data for this example are almost the same as in the example 5-10.1a. A gear drive transmits power of 4 kW at the rotational speed of the pinion $n_1 = 3{,}600$ rpm having the module $m = 2.5$ mm, $N_1 = 24$, $N_2 = 72$, and the tooth face width 38 mm. Using the safety factor $n_H = 2$, we select the material for the pinion and for the gear.

Solution:

Since the gears are made from steel, the modulus of elasticity $E = 207 \cdot 10^9$ Pa. The angular velocity of the pinion is $\omega_1 = \frac{\pi 3{,}600}{30} = 376.8\ \frac{\text{rad}}{s}$. The diameter of the pinion and the velocity ratio are $d_1 = 2.5 \cdot 24 = 60$ mm and $i = \frac{72}{24} = 3$. Using the factors $k_r = 1.25$, $k_d = 1.6$, and $k_m = 1.15$, we may now use formula (5-10.2) to calculate the surface stress:

$$\sigma_H = 1.182\sqrt{\frac{4 \cdot 10^3 \cdot 207 \cdot 10^9 \cdot 1.25 \cdot 1.6 \cdot 1.15(3+1)}{376.8 \cdot 0.06^2 \cdot 0.038 \cdot 3\sin 2 \cdot 20°}} = 327{,}161{,}317 \text{ Pa, or } \sigma_H = 327 \text{ MPa.}$$

In the example 5-10.1a the contact stress is $S_H = 47{,}219$ ksi. Hence, $S_H = 47.3 \cdot 6.9 = 325.8$ MPa, which as it is expected is almost the same as in this example.

Hence, $\frac{S_H}{n_H} = 327$ MPa, or $S_H = 2 \cdot 327 = 654$ MPa. We take $S_H = 2.45\ H_B$ MPa, therefore $H_B = \frac{S_H}{2.45} = \frac{654}{2.45} = 266.93$. We select in this case the same materials as in example 5-10.1a—namely, for the gear AISI 5160 normalized steel having $H_B = 269$ and for the pinion the AISI 4150 normalized steel having $H_B = 321$.

5-10.2. Design Calculations for Surface Durability

There is often a necessity in the industry to design a gear drive for transmitting a needed power to the output shaft that should have a required rotational speed. In these cases, the input rotational speed should be predetermined and the velocity ratio of the drive should be known. The minimum number of teeth of the pinion can be determined using formula (5-3.3), and therefore, in this situation the designer deals with the diameters of the pinion and the gear at the same time. Based on these diameters, the designer determines the center distance $C=\frac{d_2 \pm d_1}{2}=\frac{d_1(i \pm 1)}{2}$, where the negative sign is for internal engagement gears. Hence, we may write

$$d_1=\frac{2C}{(i \pm 1)} \tag{5-10.3}$$

Combining expression (5-10.2) with equation (5-10.3), we obtain

$$\sigma_H=1.182\sqrt{\frac{WEk_r k_d k_m (i \pm 1)^3}{4\omega_1 C^2 bi \sin 2\phi}}=\frac{S_H}{n_H} \tag{5-10.4}$$

Introducing an empiric coefficient for tooth face width η_c that represents the ratio between the tooth face width and the center distance

$$b=\eta_C C \tag{5-10.5}$$

we can modify the expression (5-10.4) as follows:

$$1.182\sqrt{\frac{WEk_r k_d k_m (i \pm 1)^3}{4\omega_1 C^3 \eta_C i \sin 2\phi}}=\frac{S_H}{n_H} \tag{5-10.6}$$

Squaring equation (5-10.6), we may write

$$1.182^2 \frac{WEk_r k_d k_m (i \pm 1)^3}{\omega_1 4C^3 \eta_C i \sin 2\phi}=\frac{S_H^2}{n_H^2}$$

Solving this equation for the center distance C, we obtain

$$C=0.704(i \pm 1)\sqrt[3]{\frac{WEk_r k_d k_m n_H^2}{S_H^2 \omega_1 \eta_C i \sin 2\phi}} \tag{5-10.7}$$

Because the allowable contact stresses for the gear are lower than for the pinion, we substitute into formula (5-10.7) the allowable surface stresses for the gear.

Many published sources indicate that the tooth face width coefficient η_c varies in a wide range from 0.1 to 1.2, however, for enclosed gear drives transmitting moderate power some authors recommend applying the tooth width coefficient η_c in the range 0.3 to 0.8. Upon calculation of the center distance C, we determine the pitch diameter d_1 of the pinion using the formula (5-10.3). It is recommended taking the minimum number of teeth N_1 of the pinion between 18 and 24 and calculating the diametral pitch P or module using the following formulas, respectively: $P = \frac{N_1}{d_1}$ and $m = \frac{d_1}{N_{1.}}$.

It should be emphasized that the design calculations for enclosed transmissions are, as stated above, predominantly based on surface durability. Numerous calculations show that these gears pass bending stress verifying calculations with significant margins. Nevertheless, verifying these gears with regard to bending stresses is recommended.

Example 5-10.2a: Design calculations of spur gear sets for surface durability.
Let us design a gear transmission of module system teeth using the same input data as in example 5-10.1b, which deals with verifying calculations of a module system gear set. The transmitted power is 4 kW. The numbers of teeth are $N_1 = 24$, $N_2 = 72$, and the angular velocity of the pinion is e $\omega_1 = \frac{\pi 3{,}600}{30} = 376.8\ \frac{\text{rad}}{s}$. We take $S_H = 2.45H_B$.

Solution:
From the previous example, we take the actual surface stress σ_H = 327 MPa as the allowable contact stress S_H. Now accepting the tooth face width coefficient η_c = 0.6 and the load correction factors as in the previous example, we calculate the center distance using the formula (5-10.7); while for the purpose of comparison with the previous example, the safety factor is $n_H = 1$: $C = 0.704(3+1)\sqrt[3]{\frac{4\cdot 10^{3\cdot}\cdot 207\cdot 10^{9}\cdot 1.25\cdot 1.6\cdot 1.15\cdot 1.5^{2}}{327^{2}\cdot 10^{12}\cdot 0.6\cdot 376.8\cdot 3\sin 2\cdot 20°}} = 0.128\ \text{m}$.

Using formula (5-10.3), we calculate the diameter of the pinion $d_1 = \frac{2C}{(i+1)} = \frac{2\cdot 128}{4} = 64$ mm. Therefore, the module is $m = \frac{d_1}{N_1} = \frac{64}{24} = 2.666$ mm. Comparing it with the calculated module from the previous example that is $m = \frac{25.4}{10} = 2.54$ mm, we obtain an insignificant difference in the values of these modules. This confirms that the accuracy of the design calculations that include an additional empiric coefficient η_c is adequate. Taking for the designed set of gears the closest higher standard module m = 2.75 mm, we recalculate the diameters of the gears and of the center distance. Hence, $d_1 = 2.75\cdot 24 = 66$ mm, $d_2 = 3\cdot 72 = 198$ mm, and $C = \frac{66+198}{2} = 132$ mm.

Example 5-10.2b: Design calculations based on surface durability and verifying calculations for bending stresses.
This example demonstrates the procedures for applying the results of design calculations based on surface durability to obtain the specifications of pitch or module system gears, and of performing verifying calculations based on bending stresses for repeated and completely reversed cycles.

Let us design an enclosed transmission for transmitting power of 7.5 hp at the rotational speed of the pinion of 1,800 rpm. The velocity ratio between the pinion and the gear $i = 3.2$. The safety factor $n_H = 2$.

Solution:

We select for the gear the AISI 8650 annealed steel having $S_{ut} = 103{,}750$ psi and $H_B = 212$. For the pinion we take the AISI 8740 normalized steel having $S_{ut} = 134{,}750$ psi and $H_B = 269$.

We determine the allowable contact stress for both gears based on the expression $S_H = 0.34H_B$, hence, for the gear material we have $S_H = 0.34\,H_B = 0.34{\cdot}212 = 72.08$ ksi. The angular velocity of the pinion is $\omega_1 = \dfrac{\pi 1{,}800}{30} = 188.4\ \dfrac{\text{rad}}{\text{s}}$. We take $\eta_c = 0.4$, $k_r = 1.25$, $k_d = 1.6$, and $k_m = 1.15$ and using formula (5-10.7), we calculate

$$C = 0.704(3.2+1)\sqrt[3]{\frac{7.5\cdot 550\cdot 12\cdot 30\cdot 10^6\cdot 1.25\cdot 1.6\cdot 1.15\cdot 2^2}{72.08^2\cdot 10^6\cdot 188.4\cdot 0.4\cdot 3.2\sin 2\cdot 20°}} = 7.589\text{ in.}$$

Therefore, the tooth face width is $b = \eta_c C = 0.4{\cdot}6.943 = 3.035$ in.

Now using formula (5-10.3) we determine the pitch diameter for the pinion:

$$d_1 = \frac{2C}{1+i} = \frac{2\cdot 7.589}{1+3.2} = 3.614\text{ in.}$$

Taking the number of teeth of the pinion $N_1 = 22$, we determine the diametral pitch $P = \dfrac{22}{3.614} = 6.087\ \dfrac{1}{\text{in.}}$. We take the closest smaller standard diametral pitch $P = 6\dfrac{1}{\text{in.}}$, therefore,

$$d_1 = \frac{22}{6} = 3.666\text{ in.}$$

For a module system gears we have $d_1 = 3.666{\cdot}25.4 = 93.133$ mm. Hence, the module is $m = \dfrac{93.133}{22} = 4.23$ mm. We take the closest larger standard module $m = 4.5$ mm, hence $d_1 = 4.5{\cdot}22 = 99$ mm.

Now we perform verifying calculations of the pitch system gear for a completely reversed bending loading cycle using formula (5-8.5). We need to determine the endurance limit for the material of the gear. Based on formula (1-9.2), we calculate $k_a = 0.792$, and taking $k_b = 0.95$, and $k_c = 1$, we calculate $S_e = 0.792{\cdot}0.95{\cdot}1{\cdot}0.5{\cdot}103{,}750 = 39{,}031$ psi. We calculate the tooth form factor for the pinion $y_1 = 0.52\left(1 - \dfrac{5.6}{22}\right) = 0.388$. Estimating that $K_f = 1.75$, we calculate

$$\sigma_{crc} = \frac{2\cdot 7.5\cdot 550\cdot 12\cdot 6\cdot 1.25\cdot 1.6\cdot 1.15\cdot 2}{188.4\cdot 3.333\cdot 2.78\cdot 0.374} = 4{,}185\text{ psi.}$$ Hence, $\dfrac{S_e}{n_b} = 4{,}185$ psi and we may write

$$n_b = \frac{S_e}{4{,}185} = \frac{39{,}031}{4{,}185} = 9.3260.$$ The calculated safety factor for bending stresses exceeds 4.66 times the incorporated safety factor for contact stresses $\left(\dfrac{9.326}{2} = 4.66\right)$.

Now we carry out verifying calculations of a module system for bending stresses due to a repeated cycle. Thus, we determine the amplitude stress limit using formula (5-8.2):

$$S_a = \frac{39{,}031\cdot 103.750}{39{,}031 + 103{,}750} = 28{,}361\text{ psi}$$ or $S_a = 195{,}693$ MPa. Converting the tooth face width into

the metric system, we have b = 3.035·25.4 = 77.089 mm. Applying formula (5-8.4), we calculate $\sigma_{rc} = \frac{7.5 \cdot 0.746 \cdot 10^3 \cdot 1.25 \cdot 1.6 \cdot 1.15 \cdot 2}{188.4 \cdot 0.0045 \cdot 0.090 \cdot 0.07061 \cdot 0.374} = 12{,}772{,}743$ Pa. Hence, σ_{rc} = 12,773 MPa. Therefore, $12{,}773 = \frac{S_a}{n_b}$. Hence, $n_b = \frac{195.693}{12.773} = 15.32$. In this case the safety factor for bending stresses exceeds the safety factor for contact stresses more than 7 times, as shown in following ratio: $\frac{15.32}{2} = 7.66$.

All this confirms that contact stresses are significantly more destructive for the teeth of the gears than bending stresses.

5-11. Parallel Helical Gears

Parallel helical gears represent one more type of cylindrical gears. Helical gears have their teeth inclined to the axis of rotation on an angle ψ that is called the helix angle. Helical gears have several advantages over spur gears. When compared with a spur gear set of the same diameter and the same gear blank width, the helical gear set has a wider tooth face, and it is stronger. The contact ratio of helix gears is significantly higher than that of spur gears, resulting in a smoother engagement and lower dynamic loading. However, helical gears develop axial forces that increase the load on the shaft bearings and result in additional bending moments applied to the shafts. Figure 5-11.1 shows a pair of meshing helical gears.

A special type of helical gears that is called herringbone helical gears develops equal axial forces acting in opposite directions and eliminates the axial forces applied to the shafts and the bearings. The manufacturing of helical gears is similar to spur gears, and the fabrication cost of

(a)

(b)

Figure 5-11.1. Helical gearing: (a) helical gear and pinion; (b) helical rack and pinion. (Courtesy of KHK-USA, Inc.)

helical gears is almost the same as that of spur gears. In general, using helical gears in all gear transmissions is recommended, unless there is a strong justification for using spur gears.

All geometric parameters and strength calculations presented for spur gears are applicable for helical gears with some modifications that are addressed below.

5-11.1. Geometry of Parallel Helical Gears

There are certain differences in the geometry and in the acting forces between spur and helical gears. Figure 5-11.2 presents a simplified drawing of a pair of helical gears having a helix angle ψ that usually is in range from 8° to 45°.

Two sectional planes—the transverse plane that is perpendicular to the axis of rotation and the normal plane that is perpendicular to the direction of the teeth—characterize the helix gears. A helix gear has a transverse circular pitch p_t and a normal circular pitch p_n, as shown in Figure 5-11.2. The relationship between these two circular pitches is reflected by the following expression:

$$p_n = p_t \cos\psi \tag{5-11.1}$$

Therefore, the normal diametral pitch P_n is determined from the following formula:

$$P_n = \frac{\pi}{p_t \cos\psi} \tag{5-11.2}$$

The transverse diametral pitch equals

$$P_t = \frac{\pi}{p_t} \tag{5-11.3}$$

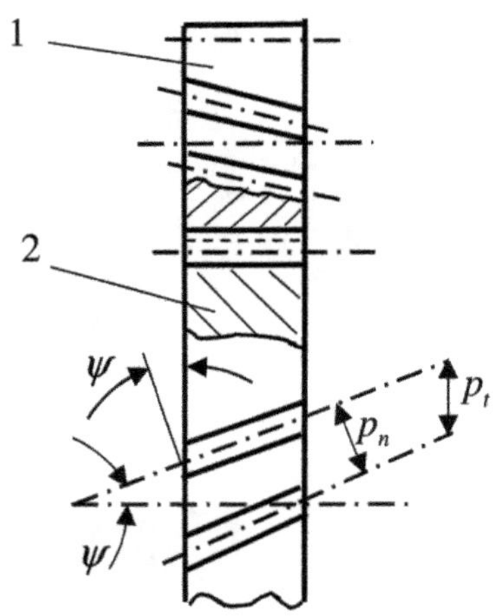

Figure 5-11.2. Helical gear geometry.

Therefore,

$$P_n = \frac{P_t}{\cos\psi} = \frac{P}{\cos\psi} \tag{5-11.4}$$

In the module system of helical gears, the relationship between the circular pitches is the same as in the pitch systems. Therefore, the normal module is determined as

$$m_n = \frac{p_n}{\pi} = \frac{p_t \cos\psi}{\pi} = m_t \cos\psi = m\cos\psi \tag{5-11.5}$$

Formulas (5-11.4) and (5-11.5) show that $P_n > P$ and $m_n < m$, which means that in the normal section of the helical gear the circumference contains more teeth than the circumference in the transverse section. The geometry of helical gears shows that the normal section of the helical gear could be defined by the diameter of a virtual (sometimes called equivalent) spur gear having teeth whose shape is very close to the shape of the teeth in the normal section of the helical gear. The virtual gear comprises a virtual number of teeth N_v. This number of teeth is accounted for in the calculation of the tooth form factor and it is computed from the following formula:

$$N_v = \frac{N}{\cos^3\psi} \tag{5-11.6}$$

It is extremely important to keep in mind that, in general, the manufacturing processes of helical gears are based on tooling producing the involute profile in the normal cross sections of the teeth. Therefore, in general, the standard diametral pitch and the standard module are designated for manufacturing helical gears. For spur gears, the circular pitch in the transverse and normal sections are the same. Therefore, for spur gears the transverse diametral pitch as well as the transverse module coincide with the normal diametral pitch and module. Actually, for spur gears they are the same. For helical gears, the transverse circular pitch exceeds the normal circular pitch, and consequently, the parameters based on these pitches are different.

It should be stressed that in the calculations of the rotational speed or angular velocity of helical gears we should use the pitch diameters based on the transverse diametral pitch (or diametral pitch, or just pitch) or module. For the pitch system helical gears, the pitch diameter equals

$$d = \frac{N}{P_n \cos\psi} \tag{5-11.7}$$

and for module system helical gears the pitch diameter is

$$d = \frac{m_n N}{\cos\psi} \tag{5-11.8}$$

The addendum and the addendum diameter for pitch system helical gears are

$$a = \frac{1}{P_n} \tag{5-11.9}$$

and

$$d_a = d + \frac{2}{P_n} \tag{5-11.10}$$

For the module system helical gears the addendum and addendum diameter are calculated using the following formulas, respectively:

$$a = m_n \tag{5-11.11}$$

and

$$d_a = d + 2m_n \tag{5-11.12}$$

During the machining processes of helical gears, the cutting tools—the rack, the shaper, and the hob—are set up in relation to the gear blank in accordance with the helix angle. This results in the form of the teeth in the normal section being identical to the form of the teeth of a spur gear.

5-11.2. Forces in Parallel Helical Gears

In comparison with spur gears where the forces between the teeth act in one plane, in parallel helical gears the forces act in the three dimensional space as it is shown in the force diagram presented in Figure 5-11.3. The pressure force F is applied to the pitch point P and is resolved into three components that are perpendicular to each other.

These components represent the tangential force F_t, the radial force F_r, and the axial force F_a. Since the tangential force is predetermined by the input parameters representing the power W, the angular velocity, and the corresponding pitch diameter of a gear, as shown in formula (5-7.1), it is reasonable to determine the relationship between the tangential force and the forces mentioned above. These relationships are apparent from the diagram in Figure 5-11.3. Therefore, the normal pressure force can be calculated from the following formula:

$$F = \frac{F_t}{\cos\psi \cos\phi} \tag{5-11.13}$$

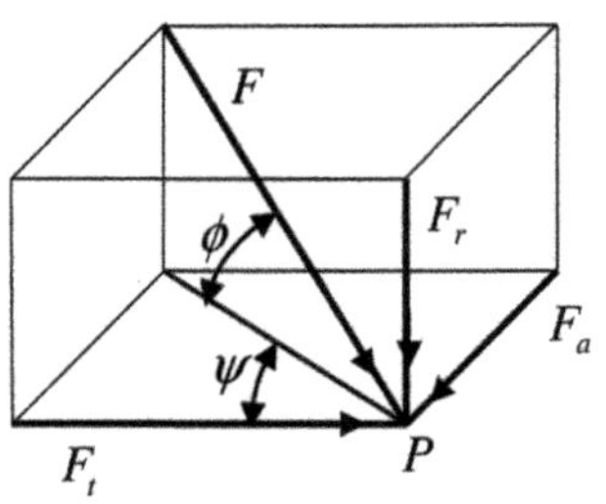

Figure 5-11.3. Forces acting in a helical gear engagement.

The formula for determining the radial force reads

$$F_r = \frac{F_t \tan\phi}{\cos\psi} \tag{5-11.14}$$

The axial force is determined using the formula

$$F_a = F_t \tan\phi \tag{5-11.15}$$

It should be emphasized that for stress calculations of helical gears the dynamic load factor for moderate operational conditions can be taken as $k_d = 1.4$, while the rest of the load factors are the same as for spur gears. In addition, the same considerations regarding the fatigue stress concentration factor for spur gears are applicable to helical gears.

5-12. Bending Stresses in Helical Gears

The basic principles of calculations of bending stresses for helical gears are similar to spur gears; however, there are a few geometry related differences. Due to the helix angle, we need to modify the tangential force, the moment of inertia, and the cross-sectional area of the tooth by accounting for this angle. All these considerations result in the corresponding change of expression (5-7.4) that obtains the following shape:

$$\sigma = \frac{F_t k_r k_d k_m \cos^2\psi}{b}\left(\frac{6l - h\tan\phi}{h^2}\right) \tag{5-12.1}$$

In addition, it should be mentioned that in helical gears the contact lines of the meshing teeth during their engagement process cause relatively less stresses in the teeth than in corresponding spur gears. This is accounted for by a so-called contact lines factor λ_b that according to some authors varies in range from 1.0 to 1.3. Modifying equation (5-12.1) similarly to equation (5-7.5) by replacing the tangential force with its value, and installing in the denominator the factor λ_b we obtain

$$\sigma = \frac{2\pi W k_r k_d k_m \cos^2\psi}{\omega d p_n b \lambda_b}\left[\frac{p_n(6l - h\tan\phi)}{\pi h^2}\right] \tag{5-12.2}$$

Based on equation (5-12.2) the helical tooth form factor y_v is defined by the following expression:

$$\frac{1}{y_v} = \frac{p_n(6l - h\tan\phi)}{\pi h^2} \tag{5-12.3}$$

Combining equations (5-12.2) and (5-12.3), we obtain the formula for the calculation of bending stresses for the helical pitch system gears:

$$\sigma = \frac{2WP_n k_r k_d k_m \cos^2\psi}{\omega d b \lambda_b y_v} \tag{5-12.4}$$

while for module system helical gears we obtain

$$\sigma = \frac{2Wk_r k_d k_m \cos^2\psi}{\omega m_n d b \lambda_b y_v} \tag{5-12.5}$$

In the formulas (5-12.4) and (5-12.5), the tooth form factor is computed from the modified formula that is similar to formula (5-7.7) for spur gears:

$$y_v = 0.52\left(1 - \frac{5.6}{N_v}\right) \tag{5-12.6}$$

where N_v is the virtual number of teeth according to formula (5-11.6).

It should be noted that the relationship between the bending stresses in the teeth of the gear and the pinion of helical gears, as in spur gears, is reflected in the following formula:

$$\sigma_g = \sigma_p \frac{y_{1v}}{y_{2v}} \tag{5-12.7}$$

The equations (5-12.4) and (5-12.5) are applicable for calculating static stresses; however, in reality the teeth are subjected to variable stresses representing repeated or completely reversed cycles. It should be emphasized again that we substitute into equations (5-12.4) and (5-12.5) and into similar formulas below ω_1, d_1, and y_{1v} or ω_2, d_2, and y_{2v} for calculating the stresses in the pinion or in the gear, respectively. As with spur gears, in order to account for the fatigue considerations it is necessary to derive corresponding expressions for calculating bending stresses acting in the teeth of helical gears.

5-13. Verifying Calculations of Bending Stresses for Helical Gears

All considerations related to the fatigue phenomenon associated with spur gears are applicable to the calculations of helical gears. The verifying calculations of bending stresses for helical gears for repeated and completely reversed cycles are presented below.

5-13.1. Verifying Calculations of Bending Stresses for Helical Gears Subjected to a Repeated Cycle

According to equation (5-12.4), the expression for calculating the bending stresses in the teeth of pitch system helical gears subjected to the action of a repeated cycle reads

$$\sigma_{rc} = \frac{WP_n k_r k_d k_m K_f \cos^2\psi}{\omega_1 d_1 b \lambda_b y_{1v}} \le \frac{S_a}{n_b} \tag{5-13.1}$$

For module system gears, bending stresses calculations for a repeated cycle according to expression (5-12.5) is as follows:

$$\sigma_{rc} = \frac{Wk_r k_d k_m K_f \cos^2\psi}{\omega_1 m_n d_1 b \lambda_b y_{1v}} \le \frac{S_a}{n_b} \tag{5-13.2}$$

where n_b is the safety factor. Use formula (5-8.2) to compute S_a and take λ_b in the range 1.0 to 1.3.

Example 5-13.1a: Verifying calculations of bending stresses for pitch system helical gears subjected to a repeated cycle.

Using the same input data as in example 5-8.1a, we determine the differences in the bending stresses between a spur and helical gear subjected to a repeated cycle. Hence, a 22-tooth pinion heaving a normal diametral pitch $P_n = 16\ \frac{1}{in.}$ transmits power of 10 hp at the rotating speed of the pinion n_1 = 1,800 rpm. The width of tooth face b = 0.8 in. The helix angle is ψ = 25°. We calculate the bending stresses in the pinion.

Solution:

Using formulas (5-11.7), (5-11.6), and (5-12.6), we determine the pitch diameter, virtual number of teeth, and tooth form factor, respectively: $d_1 = \frac{22}{16\cos 25°} = 1.517$ in., $N_{1v} = \frac{22}{\cos^3 25°} = 29.55$, and $y_{1v} = 0.52\left(1 - \frac{5.6}{29.55}\right) = 0.42$. The angular velocity is $\omega_1 = \frac{\pi 1{,}800}{30} = 188.4\ \frac{\text{rad}}{\text{s}}$. Based on Figure 5-8.1, we estimate that K_t = 1.75, therefore, using formula (1-10.2), we calculate K_f = 1.45. Taking k_r = 1.25, k_d = 1.4, k_m 1.15, λ_b = 1.15, and applying formula (5-13.1), we calculate

$$\sigma_{rc} = \frac{10\cdot 550\cdot 12\cdot 16\cdot 1.25\cdot 1.4\cdot 1.15\cdot 1.45\cos^2 25°0.821}{188.4\cdot 1.517\cdot 0.8\cdot 1.15\cdot 0.42} = 22{,}909\ \text{psi}.$$

Comparing the stress in the helical gear with the stress in the spur gear, we have $\frac{22{,}909}{43{,}799} = 0.523$ which is about 48% less than in the spur gear.

Example 5-13.1b: Verifying calculations of bending stresses for a module system subjected to a repeated cycle.

This example has the same input data as example 5-8.1b, which deals with a spur gear.

A 22-tooth pinion having a module m = 1.75 mm and a helix angle ψ = 25° transmits power of 4.0 kW at 1,800 rpm rotating in one direction. The width of the tooth face b = 20 mm. The pinion is made from the AISI 4150 normalized steel having S_{ut} = 800 MPa. We determine the safety factor of the pinion.

Solution:

From the previous example 5-13.1a, we take $\omega_1 = 188.4\ \frac{\text{rad}}{\text{s}}$, $N_{1v} = \frac{22}{\cos^3 25°} = 29.55$,

$y_{1v} = 0.52\left(1 - \frac{5.6}{29.55}\right) = 0.42$, k_r = 1.25, k_d = 1.4, λ_b = 1.15, and K_f = 1.45.

The diameter of the pinion is d_1 = 1.75·22 = 38.5 mm. Using formula (5-11.5), we calculate m_n = 1.75cos 25° = 1.59 mm. Applying formula (1-9.2), we determine k_a = 0.777 and we take k_b = 0.95 (based on the assumption that the smallest side of the tooth cross-sectional area is less than 15 mm). Since k_c = 1, we determine S_e = 0.777·0.95·1·0.5·800 = 295,288 MPa. Based on formula (5-8.2), we calculate $S_a = \frac{295.29\cdot 800}{295.29 + 800} = 215.68$ MPa. Using expression (5-13.2), we compute $\sigma_{rc} = \dfrac{4.0 \cdot 10^3 \cdot 1.25 \cdot 1.4 \cdot 1.15 \cdot 1.45 \cos^2 25°}{188.4 \cdot 0.00159 \cdot 0.0385 \cdot 0.020 \cdot 1.15 \cdot 0.42} = 86.02$ MPa, which in comparison with the stress in the spur gear $\dfrac{86.02}{135.43} = 0.645$ is less by about 35.5%.

Hence, $86.02 \cdot 10^6 = \dfrac{215.68 \cdot 10^6}{n_b}$, therefore, $n_b = \dfrac{215.68}{86.02} = 2.5$.

5-13.2 Verifying Calculations of Bending Stresses for Helical Gears Subjected to a Completely Reversed Loading Cycle

The expressions for verifying calculations of bending stresses for helical gears are based on the equations (5-12.4) and (5-12.5). Verifying calculations of the stresses in the teeth of helical gears subjected to a completely reversed cycle for the pitch systems of gears can be carried out using the following expression:

$$\sigma_{crc} = \frac{2WP_n k_r k_d k_m K_f \cos^2 \psi}{\omega_1 d_1 b \lambda_b y_{1v}} \le \frac{S_e}{n_b} \tag{5-13.3}$$

For module system helical gears the equation for verifying calculations of bending stresses with regard to a completely reversed cycle follows:

$$\sigma_{crc} = \frac{2Wk_r k_d k_m K_f \cos^2 \psi}{\omega_1 m_n d_1 b \lambda_b y_{1v}} \le \frac{S_e}{n_b} \tag{5-13.4}$$

Example 5-13.2a: Calculating the bending stresses and selecting the materials for a set of helical gears subjected to a completely reversed loading cycle.

A helical gear set having a helix angle of 25° transmits power of 15 hp at 1,800 rpm from the pinion having 18 teeth to the gear that rotates at 600 rpm. The normal diametral pitch of the teeth is $P_n = 8\dfrac{1}{\text{in.}}$ and the tooth face width is 1.0 in. Taking the safety factor $n_b = 2$, we determine the materials for the gears.

Solution:

We first determine the pitch diameter of the pinion and its angular velocity. Hence,

$d_1 = \dfrac{N_1}{P_n \cos \psi} = \dfrac{18}{8 \cos 25°} = 2.483$ in. and $\omega_1 = \dfrac{\pi n_1}{30} = \dfrac{\pi 1{,}800}{30} = 188.4 \dfrac{\text{rad}}{\text{s}}$.

Using formula (5-11.6), we compute the virtual number of teeth $N_{1v} = \dfrac{N_1}{\cos^3 \psi} = \dfrac{18}{\cos^3 25°} = 24.179$.

Now, applying formula (5-12.6), we determine the helical tooth form factor, hence,

$y_{v1} = 0.52\left(1 - \dfrac{5.6}{24.179}\right) = 0.4$.

Taking $\lambda_b = 1.15$, $k_r = 1.25$, $k_d = 1.4$, $k_m = 1.15$, and $K_f = 1.75$, we use the formula (5-13.3) to calculate the bending stress: $\sigma_{crc} = \dfrac{2\cdot 15\cdot 550\cdot 12\cdot 8\cdot 1.25\cdot 1.4\cdot 1.15\cdot 2\cos^2 25°}{188.4\cdot 2.483\cdot 1\cdot 1.15\cdot 0.4} = 24{,}325$ psi. Therefore, $\dfrac{S_e}{n_b} = 24.325$ psi.

Hence, $S_e = n_b\sigma_{crc} = 2\cdot 24{,}325 = 48{,}650$ psi. We need to determine the ultimate strength of the appropriate steel that in our case equals $S_{ut} = \dfrac{S_e}{0.5k_ak_bk_c}$. Assuming that the required steel should have the ultimate strength of approximately 150,000 psi, we estimate according to formula (1-9.2) that $k_a = 0.73$. Taking $k_b = 0.95$ and accounting for the fact that $k_c = 1$, we determine $S_{ut} = \dfrac{48{,}650}{0.5\cdot 0.73\cdot 0.95\cdot 1} = 140{,}303$ psi. Therefore we select for the pinion the AISI 4140 normalized steel having $S_{ut} = 148{,}000$ psi.

The stresses in the teeth of the gear according to formula (5-12.7) are $\sigma_g = \sigma_p \dfrac{y_{1v}}{y_{2v}}$. Thus, we need to calculate the tooth form factor for the gear. The number of teeth of the gear equals $N_2 = iN_1 = \dfrac{1{,}800}{600}\cdot 18 = 64$, therefore, we determine $N_{2v} = \dfrac{64}{\cos^3 25°} = 86$. According to formula (5-12.6), we have $y_{2v} = 0.52\left(1 - \dfrac{5.6}{86}\right) = 0.486$.

The ultimate strength for the gear material is determined in a similar way as the acting bending stresses. Therefore, $S_{utg} = 140{,}303\dfrac{0.4}{0.486} = 115{,}475$ psi. We select for the gear the AISI 9310 annealed steel having $S_{ut} = 119{,}000$ psi.

Example 5-13.2b: Verifying bending stresses calculations for a module system helical gear for a completely reversed cycle.

A 19-tooth helical pinion having a helix angle $\psi = 30°$ and a normal module 3.5 mm transmits power of 5 kW at 3,600 rpm. The tooth face width $b = 22$ mm. We determine the bending stress acting in the pinion.

Solution:

We calculate for the pinion the angular velocity, pitch diameter, and the tooth form factor as $\omega_1 = \dfrac{\pi 3{,}600}{30} = 376.8\ \dfrac{\text{rad}}{\text{s}}$, $d_1 = \dfrac{3.5\cdot 19}{\cos 30°} = 76.79$ mm, $N_{1v} = \dfrac{19}{\cos^3 30°} = 29.25$, and $y_{1v} = 0.52\left(1 - \dfrac{5.6}{29.25}\right) = 0.42$.

We take $k_r = 1.25$, $k_d = 1.4$, $k_m = 1.15$, $K_f = 1.45$, and $\lambda_b = 1.15$. Now using formula (5-13.4), we calculate the stress: $\sigma_{crc} = \dfrac{2\cdot 5\cdot 10^3\cdot 1.25\cdot 1.4\cdot 1.15\cdot 1.45\cos^2 30°}{376.8\cdot 0.0035\cdot 0.07679\cdot 0.022\cdot 1.15\cdot 0.42} = 20.338$ MPa.

5-14. Design Calculations for Bending Stresses of Helical Gears

Transforming the equations (5-12.4) and (5-12.5) in a similar manner to the corresponding equations for spur gears, we obtain the expressions for design calculations of helical gears. The equations for the design calculations with regard to bending stresses for repeated and completely reversed cycles for pitch system and module system helical gears are presented below.

5-14.1. Design Calculations Based on Bending Stresses for a Repeated Cycle

Design calculations of helical gears based on bending stresses of teeth subjected to a repeated cycle could be performed using the following equations. Hence, the equation for pitch system gears reads

$$P_n = \sqrt[3]{\frac{\omega_1 N_1 \eta_b \lambda_b y_{1v} S_a}{W k_r k_d k_m K_f n_b \cos^2 \psi}} \tag{5-14.1}$$

and for the module system the equation is

$$m_n = \sqrt[3]{\frac{W k_r k_d k_m K_f n_b \cos^2 \psi}{\omega_1 N_1 \eta_b \lambda_b y_{1v} S_a}} \tag{5-14.2}$$

where η_b is, as for spur gears, the tooth face width coefficient and n_b is the safety factor.

The stress amplitude S_a is computed from formula (5-8.2). It is recommended that λ_b be in the range of 1 to 1.3, while η_b be in the range of 10 to 25.

Example 5-14.1a: Design calculations of a pitch system helical gear for a repeated bending cycle.

In an open helical drive, the pinion transmits power of 10 hp at 1,200 rpm. The helix angle $\psi = 15°$. We determine the normal diametral pitch using the safety factor $n_b = 2$. The rotation is one directional.

Solution:

First we select the material for the pinion following the recommendations presented above. We take the AISI 5160 normalized steel having $S_{ut} = 138{,}750$ psi. In order to determine the endurance limit, we calculate based on formula (1-9.2) $k_a = 0.75$ and we take $k_b = 0.95$ assuming that the thickness of the root of the tooth is less than 0.6 in. Since $k_c = 1$, we calculate

$S_e = 0.75 \cdot 0.95 \cdot 1 \cdot 0.5 \cdot 138{,}750 = 49{,}430$ psi.

Using formula (5-8.2), we compute $S_a = \dfrac{49{,}430 \cdot 138{,}750}{49{,}430 + 138{,}750} = 36{,}446$ psi. Taking for the pinion $N_1 = 21$ and using formulas (5-11.6) and (5-12.6), we calculate $N_{1v} = \dfrac{21}{\cos^3 15°} = 23.3$ and $y_{1v} = 0.52\left(1 - \dfrac{5.6}{23.3}\right) = 0.395$, respectively.

We take k_r = 1.25, k_d = 1.4, k_m = 1.15, λ_b = 1.15, and η_b = 15. Referring to Figure 5-8.1 we estimate K_f = 1.75. The angular velocity of the pinion is $\omega_1 = \frac{\pi 1.200}{30} = 125.6 \frac{\text{rad}}{\text{s}}$. Applying formula (5-14.1), we compute $P_n = \sqrt[3]{\frac{125.6 \cdot 21 \cdot 15 \cdot 1.15 \cdot 0.395 \cdot 36{,}446}{10 \cdot 550 \cdot 12 \cdot 1.25 \cdot 1.4 \cdot 1.15 \cdot 1.75 \cdot 2\cos^2 15°}} = 14.44 \frac{1}{\text{in.}}$.

Example 5-14.1b: Design calculations of a module system helical gear for a repeated cycle.

Using same input data as in example 5-14.1a, we determine the normal module for the pinion.

Solution:

We convert the power from hp to kW and the stress from psi to MPa, hence $10 \cdot 0.746 = 7.46$ kW and $S_a = 36.446 \cdot 6.9 = 251.5$ MPa.

Using formula (5-14.2), we calculate

$$m_n = \sqrt[3]{\frac{7.46 \cdot 10^3 \cdot 1.25 \cdot 1.4 \cdot 1.15 \cdot 1.75 \cdot 2\cos^2 15°}{125.6 \cdot 21 \cdot 15 \cdot 1.15 \cdot 0.395 \cdot 251.25 \cdot 10^6}} = 0.00176 \text{ m}.$$

The numerically equivalent diametral pitch for this module is $P = \frac{25.4}{1.76} = 14.43 \frac{1}{\text{in.}}$ (as expected, it is the same as in the previous example).

5-14.2. Design Calculations of Helical Gears for a Completely Reversed Bending Cycle

The following equations allow carrying out design calculations for helical gears whose teeth are subjected to the completely reversed bending cycles. Thus, for the pitch systems the equation is

$$P_n = \sqrt[3]{\frac{\omega_1 N_1 \eta_b \lambda_b y_{1v} S_e}{2Wk_r k_d k_m K_f n_b \cos^2 \psi}} \quad (5\text{-}14.3)$$

and for the module systems the equation reads

$$m_n = \sqrt[3]{\frac{2k_r k_d k_m W K_f n_b \cos^2 \psi}{\omega_1 N_1 \eta_b \lambda_b y_{1v} S_e}} \quad (5\text{-}14.4)$$

When performing design calculations of helical gears, the tooth face width coefficient η_b should be in the range of 10 to 25 and the contact lines factor λ_b should be in the range of 1 to 1.3.

Example 5-14.2a: Design calculations of a set of helical gears subjected to a completely reversed loading cycle.
We design a module system gear set using the same input data as in example 5-13.2a, which deals with the verifying calculations of a pitch system gear subjected to a completely reversed bending cycle. Repeating the data from the example 5-13.2a, we have k_r = 1.25, k_d = 1.4, k_m = 1.15, W = 15·0.746·10³ = 11.19 kW, fatigue stress concentration factor K_f = 1.75, $\omega_1 = 188.4\ \dfrac{\text{rad}}{s}$, λ_b = 1.15, y_{1v} = 0.4, the actual stress in the pinion is $\sigma_{crc} = 24.325 \cdot 6.9 \cdot 10^6 = 167.84 \cdot 10^6$ Pa. The safety factor n_b = 2 is already accounted for in the stress σ_{crc} that is used in this example as the actual endurance limit. This allows us to compare these two examples and verify the accuracy of these design calculations. Taking η_b = 20 and using formula (5-14.4), we determine the normal module of the pinion: $m_n = \sqrt[3]{\dfrac{2 \cdot 11.19 \cdot 10^3 \cdot 1.25 \cdot 1.4 \cdot 1.15 \cdot 2 \cdot 1 \cos^2 25°}{188.4 \cdot 18 \cdot 20 \cdot 0.625 \cdot 0.4 \cdot 77.204 \cdot 10^6}} = 0.00324$ m, or m_n = 3.24 mm. The numerically equivalent pitch is $P = \dfrac{25.4}{3.24} = 7.84 \dfrac{1}{\text{in.}}$, while $P_n = \dfrac{7.84}{\cos 25°} = 8.65 \dfrac{1}{\text{in.}}$ which is close to the normal pitch of the gear in example 5-13.2a that is $P_n = 8 \dfrac{1}{\text{in.}}$. In this case, we take the closest larger standard normal module m_n = 3.5 mm.

5.15. Surface Durability, Verifying, and Design Calculations for Parallel Helical Gears

The calculations for contact stresses for helical gears are similar to the calculations for spur gears. However, due to the helix angle some differences should be accounted for in the calculations of surface durability of helical gears. The pressure force applied to the pitch point, as shown in formula (5-11.13), is influenced by the helix angle. In addition, the helix angle causes some changes in the radii of curvature of contacting profiles of the teeth, while the length of the tooth face is actually longer than the tooth face in spur gears due to the helix angle. All these changes should be accounted for in formula (5-10.2), which determines contact stresses in spur gears. Modifying this formula based on the changes mentioned above, we obtain the formula that is applicable for calculating the contact stresses of helical gears. All these changes in formula (5-10.2) are based on conventional algebraic and geometric procedures that do not represent any difficulties. A similar modification is performed for the formula for calculating bending stresses in spur gears in order to apply it to helical gears. One more change should be made in the formula mentioned above. This change deals with the length of the contact lines in the helical gears, and it is accounted for by using the contact lines coefficient λ_H that according to published data varies from 1.35 to 1.45 (while for bending stresses the similar coefficient λ_b is in the range of 1 to 1.3). Based on all these considerations, we obtain the following expression for verifying calculations of contact stresses in the teeth of helical gears:

$$\sigma_H = 1.182\sqrt{\frac{WEk_r k_d k_m (i \pm 1)\cos^2\psi}{\omega_1 d_1^2 \lambda_H b i \sin 2\phi}} \le \frac{S_H}{n_H} \tag{5-15.1}$$

where E is the modulus of elasticity, S_H is the allowable contact stress, and n_H is the safety factor with regard to contact stresses.

Applying to expression (5-15.1) transformations similar to those performed on the expression for calculating contact stresses for the spur gears, we obtain the following formula for design calculations of helical gears based on contact stresses:

$$C = 0.704(i \pm 1)\sqrt[3]{\frac{WEk_r k_d k_m n_H^2 \cos^2\psi}{S_H^2 \omega_1 \eta_C \lambda_H i \sin 2\phi}} \qquad (5\text{-}15.2)$$

The coefficient η_c and the allowable stress S_H are the same as for the spur gears. Thus, η_c varies from 0.3 to 0.8. Similar to spur gears, for preliminary calculations, the allowable contact stress S_H ranges from $0.333H_B$ in ksi to $0.377H_B$ in ksi, or from $2.3H_B$ in MPa to $2.6H_B$ in MPa, while $H_B \leq 350$ and the service life is up to 10^8 cycles. In addition, while the contact stresses in the teeth of the pinion and the gear are the same, it is recommended, as for spur gears, that the hardness of the pinion material exceed the mating gear material hardness by at least 50 units of H_B.

Example 5-15.1a: Verifying calculations of contact stresses and selecting materials for pitch system helical gears.

A helical gear set having a helix angle of 30° and a diametral normal pitch $P_n = 6\ \frac{1}{\text{in.}}$ is transmitting power of 20 hp from the pinion having the number of teeth $N_1 = 20$ to the gear having the number of teeth $N_2 = 70$ and is rotating at $n_2 = 1{,}028$ rpm. The tooth face width $b = 1.5$ in. We need to determine the stresses and select the material for the gears if the safety factor is $n_H = 2$.

Solution:

We start with determining the velocity ratio $i = \frac{70}{20} = 3.5$ and the diameter of the pinion accounting for the transverse pitch, hence, $d_1 = \frac{20}{6\cos 30°} = 3.849$ in. The angular velocity of the pinion is $\omega_1 = \frac{\pi 3.5 \cdot 1{,}028}{30} = 376.6\ \frac{\text{rad}}{\text{s}}$. We take the following values for the loading factors: $k_r = 1.25$, $k_d = 1.4$, and $k_m = 1.15$. Taking $\lambda_H = 1.4$ and using formula (5-15.1), we perform the needed calculation: $\sigma_H = 1.182\sqrt{\frac{20 \cdot 550 \cdot 12 \cdot 30 \cdot 10^6 \cdot 1.25 \cdot 1.4 \cdot 1.15(3.5+1)\cos^2 30°}{376.6 \cdot 3.849^2 \cdot 1.4 \cdot 1.5 \cdot 3.5 \sin 2 \cdot 20°}} = 37{,}782$ psi.

Therefore, $S_H = 2 \cdot 37{,}782 = 75{,}564$ psi. Based on the relationship between the stress S_H and the hardness H_B, we obtain $H_B = \frac{75.564}{0.355} = 212.85$. Hence, for the gear we take the AISI 4340 annealed steel having $H_B = 217$ and for the pinion we take the AISI 5160 normalized steel having $H_B = 269$.

Example 5-15.1b: Verifying calculations of contact stresses for module system helical gears.

An enclosed helical gear drive having $\psi = 23°$ is transmitting power of 7.5 kW at the rotational speed of the pinion $n_1 = 3{,}600$ rpm having the module $m = 2.5$ mm, $N_1 = 24$, $N_2 = 72$, and the tooth face width b = 38 mm. Taking the safety factor n_H = 2, we determine the materials for the pinion and for the gear.

Solution:
Since the gears are made from steel, the modulus of elasticity $E = 207 \cdot 10^9$ Pa. The angular velocity of the of the pinion is $\omega_1 = \frac{\pi 3{,}600}{30} = 376.8 \ \frac{\text{rad}}{s}$. The diameter of the pinion accounting for the transfer module is $d_1 = \frac{2.5 \cdot 24}{\cos 23^\circ} = 65.18$ mm. The velocity ratio is $i = \frac{72}{24} = 3$. Taking k_r = 1.25, k_d = 1.4, k_m = 1.15, and λ_H = 1.4, and using formula (5-15.1), we calculate the surface stress: $\sigma_H = 1.182\sqrt{\frac{7.5 \cdot 10^3 \cdot 207 \cdot 10^9 \cdot 1.25 \cdot 1.4 \cdot 1.15(3+1)\cos^2 23°}{376.8 \cdot 0.06518^2 \cdot 1.4 \cdot 0.038 \cdot 3\sin 2 \cdot 20°}}$ = 300,044,106 Pa.

Hence, $\frac{S_H}{n_H} = 300$ MPa, or $S_H = 2 \cdot 300 = 600$ MPa. We take $S_H = 2.45\ H_B$ MP, therefore, $H_B = \frac{S_H}{2.45} = \frac{600}{2.45} = 244.9$. We select for the gear AISI 1340 normalized steel having H_B = 248, while for the pinion the hardness should be at least 50 units higher; therefore, we take the AISI 4140 normalized steel having H_B = 302.

Example 5-15.1c: Design calculations of contact stresses for module system helical gears.
We need to design a module system helical gear drive the input shaft of which is powered by 15 kW and rotates at 3,600 rpm, while the output shaft has a rotational speed of 1,100 rpm. The safety factor is 2.

Solution:
We select for the gear the AISI 4620 normalized steel having H_B = 174, therefore, $S_H = 2.3 \cdot 174 = 400.2$ MPa. The angular velocity of the pinion is $\omega_1 = \frac{\pi 3.600}{30} = 376.8 \ \frac{\text{rad}}{\text{s}}$.

We take the helix angle ψ = 25°, the tooth width coefficient η_c = 0.5, λ_H = 1.35, k_r = 1.25, k_d = 1.4, and k_m = 1.15. The velocity ratio is $i = \frac{3{,}600}{1.100} = 3.27$ Using formula (5-15.2), we calculate

$$C = 0.704(3.27+1)\sqrt[3]{\frac{15 \cdot 10^3 \cdot 207 \cdot 10^9 \cdot 1.25 \cdot 1.4 \cdot 1.15 \cdot 2^2 \cos^2 25°}{400.2^2 \cdot 10^{12} \cdot 376.8 \cdot 1.35 \cdot 0.5 \cdot 3.27 \sin 2 \cdot 20°}} = 0.190 \text{ m}.$$

Because $d_1 = \frac{2C}{i+1} = \frac{2 \cdot 190}{3.27+1} = 89.2$ mm, and taking the number of teeth of the pinion $N_1 = 25$, we determine the module $m = \frac{89.2}{25} = 3.57$ mm. Hence, the normal module is m_n = 3.57 cos 25° = 3.234 mm. We take the closest larger standard module m_n = 3.25 mm. Now we need to recalculate the pitch diameter of the pinion: $d_1 = \frac{m_n N_1}{\cos\psi} = \frac{3.25 \cdot 25}{\cos 25°} = 89.65$ mm.

The number of teeth of the gear $N_2 = iN_1 = 3.27 \cdot 25 = 81.75$, hence, we take $N_2 = 82$. Therefore, the pitch diameter of the gear $d_2 = \frac{3.25 \cdot 82}{\cos 25°} = 294.05$ mm. The corrected center distance is $C = \frac{294.05 + 89.65}{2} = 191.85$ mm.

We select for the pinion the AISI 4820 normalized steel having $H_B = 229$.

The design calculations based on surface durability should be, as it is for spur gears, verified with regard to bending stresses.

5-16. Design Project

It is necessary to design an enclosed helical gear transmission having the power of 7.5 kW at 620 rpm on the output shaft. The drive is reversible. The design project includes calculations of the preliminary parameters of the gears, shafts, and bearings, and the preliminary engineering drawing of the assembly.

The basic steps for performing this project are presented below.

1. Design calculations of the parameters of an enclosed helical gear set.

First, we select the rotational speed for the electrical motor. As it is mentioned above for each nominal power we can select the most suitable rotational speed from the row of practical nominal rotational speeds: 3,600, 1800, 1200, and 900 rpm. For gear drives, the selection is based on the acceptable values of the velocity ratio that in our case leads to velocity ratio $i = \frac{1,800}{620} = 2.9$. Therefore, we select an electrical motor having 7.5 kW of power and the rotational speed of 1,800 rpm.

The next step is related to the selection of the material for the gear. Based on recommendations presented above, we select for the gear the AISI 5160 normalized steal having $H_B = 269$ and $S_{ut} = 138{,}750$ psi. Taking the coefficient (or multiplier) 2.45 for determining the allowable contact stress based on the H_B, we obtain $S_H = 2.45 \cdot 269 = 659.05$ MPa. We take a safety factor of 2 for all machine elements that should be calculated in the project. For the pinion we select the AISI 4150 normalized steel having $H_B = 321$ and $S_{ut} = 167{,}500$ psi.

Now we compute $\omega_1 = \frac{\pi 1,800}{30} = 188.4 \ \frac{\text{rad}}{\text{s}}$. We take $\psi = 18°$, $k_r = 1.25$, $k_d = 1.4$, $k_m = 1.15$, $\eta_c = 0.4$, $\lambda_H = 1.4$, and based on formula (5-15.2), we calculate the center distance between the gears $C = 0.704(2.9+1)\sqrt[3]{\frac{7.5 \cdot 10^3 \cdot 207 \cdot 10^9 \cdot 1.25 \cdot 1.4 \cdot 1.15 \cdot 2^2 \cos^2 18°}{659.05^2 \cdot 10^{12} \cdot 188.4 \cdot 0.4 \cdot 1.4 \cdot 2.9 \sin 2 \cdot 20°}} = 0.132$ m.

We determine the pitch diameter for the pinion $d_1 = \frac{2 \cdot 132}{1 + 2.9} = 67.69$ mm, Taking for the pinion $N_1 = 23$, we obtain $m = \frac{67.69}{23} = 2.943$ mm. Hence, the normal module is $m_n = 2.943 \cos 18° = 2.8$ mm.

We take the closest higher standard module $m_n = 3$ mm. The corrected pitch diameter of the pinion is $d_1 = \frac{3 \cdot 23}{\cos 18°} = 72.55$ mm. The number of teeth of the gear is $N_2 = 2.9 \cdot 23 = 66.7$, hence, N_2 = 67. The pitch diameter of the gear is $d_2=\frac{3\cdot 67}{\cos 18°}=211.344$ mm. The corrected center distance is $C=\frac{211.344+72.55}{2}=141.947$ mm. Hence, we take C = 142 mm. The tooth face width is $b = 0.4\cdot 142$ = 56.78 mm. We round off the width to b = 57 mm.

2. Verifying calculations for bending stresses acting in teeth of the gears.
Since the drive is reversible, the verifying calculations regarding bending stresses should be carried out for the completely reversed cycle. In order to determine the endurance limit for the material of the gear, we have to determine the surface factor using formula (1-9.2): k_a = 0.745. Assuming that the width of the root of the teeth does not exceed 15 mm, we take k_b = 0.95. Taking k_c = 1, we calculate the endurance limit, hence, $S_e = 0.745\cdot 0.95\cdot 1\cdot 0.5\cdot 138.75\cdot 6.9$ = 338.8 MPa. Based on formulas (5-11.6) and (5-12.6), we compute $N_{1v}=\frac{23}{\cos^3 18°}=26.73$ and $y_{1v}=0.52\left(1-\frac{5{,}6}{26.73}\right)=0.411$, respectively. Taking λ_b = 1.15 and estimating that K_f = 1.75, we apply formula (5-13.4) for the bending stresses in the teeth of the pinion: $\sigma_{crc}=\frac{2\cdot 7.5\cdot 10^3\cdot 1.25\cdot 1.4\cdot 1.15\cdot 1.75\cos^2 18°}{188.4\cdot 0.003\cdot 0.0726\cdot 0.057\cdot 1.15\cdot 0.411}=43{,}199{,}660$ Pa or σ_{crc} = 43.2 MPa. Therefore, $n_b=\frac{338.8}{43.2}=7.84$. Thus, $\frac{n_b}{n_H}=\frac{7.84}{2}=3.92$ the bending stress safety factor is almost 4 times higher than the contact stress safety factor. This is expected.

3. Design calculations of the shafts and bearings.
In order to determine the diameters of the shafts, we need to know the corresponding values of the torques and bending moments applied to the shafts. The bending moments could be determined if we know the distance between the centers of the bearings and the location of the gear on the shaft. We do not know the width of the bearings and their enclosures. We do not know the width of the hubs of the gears. The values of all these unknown geometric parameters could be estimated if we have the approximate dimensions of the diameters of the corresponding shafts. We can determine the preliminary diameters of the shafts by applying formula (3-3.8) that reads $d=3.65\sqrt[3]{\frac{P}{n\tau_{all}}}$, where P is the power in watt, n is the rotational speed in rpm, τ_{all} is the allowable shear stress that should be in the range from 11 to 28 MPa, and d is the diameter of the shaft in m. In order to avoid confusion, we denote the diameters of the shafts by an additional subscript s, for instance, the diameter of the gear shaft is denoted by d_{s2}. Taking τ_{all} = 25 MPa, we calculate the diameters of the shafts. Hence, the diameter of the shaft for the pinion is $d_{s1}=3.65\sqrt[3]{\frac{7.5\cdot 10^3}{1{,}800\cdot 25\cdot 10^6}}=0.0202$ m and for the gear $d_{s1}=3.65\sqrt[3]{\frac{7.5\cdot 10^3}{620\cdot 25\cdot 10^6}}=0.0288$ m. Based on the calculated values, we take the bore diameters for the bearings for the shafts of the pinion and the gear as 20 mm and 30 mm, respectively. Therefore, we select for the shaft of the pinion, on a preliminary basis, the NTN angular-contact bearing 73ZZ04 double shielded, which has the bore diameter d = 20 mm, outer diameter D = 52 mm, the width B = 15 mm, the dynamic load rating C = 18.7 kN, and the static load

rating $C_0 = 10.4$ kN. Similarly, we select for the shaft of the gear the NTN angular-contact bearing 73ZZ06 double shielded, which has the bore diameter $d = 30$ mm, the outer diameter $D = 72$ mm, the width $B = 19$ mm, the dynamic load rating $C = 33.5$ kN, and the static load rating $C_0 = 22.3$ kN.

Based on the widths and diameters of the bearings, we estimate the widths of the bearing enclosures. The diameters of the shafts allow us to determine the width of gear's hubs. Based on design considerations, we assign certain distances between the gear hubs and bearing enclosures. Now we have the needed data to perform the preliminary assembly drawing. We always make the tooth face of the pinion a few millimeters wider than the gear in order to compensate for the cumulative errors of manufacturing and assembly.

Because the bore diameter of the bearings for the shaft of the gear is 30 mm, we should increase the diameter of the shaft and the inner diameter of the hub to about 35 mm. Assuming that the length of the hub $l = 1.8 \cdot 35 = 63$ mm and taking the distance of 22 mm between the hub and the center of the bearing, we determine that the distance between the centers of the bearings of the shaft for the gear is 107 mm. Based on these design considerations, we put together a preliminary assembly drawing of the helical gear drive. The assembly drawing is shown in Figure 5-16.1 and represents the top view of the section passing through the horizontal casing joint (the bolts are removed).

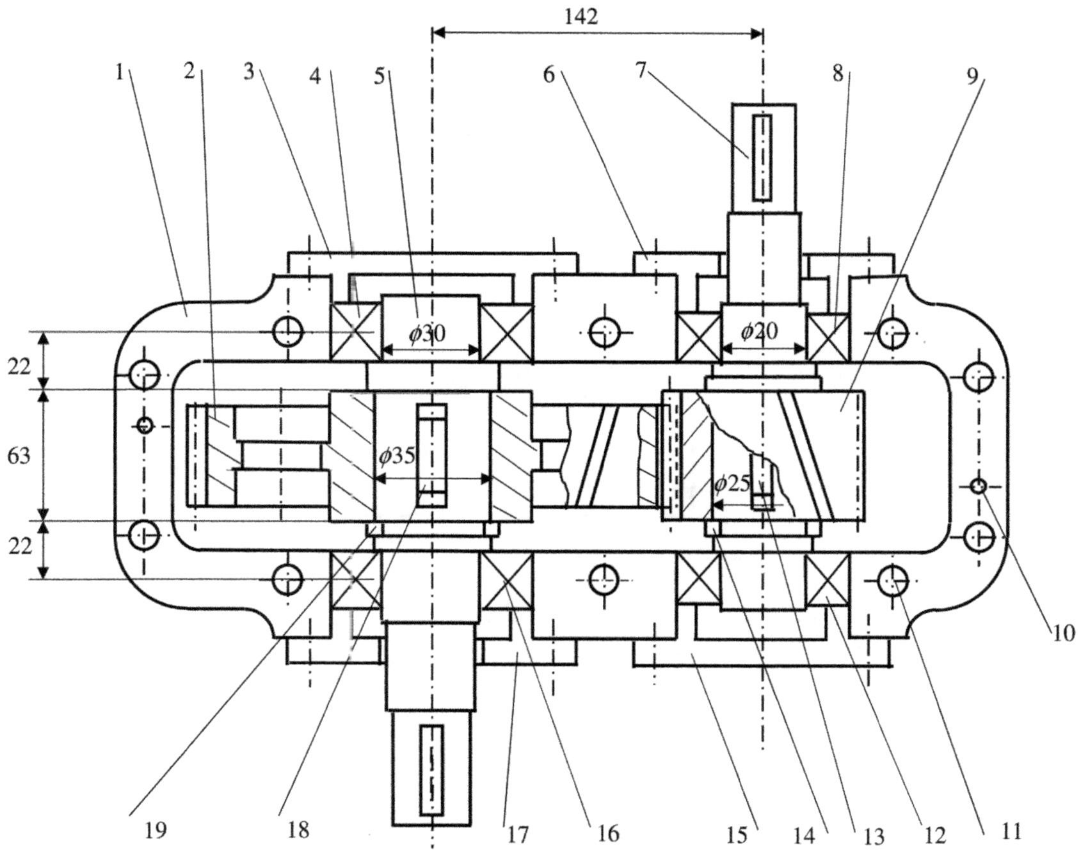

Figure 5-16.1. Top view of the assembly drawing of a helical gear drive.

The figure shows that shaft 7 supported by the bearings 8 carries pinion 9 that is engaged with gear 2 that is mounted on shaft 5 supported by bearings 4. The corresponding dimensions are shown in the drawing that is made to the approximate scale.

4. Reactions, bending and torque diagrams for the shaft of the pinion.

Now we determine the forces acting in the engagement of the gears. The tangential force applied to gear equals $F_t = \dfrac{2W}{\omega_1 d_1} = \dfrac{2 \cdot 7.5 \cdot 10^3}{188.4 \cdot 0.0726} = 1{,}096\ \text{N}$, the radial force is

$$F_r = \frac{F_t \tan\phi}{\cos\psi} = \frac{1{,}096 \tan 20^\circ}{\cos 18^\circ} = 419.6\ \text{N},$$

and the axial force equals $F_a = F_t \tan\phi = 1{,}096 \tan 20^\circ = 398.9$ N.

Based on the drawing in Figure 5-16.1, we plot for the shaft of the pinion the free-body diagram in the vertical plane that is presented in Figure 5-16.2a, where the distance between the centers of the bearings is $l = 107$ mm.

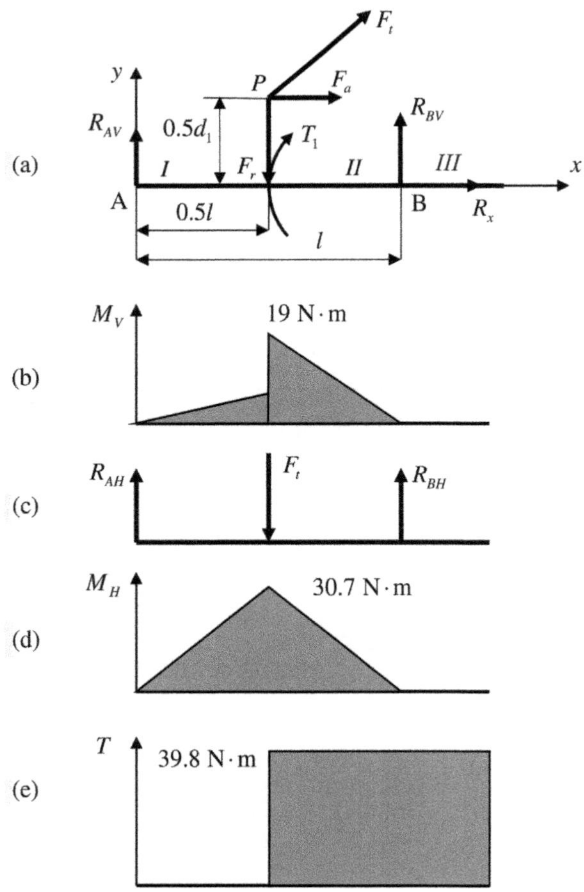

Figure 5-16.2. Free-body, bending-moments, and torque diagrams for the shaft of the pinion.

The forces mentioned above are applied to the pitch point P. Obviously, the tangential force F_t is located in the horizontal plane. Other forces are located in the vertical plane. We should consider separately the action of these forces in the vertical and horizontal planes. The free-body diagram for the vertical plane is shown in Figure 5-16.2a. Thus, projecting the forces on x axis we compute the reaction R_x that represents the axial force for the bearing. Hence, we write $F_a + R_x = 0$, therefore, $R_x = -F_a = -398.9$ N. Composing the sum of moments about point B, we determine the reaction R_{AV}, hence, $R_{AV}l - F_r\,0.5l + F_a\,0.5d_1 = 0$, and therefore, $R_{AV} = \dfrac{F_r 0.5l - F_a 0.5d_1}{l}$, while $R_{AV} = \dfrac{419.6\cdot 0.5\cdot 107 - 398.9\cdot 0.5\cdot 72.6}{107} = 74.47$ N. Projecting the forces on y axis, we calculate the reaction R_{BV}, hence, $R_{AV} - F_r + R_{BV} = 0$. Therefore, $R_{BV} = F_r - R_{AV} = 419.6 - 74.47 = 345.13$ N. As shown in Figure 5-16.2a, the shaft is divided in three intervals. The bending moment on the first interval is $M_I = R_{AV}x$. Thus, when $x = 0$, then $M_I = 0$, while when $x = 0.5l = 0.5\cdot 0.107 = 0.0535$ m the bending moment equals $M_I = 74.47\cdot 0.0535 = 3.98$ N·m. On the second interval we have $M_{II} = R_{AV}x - F_r\,(x - 0.5l) + F_a\,0.5d_1$. Therefore, when $x = 0.5l$ then $M_{II} = R_{AV}\,0.5l + F_a 0.5d_1$. Hence, $M_{II} = 3.98 + 398.9\cdot 0.5\cdot 0.0726 = 18.46$ N·m. When $x = l$ then $M_{II} = 0$. The diagram of bending moments for the vertical plane is shown in Figure 5-16.2b.

Figure 5-16.2c shows the free-body diagram for the horizontal plane.

Because the tangential force F_t is applied in the middle between the supports of the shaft the reactions in the horizontal plane are $R_{AH} = R_{BH} = 0.5\,F_t = 0.5\cdot 1{,}096 = 548$ N. Therefore, combining the reactions in both planes, we obtain that $R_A = \sqrt{R_{AV}^2 + R_{AH}^2} = \sqrt{74.47^2 + 548^2} = 553$ N and $R_B = \sqrt{R_{BV}^2 + R_{BH}^2} = \sqrt{345.13^2 + 548^2} = 647.6$ N.

For calculating the bearing, we take $F_r = 647.6$ N and $F_a = 398.9$ N.

The bending moment on the first interval in the horizontal plane is $M_I = R_{AH}x$, and when $x = 0$ then $M_I = 0$, while when $x = 0.5l$ then $M_I = 548\cdot 0.5\cdot 0.107 = 29.32$ N·m. It is obvious that at the end of the second interval the bending moment equals zero. The bending-moments diagram for the horizontal plane is shown in Figure 5-16.2d. The maximum bending moment in the horizontal plane is in the middle between the bearings of the shaft, and it equals $M_H = R_{AH}\,0.5l = M_I = 29.32$ N·m. Since the maximum bending moment in the vertical plane is also applied to the same cross-sectional area of the shaft, the maximum bending moment M_1 applied to the shaft of the pinion is determined by using the following equation: $M_1 = \sqrt{M_V^2 + M_H^2} = \sqrt{18.46^2 + 29.32^2} = 34.65\,\text{N}\cdot\text{m}$.

The torque developed by the tangent force and applied to the shaft of the pinion is $T_1 = F_t 0.5d_1 = 1{,}096\cdot 0.5\cdot 0.0726 = 39.78$ N·m. The torque diagram is shown in Figure 5-16.2e.

5. Verifying calculations of the diameter of the shaft for the pinion.
We have to calculate the diameter of the shaft in the most loaded cross-sectional area, which is in the middle between the bearings, and according to the drawing it is estimated to equal 25 mm.

For the shaft of the pinion, we select the AISI 1020 annealed steel having $S_{ut} = 394.7$ MPa and $S_{yt} = 295$ MPa. Applying formulas (1-9.2) and (1-9.3), we calculate $k_a = 0.77$ and $k_b = 0.962$, and accounting for the fact that $k_c = 1$, we determine $S_e = 0.77\cdot 0.962\cdot 1\cdot 0.5\cdot 394.5 = 146.1$ MPa. Based on the type of keyway and according to Figure 3-8.4, for the bending stresses $K_{fb} = 1.6$ and for the torsional stresses $K_{ft} = 1.3$. In addition to the keyway, stress concentration is caused due to the step in the shaft having $\frac{D}{d}=\frac{30}{25}=1.2$ and $\frac{r}{d}=0.05$. From Appendix A-1-5, we found that for bending $K_{tb}=1.92$ and for torsion $K_{tt}=1.6$. Hence, taking $q=0.6$ and using formula (1-10.2), we calculate $K'_{fb}=1.55$ and $K'_{ft}=1.36$. Taking the larger of these two fatigue stress concentrations factors, we accept $K_{fb}=1.6$, and $K_{ft}=1.36$. Now we calculate the shaft diameter of the pinion using formula (3-5.5): $d=\left\{\frac{32\cdot 2}{\pi}\left[\left(\frac{1.6\cdot 34.65\cdot 10^3}{146.1}\right)^2+\frac{3}{4}\left(\frac{1.36\cdot 39.78\cdot 10^3}{146.1}\right)^2\right]^{\frac{1}{2}}\right\}^{\frac{1}{3}}=21.546\text{ mm.}$

According to the drawing, this diameter is 25 mm, and based on the design considerations we should not make any changes.

The calculations of the shaft of the gear are not presented in this design project because they are similar to the calculations of the shaft for the pinion.

6. Calculations of the bearing for the shaft of the pinion.

For the selected bearing, the dynamic load rating is C = 18.7 kN, and the static load rating is C_0 = 10.4 kN. For calculating the equivalent static load for angular-contact bearings, we take according to the corresponding methodologies that $X_0=0.5$ and $Y_0=0.26$. In our case, $F_r=647.6$ N and $F_a=398.9$ N. Hence, according to formula (2-4.6), we calculate $F_0=0.5\cdot 647.6+0.26\cdot 398.9=427.5$ N.

Since $F_0<C_0$ (427.5 < 104,000), we may proceed with the selected bearing.

We determine the equivalent dynamic load using formula (2-4.5) that reads $F_e=XVF_r+YF_a$. Since the inner ring is rotating, then $V=1$. According to formula (2-4.7), we have $e=\frac{417.5}{10,400}=0.040$, and according to formula (2-4.8) we obtain $\frac{398.9}{647.6}=0.616>0.040$. Taking for the bearing the contact angle 5° and using the corresponding table from Appendix A-2-3, we find that $X=0.56$ and accounting for what is presented in the table of this appendix $e=0.021$; we determine by interpolating that $Y=2.06$. Therefore, $F_e=0.56\cdot 647.6+2.06\cdot 398.9=1{,}145.5$ N.

The service life of the bearing can be calculated using the formula (2-4.4), hence, $L_{10}=\left(\frac{C}{F_e}\right)^3=\left(\frac{18.7}{1.146}\right)^3=4{,}345\cdot 10^6$ revolutions or $L_{10h}=\frac{4{,}345\cdot 10^6}{60\cdot 1{,}800}=40{,}231$ hours. The calculated service life meets service life criteria.

7. Calculation of deflection and angle of rotation of the shaft for the pinion.

The shaft is subjected to the action of the tangential force $F_t=1{,}096$ N, radial force $F_r=419.6$ N, and the axial force F_a = 398.9 N. These forces are applied to the cross-sectional area of the shaft located in the middle between the bearings. The action of the forces F_t and F_r can be replaced by their resultant $F=\sqrt{F_t^2+F_r^2}=\sqrt{1{,}096^2+419.6^2}=1{,}174$ N. The force F_a creates a bending moment M_o = 398.9·0.5·0.0726 = 14.5 N·m. These loading factors should be accounted for in determining the deflection and angle of rotation of the shaft. The shaft has a stepped structure, while for the calculations of its deformation a certain uniform shaft should represent it. Using formula (3-6.11), we calculate an equivalent diameter of the shaft, hence,

$$d_e = \frac{1}{\sqrt[4]{\frac{20}{103 \cdot 20^4} + \frac{63}{103 \cdot 25^4} + \frac{20}{103 \cdot 20^4}}} = 22.4 \text{ mm.}$$

The moment of inertia of the shaft having the equivalent diameter is

$$I = \frac{\pi 22.4^4}{64} = 12{,}352 \text{ mm}^4.$$

The total deflection of the shaft consists of the vector sum of the deflection due to the force F and due to the moment M_0. Using formula (3-6.7), we calculate the deflection h of the shaft between the bearings due to the force F: $h = -\frac{Fl^3}{48EI} = -\frac{1{,}174 \cdot 107^3}{48 \cdot 207 \cdot 10^3 \cdot 12{,}574} = -0.0117$ mm. The deflection of the shaft due to the moment M_0 is calculated using formula (3-6.17). Hence, since $a = 0.5l$, we obtain $h_0 = \frac{M_0 0.5l[2 \cdot 0.5l^2 - 6 \cdot 0.5l^2 + 4(0.5l)^2]}{24lEI} = 0$. The deflection of the shaft carrying a gear should not exceed $0.0003l$, therefore, the allowable deflection for the shaft in our case is $0.0003 \cdot 107 = 0.0321$ mm that exceeds the calculated deflection $|0.0117|$ mm. In addition, the deflection of the shaft on the place where the gear is mounted should not exceed for cylindrical gears $0.03m = 0.03 \cdot 3 = 0.09$ mm that is significantly more than 0.0117 mm.

The angle of rotation is also caused by the force F and the moment M_0. The angle of rotation due to the force is calculated using the formula (3-6.10):
$\theta_F = -\frac{Fl^2}{16EI} = -\frac{1{,}174 \cdot 107^2}{16 \cdot 207 \cdot 10^3 \cdot 16{,}278} = -0.00025$ rad. The angle of rotation due to the moment M_0 can be determined based on the formula (3-6.19):

$\theta_0 = \frac{14.5 \cdot 10^3[2 \cdot 107^2 - 6 \cdot 0.5 \cdot 107 \cdot 107 + 3(0.5 \cdot 107)^2]}{6 \cdot 207 \cdot 10^3 \cdot 16{,}278 \cdot 107} = -0.0000192$ rad. These two calculated angles of rotation are oriented under an acute angle to each other, therefore, we just add them to each other. Hence, the total angle of rotation is $\theta = -0.00025 - 0.0000192 = -0.000269$ rad, which is less than the absolute value of the allowable angle of rotation for our case $|0.01|$ rad.

8. Calculation of the critical rotational speed.

According to the calculated deflection, we determine the critical rotational speed for our case using formula (3-7.15): $n_{cr} = \frac{30}{\pi}\sqrt{\frac{g}{h}} = \frac{30}{\pi}\sqrt{\frac{9.81 \cdot 10^3}{0.0117}} = 8{,}748$ rpm.

Note.

The same steps used for calculating the shaft of the pinion presented above should be performed for the shaft of the gear.

9. Assembly and disassembly.

a) Assembly

Referring to Figure 5-16.1, we begin with installing key 18 into the keyseat of shaft 5, and then we mount gear 2 and secure it with retaining ring 19. Now we mount the bearings 4 and 16 and place the shaft with the bearings into the lower casing 1. After that, we install key 13 into the keyseat of shaft 7, then mount pinion 9 and secure it with retaining ring 14. Then we place the shaft with the pinion into lower casing 1. Now we install the two centering pins 10 into lower casing 1 and mount the upper casing (not shown in the drawing). We install ten bolts into halls 11 and secure the casings by tightening the nuts. Now we install covers 3 and 17 and by using shims (not shown in the drawing), we perform the bearing preload according to the catalog specifications. During the preload, we tighten the screws of the covers. Then we install the covers 6 and 15 and perform a similar preload. After that using the special opening in the upper case, we verify the accuracy of the engagement. A simple way is to paint a tooth and to measure the spots on the meshing teeth. The spots should comply with the specifications. After that, we pour oil through this opening keeping in mind that the level of the oil should be at the middle of the lowest tooth. Then we close the opening in the upper case and allow the transmission to be used as intended.

b) Disassembly

First we open the oil plug in the lower case to empty the oil. Then we unscrew the ten nuts and take off the bolts that hold the upper and lower casings together. After that, we dismount the covers 3, 18, 6, and 15. Now we take off the shafts with the gears and bearings mounted on them. Using special pullers, we dismount the bearings from the shafts and then remove the retaining rings. After that, we dismount the gears and the keys.

Chapter 5. Problems

5-01. Determine the rotational speed of the arm 4 and the train value and its sign for the planetary gear train shown in Figure 5p-01. The rotational speed of gear 1 is $n_1 = 1{,}800$ rpm, while the ring gear 3 is braked (not rotating). The number of teeth of the gears, respectively, are $N_1 = 22$, $N_2 = 99$, and $N_3 = 220$.

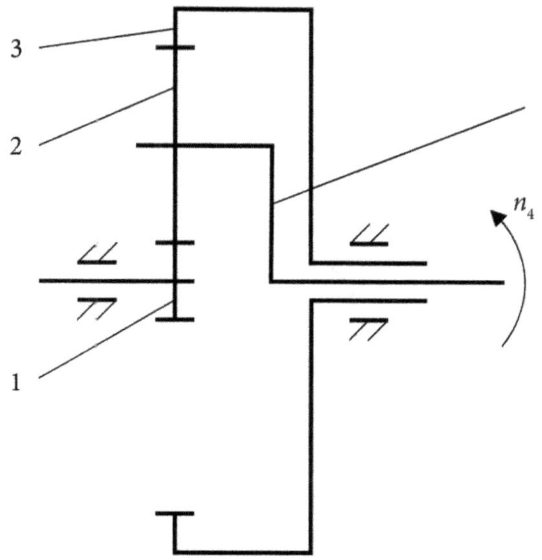

Figure 5p-01

5-02. An open one directional spur gear drive having a pressure angle of 20° and a module of 2.5 mm transmits power of 3 kW from the pinion at a rotational speed of 1,200 rpm to the gear rotating at 350 rpm. Determine the number of teeth of the pinion and of the gear, and calculate the bending stresses acting in these gears if the tooth width face is 25 mm.

5-03. An open reversible spur gear drive having a 23-tooth pinion is rotating at 1,800 rpm and transmits power of 4.0 hp to a 69-tooth gear. The diametral pitch of the teeth is $14\ \frac{1}{\text{in.}}$ and the tooth face width is 0.85 in. Determine the bending stresses in the teeth of the pinion and of the gear. In addition, calculate the safety factor of the pinion if it is made from steel having $S_{ut} = 185.5$ ksi.

5-04. Determine the module of an open one directional spur gear drive transmitting power of 7.5 kW from the 23-tooth pinion rotating at 1,800 rpm to the gear rotating at 500 rpm. Use for the pinion steel having $S_{ut} = 957$ MPa and take the safety factor with regard to bending stresses $n_b = 2$, and the tooth face width coefficient $\eta_b = 10$. Determine the required ultimate strength of the material for the gear.

5-05. An enclosed spur drive having a diametral pitch 12 $\frac{1}{\text{in.}}$ is transmitting power of 7.5 hp from a 25-tooth pinion rotating at 3,600 rpm to the 79-tooth gear. The tooth face width is 2.25 in. Determine the contact stresses in the gears and taking a safety factor with regard to contact stresses $n_H = 2$, select the materials for the gear and pinion.

5-06. Determine the center distance and the parameters of the gears for an enclosed spur drive transmitting power of 5.5 kW from a 25-tooth pinion rotating at 3,600 rpm to the 66-tooth gear. The allowable contact stress is 500 MPa. Take the tooth face width coefficient $\eta_c = 0.7$ and the safety factor with regard to contact stresses $n_H = 2$. Select the materials for the gear and pinion.

5-07. An open one directional helical gear drive having a helix angle of 22° and a transverse diametral pitch of 12 $\frac{1}{\text{in.}}$ transmits power of 10 hp from the 24-tooth pinion rotating at 1,800 rpm to the gear rotating at 450 rpm. Calculate the bending stresses acting in these gears if the tooth face width is 1.0 in.

5-08. An open one directional helical gear drive having a 23-tooth pinion that is rotating at 1,800 rpm and transmitting power of 7.5 hp to a 69-tooth gear. The helix angle is 21°, the transverse diametral pitch of the teeth is 14 $\frac{1}{\text{in.}}$ and the tooth face width is 0.75 in. Determine the bending stresses in the teeth of the pinion and of the gear. In addition, calculate the safety factor of the pinion if it is made from steel having $S_{ut} = 185.5$ ksi

5-09. Determine the normal module of an open one directional helical gear drive having a helix angle of 30° and transmitting power of 4 kW from the 23-tooth pinion rotating at 1,800 rpm to the gear rotating at 550 rpm. Use for the pinion the steel having $S_{ut} = 940$ MPa and take the safety factor with regard to bending stresses $n_b = 2$, and the tooth face width coefficient $\eta_b = 15$. Determine the required ultimate strength of the material for the gear.

5-10. An enclosed helical drive having a transverse diametral pitch of 12 $\frac{1}{\text{in.}}$ is transmitting power of 7.5 hp from a 25-tooth pinion rotating at 3,600 rpm to the 75-tooth gear. The helix angle is 22° and the tooth face width is 1.75 in. Taking $\lambda_H = 1.4$ and the safety factor with regard to contact stresses $n_H = 2$, determine the contact stresses in the gears and select the materials for the pinion and gear.

CHAPTER 6

Bevel Gears

Bevel gears resemble a frictional drive of two cones transmitting power between intersecting shafts. The commonly used angle of intersection is 90°. Bevel gears might have straight, inclined, or curved teeth. The manufacturing of bevel gears is associated with highly specialized metal cutting equipment and represents a more complicated and more expensive process compared with the processes used for cylindrical gears. In spite of that, bevel gears have a very widespread application in industry. Usually, bevel gears with straight teeth are used in transmissions that work at velocities below 3,000 fpm $\left(4\frac{\text{m}}{\text{s}}\right)$ and velocity ratios on the order of $i \leq 5$. Curved tooth bevel gears allow velocities up to 6,000 fpm $\left(8\frac{\text{m}}{\text{s}}\right)$ and velocity ratios up to $i = 10$. Bevel gears are made from the same materials as cylindrical gears. In this book, we consider bevel gears with straight teeth.

6-1 Bevel Gears: Their Geometry and Acting Forces

Figure 6-1.1 shows a pair of meshing bevel gears.

Figure 6-1.1. A pair of meshing bevel gears. (Courtesy of KHK-USA, Inc.)

Figure 6-1.2 shows a bevel pinion and a bevel gear. The gears shown in the figures represent off-the-shelf products that allow machining the holes in the hubs to the required diameters.

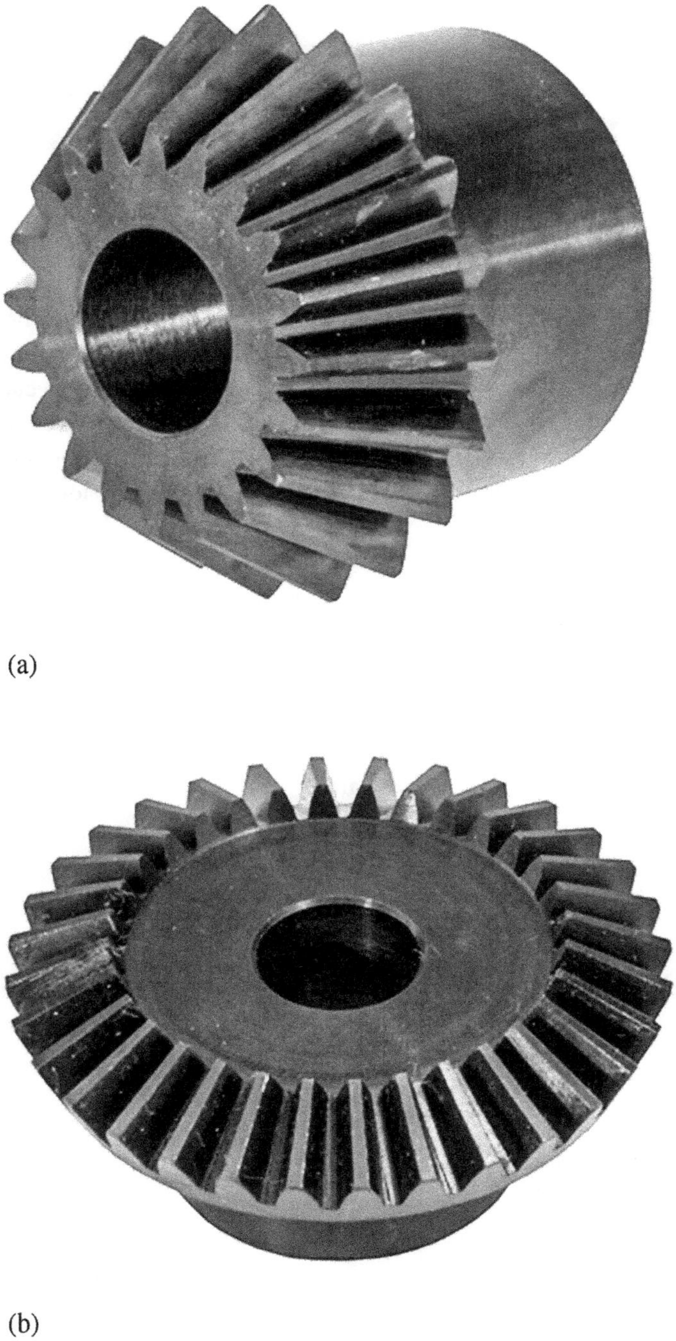

(a)

(b)

Figure 6-1.2. (a) Bevel pinion; (b) bevel gear. (Courtesy of Martin Sprocket & Gear, Inc.)

Figure 6-1.3 shows two cones *OAB* and *OBC* having a common line *OB*, and representing a schematic of a pair of bevel gears while their axes intersect under 90°.

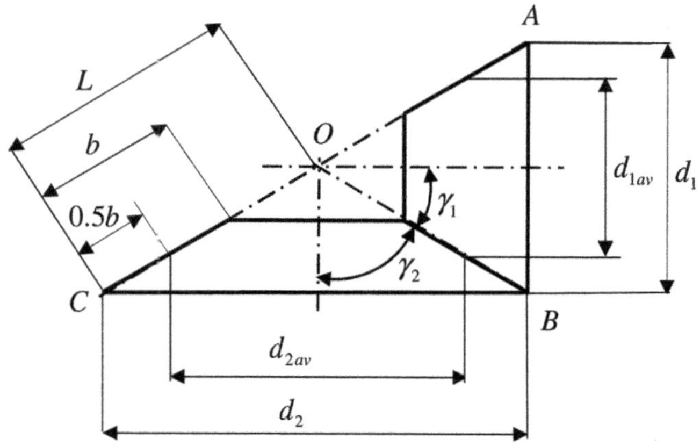

Figure 6-1.3. Bevel gear geometry.

Point *O* represents the common pitch apex, while the two cones are the pitch cones. The angles γ_1 and γ_2 are, respectively, the pitch angles of the pinion and of the gear. The relationship between the circular pitch, the diametral pitch, and the pitch diameter are the same as for spur gears; however, all these parameters are measured at the large end of the teeth. The pitch diameters for the pinion d_1 and for the gear d_2, as well as the tooth face width *b* and cone distance *L*, are shown in Figure 6-1.3. The velocity ratio, as it is for the spur gears, is the ratio between the diameters d_2 and d_1, as well as between the number of teeth N_2 and N_1, respectively:

$$i=\frac{d_2}{d_1}=\frac{N_2}{N_1} \tag{6-1.1}$$

From Figure 6-1.3 we have,

$$i=\frac{d_2}{d_1}=\frac{N_2}{N_1}=\tan\gamma_2=\frac{\sin\gamma_2}{\sin\gamma_1} \tag{6-1.2}$$

Applying some additional conventional transformations to the geometric relationships between the parameters in Figure 6-1.3, we obtain

$$\sin\gamma_1=\cos\gamma_2=\frac{1}{\sqrt{i^2+1}} \tag{6-1.3}$$

$$\sin\gamma_2=\cos\gamma_1=\frac{i}{\sqrt{i^2+1}} \tag{6-1.4}$$

$$L = \frac{d_1\sqrt{i^2+1}}{2} = \frac{d_2\sqrt{i^2+1}}{2i} \tag{6-1.5}$$

The analysis of forces acting between the teeth of straight tooth bevel gears, which are called spur bevel gears, is usually associated with the average pitch diameters. From Figure 6-1.3 we determine the average diameters of the pinion and the gear:

$$d_{1av} = \frac{d_1\,(L-0.5b)}{L} = d_1 - b\cos\gamma_2 \tag{6-1.6}$$

and

$$d_{2av} = \frac{d_2\,(1-0.5b)}{L} = d_2 - b\cos\gamma_1 \tag{6-1.7}$$

Figure 6-1.4 shows a schematic of two engaged pitch cones 1 and 2 having their common pitch apex at the point O. At the intersection of the average pitch diameters in pitch point P (in the middle of the tooth face width) are conventionally applied forces acting on the bevel gear engagement. On pitch cone 1 is drawn a space diagram that allows us to analyze the relationships between the forces applied to the pinion. Obviously, similar relationships apply to the gear.

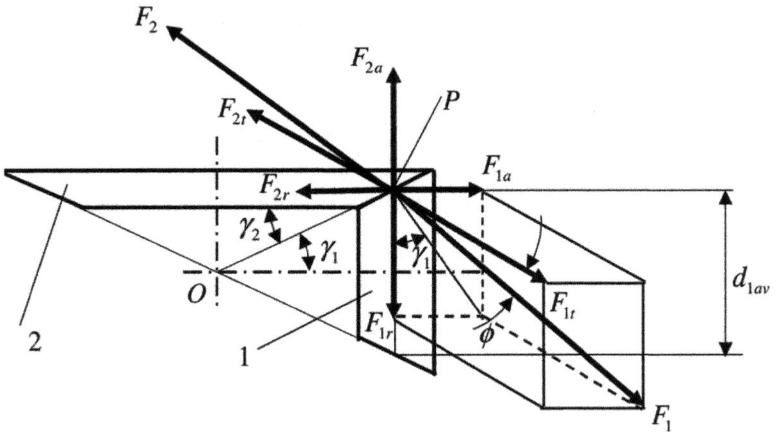

Figure 6-1.4. Forces acting on the bevel gear engagement.

Pressure force F_1 acting along the pressure line is equal and opposite to pressure force F_2; both are applied to pitch point P. Each of the pressure forces are resolved into three components: the tangential forces, the radial forces, and the axial forces. However, as shown in Figure 6-1.4, among the resolved forces only the tangential forces acting in the pinion and in the gear have the same values. Hence, the radial force of the pinion and the axial force of the gear have the same absolute values, and the axial force of the pinion is equal to the radial force of the gear in terms of their values. The tangential force is determined from the following expression:

$$F_t = F_{1t} = F_{2t} = \frac{2W}{\omega_1 d_{1av}} = \frac{2W}{\omega_2 d_{2av}} \tag{6-1.8}$$

Hence,

$$F = F_1 = F_2 = \frac{F_t}{\cos\phi} \tag{6-1.9}$$

Based on Figure 6-1.4, we determine the radial and axial forces applied to the pinion:

$$F_{1r} = F_t \tan\phi \sin\gamma_1 \tag{6-1.10}$$

and the axial force is

$$F_{1a} = F_t \tan\phi \cos\gamma_1 \tag{6-1.11}$$

We obtain similar relationships for the gear:

$$F_{2r} = F_t \tan\phi \sin\gamma_2 \tag{6-1.12}$$

and

$$F_{2a} = F_t \tan\phi \cos\gamma_2 \tag{6-1.13}$$

Combining equations (6-1.3) and (6-1.4) with equations (6-1.10), (6-1.11), (6-1.12) and (6-1.13), we may write, respectively,

$$F_{1r} = F_{2a} = \frac{F_t \tan\phi}{\sqrt{i^2+1}} \tag{6-1.14}$$

$$F_{1a} = F_{2r} = \frac{F_t i \tan\phi}{\sqrt{i^2+1}} \tag{6-1.15}$$

6-2. Bending Stress Calculations for Bevel Gears

The stress analysis of bevel gears is based on considering the stresses in the teeth of a virtual spur gear whose teeth have a form like a bevel gear's teeth. Figure 6-2.1 shows that the center of the virtual spur gear coincides with the apex of the back cone at point O_v, and the diameters of the virtual pinion and virtual gear are

$$d_{1v} = \frac{d_1}{\cos\gamma_1} \tag{6-2.1}$$

and

$$d_{2v} = \frac{d_2}{\cos \gamma_2} \qquad (6\text{-}2.2)$$

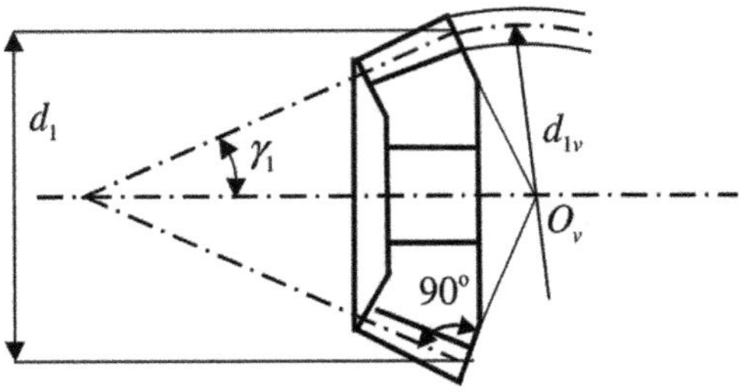

Figure 6-2.1. Imaging of the virtual gear.

Accordingly, the virtual number of teeth of the virtual pinion and virtual gear are determined from the following formulas:

$$N_{1vb} = \frac{N_1}{\cos \gamma_1} \qquad (6\text{-}2.3)$$

and

$$N_{2vb} = \frac{N_2}{\cos \gamma_2} \qquad (6\text{-}2.4)$$

The tooth form factor for the virtual gears is determined as it is for helical gears using formula (5-12.6).

Based on appropriate experimental investigations, several sources recommend increasing the load correction factors for bevel gears 10 to 15% over spur gears. Therefore, we take $k_r = 1.3$, $k_d = 1.8$, and $k_m = 1.2$. The fatigue stress concentration factor is the same as for spur and helical gears.

The calculations of bending stresses for bevel gears are related to the virtual spur gear that is generated by the normal section passing through the midpoint of the tooth face width. This section is associated with the average diameter and average normal pitch or average normal module of the real bevel gear.

Based on these considerations, it is possible to present the following formula for calculating bending stresses in the pitch system bevel gears. Hence, for pitch system pinion the formula reads

$$\sigma = \frac{2WP_{av}k_r k_d k_m}{\omega_1 d_{1av} b y_{1v}} \qquad (6\text{-}2.5)$$

where P_{av} is the average value of the diametral pitch, d_{1av} is the average value of the pinion pitch diameter, and y_{1v} is the virtual tooth form factor for the pinion, while $P_{av} = \frac{N}{d_{av}}$. For the module system pinion we obtain

$$\sigma = \frac{2Wk_r k_d k_m}{\omega_1 m_{av} d_{1av} b y_{1v}} \tag{6-2.6}$$

where m_{av} is the average value of the module, and $m_{av} = \frac{d_{av}}{N}$.

The stresses in the gear can be determined, as for helical gears, using formula (5-7-11).

Equations (6-2.5) and (6-2.6) are applicable for calculating stresses in cases of static loading. Because the teeth of bevel gears, like the teeth of spur and helical gears, are subjected to variable loading, these equations should be modified to account for the fatigue considerations. The modification of these equations consists in accounting for the stress concentration factor and the characteristics of the variable loading cycles. As discussed earlier for spur and helical gears, the fatigue stress concentration factor K_f for the teeth is estimated to be in the range of 1.45 to 2.05. The teeth are subjected to a repeated loading cycle if the drive is not reversible, or to a completely reversed loading cycle if the transmission is reversible. The modified formulas are presented below.

6-2.1. Verifying Calculations for a Repeated Cycle of Bending Stresses in Bevel Gears

Similar to spur and helical gears, bevel gears under a repeated loading cycle should be calculated based on the amplitude value of the bending moment according to equation (5-8.1) and the stress amplitude limit S_a that should be determined from equation (5-8.2). Therefore, we obtain for the pitch system pinion the following expression for the verifying calculations:

$$\sigma_{rc} = \frac{WP_{av} k_r k_d k_m K_f}{\omega_1 d_{1av} b y_{1v}} \le \frac{S_a}{n_b} \tag{6-2.7}$$

and for the module system pinion the expression reads

$$\sigma_{rc} = \frac{Wk_r k_d k_m K_f}{\omega_1 m_{av} d_{1av} b y_{1v}} \le \frac{S_a}{n_b} \tag{6-2.8}$$

It should be emphasized that for stress calculations of bevel gears the velocity ratio or the pitch angle of one of the gears should be given. As mentioned above, the axes of the gears intersects under a straight angle.

Example 6-2.1a: Verifying calculations for a repeated cycle of bending stresses for a pitch system bevel pinion.

An open bevel gear drive transmits 1.5 hp of power from an 18-tooth bevel pinion having a diametral pitch $P = 14\frac{1}{\text{in.}}$ on the large end and is rotating at a speed of n_1 = 1,800 rpm to the gear having a rotational speed of 600 rpm. The width of the tooth face b = 0.6 in. The pinion is made from AISI 8650 normalized steel having S_{ut} = 148,500 psi. We determine the safety factor of the pinion with regard to bending stresses. The rotation is one directional.

Solution:

Calculating the angular velocity and the pitch diameter on the large end we obtain

$$\omega_1 = \frac{\pi 1{,}800}{30} = 188.4\ \frac{\text{rad}}{\text{s}},\quad d_1 = \frac{18}{14} = 1.285\ \text{in.}$$

Since the velocity ratio is $i = \frac{1{,}800}{600} = 3$ and considering formula (6-1.2), we have $\tan\gamma_2 = 3$, therefore, $\tan^{-1} 3 = \gamma_2 = 71.565°$, hence, $\gamma_1 = 90° - 71.565° = 18.435°$. The virtual number of teeth according to formula (6-2.1) and tooth form factor according to formula (5-12.6), respectively, are $N_{1v} = \frac{18}{\cos 18.435°} = 18.97$ and $y_{1v} = 0.52\left(1 - \frac{5.6}{18.97}\right) = 0.366$. We take $k_r = 1.3$, $k_d = 1.8$, and $k_m = 1.2$ and estimate that $K_f = 1.7$. Using formula (6-1.6), we calculate the average diameter of the pinion: $d_{1av} = d_1 - b\cos\gamma_2 = 1.285 - 0.6\cos 71.565° = 1.095$ in. The average diametral pitch equals $P_{av} = \frac{18}{1.095} = 16.44\frac{1}{\text{in.}}$. In order to determine the endurance limit, we calculate using formula (1-9.2) $k_a = 0.74$, and we take $k_b = 0.95$. Because $k_c = 1$, we determine $S_e = 0.74 \cdot 0.95 \cdot 1 \cdot 0.5 \cdot 148{,}500 = 52{,}198$ psi.

Based on formula (5-8.2), we calculate $S_a = \frac{52{,}198 \cdot 148{,}500}{52{,}198 + 148{,}500} = 38{,}622$ psi.

Now, based on formula (6-2.5) we calculate $\sigma_{rc} = \frac{1.5 \cdot 550 \cdot 12 \cdot 16.44 \cdot 1.3 \cdot 1.8 \cdot 1.2 \cdot 1.7}{188.4 \cdot 1.095 \cdot 0.6 \cdot 0.366} = 17{,}150$ psi.

Therefore, $22{,}866 = \frac{S_a}{n_b}$, hence, $n_b = \frac{38{,}622}{17{,}150} = 2.25$.

6-2.2. Verifying Calculations for a Completely Reversed Cycle of Bending Stresses of Bevel Gears

For this cycle the calculations are based on the maximum value of the bending moment (having in the numerator the multiplier 2) and the endurance limit. The same approach is applied to the verifying calculations of spur and helical gears. Based on formula (6-2.5), we obtain the expression of the verifying calculations for a pitch system pinion:

$$\sigma_{crc} = \frac{2WP_{av}k_r k_d k_m K_f}{\omega_1 d_{1av} b y_{1v}} \le \frac{S_e}{n_b} \tag{6-2.9}$$

Modifying formula (6-2.6), we write the expression for verifying calculations of a module system of bevel gear:

$$\sigma_{crc} = \frac{2Wk_r k_d k_m K_f}{\omega_1 m_{av} d_{1av} b y_{1v}} \le \frac{S_a}{n_b} \tag{6-2.10}$$

Example 6-2.2a: Verifying calculations for a completely reversed cycle of bending stresses for module system bevel gears.

In an open drive, a 19-tooth bevel pinion having a pitch angle $\gamma_1 = 32°$ and module 4 mm is transmitting power of 11 kW at 3,600 rpm. The tooth face width $b = 30$ mm. The pinion is made from the AISI 4820 normalized steel having S_{ut} = 755 MPa. We determine the bending stress safety factor if the drive is reversible.

Solution:
Using formula (1-9.2), we calculate k_a = 0, 784. Taking k_b = 0.95 and accounting for k_c = 1, we calculate $S_e = 0.784 \cdot 0.95 \cdot 1 \cdot 0.5 \cdot 755 = 281.353$ MPa. Now we determine for the pinion the angular velocity, pitch diameter at the larger end, and the tooth form factor, respectively:
$\omega_1 = \dfrac{\pi 3,600}{30} = 376.8\ \dfrac{1}{s}$, $d_1 = 4 \cdot 19 = 76$ mm, $N_{1v} = \dfrac{19}{\cos 32°} = 22.4$ and $y_{1v} = 0.52\left(1 - \dfrac{5.6}{22.4}\right) = 0.39.$
We take k_r = 1.3, k_d = 1.8, k_m = 1.2, and K_f = 1.6. We calculate γ_2 = 90° − 32° = 58°, hence, d_{1av} = 76 − 30 cos 58° = 60.1 mm and $m_{av} = \dfrac{60.1}{19} = 3.163$ mm. Now using formula (6-2.10), we determine $\sigma_{crc} = \dfrac{2 \cdot 11 \cdot 10^3 \cdot 1.3 \cdot 1.8 \cdot 1.2 \cdot 1.6}{376.8 \cdot 0.003163 \cdot 0.0601 \cdot 0.03 \cdot 0.39} = 117{,}942$ MPa. Hence, $117{,}942 = \dfrac{S_e}{n_b}$, therefore, $n_b = \dfrac{281.353}{117{,}942} = 2.386.$

6-3. Design Calculations of Bevel Gears for Bending Stresses

Based on expressions (6-2.5) and (6-2.6), we derive formulas for design calculations of a bevel pinion for repeated and completely reversed cycles of bending stresses. These calculations are associated with the empiric tooth face width coefficient η_L that represents the ratio between the width and the cone distance:

$$\eta_L = \frac{b}{L} \tag{6-3.1}$$

Numerous experimental investigations suggest that η_L varies in the range from 0.25 to 0.35.

6-3.1. Design Calculations of Bevel Gears for a Repeated Cycle of Bending Stresses

Combining formula (6-3.1) with expressions (6-2.7) and (6-2.8) and accounting for expressions (6-1.5), (6-1.6), and that $d_{1av} = \dfrac{N_1}{P_{av}}$, we obtain $\dfrac{2P_{av}^3 W k_r k_d k_m K_f (1-0.5\eta_L)}{\omega_1 N_1^2 \eta_L y_{1v} \sqrt{i^2+1}} = \dfrac{S_a}{n_b}$. Solving this equation for P_{av}, we derive the formula for calculating the average diametral pitch of the pitch system bevel gears experiencing a repeated cycle of bending stresses:

$$P_{av} = \sqrt[3]{\frac{\omega_1 N_1^2 \eta_L y_{1v} S_a \sqrt{i^2+1}}{2Wk_r k_d k_m K_f n_b (1-0.5\eta_L)}} \tag{6-3.2}$$

where P_{av} in $\dfrac{1}{\text{in.}}$, S_a in psi, and W in hp.

Similarly, using expression (6-2.6) and having that $d_{1av} = m_{av}N_1$, we obtain the formula for calculating the average module for the module system bevel gears subjected to a repeated cycle of bending stresses:

$$m_{av} = \sqrt[3]{\frac{2Wk_r k_d k_m K_f n_b (1-0.5\eta_L)}{\omega_1 N_1^2 \eta_L y_{1v} S_a \sqrt{i^2+1}}} \tag{6-3.3}$$

where m_{av} in meter (m), S_a in pascal (Pa), and W in watt (W).

Example 6-3.1a: Design calculations of a pitch system bevel pinion for a repeated cycle of bending stresses.

The pinion of an open bevel gear drive transmits 5.5 hp of power at 1,800 rpm. The number of teeth in the pinion is N_1 = 18, the velocity ratio is i = 2.6, and the rotation is one directional. The pinion is made of AISI 1340 normalized steel having S_{ut} = 121, 250 psi. We determine the diametral pitch for the safety factor n_b = 2.

Solution:

First we determine the endurance limit. Based on formula (1-9.2), we calculate $k_a = 0.77$. We take $k_b = 0.95$ because the thickness of the tooth is less than 0.75 in. Accounting for the fact that $k_c = 1$, we calculate $S_e = 0.77 \cdot 0.95 \cdot 1 \cdot 0.5 \cdot 121{,}250 = 44{,}347$ psi. Using formula (5-8.2), we determine $S_a = \dfrac{44{,}347 \cdot 121{,}250}{44{,}347 + 121{,}250} = 32{,}471$ psi. Since $i = \tan\gamma_2$ we found $\tan\gamma_2 = 2.6$, therefore, $\gamma_2 = 69°$ and $\gamma_1 = 21°$. Applying formulas (6-2.2) and (5-12.6), we calculate, respectively, $N_{1v} = \dfrac{18}{\cos 21°} = 19{,}28$ and $y_{1v} = 0.52\left(1 - \dfrac{5.6}{19.28}\right) = 0.368$. We take k_r = 1.3, k_d = 1.8, k_m = 1.2, K_f = 1.7, and η_L = 0.3. The angular velocity of the pinion $\omega_1 = \dfrac{\pi 1{,}800}{30} = 188.4\ \dfrac{1}{\text{s}}$. Based on formula (6-3.2), we calculate

$$P_{av} = \sqrt[3]{\frac{188.4 \cdot 18^2 \cdot 0.3 \cdot 0.368 \cdot 32{,}471\sqrt{2.6^2+1}}{2 \cdot 5.5 \cdot 550 \cdot 12 \cdot 1.3 \cdot 1.8 \cdot 1.2 \cdot 1.7 \cdot 2(1-0.5 \cdot 0.3)}} = 10.9\ \frac{1}{\text{in.}}$$

Hence, $d_{1av} = \dfrac{18}{10.09} = 1.784$ in. Based on the equation (6-1.6), we have

$d_1 = \dfrac{d_{1av}}{1-0.5\eta_L} = \dfrac{1.784}{1-0.5 \cdot 03} = 2.099$ in. Therefore, $P = \dfrac{18}{2.099} = 8.58\ \dfrac{1}{\text{in.}}$.

We take the closest smaller standard pitch $P = 8\ \dfrac{1}{\text{in.}}$.

6-3.2. *Design Calculations of Bevel Gears for a Completely Reversed Cycle of Bending Stresses*

Modifying formulas (6-3.2) and (6-3.3) by replacing S_a with S_e and applying the multiplier 2 to the power W, we obtain the formulas for calculating the average diametral pitch and average module for bevel gears subjected to a completely reversed loading cycle. These formulas read

$$P_{av} = \sqrt[3]{\frac{\omega_1 N_1^2 \eta_L y_{1v} S_e \sqrt{i^2+1}}{4Wk_r k_d k_m K_f n_b (1-0.5\eta_L)}} \tag{6-3.4}$$

where P_{av} in $\frac{1}{\text{in.}}$, S_a in psi, and W in hp

and

$$m_{av} = \sqrt[3]{\frac{4Wk_r k_d k_m K_f n_b (1-0.5\eta_L)}{\omega_1 N_1^2 \eta_L y_{1v} S_e \sqrt{i^2+1}}} \tag{6-3.5}$$

where m_{av} in meter (m), S_a in pascal (Pa), and W in watt (W).

As emphasized earlier, it is recommended that you use the U. S. units and SI units in the formulas for pitch system and module system gears.

Example 6-3.2a: Design calculations of a module system bevel gear for a completely reversed cycle of bending stresses.

A 19-tooth pinion of an open bevel gear drive transmits 7.5 kW of power at 1,800 rpm. The velocity ratio $i = 2.7$. The pinion is made from AISI 5160 normalized steel having $S_{ut} = 960$ MPa. We determine the module for a safety factor $n_b = 2$. The rotation is reversible.

Solution:

According to formula (1-9.2), we calculate that $k_a = 0.75$, and we take $k_b = 0.95$ assuming that the thickness of the tooth is less than 20 mm. Since $k_c = 1$, we calculate

$S_e = 0.75 \cdot 0.95 \cdot 1 \cdot 0.5 \cdot 860 = 306.4$ MPa.

Considering that $i = \tan\gamma_2 = \cot\gamma_1 = 2.7$, therefore, $\gamma_1 = 20.32°$.

Using formulas (6-2.2) and (5-12.6), we calculate $N_{1v} = \dfrac{19}{\cos 20.32°} = 20.26$ and $y_{1v} = 0.52\left(1 - \dfrac{5.6}{20.26}\right) = 0.376$, respectively. We take $k_r = 1.3$, $k_d = 1.8$, $k_m = 1.2$, and $\eta_L = 0.3$. Referring to Figure 5-8.1, we estimate $K_f = 1.70$. The angular velocity of the pinion is $\omega_1 = \dfrac{\pi 1{,}800}{30} = 188.4\,\dfrac{\text{rad}}{\text{s}}$.

Applying formula (6-3.5), we compute

$$m_{av} = \sqrt[3]{\frac{4 \cdot 7.5 \cdot 10^3 \cdot 1.3 \cdot 1.8 \cdot 1.2 \cdot 1.7 \cdot 2(1-0.5 \cdot 0.3)}{188.4 \cdot 19^2 \cdot 0.3 \cdot 0.376 \cdot 306.4 \cdot 10^6 \sqrt{2.7+1}}} = 0.0038 \text{ m}.$$

Hence, $d_{1av} = 3.8 \cdot 19 = 72.14$ mm. Using equation (6-1.6), we have

$$d_1 = \frac{d_{1av}}{1-0.5\eta_L} = \frac{72.14}{1-0.5\cdot 03} = 84.87 \text{ mm.}$$

Therefore, $m = \frac{84.87}{19} = 4.467$ mm. We take the closest standard module $m = 4.5$ mm.

6-4. Surface Durability of Bevel Gears

As mentioned above, the stress analysis of bevel gears is associated with the consideration of a virtual spur gear that is characterized by a virtual average pitch diameter and a virtual number of teeth. Therefore, in order to use the basic formula (1-4.30) for contact stresses, we have to determine the virtual radii of curvature of the virtual teeth profiles on the pitch point. Hence, the formulas for the virtual radii for the pinion and gear, respectively, read

$$r_{1v} = \frac{d_{1av}\sin\phi}{2\cos\gamma_1} \tag{6-4.1}$$

and

$$r_{2v} = \frac{d_{2av}\sin\phi}{2\cos\gamma_2} \tag{6-4.2}$$

where the pressure angle $\phi = 20°$.

The tangential force applied in the midpoint of the teeth is determined from the formula (6-1.8) that reads $F_t = \frac{2W}{\omega_1 d_{1av}}$. Combining this formula with formulas (1-4.30), (6-4.1), and (6-4.2), and replacing l with b, we may write

$$\sigma_H = 0.418\sqrt{\frac{4WE(d_{2av}\cos\gamma_1 + d_{1av}\cos\gamma_2)}{\omega_1 d_{1av}^2 b d_{2av}\sin\phi}} \tag{6-4.3}$$

Applying to equation (6-4.3) formulas (6-1.3) and (6-1.4), and since $d_{2av} = id_{1av}$ we have

$$\sigma_H = 0.418\sqrt{\frac{4WE\sqrt{i^2+1}}{0.342\omega_1 d_{1av}^2 bi}} \tag{6-4.4}$$

Appropriately modifying formula (6-4.4), we obtain expressions for verifying and design calculations of bevel gears with regard to contact stresses.

6-4.1. Verifying Calculations of Bevel Gears Based on Surface Durability

Accounting for the load factors in formula (6-4.4), we obtain an expression for verifying calculations of bevel gears with regard to contact stresses:

$$\sigma_H = 1.43\sqrt{\frac{WEk_r k_d k_m \sqrt{i^2+1}}{\omega_1 d_{1av}^2 bi}} \le \frac{S_H}{n_H} \qquad (6\text{-}4.5)$$

The allowable contact stresses S_H for bevel gears are based on the hardness of the material—from $0.333H_B$ to $0.377H_B$ expressed in ksi and from $2.3H_B$ to $2.6H_B$ expressed in MPa. The hardness of the material of the pinion should be at least $50H_B$ higher than the hardness of the gear, as it is for spur and helical gears.

Example 6-4.1a: Verifying calculations of contact stresses for module system bevel gears.

An enclosed bevel gear drive transmits 2.2 kW of power; the rotational speed of the pinion n_1 = 1,800 rpm. The module of the larger end of the gears is m = 3.5 mm. The number of teeth of the pinion N_1 = 19, and of the gear N_2 = 57. The tooth face width b = 40 mm. Taking the safety factor n_H = 2, we will determine the materials for the pinion and for the gear.

Solution:

We calculate the angular velocity of the pinion $\omega_1 = \frac{\pi 1{,}800}{30} = 188.4\ \frac{\text{rad}}{s}$. The diameter of the pinion on the larger end is d_1 = 3.5·19 = 66.5 mm. The velocity ratio is $i = \frac{57}{19} = 3$. Because $\tan\gamma_2 = i = 3$, we determine that $\gamma_2 = 71.56°$. Hence, using formula (6-1.6), we compute the average diameter of the pinion $d_{1av} = d_1 - b\cos\gamma_2 = 65.5 - 40\cos 71.56° = 52.85$ mm. Taking $k_r = 1.3$, $k_d = 1.8$, and $k_m = 1.2$, we use formula (6-4.5) to calculate the surface stress:

$$\sigma_H = 1.43\sqrt{\frac{2.2\cdot 10^3\cdot 207\cdot 10^9\cdot 1.3\cdot 1.8\cdot 1.2\sqrt{3^2+1}}{188.4\cdot 0.05285^2\cdot 0.04\cdot 3}} = 362\cdot 10^6 \text{ Pa. Hence, } \sigma_H = 362 \text{ MPa.}$$

Therefore, $\frac{S_H}{n_H} = 362$ MPa, or S_H = 2.362 = 724 MPa. Taking $S_H = 2.45\ H_B$ (MPa), we calculate $H_B = \frac{S_H}{2.45} = \frac{724}{2.45} = 295.4$. Therefore, we select for the gear AISI 3140 normalized steel having H_B = 262, while for the pinion the hardness should be at least 50 units higher, therefore, we take AISI 4140 normalized steel having H_B = 302.

6-4.2. Design Calculations of Bevel Gears Based on Surface Durability

Combining formula (6-3.1) with expression (6-4.6), we may write

$$1.43\sqrt{\frac{WEk_r k_d k_m \sqrt{i^2+1}}{\omega_1 d_{1av}^2 \eta_L Li}} = \frac{S_H}{n_H} \qquad (6\text{-}4.6)$$

Linking formulas (6-1.5), (6-1.6), and (6-3.1), we have

$$L = \frac{d_{1av}\sqrt{i^2+1}}{2(1-0.5\eta_L)} \quad (6\text{-}4.7)$$

Combining equations (6-4.6) and (6-4.7) yields

$$1.43\sqrt{\frac{2WEk_r k_d k_m(1-0.5\eta_L)}{\omega_1 d_{1av}^3 \eta_L i}} = \frac{S_H}{n_H} \quad (6\text{-}4.8)$$

Squaring equation (6-4.8), we write

$$\frac{1.43^2 \cdot 2WEk_r k_d k_m(1-0.5\eta_L)}{\omega_1 d_{1av}^3 \eta_L i} = \frac{S_H^2}{n_H^2} \quad (6\text{-}4.9)$$

Solving equation (6-4.9) for d_{1av}, we obtain the formula for design calculations of bevel gears with respect to surface durability:

$$d_{1av} = 1.6\sqrt[3]{\frac{WEk_r k_d k_m n_H^2(1-0.5\eta_L)}{\omega_1 \eta_L i S_H^2}} \quad (6\text{-}4.10)$$

Example 6-4.2a: Design calculations of bevel gears based on surface durability.
It is necessary to design an enclosed bevel gear drive that transmits 7.5 hp of power to the output shaft that rotates at 430 rpm. The input shaft rotates at the angular velocity of 1,200 rpm. The safety factor is 2. We will determine the pitch diameters of the gears, the tooth face width, and the cone distance.

Solution:
When selecting the materials for the gear and pinion, the hardness for the pinion should exceed the hardness of the gear by at least 50 units of H_B. Therefore, for the gear we select AISI 6150 normalized steel having $H_B = 269$, and for the pinion we select AISI 4150 normalized steel having $H_B = 321$. We substitute into the formula for design calculations the data for the steel of the gear. Hence, $S_H = 0.35 \cdot 269 = 94.15$ ksi. The angular velocity of the pinion is $\omega_1 = \frac{\pi 1{,}200}{30} = 125.6\ \frac{\text{rad}}{\text{s}}$.

We take $k_r = 1.3$, $k_d = 1.8$, $k_m = 1.2$ and $\eta_L = 0.3$. The velocity ratio is $i = \frac{1{,}200}{430} = 2.79$. Using formula (6-4.10) we calculate $d_{1av} = 1.6\sqrt[3]{\frac{7.5 \cdot 550 \cdot 12 \cdot 30 \cdot 10^6 \cdot 1.3 \cdot 1.8 \cdot 1.2 \cdot 2^2(1-0.5 \cdot 0.3)}{125.6 \cdot 0.3 \cdot 2.79 \cdot 94.5^2 \cdot 10^6}} = 3.948$ in.

Combining formulas (6-1.6) and (6-3.1), we have $d_1 = \frac{d_{1av}}{1-0.5\eta_L} = \frac{3.948}{1-0.5 \cdot 0.3} = 4.645$ in. Taking the number of teeth for the pinion N_1 = 20, we compute $P=\frac{20}{4.645}=4.3\frac{1}{\text{in.}}$. We select the closest smaller standard diametral pitch $P=4\frac{1}{\text{in.}}$. We recalculate $d_1=\frac{20}{4}=5.0$ in. Hence, $d_2=\frac{2.79\cdot 20}{4}=13.95$ in. Since $\tan\gamma_2 = i = 2.79$, we determine $\gamma_2 = 70.28°$ and $\gamma_1 = 19.72°$.

Using formula (6-1.5), we determine $L=\frac{5\sqrt{2.79^2+1}}{2}=7.41$ in., therefore, $b = 0.3\cdot 7.41 = 2.223$ in.

6-5. Design Project

We need to design an enclosed bevel gear drive that transmits 2.2 kW of power from the pinion that rotates at 1,800 rpm to the output shaft that rotates at 500 rpm. The safety factor is 2. The drive is reversible.

1. Design calculations for the pinion.

The design project requirements emphasize that the drive should have an outboard structure for the pinion subassembly, allowing a simplified possibility of axial adjustment of the pinion in order to bring both pitch apexes to one point.

We select for the gear AISI 4340 annealed steel having H_B = 217 and for the pinion AISI 4340 normalized steel heaving H_B = 269. We use the allowable stress of the gear in the calculations; therefore, we determine this stress $S_H = 2.45\cdot 217 = 532$ MPa. The angular velocity of the pinion and the velocity ratio are $\omega_1=\frac{\pi 1{,}800}{30}=188.4\frac{\text{rad}}{\text{s}}$ and $i=\frac{1{,}800}{500}=3.6$. We take k_r = 1.3, k_d = 1.8, k_m = 1.2, and η_L = 0.3. Applying formula (6-4.10), we calculate $d_{1av}=1.6\sqrt[3]{\frac{2.2\cdot 10^3\cdot 207\cdot 10^9\cdot 1.3\cdot 1.8\cdot 1.2\cdot 2^2(1-0.5\cdot 03)}{188.4\cdot 0.3\cdot 3.6\cdot 532^2\cdot 10^{12}}}=0.0676$ m. Combining formulas (6-1.6) and (6-3.1), we write $d_1=\frac{d_{1av}}{1-0.5\eta_L}=\frac{67.6}{1-0.5\cdot 0.3}=79.57$ mm. We take for the pinion N_1 = 18 and calculate $m=\frac{79.57}{18}=4.42$ mm. We select the closest higher standard module m = 4.5 mm. Recalculating the diameter of the pinion at the larger end, we have $d_1 = 4.5\cdot 18 = 81$ mm. Therefore, $d_2 = 4.5\cdot 3.6\cdot 18 = 291.6$ mm. Since $\tan\gamma_2 = i = 3.6$, we calculate $\gamma_2 = 74.47°$ and $\gamma_1 = 15.53°$. Using formula (6-1.5), we determine $L=\frac{81\sqrt{3.6^2+1}}{2}=151.32$ mm, hence, $b = 0.3\cdot 151.32 = 45.4$ mm. Therefore, we recalculate $d_{1av} = 81 - 45.4\cos 74.47° = 68.84$ mm and $m_{av}=\frac{68.84}{18}=3.60$ mm.

2. Verifying calculations for bending stresses.
Because the drive is reversible, we carry out the verifying calculations for bending stresses with regard to the completely reversed loading cycle.

The material of the gear has S_{ut} = 745 MPa. Applying formula (1-9.2), we calculate k_a = 0, 790. Taking k_b = 0.95 and accounting that k_c = 1, we calculate $S_e = 0.786 \cdot 0.95 \cdot 1 \cdot 0.5 \cdot 745 = 279.693$ MPa.

Now we calculate $N_{1v} = \dfrac{18}{\cos 15.53°} = 18.68$ and $y_{1v} = 0.52\left(1 - \dfrac{5.6}{18.68}\right) = 0.364$.

We take k_r = 1.3, k_d = 1.8, k_m = 1.2, and K_f = 1.6. Using formula (6-2.10), we calculate $\sigma_{crc} = \dfrac{2 \cdot 3 \cdot 10^3 \cdot 1.3 \cdot 1.8 \cdot 1.2 \cdot 1.6}{188.4 \cdot 0.003825 \cdot 0.06884 \cdot 0.0454 \cdot 0.364} = 32{,}881{,}975.42$ Pa. Hence, σ_{crc} = 32.9 MPa and $32.9 = \dfrac{S_e}{n_b}$. Therefore, $n_b = \dfrac{279.693}{32.9} = 8.50$ which is more than four times larger than the safety factor for the contact stresses $\left(\dfrac{8.5}{2} = 4.25\right)$.

3. Calculation of the bore diameter of the bearings for the shaft of the pinion.
To make a preliminary selection of the bearings for the shaft of the pinion, we can determine the bore diameter of the bearings by calculating the diameter of the shaft based on formula (3-3.8) that reads $d = 3.65\sqrt[3]{\dfrac{P}{\tau_{all}}}$, where P is the power in watt, n in rpm, τ_{all} is in the range of 11 to 28 MPa, and d is in m. In order to avoid confusion, we denote the diameters of shafts by an additional subscript s: for example, the diameter of the shaft for the pinion is denoted by d_{s1}. Taking τ_{all} = 11 MPa, we calculate the diameters of the shaft for the pinion. Hence, the diameter of shaft for the pinion is $d_{s1} = 3.65\sqrt[3]{\dfrac{2.2 \cdot 10^3}{1.800 \cdot 11 \cdot 10^6}} = 0.0175$ m. We round up to 20 mm and, since the pinion is mounted on the shaft's cantilever, we assign this diameter for the pinion and take the next standard bore diameter for the bearings. Thus, we take the bore diameter of the bearing for the shaft of the pinion d = 25 mm. Often the shafts of bevel gears are supported by angular-contact ball or taper roller bearings; therefore, we make a preliminary selection of angular-contact ball bearing SKF 7205BEP having the outside diameter of 52 mm, the width 15 mm, the dynamic load rating C = 14.8 kN, and static load rating C_0 = 9.3 kN.

Obviously, since we determined the geometric parameters of the pinion, we can calculate the dimensions of the gear, and, in addition, we can determine the bore diameter of the bearings for the shaft that carries the gear and begin to compose the assembly drawing of the bevel gear transmission. However, the design of cylindrical and bevel gear transmissions differ from one another in two ways. The first is the necessity of providing for the adjustment of the bevel gears with the purpose of combining at one point the apexes of the pinion and the gear. The second difference consists in the impossibility of mounting both gears between the bearings; therefore, one of the gears should be mounted on the cantilever of the shaft (outboard), and, in most cases, the pinion is placed on the cantilever. Usually, the pinion with the associated parts is a subassembly that allows for the adjustments of the apexes mentioned above. Hence, this design project is focused on the design peculiarities of the subassembly of the pinion. The design considerations related to the bevel gear are similar to the helical gear with the difference being that the bevel gear cannot be mounted in the middle between the bearings, and it is placed closer to one of the bearings. Figure 6-5.1 shows the partial view of the transmission and the pinion subassembly. The distance between the pinion and the bearing and the distance between the bearings conform to conventional design considerations.

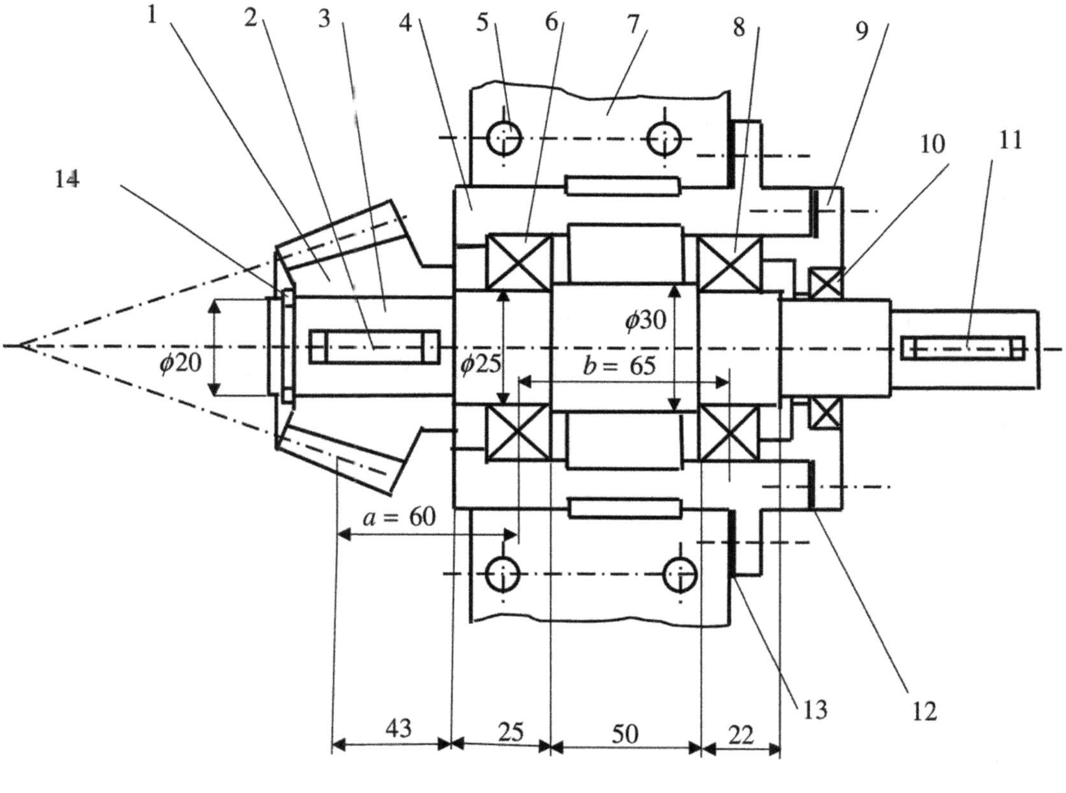

Figure 6-5.1. Partial top view of the bevel gear drive.

In this figure, pinion 1 is mounted on the cantilever of shaft 3 that is supported by angular-contact ball bearings 6 and 8 that are accommodated by sleeve 4.

Based on appropriate design considerations associated with the structure of the pinion and the dimensions of the bearing, the distance between the midpoint of teeth of the pinion and the center of the bearing $a = 60$ mm and the distance between the centers of the bearing $b = 65$ mm. The drawing is not made to the scale; however, we can now draw the free-body diagram and continue the engineering calculations.

4. Free-body diagram, reactions, bending and torque diagrams for the shaft of the pinion.

Based on Figure 6-5.1, we draw the free-body diagram of the shaft of the pinion in order to calculate the reactions in the supports and determine the bending moments and the torques applied to the shaft. The free-body diagram for the vertical plane is shown in Figure 6-5.2a (the force F_t is acting in the horizontal plane).

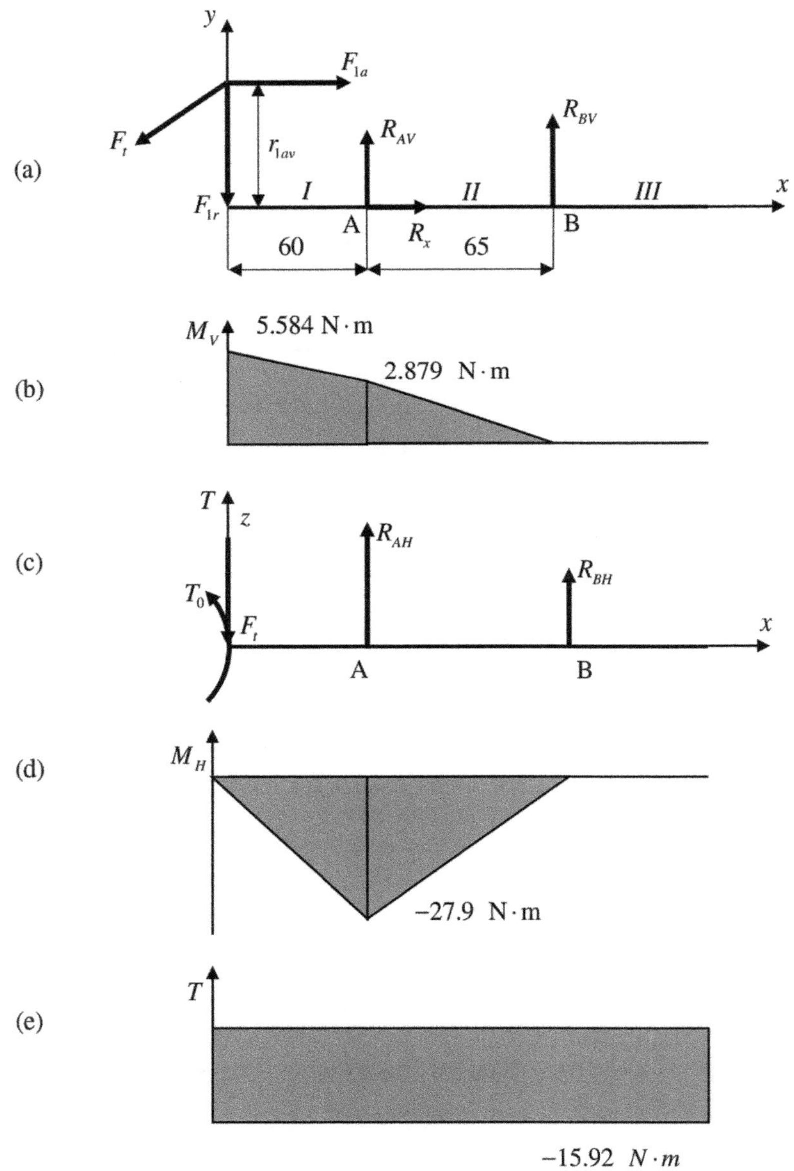

Figure 6-5.2. Free-body diagram, reactions, bending moments, and torque diagrams.

Using formula (6-1.8), we determine the tangential force applied to the gears

$$F_t = \frac{2W}{\omega_1 d_{1av}} = \frac{2 \cdot 3 \cdot 10^3}{188.4 \cdot 0.06884} = 462.63 \text{ N}.$$

The radial force applied to the pinion according to formula (6-1.10) equals $F_{1r} = F_t \tan\phi \sin\gamma_1 = 462.63 \tan 20° \sin 15.53° = 45.08$ N and the axial force according to formula (6-1.11) is $F_{1a} = F_t \tan\phi \cos\gamma_1 = 462.63 \tan 20° \cos 15.53° = 162.23$ N.

Projecting the forces on the x axis, we determine the reaction R_x, hence, $F_{1a} + R_x = 0$. Therefore, $R_x = -F_{1a} = -162.23$ N. Writing the sum of moments about point B, we determine the reaction R_{AV}. Hence, we have $F_{1a} r_{1av} - F_{1r}(60 + 65) + R_{AV} 65 = 0$, where $r_{1av} = 0.5 \cdot 68.84 = 34.42$ mm. Therefore, $R_{AV} = \dfrac{45.08(60+65) - 162.23 \cdot 34.42}{65} = 0.785$ N. Projecting the forces on the y axis, we determine the reaction R_{BV}. Therefore, $-F_{1r} + R_{AV} + R_{BV} = 0$, hence, we have $R_{BV} = F_{1r} - R_{AV} = 45.08 - 0.785 = 44.295$ N. Now we can determine the bending moments in the vertical plane. The shaft is divided into three intervals. On the first interval we have $M_I = F_{1a} r_{1av} - F_{1r} x$. When $x = 0$ then $M_I = 162.23 \cdot 0.03442 = 5.584$ N·m. When $x = 0.06$ m then $M_I = 5.584 - 45.08 \cdot 0.06 = 2.879$ N·m. On the second interval, when $x = 0.06 + 0.065 = 0.125$ m then $M_{II} = 5.584 - 5.635 + 0.785 \cdot 0.065 = 0$. The diagram of moments in the vertical plane is shown in Figure 6-5.2b. Figure 6-5.2c shows the free-body diagram in the horizontal plane.

Taking the sum of moments about point B, we determine the reaction R_{AH}. Hence, $-F_t(60 + 65) + R_{AH} 65 = 0$. Thus, $R_{AH} = \dfrac{462.43\,(60+65)}{65} = 889.67$ N. Projecting the forces on the z axis, we determine the reaction R_{BH}. Hence, $-F_t + R_{AH} + R_{BH} = 0$. Therefore, $R_{BH} = F_t - R_{AH} = 462.63 - 889.67 = -427.04$ N. Now we determine the forces applied to the bearings A and B. The radial force for bearing A is $F_{Ar} = \sqrt{R_{AV}^2 + R_{AH}^2} = \sqrt{0.785^2 + 889.67^2} = 889.67$ N and the radial force for bearing B is $F_{Br} = \sqrt{R_{BV}^2 + R_{BH}^2} = \sqrt{44.295^2 + 427.04^2} = 429.33$ N. The axial force is $F_a = R_x = 162.23$ N. Hence, for the bearing calculations we take the most loaded bearing A having the radial force $F_r = 889.67$ N and the axial force $F_a = 162.23$ N.

Now we determine the bending moments in the horizontal plane. On the first interval we have $M_I = -F_t x$, therefore, when $x = 0$ then $M_I = 0$ and when $x = 0.06$ m then $M_I = -27.76$ N·m.

On the second interval $M_{II} = -F_t x + R_{AH}(x - 0.06)$. When $x = 0.06 + 0.065 = 0.125$ m, then $M_{II} = 0$. The diagram of bending moments on the second interval is shown in Figure 6-5.2d. From this figure and from Figure 6-5.2b it becomes obvious that the maximum bending moment is applied to the cross-section passing through point A and, therefore, $M_{\max} = \sqrt{M_V^2 + M_H^2} = \sqrt{2.879^2 + 27.76^2} = 27.9$ N·m.

The torque applied to the shaft of the pinion equals $T_0 = F_t r_{1av} = 462.63 \cdot 0.03442 = 15.92$ N·m. The Figure 6-5.2e shows the torque diagram for all three intervals.

5. Verifying calculations of the diameter of the shaft for the pinion.

The most loaded cross-sectional area of the shaft passes through point A in Figure 6-5.2. Near this cross-sectional area, the cantilever has a stepped structure and a keyseat. This causes a stress

concentration that should be taken into consideration for the corresponding calculations of the diameter of the shaft. We select for the shaft of the pinion AISI 1030 annealed steel having $S_{ut} = 450$ MPa. Applying formula (1-9.2), we calculate $k_a = 0.92$. Taking the diameter $d = 20$ mm and using formula (1-9.3), we have $k_b = 0.912$. Since $k_c = 1$, we determine
$S_e = 0.92 \cdot 0.912 \cdot 1 \cdot 0.5 \cdot 450 = 188.78$ MPa.
According to the type of the keyseat, we take, based on Figure 3-8.4, that for bending $K_{fb} = 1.6$ and for torsion $K_{ft} = 1.3$. Accounting for the stress concentration that is caused by the step of the shaft, we estimate that $\frac{D}{d} = \frac{25}{20} = 1.25$ and $\frac{r}{d} = 0.1$. Hence, from Appendix A-1-5 we determine that for bending $K_{tb} = 1.65$ and for torsion $K_{tt} = 1.35$. Therefore, using formula (1-10.2) and taking $q = 0.6$ we calculate for bending $K'_{fb} = 1.39$ and for torsion $K'_{ft} = 1.21$. However, it is recommended that we take the larger values of these coefficients: $K_{fb} = 1.6$ and $K_{ft} = 1.3$. Using formula (3-5.9) for alternative bending and torsion and applying the fourth theory of failure, we determine $d = \left\{ \frac{32 \cdot 2}{\pi} \left[\left(\frac{1.6 \cdot 27.9 \cdot 10^3}{188.78} \right)^2 + \frac{3}{4} \left(\frac{1.3 \cdot 15.92 \cdot 10^3}{188.78} \right)^2 \right]^{\frac{1}{2}} \right\}^{\frac{1}{3}} = 17.268$ mm.

The diameter of the cantilever is 20 mm, and there is no need for changes.

6. Calculations of the bearing for the shaft of the pinion.

It was determined above that for the bearing calculations we have $F_r = 889.67$N and $F_a = 162.23$ N. For the selected angular-contact ball bearing SKF 7202BEP the dynamic load rating is $C = 14.8$ kN, and the static load rating is $C_0 = 9.3$ kN. In order to calculate the equivalent static load of the angular-contact bearings, we take the recommended values of the load factors $X_0 = 0.5$ and $Y_0 = 0.26$. According to formula (2-4.6), the equivalent static load is $F_0 = 0.5 \cdot 889.67 + 0.26 \cdot 162.23 = 487.01$ N. Since $F_0 < C_0$ (487.01<9,300), we may proceed with the bearing calculations.

For the selected bearing SKF 7202BEP according to formula (2-4.7), we have
$e = \frac{F_a}{C_0} = \frac{162.23}{9{,}300} = 0.0174.$

Because $\frac{F_a}{F_r} = \frac{162.23}{889.67} = 0.182 > e = 0.0174$, it is recommended to take the values of the corresponding load factors $X = 1$ and $Y = 0$. The equivalent dynamic load is determined according to formula (2-4.5) that reads: $F_e = XVF_r + YF_a$. Since the inner ring is rotating, we take $V = 1$. Therefore, $F_e = 1 \cdot 889.67 = 889.67$ N. The service life of the bearing according to formula (2-4.4) is $L_{10} = \left(\frac{C}{F_e} \right)^3 = \left(\frac{14{,}800}{889.67} \right)^3 = 4{,}604 \cdot 10^6$ revolutions or $L_{10h} = \frac{4{,}604 \cdot 10^6}{60 \cdot 1{,}800} = 42{,}626$ hours.
The calculated duration of the service life seems to be higher than average, however, we have taken the lowest dynamic load rating for this bearing in the SKF catalog, and in some cases the required duration of the service life may exceed 100,000 hours; therefore, we accept the selected bearing.

7. Calculation of the deflection and angle of rotation of the shaft for the pinion.
The following forces are applied to the shaft: the tangential force $F_t = 462.63$ N; the radial force $F_r = 45.08$ N; and the axial force $F_a = 162.23$ N, while these forces are acting on the cross-sectional area of the shaft located at the midpoint of the teeth of the pinion. The action of the forces F_t and F_r can be determined by using their resultant $F=\sqrt{F_t^2+F_r^2}=\sqrt{462.63^2+45.08^2}=465$ N. The force F_a causes a bending moment $M_o = 162.23\cdot 0.5\cdot 0.06884 = 5.584\ \text{N}\cdot\text{m}$. Because the shaft of the pinion has a stepped structure, it is reasonable for the calculations of the shaft deformations to determine the equivalent diameter of the uniform shaft based on the dimensions of the actual stepped shaft of the pinion. According to the drawing in Figure 6-5.1, we estimate the lengths of the following steps of the shaft: step $l_1 = 43$ mm (the cantilever has the diameter of 20 mm), step $l_2 = 25$ mm and has the diameter of 25 mm, the length of step $l_3 = 50$ mm between the bearings and has the diameter of 30 mm, and, finally, step $l_4 = 22$ and has the diameter 25 mm. We take the total length of the shaft $l = 43 + 25 + 50 + 22 = 140$ mm. Based on formula (3-6.11), we calculate

$$d_e=\frac{1}{\sqrt[4]{\dfrac{43}{140\cdot 20^4}+\dfrac{25}{140\cdot 25^4}+\dfrac{50}{140\cdot 30^4}+\dfrac{22}{140\cdot 25^4}}}=23.6\ \text{mm}.$$

The moment of inertia of the shaft having the equivalent diameter is

$$I=\frac{\pi 23.6^4}{64}=15{,}237\ \text{mm}^4.$$

In this case, the deflection of the shaft occurs in the horizontal plane due to the force F and in the vertical plane due to the moment M_0. The total deflection represents a vector sum of these two deflections. Using formula (3-6.14), we calculate the deflection h_F of the shaft in the horizontal plain at the midpoint of the teeth of the pinion. According to Figure 6-5.1, we take $l_C = 43 + 25 + 50 = 118$ mm and $l = 43$ mm, and applying formula (6-3.20), we calculate $h_F=-\dfrac{465\cdot 43^2(43+118)}{3\cdot 207\cdot 10^3\cdot 15{,}237}=-0.0146$ mm. The deflection of the midpoint of teeth of the pinion due to the moment M_0 is calculated using formula (6-3.21):

$$h_0=\frac{5{,}584\cdot 60}{6\cdot 207\cdot 10^3\cdot 15{,}237}\left[3\cdot 60-2(60+65)\right]=-0.00124\ \text{mm}.$$

The total deflection is $h=\sqrt{h_F^2+h_0^2}=\sqrt{0.0146^2+0.00124^2}=0.01465$ mm. The allowable deflection is $0.007m = 0.007\cdot 4.5 = 0.0315$ mm, which exceeds the actual deflection.

The angle of rotation of the shaft in the cross-section passing through point A also depends on the force F acting in the horizontal plane and bending moment M_0 acting in the vertical plan. The total angular rotation is determined as a vector sum of these two deflections.

Therefore, $\theta_F=-\dfrac{Fa(a+b)}{3EI}=-\dfrac{465\cdot 60(60+65)}{3\cdot 207\cdot 10^3\cdot 15{,}237}=-0.0003686$ rad, and

$$\theta_0=-\frac{M_0(a+b)}{3EI}=-\frac{5{,}584(60+65)}{3\cdot 207\cdot 10^3\cdot 15{,}237}=-0.0000738\ \text{rad}.$$

Hence, $\theta=\sqrt{\theta_F^2+\theta_0^2}=\sqrt{0.00036866^2+0.0000738^2}=0.00076$ rad, which is less than the allowable angle of rotation 0.01 rad.

7. Calculation of the critical rotational speed.

Based on the calculated deflection, we determine the critical rotational speed of the pinion using the following formula: $n_{cr} = \frac{30}{\pi}\sqrt{\frac{g}{h}} = \frac{30}{\pi}\sqrt{\frac{9.8 \cdot 10^3}{0.01465}} = 7{,}814 \text{ rpm}.$

8. Assembly and disassembly.

a) Assembly.

Referring to Figure 6-5.1, we begin with mounting the bearings 6 and 8 onto shaft 3, and then we place the shaft with the bearings into sleeve 4 creating a subassembly. After that, we install key 2 into the keyseat of shaft 3, and mount pinion 1, securing it with retaining ring 14. Now we place the subassembly into the lower casing 7. Then we place the upper casing securing both halves of the casings by bolts that fit into the halls 5. After that, we adjust the preload of the bearings by placing cover 9 and the set of shims 12 and tighten the screws of this cover (the axes of the screws are shown in the drawing). Now we adjust the apexes of the gears by using the set of shims 13 and tighten the screws of the sleeve 4. After that, we install the seal 10 and key 11. Now we examine the engagement of the gears, applying appropriate procedures. After that, we pour oil through the special opening into the casing, keeping in mind that the oil should cover just one-half of a tooth of the lower gear. After closing the opening, the transmission is ready to be used.

It should be emphasized that adjustments of the bearings and gears and also the testing of the teeth engagement should be performed in accordance with the appropriate procedures and instruments.

b) Disassembly.

We unscrew the oil plug in the lower case to empty the oil. After that, we unscrew the screws that hold cover 9 and sleeve 4. We remove cover 9 and key 11. Now, we unscrew the bolts and nuts that hold together the lower and upper halves of the casings and remove the upper casing. We take off the subassembly and remove the retaining ring 14, the pinion 1, and the key 2. Now we remove from sleeve 4 the shaft 3 with the bearings 6 and 8. It could happen that the left outer ring of the angular-contact bearing 6 will remain in the sleeve. Accounting for that, the outer rings of the bearings are installed into the sleeve with a sliding fit, and it is easy to push them out manually with an appropriate tool. After that, the bearings are removed from the shaft using appropriate pullers. If the seal 10 must be replaced, it can be removed using the appropriate manual tools.

Chapter 6. Problems

6-01. An open one directional bevel gear drive transmits 4 kW of power from the 25-tooth pinion at the rotational speed of n_1 = 1,200 rpm to the 75-tooth gear. The module is 3 mm the tooth face width is 20 mm, the pressure angle is 20°. Using the recommended load, factors, determine the bending stresses in the pinion and in the gear.

6-02. An open one directional bevel gear drive transmits 5.5 kW of power from the 21-tooth pinion at the rotational speed of n_1 = 900 rpm to the gear. The module is 5 mm, the pitch angle of the pinion is 21°, the tooth face width is 30 mm, and the pressure angle is 20°. The pinion and the gear are made from steel having S_{ut} = 794 MPa. Using the recommended load factors, determine the bending stress safety factor for the gears.

6-03. An open reversible bevel gear drive transmits 3.0 hp of power from the 21-tooth pinion at the rotational speed of n_1 = 900 rpm to the gear. The diametral pitch on the large end is $P = 5\frac{1}{\text{in.}}$, the tooth face width is 1.6 in., the pressure angle is 20°, and the pitch angle of the pinion is 20°. The pinion and the gear are made from steel having S_{ut} = 112, 500 psi. Using the recommended load factors, determine the bending stress safety factor of the teeth of the gears.

6-04. An open reversible bevel gear drive transmits 7.5 kW of power from the 25-tooth pinion at the rotational speed of n_1 = 1, 200 rpm to the gear. The module is 6 mm, the pitch angle of the gear is 71°, the tooth face width is 20 mm, and the pressure angle is 20. The pinion and the gear are made from steel having S_{ut} = 527 MPa. Using the recommended load factors, determine the bending stress safety factors for the gears.

6-05. A one directional open bevel gear drive transmits 7.5 hp of power from the 25-tooth pinion at the rotational speed of n_1 = 1,800 rpm to the gear. The velocity ratio is 3.1, the pinion is made from steel having S_{ut} = 129,250 psi, and the bending stress safety factor is 2. Using the recommended load factors and taking the face width coefficient η_L = 0.3, determine the diametral pitch of the teeth.

6-06. A one directional open bevel gear drive transmits 3 hp of power from the 21-tooth pinion at the rotational speed of the n_1 = 1,800 rpm to the gear that has the pitch angle equal to 68°. The pinion is made from steel having S_{ut} = 115,000 psi , and the bending stress safety factor is 2. Using the recommended load factors and taking the face width coefficient η_L = 0.3, determine the diametral pitch of the teeth.

6-07. A reversible open bevel gear drive transmits 5.5 hp of power from the 25-tooth pinion at the rotational speed of the n_1 = 3,600 rpm to the 68-tooth gear. The pinion is made from steel having S_{ut} = 129,250 psi, and the bending stress safety factor is 2. Using the recommended load factors and taking the face width coefficient η_L = 0.3, determine the diametral pitch of the teeth.

6-08. A reversible open bevel gear drive transmits 3 kW of power from the 21-tooth pinion at the rotational speed of the n_1 = 1,800 rpm to the gear that has the pitch angle equal to 70°. The pinion is made from steel having S_{ut} = 873 MPa, and the bending stress safety factor is 2. Using the recommended load factors and taking the face width coefficient η_L = 0.3, determine the normal module of the teeth.

6-09. An enclosed bevel gear drive transmits 4.5 kW of power from the 23-tooth steel pinion at the rotational speed of the n_1 = 1,800 rpm to the 73-tooth steel gear. The module is 4 mm and the tooth face width is 42 mm. Using the recommended load factors , determine the contact stresses in the teeth.

6-10. An enclosed bevel gear drive transmits 2.2 kW of power at the rotational speed of the 23-tooth pinion n_1 = 1,800 rpm to the gear rotating at the rotational speed n_2 = 550 rpm. Take the tooth face width coefficient $\eta_L = 0.3$, the allowable contact stress S_H = 573 MPa, and the safety factor $n_H = 2$. Using the recommended load factors, determine the normal module, the pitch diameters of the gears, the tooth face width, and the materials for the gears.

CHAPTER 7

Worms and Worm Gears

Worm gear drives transmit power between crossing (nonintersecting) axes. They consist of a worm that resembles threshed rods and worm gears that are similar to helix gears.

7-1. Worms and Worm Gears and Their Geometry

Figure 7-1.1 provides a view of a worm gear drive.

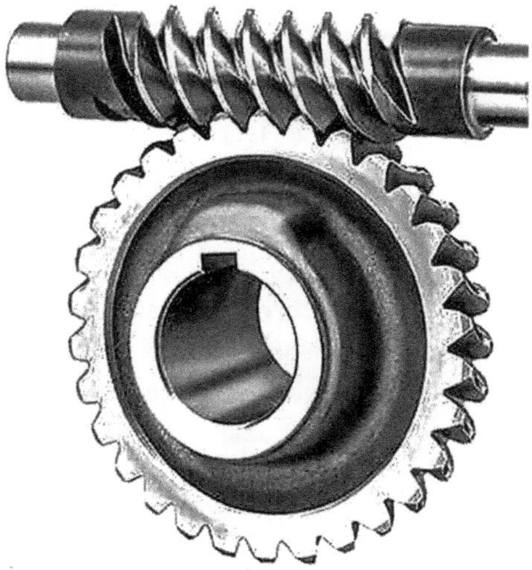

Figure 7-1.1. A meshing pair of a worm and a worm gear. (Courtesy of Emerson Power Transmission Corporation)

The crossing angle between axes of a worm and worm gear is predominantly 90°; however, it is possible to implement other angles. Figure 7-1.2 shows a worm gear set, where worm 1 having a pitch diameter d_1 is rotating at the angular velocity ω_1, and it is engaged with worm gear 2 that has a pitch diameter d_2 and rotates at the angular velocity ω_2. The center distance of this drive is C.

One of the most distinguishing features of worm gear transmissions is their ability to obtain large velocity ratios in one gearbox. Normally, the velocity ratio of worm gear transmissions ranges from 8 to 80; however, 100 and even 1,000 is also achievable. A worm gear drive

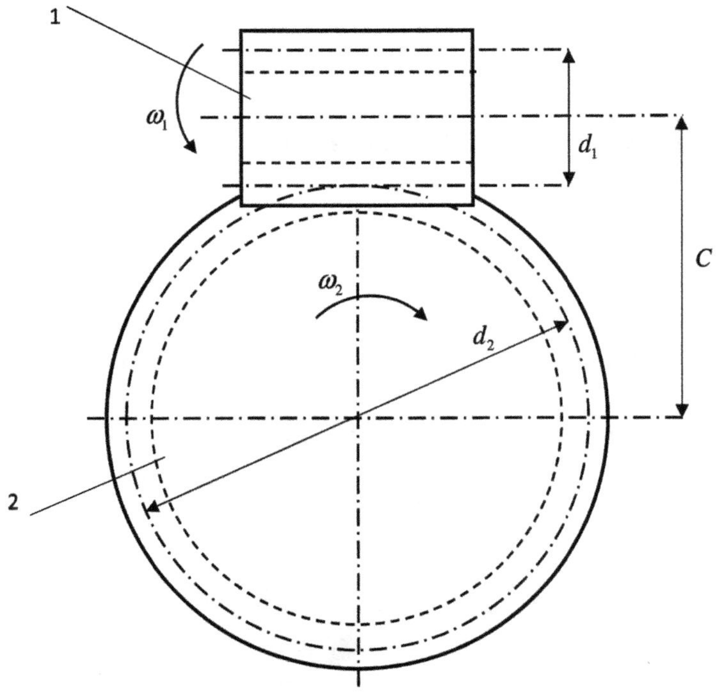

Figure 7-1.2. Worm gear set.

consists of a worm and a worm gear. The worm is usually the driving element, and the worm gear is the driven. The principle action of a worm gear drive resembles the interaction between a threaded screw and a nut. In other words, when the screw makes one revolution the nut makes an axial displacement that corresponds to the axial pitch of the screw-nut pair. The axial pitch represents the distance between two identical points of adjacent threads. Actually, widely used worms are structurally similar to threaded screws and in their axial sections resemble the shape of an involute rack. The worm gear resembles a helical gear interacting with the worm as with a rack, while the teeth of the worm gear wrap partially around the worm. This type of a worm gear set is called the single enveloping worm gear drive. Thus, a worm gear transmission possesses certain properties of threaded elements and helical gears.

The next distinguishing feature of worm gear transmissions is that at certain geometric parameters of the worm and the worm gear the system becomes self-locking, making the transfer of the rotation from the gear to the worm impossible. This feature eliminates the need of applying dedicated braking mechanisms in some load lifting equipment. However, it should be stressed that in cases of incorporating worm gear transmissions into industrial cranes, it is strongly recommended to use dedicated corresponding braking devices.

Worm gear transmissions possess an essential disadvantage: high friction between the worm and worm gear. As a result of this, these drives are characterized by friction loses that could exceed 50% of the transmitted power. This leads to generating a significant amount of heat that may result in critical dilution of the lubricating oil that in turn may cause the failure of the worm gear drive. This is the reason why some worm gear transmissions require special cooling devices. In addition, it is highly desirable that the worm gear set is manufactured from materials that possess low friction coefficients.

Usually, the worm is made from steel and the worm gear from bronze; however, sometimes the gear is made from cast iron or plastics. Bronze is an expensive material that increases the cost of the transmission and has lower strength characteristics than steel. The friction loses and the strength issues of worm gear material are the reason for the limited level of power transmission by worm gear drives. Usually, they are used for transmitting power that does not exceed 80 hp (60 kW), but there are worm gear drives that transmit 100 kW and more. In addition, worm gear drives are limited in the operational level of linear velocity that should not exceed 3,000 fpm $\left(15\frac{\mathrm{m}}{\mathrm{s}}\right)$.

It should be mentioned that manufacturing worm gear transmissions is more complicated and more expensive than cylindrical toothed transmissions.

Because as shown in Figure 7-1.3, the structure of a worm is similar to the structure of an external threaded cylinder, resembling a screw, we introduce some related parameters of threads.

Figure 7-1.3. A single-start worm. (Courtesy of Martin Sprocket & Gear, Inc.)

Figure 7-1.4 shows a worm having a trapezoidal thread that is characterized by a profile angle 2α, a lead angle λ, an axial pitch p_a, and a lead L.

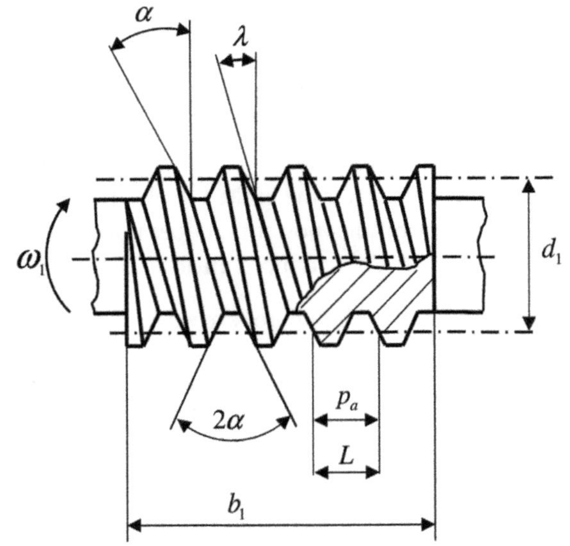

Figure 7-1.4. Geometry of the worm.

It should be mentioned that the details related to the geometry of threads are presented in chapters 8 and 9. This figure shows a particular case where the cutting process of the thread is characterized by a single starting point. In this case, the screw has one thread and one starting point (one start). For a one-start thread, as shown in Figure 7-1.4, the axial pitch P_a is equal to the lead L.

Looking at the front view of a screw or worm, it is impossible to determine the number of threads (or starts). However, we may clearly see the number of thread starts when we look at the face of the screw or worm (the side view). Figure 7-1.5 presents a view of a four-start worm.

Figure 7-1.5. A four-start worm. (Courtesy of Martin Sprocket & Gear, Inc.)

Thus, in cases of a single thread, we see one place on the periphery of the circle where the tool begins to cut. In cases of two starts, we see two starting places on each side of the diameter; in cases of four starts the beginning cutting places are located on the circle 90° apart (three starts are seldom used).

Figure 7-1.6 illustrates cases with one and two starts. In cases of the engagement of a one-start worm with a worm gear, during one revolution of the worm the gear will rotate on an angle that corresponds to one tooth of the worm gear (the worm does not have teeth). Therefore, when the worm or screw has one thread, the lead equals the axial pitch. Figure 7-1.6a schematically shows the structure of a one-start screw, where the start of the tread begins in area 1 on the front and side views.

Figure 7-1.6b shows the beginning of cutting a two-start screw or worm and makes it clear that in a two-start screw or worm the lead equals two axial pitches—$L = 2p_a$.

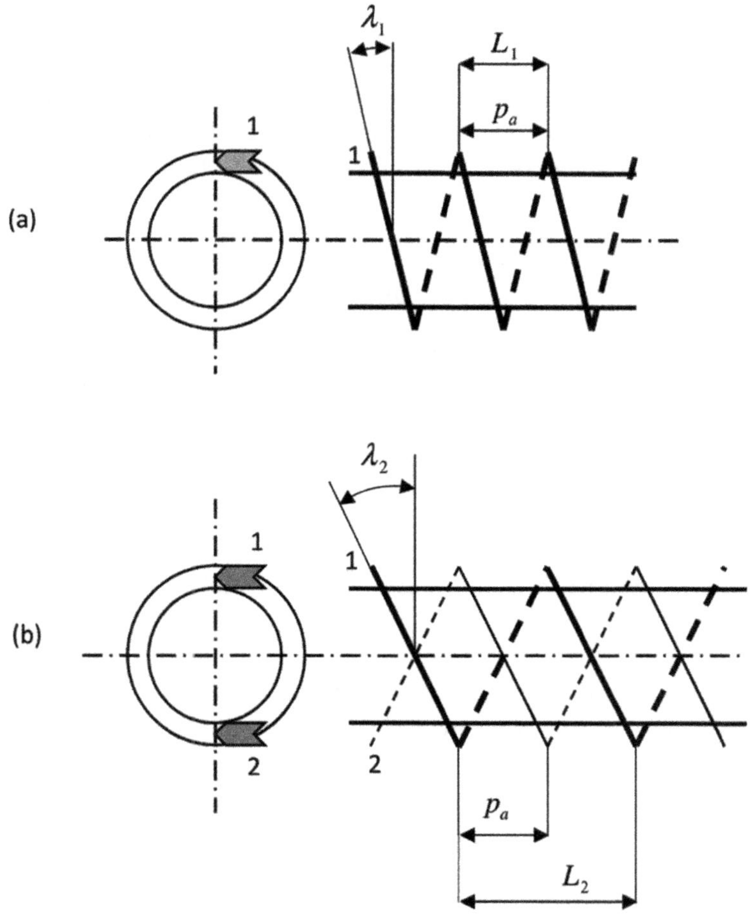

Figure 7-1.6. One- and two-start screw or worm.

Now we can clarify the definitions of the axial pitch and the lead and emphasize the difference between them. The axial pitch, as mentioned above, is the distance along the axis of the worm (screw) between two identical points of two adjacent threads. In cases where the screw has one thread that is cut from one start, these two points actually belong to the same thread, which means that if we move the point from its initial position along the thread, then after one revolution we will come to the end of the axial pitch. Defining the axial pitch, we consider the distance between two adjacent identical points of the thread. The lead is the distance that the nut moves axially during one revolution of the screw, in other words, it is the distance between the identical points of the same thread. Obviously, depending on the number of starts, this distance could be between two identical closest points (one start) or between several threads (several starts). Therefore, when the worm or screw has one thread the lead equals the axial pitch.

Figure 7-1.6b shows that the first start begins in the area 1 while the second start begins 180° from the first start. The second start is presented on the front view by a thinner line. These figures are drawn in the same scale that allows us to see that the axial pitch in both cases is the same; however, $\lambda_2 > \lambda_1$ and $L_2 = 2L_1$. Therefore, during one revolution of a worm with two starts, the worm gear makes an angular displacement that is equivalent to two circular pitches and, consequently, to two teeth of the gear. This makes it clear that the velocity ratio in a worm gear set is associated with the number of teeth of the worm gear and with the number of starts of the worm. Hence, we may write

$$i = \frac{\omega_1}{\omega_2} = \frac{N_2}{N_1} \qquad (7\text{-}1.1)$$

where ω_1 and ω_2 are the angular velocities of the worm and the worm gear, respectively; N_1 is the number of starts of the worm; and N_2 is the number of teeth of the worm gear. Usually in worm gear drives, the number of starts is taken 1, 2, and 4. Hence, accounting for the above considerations, we can see that the lead length can be calculated using the following formula:

$$L = N_1 p_a \qquad (7\text{-}1.2)$$

The engagement of a worm with a worm gear is possible if the three following requirements are met:

1) The axial pitch of the worm equals the circular pitch of the worm gear ($p_a = p$).
2) The lead angle of the worm equals the helix angle of the worm gear ($\lambda = \psi$).
3) The pressure angle of the worm gear equals one-half of the worm thread profile angle ($\phi = \alpha$). Since the pressure angle of gears in this book is taken $\phi = 20°$, we have $2\alpha = 40°$. All these angles and the pitch diameter d_1 are shown in Figure 7-1.4.

Figure 7-1.7 shows cast iron and bronze work gears.

Figure 7-1.7. (a) Cast iron and (b) bronze worm gears. (Courtesy of Martin Sprocket & Gear, Inc.)

Figure 7-1.8 shows a triangle that represents the development of one turn of the thread, where the horizontal side equals the circumference of the pitch circle, the vertical side equals the lead, and the hypotenuse equals the length of the thread.

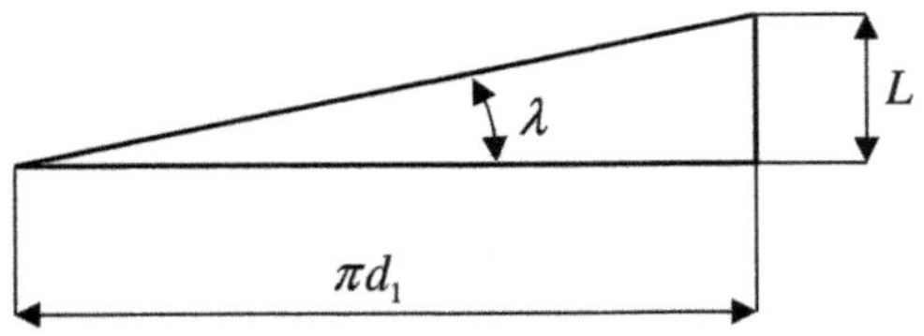

Figure 7-1.8. The relationship among the pitch diameter, lead, and lead angle.

This triangle demonstrates the relationship among the lead L, the lead angle γ, and the pitch diameter d_1 of the worm. Hence, we may write

$$\tan\lambda = \frac{L}{\pi d_1} = \frac{p_a N_1}{\pi d_1} = \frac{pN_1}{\pi d_1} = \frac{N_1}{Pd_1} = \frac{mN_1}{d_1} \tag{7-1.3}$$

where p_a is the axial pitch of the worm, p is the circular pith of the worm gear, P is the diametral pitch of the worm gear, and m is the module of the worm gear.

The pitch diameter of the worm is not associated with the diametral pitch or module and is determined based on the strength and stiffness of the worm shaft. Smaller worms are cut as one component with the shaft. During the design procedures, the pitch diameter of the worm is determined based on corresponding experimental design considerations. For the similarity of calculation, the pitch diameter of a worm is determined using the following empiric formula for the pitch system:

$$d_1 = \frac{w}{P} \tag{7-1.4}$$

where w is an empiric coefficient of the worm pitch diameter, and P is the transverse diametral pitch of the worm gear, while for module system gears

$$d_1 = wm \tag{7-1.5}$$

where m is the transverse module of the worm gear.

Based on published sources, it is recommended that the coefficient w vary from 8 to 25; however, usually this coefficient ranges from 8 to 13, while the larger values are used for smaller teeth. Unlike the number of teeth, w can be a fraction or an integer number. In addition, there are recommendations that the pitch diameter of the worm be in a certain relation with the center distance C between the gears, and $C = 0.5(d_1 + d_2)$. According to the widely published recommendation, the pitch diameter of the worm for pitch system gears should be in the following range:

$$\frac{C^{0.875}}{3.0} \le d_1 \le \frac{C^{0.875}}{1.7} \tag{7-1.6}$$

where C and d_1 are in inches. Modifying this formula for the module system gears, we may write

$$\frac{(0.0394C)^{0.875}}{3.0} \leq 0.0394d_1 \leq \frac{(0.0394C)^{0.875}}{1.7} \tag{7-1.7}$$

where C and d_1 are in millimeters.

Figure 7-1.9 represents a front view of the vertical axial section of a worm gear set, where worm 1 is meshing with bronze worm gear rim 2 that is rigidly attached to the steel wheel 3.

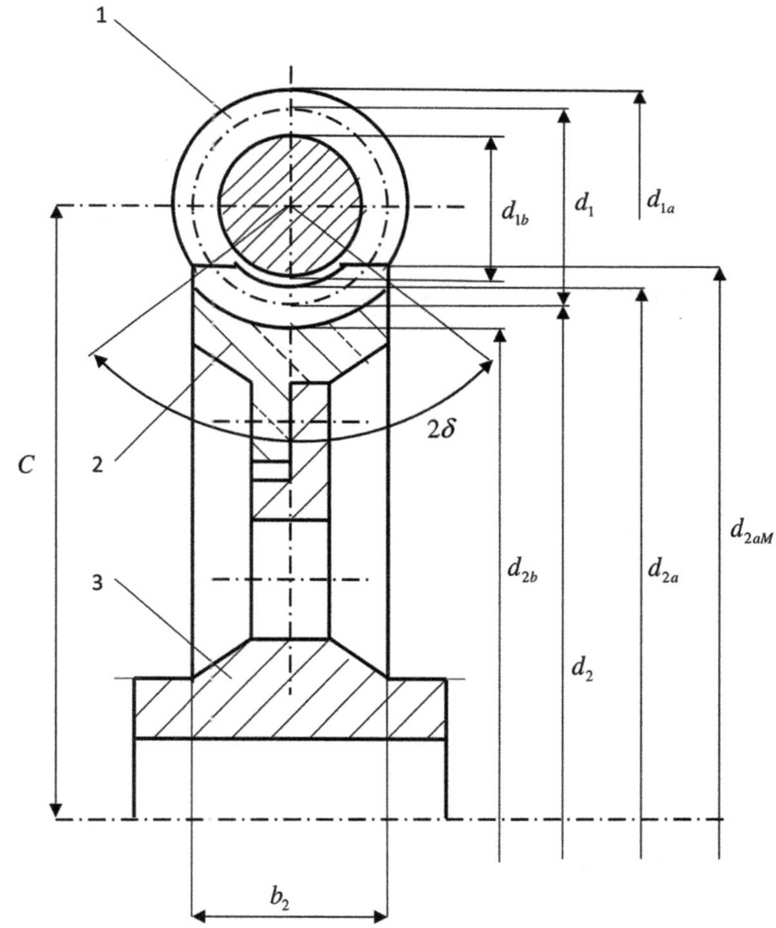

Figure 7-1.9. Design of the worm and worm gear.

The rim is pressed onto the wheel and is secured with bolts and nuts (the axes of these bolts are shown in the drawing). The basic parameters of the worm comprise the pitch diameter, the addendum diameter, the dedendum circle diameter, and the width (sometimes called the length) of the worm. For pitch system gears, the pitch diameter of the worm is determined using formula (7-1.4). The addendum diameter and the diameter of the dedendum circle are determined from the following formulas:

$$d_{1a} = d_1 + \frac{2}{P} \tag{7-1.8}$$

$$d_{1b} = d_1 - \frac{2.4}{P} \tag{7-1.9}$$

The width b_1 (see Figure 7-1.2) of the worm in cases of one or two starts ($N_1 = 1$ or $N_1 = 2$) is computed from the following empiric formula:

$$b_1 = \frac{11 + 0.06N_2}{P} \text{ (in.)} \tag{7-1.10}$$

and in cases of four starts ($N_1 = 4$) can be determined from a similar empiric formula that reads

$$b_1 = \frac{12.5 + 0.09N_2}{P} \text{ (in.)} \tag{7-1.11}$$

The pitch diameter d_1 of the worm for a module system is given by formula (7-1.5), while the addendum diameter and the diameter of the dedendum circle, respectively, are

$$d_{1a} = d_1 + 2m \tag{7-1.12}$$

$$d_{1b} = d_1 - 2.4m \tag{7-1.13}$$

The width of the worm for the module system in cases of one or two starts is calculated using the following empiric formula:

$$b_1 = (11 + 0.06N_2)m \text{ (mm)} \tag{7-1.14}$$

and in cases of four starts the width can be determined from the following formula:

$$b_1 = (12.5 + 0.09N_2)m \text{ (mm)} \tag{7-1.15}$$

The formula for calculating the pitch diameter of the worm gear in the pitch system is

$$d_2 = \frac{N_2}{P} \tag{7-1.16}$$

The diameters of the addendum and dedendum circles, respectively, are calculated using the formulas

$$d_{2a} = d_1 + \frac{2}{P} \tag{7-1.17}$$

$$d_{2b} = d_2 - \frac{2.4}{P} \qquad (7\text{-}1.18)$$

The maximum outside diameter of the worm gear is determined based on the following empiric formula:

$$d_{2M} = d_{2a} + \frac{6}{(N_1 + 2)P} \qquad (7\text{-}1.19)$$

The approximate width of the worm gear tooth face for one and two starts worms is $b_2 = 0.75d_{1a}$, while for four starts worms $b_2 = 0.67d_{1a}$. The wrapping angle 2δ is determined using the formula that reads

$$\sin\delta = \frac{b_2}{d_{1a} - \frac{0.5}{P}} \qquad (7\text{-}1.20)$$

Formulas applicable for similar calculations of the geometric parameters for module system gears are presented below. Hence, for calculating the pitch diameter of the worm gear the following is used:

$$d_2 = mN_2 \qquad (7\text{-}1.21)$$

To determine the diameters of the addendum and dedendum circles, use the following formulas:

$$d_{2a} = d_2 + 2m \qquad (7\text{-}1.22)$$

$$d_{2b} = d_2 - 2.4m \qquad (7\text{-}1.23)$$

The maximum outside diameter of the worm gear is calculated using the following formula:

$$d_{2M} = d_{2a} + \frac{6m}{N_1 + 2} \qquad (7\text{-}1.24)$$

The width of the worm gear tooth face is taken based on the same empiric approach as used with the pitch system gears. The wrapping angle 2δ is calculated based on the formula that reads

$$\sin\delta = \frac{b_2}{d_{1a} - 0.5m} \qquad (7\text{-}1.25)$$

Based on design considerations some authors recommend determining the number of worm starts depending on the velocity ratio. Thus, for the velocity ratio from $i = 10$ to $i = 18$, the number of starts should be $N_1 = 4$; for velocity ratios from $i > 18$ to $i = 40$, the number should be $N_1 = 2$; and for $i > 40$ the number of starts should be $N_1 = 1$.

It is recommended to consider the minimum number of teeth for the gear $N_{2\min} = 28$.

7-2. Materials and Manufacturing of Worms and Worm Gears

Worms are made from carbon and alloyed steels subjected to corresponding thermal treatment in order to achieve relatively high surface hardness. Due to the intensive friction between the worm threads and worm gear teeth, the gears are usually made of materials that possess low friction coefficients. Most often, the rims of the worm gears are made of bronze and rigidly secured to steel wheels. In this case, the worm gear has a compound structure that is shown in Figure 7-1.9. In addition, worm gears are made from aluminum, cast iron, and plastics.

Numerous investigations report that the surface finish and the surface hardness of the worm play a significant role in the operational effectiveness of a worm gear set. There are several types of configurations of worms in the axial sections. In these sections, the form of the teeth resembles a rack. One of the types of the usually utilized worms has in the axial section trapezoidal shaped teeth (see Figure 7-1.4). When these teeth mesh with the involute teeth of the worm gears, it creates a common engagement of a gear with a rack. In addition, these worms can be machined on a conventional lathe using a single-point cutting tool. However, in order to grind this type of worm, special and relatively costly machining equipment and grinding tools are required. Usually, these worms do not go through grinding operations and thermal treatment. Often these worms are made from alloyed steels that possess relatively high hardness, approaching 350 H_B. The teeth of other types of worms have an involute profile in the section perpendicular to the axis of the worm and a trapezoid profile in the section perpendicular to the direction of the tread. These types of worms resemble a helical gear having a small number of teeth at a large helix angle. The grinding operation of these worms is less complicated than for the worms considered above, and this advantage provides the opportunity to utilize these worms in mass production. This type of worms is widely used in applications requiring surface hardness higher than 350 H_B and high quality surface finish.

There are other types of worms having curved thread profiles in the axial section and trapezoidal profiles in the section normal to the direction of the thread; however, they have limited application. Usually, manufacturing methods play an important role in the selection of the types of worms.

The machining processes for manufacturing worm gears are based on the method of generating using the hobbing tool whose shape is a copy of the worm that will be meshed with the gear. This produces the conjugate profiles of the worm and worm gear.

7-3. Efficiency and Heat Calculations for Worm Gear Drives

Figure 7-3.1a shows a worm gear set, where the worm 1 is distanced from worm gear 2 for the clarity of the drawing.

The vectors of the linear velocities of the worm V_1 and the worm gear V_2 in the pitch point P are shown in Figure 7-3.1b. These vectors are not equal to each other and are directed perpendicularly to each other. This causes continual sliding between the threads of the worm and the teeth of the worm gear, resulting in power loses and heat generation. The sliding

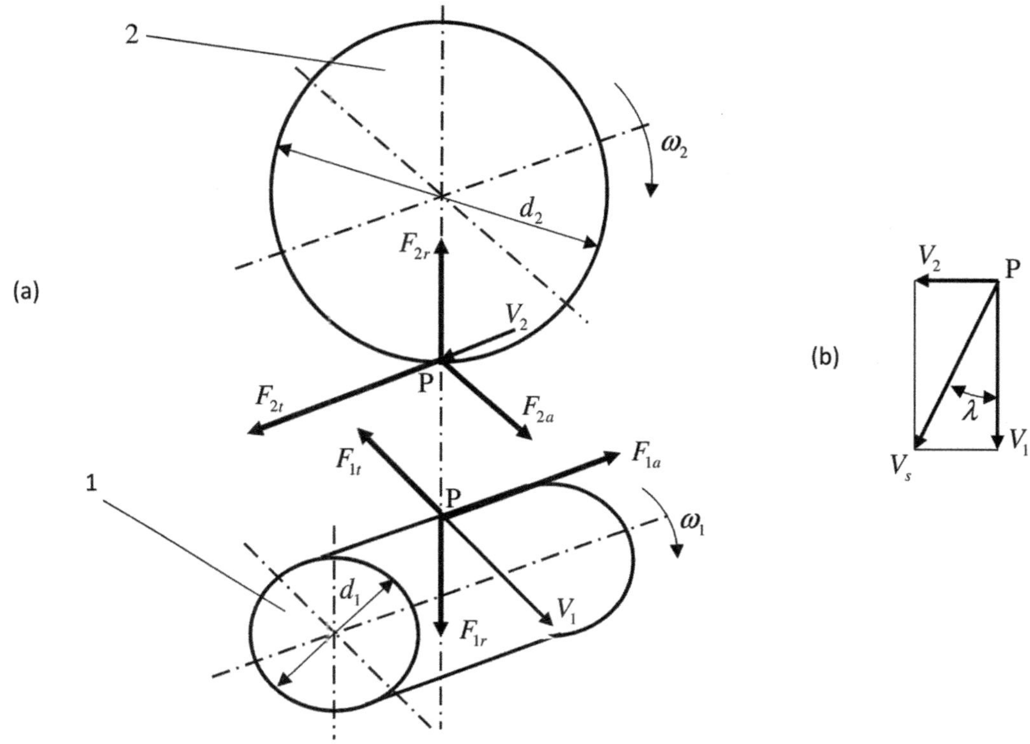

Figure 7-3.1. Forces and linear velocities in a worm gear set.

velocity V_s represents the relative velocity between the worm and worm gear and, as shown in Figure 7-3.1b, equals to $V_s = \sqrt{V_1^2 + V_2^2}$. Since $\frac{V_2}{V_1} = \tan\lambda$, we may write $V_s = V_1\sqrt{1+\tan^2\lambda}$ and finally

$$V_s = \frac{V_1}{\cos\lambda} \tag{7-3.1}$$

Replacing in formula (7-3.1) the linear velocity of the worm V_1 with its value, we obtain

$$V_s = \frac{d_1\omega_1}{2\cos\lambda} \tag{7-3.2}$$

In a worm gear set, the sliding velocity exceeds the linear velocity of the worm ($V_s > V_1$), resulting in power losses. The power losses in a worm gear transmission are mostly associated with friction in the engagement; however, some power losses are related to friction in the bearings and to oil mixing in the enclosed drives. Because the engagement of a worm gear set is similar to the interaction between the threads of a screw and a nut, it is justifiable to apply the formula for calculating the efficiency of a threaded pair to a worm gear set, therefore we may write

$$e = \frac{\tan \lambda}{\tan(\lambda + \mu)} \tag{7-3.3}$$

where e is the efficiency and μ is the friction angle for a pair of interacting materials, while $\tan \mu = f$ where f is the friction coefficient for the mentioned pair of materials. (The details related to the analysis of a threaded pair are presented in Chapter 8.)

In cases when the worm gear is the driving link, the formula for calculating the efficiency obtains the following shape:

$$e' = \frac{\tan(\lambda - \mu)}{\tan \lambda} \tag{7-3.4}$$

In cases when $\lambda \leq \mu$, then the efficiency $e' = 0$ and the worm gear set becomes self-locking—the rotation of the worm cannot be performed by the worm gear. Increasing the number of starts results in increasing the lead angle λ and, consequently, increasing the efficiency of the gear set.

Taking realistic values of λ and μ, such as $\lambda = 7°$ and $\mu = 5°$, we obtain according to formula (7-3.3) that $e = 0.578$. In this case, about 42% of the energy lose is associated with the generation of heat that raises the temperature of the lubricating oil in the worm gear enclosure. This leads to a drastic decline of the lubricating properties of the oil. In cases when the natural dissipation of heat through the outside surfaces of the worm gear drive is not sufficient to keep the oil at the required operational temperature, it is necessary to apply a dedicated cooling device to the enclosed worm gear transmission. Therefore, for enclosed transmissions it is necessary to calculate the level of natural heat dissipation in order to determine if a dedicated cooling device is needed for the transmission. The amount of power Q_e that is converted into heat due to friction depends on the efficiency of the worm gear set and it is equal to

$$Q_e = 859.85W(1 - e) \tag{7-3.5}$$

where W is the input power in kW and Q_e is in $\frac{\text{kcal}}{\text{h}}$. The heat flow Q_d that naturally dissipates into the environment through the outside (active) surfaces of the enclosure of the worm gear drive can be determined based on the following expression:

$$Q_d = C_H A_w (t_{all} - t_a) \tag{7-3.6}$$

where Q_d is in $\frac{\text{kcal}}{\text{h}}$, while C_H is the heat transfer coefficient in $\frac{\text{kcal}}{°\text{Cm}^2\text{h}}$, A_w is the effective area of the enclosure of the worm gear transmission in m^2, t_{all} is the maximum allowable temperature of the oil in °C, and t_a is the ambient temperature in °C. Denoting by q_e the ratio between Q_e and Q_d we may write

$$q_e = \frac{Q_e}{Q_d} = \frac{859.85W(1 - e)}{C_H A_w (t_{all} - t_a)} \tag{7-3.7}$$

In order to determine the efficiency, we need to estimate the friction coefficient that depends on the sliding velocity between the worm and worm gear. The experimental investigations indicate that the friction coefficient drops from about 0.1 at the sliding velocity of 20 fpm $\left(0.1\dfrac{\text{m}}{\text{s}}\right)$ to about 0.018 at 3,000 fpm $\left(15\dfrac{\text{m}}{\text{s}}\right)$. Here is an empiric formula, recommended by several authors, for estimating the approximate value of the friction coefficient:

$$\begin{aligned} &f = 0.05 - 2.5\cdot 10^{-5}(V_s - 200) && 200\text{ fpm} \le V_s \le 1{,}400\text{ fpm} \\ &f = 0.05 - 5\cdot 10^{-3}(V_s - 1.0) && 1.0\ \frac{\text{m}}{\text{s}} \le V_s \le 7.0\ \frac{\text{m}}{\text{s}} \end{aligned} \tag{7-3.8}$$

If the sliding velocity is in the range between 1,400 fpm $\left(7.0\dfrac{\text{m}}{\text{s}}\right)$ and 3,000 fpm $\left(15\dfrac{\text{m}}{\text{s}}\right)$, we determine the value of the friction coefficient by interpolation in the range from 0.025 to 0.018. Obviously, the value of the friction coefficient that represents the friction angle should be converted into degrees in order to be substituted into formula (7-3.4).

Referring now to formula (7-3.7), if $q_e \le 1$, then the natural cooling of the transmission is adequate and there is no need for additional cooling arrangements. The maximum allowable temperature for the oil is 200*F*° (95° *C*). The value of the heat transfer coefficient can be found in published sources, while for approximate calculations it could be taken $C_H = 7.0\ \dfrac{\text{kcal}}{{}^\circ\text{Cm}^2\text{h}}$. It is recommended that the structure of the housing of the worm gear transmission includes, as much as possible, cooling ribs to intensify the heat dissipation.

7-4. Forces Acting on Worm Gear Sets

The expression for calculating the normal force *F* applied to the pitch point *P* of a worm gear set is similar to formula (5-11.13) that is used for helical gears, and it reads

$$F = \frac{F_{2t}}{\cos\phi\,\cos\lambda} \tag{7-4.1}$$

where F_{2t} is, as shown in Figure 7-3.1a, the tangential force applied to the worm gear, ϕ is the pressure angle, and λ is the lead angle of the worm that is the helix angle of the worm gear. Unlike in helical gears, the tangential forces do not equal each other ($F_{1t} \ne F_{2t}$), while the tangential force of the gear equals the axial force of the worm ($F_{2t} = F_{1a}$). The tangential force of the worm gear is computed from the following formula:

$$F_{2t} = \frac{2W}{\omega_2 d_2} = F_{1a} \tag{7-4.2}$$

where W is the input power and F_{1a} is the axial force of the worm. Combining formulas (7-4.1) and (7-4.2) we obtain

$$F = \frac{2W}{\omega_2 d_2\cos\phi\,\cos\lambda} \tag{7-4.3}$$

The axial force of the gear equals the tangential force of the worm, hence, we write

$$F_{2a} = F_{1t} = F\cos\phi\sin\lambda = \frac{2W\tan\lambda}{\omega_2 d_2} \tag{7-4.4}$$

The radial forces of the gear and of the worm equal each other:

$$F_{2r} = F_{1r} = F\sin\phi = \frac{2W\tan\phi}{\omega_2 d_2\cos\lambda} \tag{7-4.5}$$

7-5. Bending Stress Calculations for Worm Gears

Usually, the worms are made from steel and the worm gears are made from bronze, while the worm threads are significantly stronger than the teeth of the worm gear. This is the reason why in a worm gear transmission the strength calculations are performed on the teeth of the worm gear only. Like other gear transmissions, the worm gear drive might have an open or enclosed structure. For open worm gear drives, strength calculations are based on bending stresses. However, it should be emphasized that the majority of worm gear transmissions are enclosed transmissions.

The calculations of bending stresses for the teeth of the worm gears are basically similar to the calculations used for helical gears. However, there are some differences that should be accounted for in the stress calculations. The teeth of worm gears wrap the worm threads and have a shape that resembles an arc. Due to this shape, the worm gear teeth are relatively stronger than the teeth of helical gears if they are made from the same material. This is reflected by the adjusted value of the virtual tooth form coefficient y_{2v}, for worm gears that is about 20% to 40% larger than for helical gears. This coefficient is computed using the following formula:

$$y_{2vw} = 0.52\left(1 - \frac{5.6}{N_{2vw}}\right)(1.18 + 0.001N_{2vw}) \tag{7-5.1}$$

where N_{2vw} is the virtual number of teeth of the worm gear and is calculated using a formula that is similar to formula (5-11.6):

$$N_{2vw} = \frac{N_2}{\cos^3\psi} \tag{7-5.2}$$

where ψ is, as mentioned earlier, the helix angle of the worm gear and $\psi = \lambda$, while λ is the lead angle of the worm.

By analogy with formulas (5-12.4) and (5-12.5), we write the formulas for calculating the bending stresses of the pitch and module systems worm gears, respectively:

$$\sigma_{wg} = \frac{2WP_n k_r k_d k_m \cos^2\psi}{\omega_2 d_2 b_2 \lambda_b y_{2vw}} \tag{7-5.3}$$

and

$$\sigma_{wg} = \frac{2Wk_r k_d k_m \cos^2 \psi}{\omega_2 m_n d_2 b_2 \lambda_b y_{2vw}} \quad (7\text{-}5.4)$$

where P_n and m_n are the normal pitch and normal module, λ_b is the contact line factor, and k_r, k_d, and k_m are the loading factors. For worm gears, as for helical gears, λ_b is in range from 1.1 to 1.3. Based on considerations presented in several sources, it is acceptable for preliminary calculations of worm gears to take the following values of the loading factors: the random loading factor is $k_r = 1.2$; the dynamic factor is $k_d = 1.3$; and the misalignment factor is $k_m = 1.1$.

Formulas (7-5.3) and (7-5.4) are applicable for determining bending stresses in worm gear teeth subjected to static loading. However, worm gears, as other gears, experience variable stresses, and the strength of their teeth should be calculated accounting for fatigue considerations. But if the number of loading cycles does not exceed 1,000, as for other gears, we may use the calculations for static stresses. Thus, for variable stress calculations, it is necessary to account for the fatigue stress concentration factor and the influence of the fatigue phenomenon on the strength of the worm gear materials. Some sources recommend using for approximate calculations of worm gears the fatigue stress concentration factor $K_f = 1.3$.

In cases of reversible drives, the teeth of the worm gears are subjected to a completely reversed loading cycle, while in cases of one directional drives the loading is characterized by a repeated cycle. In the majority of cases, worm gears or their rims are made of bronze. Bronze like other nonferrous alloys does not have an endurance limit like steel, and selecting allowable bending stresses is based on appropriate experimental data related to the bronze specimen. The experimental investigations are based on applying to the bronze material completely reversed cycles of bending stresses, usually in the amount of 10^7 cycles. According to these investigations, rather than an endurance limit a so-called fatigue strength limit S_{fe} for bronze should be used. Based on the appropriate investigations, we calculate the fatigue strength limit using the following empiric formula:

$$S_{fe} = 0.16 S_{ut} \quad (7\text{-}5.5)$$

Commonly, worm gears are made from sand cast bronze that, according to the American Gear Manufacturers Association (AGMA) records, has S_{fe} = 39 MPa (5.65 ksi). Following the above recommendation, we obtain for this bronze that the ultimate limit $S_{ut} = \frac{39}{0.16} = 244$ MPa (35.31 ksi) that is very close to the ultimate stresses of bronzes and leaded tin bronzes designated by UNS C9XXXX (unified numbering system of materials). These bronzes are recommended for manufacturing worm gears.

In this book, bending stress calculations of bronze worm gears are presented for completely reversed loading cycles only.

7-5.1. Verifying Worm Gear Bending Stress Calculations

Modifying equations (7-5.3) and (7-5-4) by accounting for the fatigue stress concentration factor, the fatigue strength limit S_{fe}, and the safety factor, we obtain formulas for verifying calculations of pitch and module systems of worm gears with regard to bending stresses for completely reversed cycles:

$$\sigma_{wg} = \frac{2WP_n k_r k_d k_m K_f \cos^2\psi}{\omega_2 d_2 b_2 \lambda_b y_{2vw}} \leq \frac{S_{fe}}{n_b} \tag{7-5.6}$$

and

$$\sigma_{wg} = \frac{2Wk_r k_d k_m K_f \cos^2\psi}{\omega_2 m_n d_2 b_2 \lambda_b y_{2vw}} \leq \frac{S_{fe}}{n_b} \tag{7-5.7}$$

Example 7-5.1a: Verifying calculations of bending stresses for an open worm gear set. For an open worm gear transmission, we determine the safety factor of the worm gear having a rotational speed of 260 rpm, a transverse module $m = 5$ mm, the number of teeth $N_2 = 45$, a helix angle $\psi = 10°$, and the width of the rim $b_2 = 35$ mm. The input power $W = 2.2$ kW. The worm gear is made of bronze casted in sand having $S_{fe} = 39$ MPa.

Solution:

We take $k_r = 1.2$, $k_d = 1.3$, $k_m = 1.1$, $K_f = 1.3$, and $\lambda_b = 1.15$. The angular velocity of the gear $\omega_2 = \frac{\pi 260}{30} = 27.21\ \frac{\text{rad}}{\text{s}}$. The pitch diameter $d_2 = mN_2 = 5 \cdot 45 = 225$ mm. The normal module $m_n = m\cos\psi = 5\cos 10° = 4.92$ mm. Using formula (7-5.2), we determine the virtual number of teeth of the gear $N_{2vw} = 47.1$, while using formula (7-5.1) we calculate the gear tooth form factor $y_{2vw} = 0.562$. Based on the formula (7-5.7), we determine the bending stress in the teeth:

$$\sigma_{wg} = \frac{2 \cdot 2.2 \cdot 10^3 \cdot 1.2 \cdot 1.3 \cdot 1.1 \cdot 1.3\cos^2 10°}{16.75 \cdot 0.00492 \cdot 0.225 \cdot 0.035 \cdot 1.15 \cdot 0.562} = 17{,}116{,}524 \text{ Pa. Therefore } \sigma_{wg} = 17.117 \text{ MPa.}$$

Hence, $17.117 = \frac{39}{n_b}$. Thus, $n_b = \frac{39.0}{17.117} = 2.28$, which is acceptable.

7-5.2. Design Calculations of Worm Gear Sets Based on Bending Stresses

The goal of design calculations is to derive the expressions for determining the diametral pitch or module of the worm gear based on appropriate input data. Thus, we modify the expressions (7-5.6) and (7-5.7) on the basis of the relationship between the pitch diameter d_1 of the worm and the width b_2 of the worm gear, which is characterized by the following formula:

$$b_2 = \kappa d_1 \tag{7-5.8}$$

where κ is an empiric coefficient of the worm gear width. The value of this coefficient, according to some authors, can be in the range of 0.67 to 0.75. Now, since for pitch system gears $d_1 = \frac{w}{P}$

and $d_2 = \frac{N_2}{P}$, we may write that $P = \frac{w}{d_1}$ and $P = \frac{N_2}{d_2}$. Similarly, we obtain for module system gears that $m = \frac{d_1}{w}$ and $m = \frac{d_2}{N_2}$. Using these two ratios, we may write $d_2 = \frac{d_1 N_2}{w}$. Combining this formula with expression (7-5.6) and formula (7-5.8), and recalling that $P_n = \frac{P}{\cos\psi}$ and $d_1 = \frac{w}{P}$, we have

$$\sigma_{wg} = \frac{2WP^3 k_r k_d k_m K_f \cos\psi}{\omega_2 N_2 w\kappa\lambda_b y_{2vw}} \tag{7-5.9}$$

Solving equation (7-5.9) for the diametral pitch, we obtain

$$P = \sqrt[3]{\frac{\omega_2 N_2 w\kappa\lambda_b y_{2vw}\,\sigma_{wg}}{2Wk_r k_d k_m K_f \cos\psi}} \tag{7-5.10}$$

Denoting

$$\eta_w = w\kappa \tag{7-5.11}$$

and following the recommendation based on actual data that the average value of the coefficient w may be taken as $w = 10$, we determine that the coefficient η_w may vary from 6.7 to 7.5, while 7.5 should be taken for worms having one or two starts, and 6.7 for worms having four starts. Combining equation (7-5.10) with formula (7-5.11) and replacing σ_{wg} with the ratio between the fatigue strength limit S_{fe} and the safety factor n_b, we obtain the design calculation formula for pitch system worm gears with regard to bending stresses:

$$P = \sqrt[3]{\frac{\omega_2 N_2 \eta_w \lambda_b y_{2vw}\,S_{fe}}{2Wk_r k_d k_m K_f n_b \cos\psi}} \tag{7-5.12}$$

Performing similar algebraic procedures, we derive the formula for design calculations for module system gears based on bending stresses:

$$m = \sqrt[3]{\frac{2Wk_r k_d k_m K_f n_b \cos\psi}{\omega_2 N_2 \eta_w \lambda_b y_{2vw} S_{fe}}} \tag{7-5.13}$$

Example 7-5.2a: Design calculations of a worm gear set with regard to bending stresses. We need to design an open worm gear transmission in which the worm is getting power of 1.0 hp at the rotational speed of 1,800 rpm, while the worm gear has a rotational speed of 60 rpm. The safety factor $n_b = 2$.

Solution:

According to the input data, the velocity ratio is $i = \frac{1{,}800}{60} = 30$. Accounting for the rotational speed of the worm and the velocity ratio we take $N_1 = 2$ and, consequently, $N_2 = iN_1 = 30{\cdot}2 = 60$.

We preliminarily select $\psi = \lambda = 12°$ and take $k_r = 1.2$, $k_d = 1.3$, $k_m = 1.1$, $K_f = 1.3$, and $\lambda_b = 1.15$. For the worm, we select AISI normalized steel having $S_{ut} = 185$ ksi and $H_B = 363$. The rim of the gear should be made from the C92600 bronze having $S_{ut} = 44$ ksi, and consequently, $S_{fe} = 0.16 \cdot 44 = 7.04$ ksi. The angular velocity of the gear $\omega_2 = \frac{\pi 60}{30} = 6.28 \frac{\text{rad}}{\text{s}}$. The virtual number of teeth of the worm gear according to formula (7-5.2) equals $N_{2vw} = 64.1$. Using formula (7-5.1), we calculate the gear tooth form factor $y_{2vw} = 0.589$. Since $N_1 = 2$ we take $\eta_w = 7.5$. Now, using formula (7-5.12) we calculate the diametral pitch of the worm gear: $P = \sqrt[3]{\frac{6.28 \cdot 60 \cdot 7.5 \cdot 1.15 \cdot 0.589 \cdot 7.04 \cdot 10^3}{2 \cdot 1 \cdot 550 \cdot 12 \cdot 1.2 \cdot 1.3 \cdot 1.1 \cdot 1.3 \cdot 2\cos 12°}} = 6.23 \frac{1}{\text{in.}}$. We have to select the closest smaller standard diametral pitch, which is $P = 6 \frac{1}{\text{in.}}$. Therefore, $d_1 = \frac{w}{P} = \frac{10}{6} = 1.667$ in. and $d_2 = \frac{N_2}{P} = \frac{60}{6} = 10$ in. Based on formula (7-1.3), we define the helix angle, hence, $\tan \lambda = \tan \psi = \frac{N_1}{Pd_1} = \frac{2}{6 \cdot 1.667} = 0.2$ and $\psi = 11.3°$. The center distance is $C = \frac{1.667 + 10}{2} = 5.833$ in. The pitch diameter of the worm depending on the center distance should be in the range that is determined by expression (7-1.6). The calculations show that $1.558 < 1.667 < 2.7$; therefore, the value of the worm pitch diameter satisfies the conditions mentioned above. The length of the worm according to formula (7-1.10) equals $b_1 = 2.433$ in. According to formula (7-5.8), the width of the worm gear $b_2 = 0.75 d_1 = 0.75 \cdot 1.667 = 1.25$ in. Hence, the basic parameters of the gear set are determined.

7-6. Calculations of Worm Gear Surface Durability

For enclosed worm gear transmissions as with other gear drives, contact stresses represent the most destructive factor that influences the serviceability of the drive. The calculations of contact stresses acting between the worm and worm gear are similar to the calculations of helical gears. However, we should take into consideration certain differences related to the moduli of elasticity of the steel worm and the bronze worm gear and to the infinitely large radius of the curvature of the worm profile. These differences result in a reduced modulus of elasticity and a reduced radius of curvature at the contact point of the worm and the worm gear. The interaction between two elements that are made from different materials is characterized by the expression for the reduced modulus of elasticity E_{red} that reads

$$E_{red} = \frac{2E_1E_2}{E_1 + E_2} \tag{7-6.1}$$

where E_1 and E_2 are the moduli of elasticity of the material of the worm and the worm gear, respectively.

The reduced radius of curvature for the engagement of the worm and the worm gear is determined from the following formula:

$$\frac{1}{\rho_{red}}=\frac{1}{\rho_1}+\frac{1}{\rho_2} \quad (7\text{-}6.2)$$

where ρ_1 and ρ_2 are the radii of curvature of the worm and the worm gear at the pitch point. Because the radius of curvature of the thread of the worm at the pitch point tends to infinity ($0.5d_1 = \rho_1 \rightarrow \infty$), therefore, $\frac{1}{\rho_1}\rightarrow 0$, and the reduced radius of curvature according to formula (7-6.2) becomes equal to

$$\frac{1}{\rho_{red}}=\frac{1}{\rho_2}=\frac{2\cos^2\psi}{d_2\sin\phi} \quad (7\text{-}6.3)$$

Combining formulas (1-4.30), (5-15.1), (7-4.3), (7-6.1), and (7-6.3), we obtain the formula for calculating the contact stresses in the worm gear set:

$$\sigma_H = 0.418\sqrt{\frac{4Wk_r k_d k_m E_{red}\cos^2\psi}{\omega_2 d_2^2 \lambda_{Hw} b_2 \cos\phi\sin\phi}} \quad (7\text{-}6.4)$$

where λ_{Hw} is the coefficient of the length of the contact lines that for worm gears may be taken $\lambda_{Hw} \approx 1.6$.

Based on formula (7-6.1), we determine the reduced modulus of elasticity for the interacting pair of materials—a steel worm having $E_1 = 30\cdot 10^6$ psi ($E_1 = 207\cdot 10^9$ Pa) and a bronze gear having $14.5\cdot 10^6$ psi ($E_2 = 100\cdot 10^9$ Pa), hence, $E_{red}=\frac{2\cdot 30\cdot 10^6\cdot 14.5\cdot 10^6}{30\cdot 10^6+14.5\cdot 10^6}=19.55\cdot 10^6$ psi or $E_{red}=\frac{2\cdot 207\cdot 10^9\cdot 100\cdot 10^9}{207\cdot 10^9+100\cdot 10^9}=135.853\cdot 10^9$ Pa. Therefore, the value of the reduced moduli of elasticity in psi and in MPa, respectively, are

$$\begin{aligned} E_{red} &= 19.55\cdot 10^6\,\text{psi} \\ E_{red} &= 136\cdot 10^9\ \text{GPa} \end{aligned} \quad (7\text{-}6.5)$$

This allows modifying formula (7-6.4) to the following shape:

$$\sigma_H = 1.182\sqrt{\frac{Wk_r k_d k_m E_{red}\cos^2\psi}{\omega_2 d_2^2 \lambda_{Hw} b_2 \sin 2\phi}} \quad (7\text{-}6.6)$$

According to sources based on corresponding experimental data, the allowable contact stresses S_{Hw} of bronze for the number of loading cycles 10^7 can be determined using the following empiric formula:

$$S_{Hw} = 0.6S_{ut} \quad (7\text{-}6.7)$$

This coincides with the recommendation of AGMA indicating that the allowable contact stress for sand cast bronze may be equal to 163 MPa. Since the ultimate strength of this type of bronze according to reliable sources equals 275 MPa, we have $S_{Hw} = 0.6 \cdot 275 = 165$ MPa, which is practically the same as the value recommended by AGMA.

7-6.1. Verifying Calculations of Worm Gears that Experience Contact Stresses

Based on formula (7-6.6), we can write the formula for verifying calculations of contact stresses for the worm gears:

$$\sigma_H = 1.182\sqrt{\frac{Wk_r k_d k_m E_{red}\cos^2\psi}{\omega_2 d_2^2 \lambda_{Hw} b_2 \sin 2\phi}} \le \frac{S_{Hw}}{n_{Hw}} \tag{7-6.8}$$

where n_{Hw} is the safety factor for contact stresses.

Example 7-6.1a: Verifying calculations of a worm gear that experiences contact stresses.

An enclosed worm gear drive transmits 2.0 kW of power at the rotational speed of the worm $n_1 = 3{,}600$ rpm to the worm gear having a rotational speed $n_2 = 150$ rpm. The worm gear is made from the UNS C92000 bronze having $S_{ut} = 275$ MPa. The module $m = 7$ mm, the number of starts of the worm $N_1 = 2$, and the width of the gear $b_2 = 52$ mm. We determine the safety factor with regard to contact stresses.

Solution:

The allowable contact stress according to formula (7.6.7) is $S_{Hw} = 0.6 \cdot 275 = 165$ MPa. The angular velocity of the gear is $\omega_2 = \frac{\pi 150}{30} = 15.7\ \frac{\text{rad}}{\text{s}}$. The velocity ratio is $i = \frac{3{,}600}{150} = 24$ and since $N_1 = 2$, therefore, the number of teeth of the gear is $N_2 = 2 \cdot 24 = 48$. Hence, $d_2 = 7 \cdot 48 = 336$ mm. Now we take $\lambda_{Hw} = 1.6$ and $k_r = 1.2$, $k_d = 1.3$, and $k_m = 1.1$. Combining formulas (7-6.5) and (7-6.7), we calculate $\sigma_H = 1.182\sqrt{\frac{2 \cdot 10^3 \cdot 1.2 \cdot 1.3 \cdot 1.1 \cdot 136 \cdot 10^9 \cdot \cos^2 14^\circ}{15.7 \cdot 0.336^2 \cdot 1.6 \cdot 0.052 \sin 2 \cdot 20^\circ}} = 80{,}445{,}013$ Pa.

Hence, $\sigma_H = 80.445$ MPa. Therefore, $n_{Hw} = \frac{165}{80.445} = 2.05$, which is acceptable.

7-6.2. Design Calculations of Worm Gears that Experience Contact Stresses

Combining expressions (7-5.7), (7-5.8), and (7-6.8) to derive design calculation formulas related to bending stresses, we may write

$$\sigma_H = 1.182\sqrt{\frac{Wk_r k_d k_m E_{red} N_2 \cos^2\psi}{\omega_2 d_2^3 \lambda_{Hw} \kappa w \sin 2\phi}} \le \frac{S_{Hw}}{n_{Hw}} \tag{7-6.9}$$

Modifying and squaring the expression (7-6.9) and accounting for $N_2 = iN_1$, we have

$$1.182^2 \frac{Wk_r k_d k_m E_{red} i N_1 \cos^2 \psi}{\omega_2 d_2^3 \lambda_{Hw} \kappa w \sin 2\phi} = \left(\frac{S_{Hw}}{n_{Hw}}\right)^2 \tag{7-6.10}$$

Solving equation (7-6.10) for d_2 we write

$$d_2 = \sqrt[3]{\frac{1.182^2 Wk_r k_d k_m E_{red} i N_1 n_{Hw}^2 \cos^2 \psi}{\omega_2 \lambda_{Hw} \kappa w S_{Hw}^2 \sin 2\phi}} \tag{7-6.11}$$

Modifying this equation and combining it with formula (7-5.11), we obtain the formula for design calculations of worm gear pitch diameters based on contact stresses:

$$d_2 = 1.118\sqrt[3]{\frac{Wk_r k_d k_m E_{red} i N_1 n_{Hw}^2 \cos^2 \psi}{\omega_2 \lambda_{Hw} \eta_w S_{Hw}^2 \sin 2\phi}} \tag{7-6.12}$$

where, as with bending stresses, $\eta_w = 7.5$ in cases when the number of starts of the worm is $N_1 = 1$ or $N_1 = 2$ and $\eta_w = 6.7$ in cases when the number of starts $N_1 = 4$.

Example 7-6.2a: Design calculations of a worm gear set that experiences contact stresses. It is required to design an enclosed pitch system worm gear drive that reduces the rotational speed from 1,800 rpm to 90 rpm. The input power is 1.0 hp. We take the safety factor $n_{Hw} = 2$.

Solution:

We select for the worm AISI 4340 normalized steel having $S_{ut} = 185.5$ ksi and $H_b = 363$, while for the worm gear we take UNS C90900 bronze having $S_{ut} = 40$ ksi. According to formula (7-6.7), we determine $S_{Hw} = 24$ ksi. Since the velocity ratio is $i = \frac{1{,}800}{90} = 20$, we take $N_1 = 2$ and, therefore, $N_2 = iN_1 = 2 \cdot 20 = 40$. We take the helix angle for the worm gear $\psi = 20°$. The angular velocity of the gear is $\omega_2 = \frac{\pi 90}{30} = 9.42 \frac{\text{rad}}{\text{s}}$. Now we take $k_r = 1.2$, $k_d = 1.3$, k_m 1.1, $\lambda_{Hw} = 1.6$, and $\eta_w = 7.5$. From formula (7-6.5), we have $E_{red} = 19.55 \cdot 10^6$ psi. Substituting the data presented above into formula (7-6.12), we calculate $d_2 = 1.118\sqrt[3]{\frac{1.0 \cdot 550 \cdot 12 \cdot 1.2 \cdot 1.3 \cdot 1.1 \cdot 19.55 \cdot 10^6 \cdot 20 \cdot 2 \cdot 2^2 \cos^2 12°}{9.42 \cdot 1.6 \cdot 7.5 \cdot 24^2 \cdot 10^6 \sin 2 \cdot 20°}} = 12.022$ in.

The diametral pitch is $P = \frac{N_2}{d_2} = \frac{40}{12.022} = 3.327 \frac{1}{\text{in.}}$. Taking the closest smaller standard pitch, we have $P = 3 \frac{1}{\text{in.}}$; therefore, recalculating the gear pitch diameter we obtain $d_2 = 13.333$ in. Taking the average value for the coefficient $w = 10$, we determine the pitch diameter of the worm $d_1 = \frac{w}{P} = \frac{10}{3} = 3.333$ in. Hence, the width of the worm gear is $b_2 = 0.75 \cdot 3.333 = 2.5$ in. Using formula (7-1.10), we calculate the length of the worm $b_1 = \frac{11 + 0.06 \cdot 40}{3} = 4.467$ in. The center distance is $C = \frac{3.333 + 13.333}{2} = 9.833$ in. Verifying the value of the worm pitch diameter according to expression (7-1.6), we obtain $2.46 \le 3.333 \le 4.34$, which is in agreement with the recommendations. Thus, the basic parameters of the worm gear are determined.

7-7. Design Project

It is necessary to design an enclosed worm gear reducer having input power of 3 kW and output rotational speed of 150 rpm.

1. Calculating the basic parameters of the worm gear set.

We start with selecting the materials for the worm and worm gear. For the worm we take AISI 4340 normalized steel having S_{ut} = 1,280 MPa and H_b = 363, and for the gear we select UNS C90900 bronze having S_{ut} = 275 MPa. The fatigue strength limit of bronze is $S_{fe} = 0.16 \cdot 275 = 44$ MPa, and the allowable contact stress is $S_{Hw} = 0.6 \cdot 275 = 165$ MPa. We assign the safety factor $n_{Hw} = 2$ and take preliminarily the helix angle of the gear $\psi = 15°$. We choose an electrical motor having the output power of 3 kW and rotational speed of 3,600 rpm. Therefore, $i = \frac{3{,}600}{150} = 24$ that leads to the number of starts $N_1 = 2$. Hence, the number of teeth of the gear is $N_2 = N_1 i = 2 \cdot 24 = 48$. Taking the recommended values of loading factors and of other appropriate coefficients, we write $k_r = 1.2$, $k_d = 1.3$, $k_m = 1.1$, $\lambda_{Hw} = 1.6$, and $\eta_w = 7.5$. The angular velocity of the worm gear is $\omega_2 = \frac{\pi 150}{30} = 15.7 \frac{\text{rad}}{\text{s}}$. We take the reduced modulus of elasticity according to formula (7-6.5). Now using formula (7-6.11), we calculate the pitch diameter of the worm gear: $d_2 = 1.118 \sqrt[3]{\frac{3 \cdot 10^3 \cdot 1.2 \cdot 1.3 \cdot 1.1 \cdot 136 \cdot 10^9 \cdot 24 \cdot 2 \cdot 2^2 \cos^2 15°}{15.7 \cdot 1.6 \cdot 7.5 \cdot 165^2 \cdot 10^{12} \sin 2 \cdot 20°}} = 0.376$ m. Hence, the module is $m = \frac{d_2}{N_2} = \frac{376}{48} = 7.832$ mm. The closest larger standard module is 8 mm. Recalculating the pitch diameter of the gear $d_2 = 8 \cdot 48 = 384$ mm. Taking the average value of the coefficient of the worm pitch diameter $w = 10$, we calculate $d_1 = 10 \cdot 8 = 80$ mm. Therefore, the center distance is $C = \frac{80 + 384}{2} = 232$ mm. Using expression (7-1.7), we calculate $2.309 \le 3.15 \le 4.075$, which is in compliance with the above recommendation.

The width of the worm gear is $b_2 = 0.75 \cdot 80 = 60$ mm and the length of worm according to formula (7-1.14) is $b_1 = (11 + 0.06 \cdot 48)8 = 111.04$ mm. Now we define the helix angle according to formula (7-1.3), hence, $\tan \psi = \frac{8 \cdot 2}{80} = 0.2$, therefore, $\psi = 11.31°$ which is noticeably less than the preliminary accepted helix angle of 15°, and we are required to recalculate the pitch diameter of the worm gear. The recalculated value of the worm gear pitch diameter $d_2 = 0.372$ m leads to the same module $m = 8$ mm.

2. Verifying calculations of the bending stresses in the teeth of the worm gear.
As part of the design process of enclosed gear drives, it is recommended that we perform verifying calculations with regard to bending stresses. We take K_f = 1.3 and λ_b = 1.15. In addition, we calculate $N_{2vw} = \dfrac{48}{\cos^3 11.31°} = 50.23$ and using formula (7-5.1), we compute $y_{2vw} = 0.52\left(1 - \dfrac{5.6}{50.23}\right)(1.18 + 0.001 \cdot 50.23) = 0.568$. Now using formula (7-5.7) and accounting that m_n = 8 cos 11.31° = 7.844 mm, we calculate the bending stresses: $\sigma_{wg} = \dfrac{2 \cdot 3 \cdot 10^3 \cdot 1.2 \cdot 1.3 \cdot 1.1 \cdot 1.3\cos^2 11.31°}{16.74 \cdot 0.007844 \cdot 0.384 \cdot 0.06 \cdot 1.15 \cdot 0.568} = 6{,}947{,}380$ Pa. Hence, σ_{wg} = 6.95 MPa. Based on the fatigue strength limit and assigned safety factor, we may write $\dfrac{44}{2 \cdot 6.95} = 3.165$. Hence, the actual safety factor with regard to bending stresses is 3.165, which is acceptable.

3. Heat calculations.
The determined values of the center distance and the dimensions of the worm and worm gear allow us to estimate the volume of the gear drive enclosure and, consequently, calculate the approximate value of the active area of the enclosure that plays an important role in the dissipation of heat. Assuming that the center distance is measured in the vertical direction, then accounting for the distances between the gear and walls and accounting for the volume of needed oil, we estimate that the height of the enclosure is about 2.75*C* (*C* is the center distance). The length of the enclosure depends on the diameter of the worm gear and could be estimated at about $1.25d_2$. The width of the enclosure depends on the widths of the gear's hub, the bearings, and the required distances between the gear and the walls of the enclosure and might be estimated as equal to 0.75*C*. We do not include the area of the bottom of the enclosure. These considerations allow us to calculate the approximate area of the active surface of the enclosure. Hence, we have $A_w = 2 \cdot 2.75 \cdot 0.232 \cdot 1.25 \cdot 0.384 + 2 \cdot 2.75 \cdot 0.232 \cdot 0.75 \cdot 0.232 + 1.25 \cdot 0.384 \cdot 0.75 \cdot 0.232 = 0.918$ m². In order to determine the efficiency of the worm gear drive, we need to determine the corresponding friction coefficient that depends on the sliding velocity between the worm and gear. The angular velocity of the worm is $\omega_1 = \dfrac{\pi 3{,}600}{30} = 376.8\ \dfrac{\text{rad}}{\text{s}}$. Applying formula (7-3.2) and considering that $\lambda = \psi = 11.31°$, we calculate the sliding velocity: $V_s = \dfrac{0.08 \cdot 376.8}{2\cos 11.31°} = 15.37\ \dfrac{\text{m}}{\text{s}}$. As discussed above, for this value of sliding velocity we take f = 0.018, hence, the friction angle is μ = 57.3·0.018 = 1.03°. Using formula (7-3.3), we calculate the efficiency $e = \dfrac{\tan 11.31°}{\tan(11.31° + 1.03°)} = 0.914$. Applying formula (7-3.5), we determine the amount of heat generated per hour $Q_e = 859.85 \cdot 3(1 - 0.914) = 221.84\ \dfrac{\text{kcal}}{\text{h}}$. Taking the heat transfer coefficient $C_H = 7.0\ \dfrac{\text{kcal}}{°\text{Cm}^2\text{h}}$, the maximum allowable temperature of the oil 95°C, the ambient temperature 20°C, and applying formula (7-3.6), we determine the amount of dissipated heat per hour
$Q_d = 7 \cdot 0.918(95°C - 20°C) = 482\ \dfrac{\text{kcal}}{\text{h}}$.

Using formula (7-3.7), we determine $q_e = \dfrac{Q_e}{Q_d} = \dfrac{221.84}{482} = 0.46 < 1$. In our case, the natural heat dissipation is sufficient, and a dedicated cooling device is not needed.

4. Preliminary calculations of the diameter for the worm gear shaft and selecting the bearings.

Using formula (3-3.8) and taking $\tau_{all} = 16$ MPa, we calculate the diameter of the worm gear shaft that is considered the same as the bore diameter of the bearing: $d = 3.65\sqrt[3]{\dfrac{3 \cdot 10^3}{150 \cdot 16 \cdot 10^6}} = 0.0394$ m.

We take d = 40 mm and select preliminarily SKF 7208 BECBP angular-contact bearing having the bore diameter d = 40 mm, the outer diameter D = 80 mm, the width B = 18 mm, the dynamic load rating C = 36.5 kN, and static load rating C_0 = 26 kN.

5. Reactions, bending moments, and torque diagrams for the gear shaft.

Based on the determined geometric parameters of the gear, the shaft, and the bearings it became possible to compose the preliminary assembly drawing of the worm gear reducer that is presented in Figure 7-7.1. This figure shows that the width of the hub of the worm gear is 84 mm and the distance between the hub and the walls is equal to 24 mm on each side.

Accounting for the width of the bearing, we determine that the distance between the centers of the bearings l = 150 mm. Now, based on Figure 7-7.1 and the considerations above, we can draw the free-body diagram of the gear shaft that is shown in Figure 7-7.2a. This free-body diagram is combined with the free-body diagram of the vertical plane and indicates all forces applied to the pitch point P.

Obviously, the tangential force F_{2t} acts in the horizontal plane, and it is considered in Figure 7-7.2c. Using formulas (7-4.4) and (7-4.5), we calculate the axial and radial forces applied to the shaft of the gear: $F_{2a} = F_{1t} = \dfrac{2 \cdot 3 \cdot 10^3 \cdot \tan 11.31°}{15.7 \cdot 0.384} = 205.66$ N and $F_{2r} = F_{1r} = \dfrac{2 \cdot 3 \tan 20°}{15.7 \cdot 0.384 \cos 11.31°} = 359.05$ N. Projecting the forces onto the x axis, we write $R_x + F_{2a} = 0$, therefore $R_x = -F_{2a} = 205.66$ N. Now taking the sum of moments about point B, we calculate the reaction R_{AV}. $R_{AV}l - 0.5F_{2r}l + 0.5Fd_2 = 0$, therefore, $R_{AV} = \dfrac{0.5 \cdot 359.05 \cdot 150 - 0.5 \cdot 384 \cdot 205.66}{150} = -83.72$ N. Projecting the forces onto the y axis, we calculate the reaction R_{BV}. $R_{AV} - F_{2r} + R_{BV} = 0$. Hence, $R_{BV} = 83.72 + 359.05 = 442.77$ N. Now we determine the bending moments in the vertical plane. The free-body diagram is divided into two intervals, while the bending moment on the first interval is $M_I = R_{AV}x_1$. When $x_1 = 0$ then $M_I = 0$; and when $x_1 = 0.5l = 0.5 \cdot 0.15 = 0.075$ m then $M_I = -83.72 \cdot 0.075 = -6.28$ N·m. On the second interval the bending moments equal $M_{II} = R_{AV}x_2 - F_{2r}(x_2 - 0.5l) + 0.5F_{2a}d_2$. When $x_2 = 0.5l = 0.075$ m then $M_{II} = -83.72 \cdot 0.075 + 0.5 \cdot 0.384 \cdot 205.66 = 33.2$ N·m, and when $x = l$

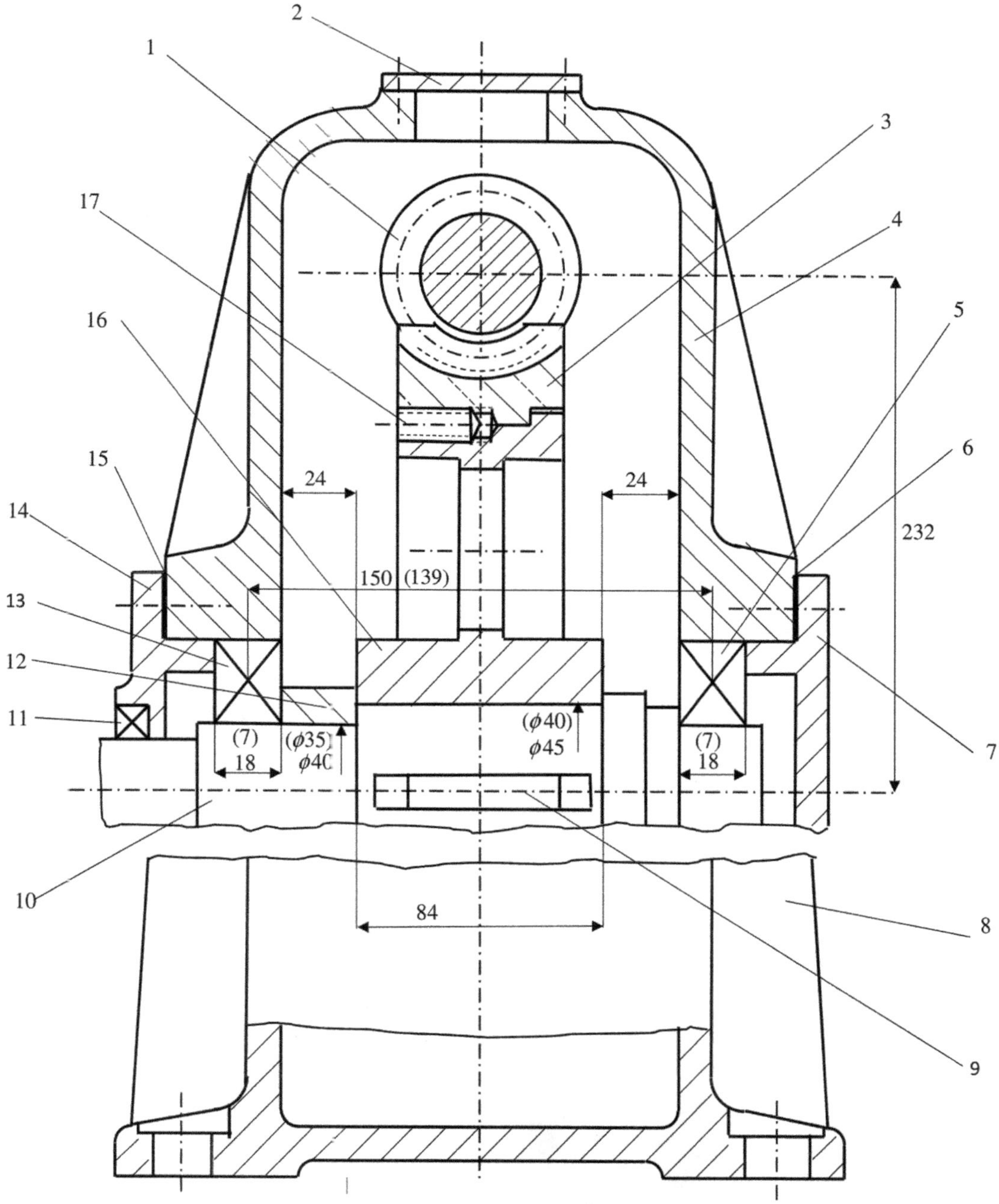

Figure 7-7.1. Worm gear reducer.

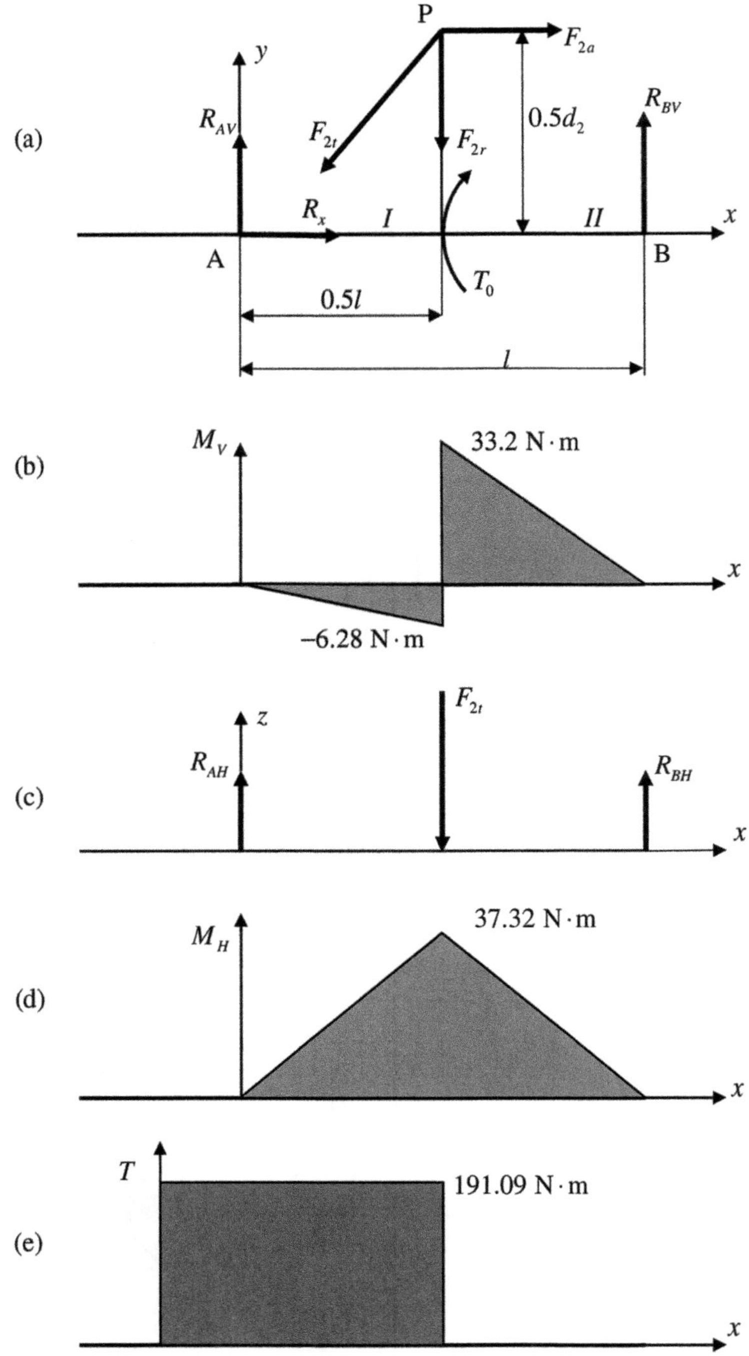

Figure 7-7.2. Reactions, bending moments and torque diagrams.

then $M_{II} = -83.72 \cdot 0.15 - 359.05 \cdot 0.075 + 205.66 \cdot 0.5 \cdot 0.384 = 0$. The bending-moment diagram in the vertical plane is presented in Figure 7-7.2b that shows that $M_{V\max} = 33.2$ N·m, and it is applied in the section located in the middle between the supports. The free-body diagram in the horizontal plane is shown in Figure 7-7.2c. The tangential force that is applied in the middle between the bearings equals $F_{2t} = \dfrac{2 \cdot 3 \cdot 10^3}{15.7 \cdot 0.3840} = 995.24$ N, therefore, $R_{AH} = R_{BH} = 0.5F_{2t} = 0.5 \cdot 995.22 = 497.6$ N.

The bending moment on the first interval $M_I = R_{AH}x_1$, and when $x_1 = 0$ then $M_I = 0$, while when $x_1 = 0.5l$ then $M_I = 497.6 \cdot 0.075 = 37.32$ N·m. On the second interval, the bending moment equals $M_{II} = R_{AH}x_2 - F_{2t}(x_2 - 0.5l)$. Hence, when $x_2 = 0.5l$ then $M_{II} = 37.32$ N·m, while when $x_2 = l$ then $M_{II} = 0$. The bending-moment diagram in the horizontal plane is presented in Figure 7-7.2d, indicating that $M_{H\max} = 37.32$ N·m, and it is also acting in the section located in the middle between the supports. The resultant bending moment applied to the shaft is $M = \sqrt{M_{V\max}^2 + M_{H\max}^2} = \sqrt{33.2^2 + 37.32^2} = 49.95$ N·m. The torque applied to the shaft equals $T_0 = 0.5F_{2t}d_2 = 0.5 \cdot 995.24 \cdot 0.384 = 191.09$ N·m. The torque diagram is shown in Figure 7-7.2e. The bending moment and torque diagrams show that the most loaded section of the shaft is located in the middle between the supports. Now we determine the resultants of the reactions in the supports: $R_A = \sqrt{R_{AV}^2 + R_{AH}^2} = \sqrt{(-83.72)^2 + 497.62^2} = 504$ N and $R_B = \sqrt{R_{BV}^2 + R_{BH}^2} = \sqrt{442.77^2 + 497.62^2} = 666.1$ N. Therefore, for the verifying calculations of the selected bearings, we take $F_r = 666.1$ N and $F_a = 205.66$ N.

6. Verifying calculations of the diameter of the shaft for the worm gear.

We select for the shaft AISI 1015 annealed steel having $S_{ut} = 385$ MPa. The shaft is subjected to a completely reversed loading cycle. Using formulas (1-9.2) and (1-9.3), we calculate k_a 0.948 and $k_b = 0.816$, respectively, and taking $k_c = 1$, we calculate $S_e = 0.948 \cdot 0.816 \cdot 1 \cdot 0.5 \cdot 385 = 149$ MPa. Due to the keyseat, we take according to Figure 3-8.4 the fatigue stress concentration factor $K_{fb} = 1.6$ for bending and $K_{ft} = 1.3$ for torsion. The shaft has a step that is characterized by $\dfrac{D}{d} = 1.1$ and $\dfrac{r}{d} = 0.05$. According to Appendix A-1-5, we determine the theoretical stress concentration factor for bending $K_{tb} = 1.85$ and for torsion $K_{tt} = 1.4$. Taking the notch sensitivity factor $q = 0.6$ and using formula (1-10.2), we compute for bending $K'_{fb} = 1.51$ and for torsion $K'_{ft} = 1.24$. Taking the correspondingly larger factors, we apply $K_{fb} = 1.6$ and $K_{ft} = 1.3$. Considering for the shaft the safety factor $n = 2$ and applying formula (3-5.9) for alternative bending and torsion according to the fourth theory of failure, we calculate $d = \left\{ \dfrac{32 \cdot 2}{\pi} \left[\left(\dfrac{1.6 \cdot 49.95 \cdot 10^3}{149} \right)^2 + \dfrac{3}{4} \left(\dfrac{1.3 \cdot 191.09 \cdot 10^3}{149} \right)^2 \right]^{\frac{1}{2}} \right\}^{\frac{1}{3}} = 31.36$ mm. Hence, we may take the bore diameter as 35 mm, which will result in the reduction of the weight and of the dimensions of the transmission.

7. Calculations of the bearings for the shaft of the worm gear.

Now we select SKF 7207 BEP angular-contact bearing having the bore diameter of 35 mm, the outer diameter of 72 mm, the width of 17 mm, the dynamic load rating $C = 29.1$ kN, and the static load rating $C_0 = 19$ kN.

For this type of bearings the load factors for calculating the equivalent static load are $X_0 = 0.5$ and $Y_0 = 0.26$. Using formula (2-4.6), we calculate the equivalent static load:

$F_0 = 0.5 \cdot 666.1 + 0.26 \cdot 205.66 = 386.52$ N.

Since $F_0 < C_0$ (4386.52 N < 19,000 N), we may proceed with the newly selected bearing. According to formula (2-4.7), $e = \frac{F_a}{C_0} = \frac{205.66}{19{,}000} = 0.0108$. Because the inner ring rotates, we take $V = 1$ and using formula (2-4.8), we determine $\frac{F_a}{F_r} = \frac{205.66}{666.1} = 0.308 > e = 0.0108$. For this case, considering that $\frac{F_a}{F_r} = 0.308 < 1.14$, the SKF catalog recommends for calculating the equivalent dynamic load to use in formula (2-4.5) the following values of loading factors: $X = 1$ and $Y = 0.55$. Therefore, we have $F_e = XVF_r + YF_a = 1 \cdot 1 \cdot 666.1 + 0.55 \cdot 205.66 = 779.2$ N. The service life of the bearing is $L_{10} = \left(\frac{C}{F_e}\right)^3 = \left(\frac{29100}{779.2}\right)^3 = 52{,}087 \cdot 10^6$ revolutions or $L_{10h} = \frac{52{,}087 \cdot 10^6}{60 \cdot 150} = 5{,}787{,}444$ hours, which significantly exceeds the conventional service life. We select now SKF 61807 radial ball bearing that has the same bore diameter as the previously selected bearing; however, the outer diameter is 47 mm, the width is 7 mm, the dynamic load rating is $C = 4.36$ kN, and the static load rating $C_0 = 3.35$ kN. In this case, $e = \frac{205.66}{3{,}350} = 0.061$ which is still less than $\frac{F_a}{F_r} = 0.308$, therefore, from Appendix A-2-2 we found that for this case $X = 0.56$ and $Y = 1.65$. Hence, the equivalent load is $F_e = 0.56 \cdot 666.1 + 1.65 \cdot 205.66 = 712.36$ N. Calculating the service life of the bearing, we have $L_{10} = \left(\frac{4.36 \cdot 10^3}{712.36}\right)^3 = 229.27 \cdot 10^6$ revolutions or $L_{10h} = \frac{229{,}27 \cdot 10^6}{60 \cdot 150} = 25{,}475$ hours, which is acceptable.

8. Calculations of deflection and angle of rotation of the worm gear shaft.

The following forces should be accounted for in the calculations of the deflection and the angle of rotation of the shaft—namely, the tangential force $F_{2t} = 995.24$ N, the radial force $F_{2r} = 364$ N, and the axial force $F_{2a} = 267$ N. The forces F_{2t} and F_{2r} are perpendicular to each other, and they can be replaced by their resultant $F = \sqrt{F_{2t}^2 + F_{2r}^2} = \sqrt{995.24^2 + 359.05^2} = 1{,}058$ N. The axial force F_{2a} causes a bending moment $M_o = 205.66 \cdot 0.5 \cdot 0.384 = 39{,}487$ N·m. The shaft has a stepped shape, while the diameters of the steps due to the newly selected bearings are reduced to 35 and to 40 mm (Figure 7-7.1). The width of the newly selected bearings is 7 mm; therefore, the length of shaft should be corrected in these calculations $l' = 7 + 2 \cdot 24 + 84 = 139$ mm (Figure 7-7.1). It should be mentioned that due to this correction there is no need to perform a new iteration of the previous calculations. Now, based on formula (3-6.11), we determine the equivalent diameter of the shaft: $d_e = \frac{1}{\sqrt[4]{\sum_1^k \frac{l_j}{l d_j^4}}} = \frac{1}{\sqrt[4]{\frac{2 \cdot (7+24)}{139 \cdot 35^4} + \frac{84}{139 \cdot 40^4}}} = 37$ mm.

The moment of inertia of the shaft having the equivalent diameter equals
$I=\frac{\pi d^4}{64}=\frac{\pi 37^4}{64}=91{,}951\ \text{mm}^4$.
Applying formula (3-6.8), we determine the deflection h of the shaft between the bearings due to the force F, therefore, $h_F=-\frac{Fl^3}{48EI}=-\frac{1{,}058\cdot 146^3}{48\cdot 207\cdot 10^3\cdot 91{,}951}=-0.00359$ mm. The deflection of the shaft caused by the moment M_0 according to formula (3-6.17) equals zero. According to recommendations presented above, the deflection of the shaft carrying a gear should not exceed $0.0003l$. In our case, $0.0003l = 0.0003\cdot 139 = 0.0417$ mm that exceeds the determined deflection 0.00359 mm.

The angles of rotation caused by radial and axial forces occur in the same plane and the resultant of these two angles represents their superposition. The angle of rotation due to the radial force is calculated using formula (3-6.): $\theta_{Fr}=-\frac{359.05\cdot 139^2}{16\cdot 207\cdot 10^3\cdot 91{,}951}=-0.0000228$ rad. The angle of rotation due to the axial force that creates the moment M_0 = 51.26 N·m is calculated using formula (3-6.19). After applying to this formula the related transformations, we obtain $\theta_M=-\frac{M_0 l}{24EI}=-\frac{51.26\cdot 10^3\cdot 139}{24\cdot 207\cdot 10^3\cdot 91{,}951}=-0.0000156$ rad. The superposition of the calculated angles of rotation equals $\theta_1=\theta_{Fr}+\theta_M=-0.0000384$ rad. The angle of rotation caused by the tangential force is calculated applying formula (3-6.10):
$\theta_{Ft}=-\frac{995.24\cdot 139^2}{16\cdot 207\cdot 10^3\cdot 91{,}951}=-0.0000631$ rad. The total angle of rotation θ_A represents a vector sum of θ_1 and θ_{Ft}, hence, $\theta_A=\sqrt{\theta_1^2+\theta_{Ft}^2}=\sqrt{0.0000384^2+0.0000631^2}=0.0000739$ rad. The allowable angle of rotation for radial ball bearings is 0.01 rad that exceeds the calculated value of the angle of rotation.

9. Calculation of the rotational speed.
Accounting for the calculated deflection, we determine the critical rotational speed using formula (3-7.12): $n_{cr}=\frac{30}{\pi}\sqrt{\frac{9.81\cdot 10^3}{0.00359}}=15{,}793$ rpm.

10. Assembly and disassembly.

a) Assembly.
Referring to Figure 7-7.1, it should be mentioned that the bronze rim 3 is permanently pressed onto the wheel 16 and secured with four set screws 17.

We start with inserting key 9 into the keyseat of the shaft 10 and placing the wheel 16 that became a compound gear on the shaft and then placing distance bushing 12. After that, we install bearings 5 and 13, and then we place the subassembly into the lower casing 8. The worm subassembly is not shown in drawing, but it is structured similar to the worm gear subassembly. With regard to the worm subassembly, we start by inserting the key into keyseat of the shaft of the worm 1, and we place the worm on its shaft and then install the distance bushing and press in the bearings on the worm shaft. After that, the worm subassembly is placed into the upper casing 4. Then the upper casing is put in place covering the worm gear subassembly, and then it is secured with the lower casings using appropriate bolts and nuts (not shown in the drawing). Then we put in place the bearing covers with the shims of the worm subassembly. We adjust the preload of the bearings of the worm shaft and secure the covers with screws. After that, we put in place covers 7 and 14 with the sets of shims 6 and 15. Then we adjust the bearings as well as the worm gear engagement and secure the covers with screws. We install the seal 11 and pour the oil through the opening in the upper casing, keeping in mind that the level of oil should cover one-half of a tooth of the worm gear. Then we close the opening with cover 2. Now the worm reducer is ready to be put to work. It should be mentioned that all adjustments are performed according to appropriate specifications.

b) Disassembly.

We start with opening the oil plug and draining the oil from the worm gear reducer. Then we unscrew the screws holding the covers 7 and 14 in place. After that, we unscrew the screws holding the covers of the worm shaft bearings in place. We also unscrew the screws keeping the cover 2 in place. Now we unscrew the bolts that hold together upper casing 4 and lower casing 8 and remove the upper casing. We take off the worm subassembly from the upper casing and using special pullers remove the bearings. Then, we remove the distance bushing and dismount the worm and the key. We also take off the worm gear subassembly and using special pullers uninstall the bearings. Then we remove the distance bushing and take off the worm gear and the key. We also remove the seal.

Chapter 7. Problems

7-01. An enclosed worm gear drive transmits 3 kW of power at the rotational speed of the worm $n_1 = 3{,}600$ rpm to the 52-tooth worm gear. The number of starts of the worm $N_1 = 1$, the module $m = 6$ mm, the active surface of the drive enclosure $A_W = 1.1$ m^2, and the pitch diameter of the worm $d_1 = 60$ mm. Determine if the drive needs a dedicated cooling device considering that the allowable operational temperature of oil is 95 °C, the ambient temperature is 20°C, and the heat transfer coefficient $C_H = 7\frac{kcal}{°Cm^2h}$.

7-02. An open worm gear drive has a 56-tooth worm gear that transmits 1.5 hp of power at the rotational speed $n_2 = 130$ rpm. The diametral pitch $P = 6\ \frac{1}{\text{in.}}$, the number of starts of the worm $N_1 = 2$, the worm pitch diameter $d_1 = 1.667$ in., and the width of the worm gear $b_2 = 1.3$ in. The worm gear is made from bronze having $S_{ut} = 35.31$ ksi. Determine the safety factor with regard to bending stresses.

7-03. An open worm gear drive transmits 3 hp of power at the rotational speed of the worm $n_1 = 3{,}600$ rpm to the worm gear that has a rotational speed $n_2 = 120$ rpm. The number of starts of the worm $N_1 = 2$. The worm gear is made from bronze having $S_{ut} = 44$ ksi. Taking the preliminary value of the helix angle $\psi = 12°$ and the safety factor $n_b = 2$, determine the diametral pitch of the worm gear.

7-04. An open worm gear drive transmits 4 kW at the rotational speed of the worm $n_1 = 1{,}800$ rpm to the worm gear that has a rotational speed $n_2 = 60$ rpm. The number of starts of the worm $N_1 = 2$. The worm gear is made from bronze having $S_{ut} = 303$ MPa. Taking the preliminary value of the helix angle $\psi = 12°$ and the safety factor $n_b = 2$, determine the module of the worm gear.

7-05. An enclosed worm gear drive transmits 4 hp at the rotational speed of the worm $n_1 = 1{,}800$ rpm to the worm gear that has a rotational speed $n_2 = 75$ rpm. The number of starts of the worm $N_1 = 2$, the diametral pitch of the worm gear $P = 4\ \frac{1}{\text{in.}}$, and the width of the worm gear $b_2 = 2.0$ in. Taking the value of the helix angle $\psi = 12°$, determine the contact stresses in the teeth of the worm gear.

7-06. An enclosed worm gear drive transmits 5 kW at the rotational speed of the worm $n_1 = 1{,}200$ rpm to the worm gear that has a rotational speed $n_2 = 40$ rpm. The number of starts of the worm $N_1 = 2$. The module of the worm gear $m = 6$ mm and the width of the gear $b = 52$ mm. Taking the value of the helix angle $\psi = 12°$, determine the contact stresses in the teeth of the worm gear.

7-07. An enclosed worm gear drive transmits 2 hp of power at the rotational speed of the worm $n_1 = 3{,}600$ rpm to the worm gear that has a rotational speed $n_2 = 360$ rpm. The number of starts of the worm $N_1 = 4$, the diametral pitch $P = 4\ \frac{1}{\text{in.}}$, and the width of the worm gear $b_2 = 2.2$ in. The worm gear is made from bronze having $S_{ut} = 44$ ksi. Taking the preliminary value of the helix angle $\psi = 22°$, determine the safety factor with regard to contact stresses.

7-08. An enclosed worm gear drive transmits 2.2 kW at the rotational speed of the worm $n_1 = 3{,}600$ rpm to the worm gear that has a rotational speed $n_2 = 240$ rpm. The number of starts of the worm $N_1 = 4$. The module $m = 6$ mm and the width of worm gear $b_2 = 54$ mm. The worm gear is made from bronze having $S_{ut} = 303$ MPa. Taking the preliminary value of the helix angle $\psi = 22°$, determine the safety factor with regard to contact stresses.

7-09. An enclosed worm gear drive transmits 2 hp at the rotational speed of the worm $n_1 = 1{,}800$ rpm to the worm gear that has a rotational speed $n_2 = 72$ rpm. The number of starts of the worm $N_1 = 2$. The worm gear is made from bronze having $S_{ut} = 40$ ksi. Taking the value of the helix angle $\psi = 12°$ and safety factor $n_{Hw} = 2$, determine the pitch diameter of the worm gear.

7-10. An enclosed worm gear drive transmits 1.1 kW of power at the rotational speed of the worm $n_1 = 1{,}200$ rpm to the worm gear that has a rotational speed $n_2 = 45$ rpm. The number of starts of the worm $N_1 = 2$. The worm gear is made from bronze having $S_{ut} = 265$ MPa. Taking the value of the helix angle $\psi = 12°$ and the safety factor $n_{Hw} = 2$, determine the pitch diameter of the worm gear.

CHAPTER 8

Power Screw Drives

Power screw drives consist of a threaded pair comprising a screw and nut. In its simplest form, either the screw or the nut is rigidly secured to a non-movable support, and the other mating element is subjected to an external torque enabling it to perform rotational and translational motion simultaneously. In this case, because of the specific interaction between the threads of the threaded pair, the movable element develops a significant force in the direction of its linear translational motion. The main advantages of power screw drives are their ability to develop relatively large forces from relatively small torques, structural and manufacturing simplicity, and the highly accurate operation.

These features explain the extremely wide use of power screws in an endless variety of human activities and industrial applications. Power screw drives have widespread use in a variety of lifting mechanism, such as manual and power jacks, lead screws in milling machines, lathes, assembly and disassembly devices, measuring and control instrumentation, and in numerous other mechanisms.

8-1. Power Screw Mechanics

Figure 8-1.1 shows an example of a manual power screw jack comprising a power screw 1 that interacts with the nut 2. Handle 3 enables the screw rotating. The nut 2 is rigidly secured to the body 3. The threaded pair in this figure has square profile threads.

Very often, the threads of power screw drives have a trapezoidal or triangular profile. Figure 8-1.2b schematically shows the square and trapezoidal thread profiles of the UNS (Unified Numbering System, in inches) and of the ISO (International Standardization Organization, in millimeters).

Square and trapezoidal profiles have the same pitch p (axial pitch) that according to the American National Standard represents the distance between identical points of two adjacent threads measured parallel to the axis of the treads. The definition of pitch in the ISO system is the same. The profile angle of the Acme (UNS) trapezoidal thread is $2\alpha = 29°$, while the ISO trapezoidal profile is $2\alpha = 30°$. In the American National Standard of threads, the number of threads per inch N_t represents the reciprocal of the pitch; therefore, $N_t = \dfrac{1}{p}$ and, consequently, $p = \dfrac{1}{N_t}$ (measured in inches). The mean diameter of the thread, also called pitch diameter, and the minor diameter of the threads are

$$d_m = d - 0.5p = d - \frac{0.5}{N_t} \qquad (8\text{-}1.1)$$

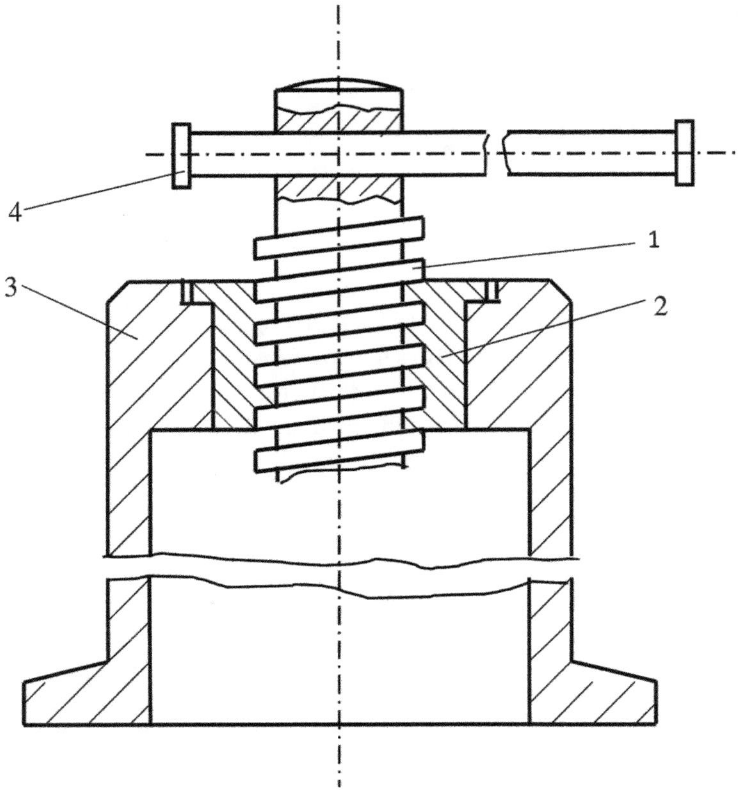

Figure 8-1.1. Power screw jack.

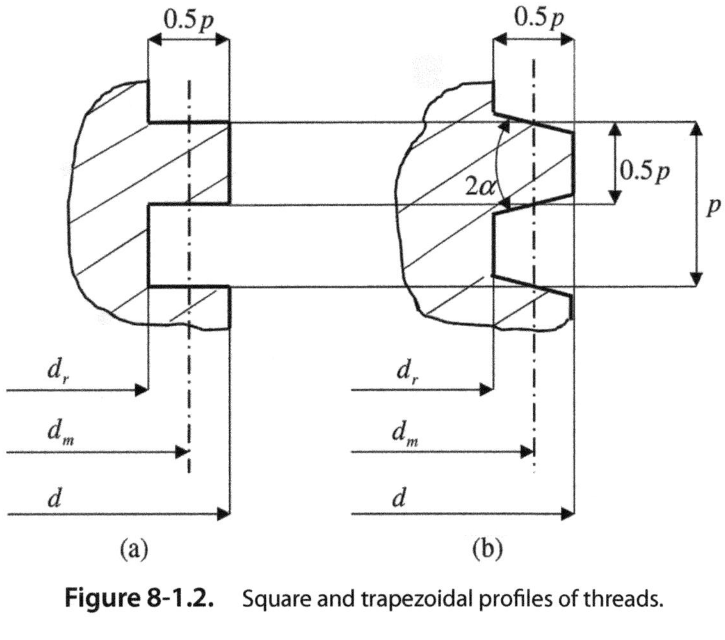

Figure 8-1.2. Square and trapezoidal profiles of threads.

where d is the major diameter and d_m is the mean (pitch) diameter and

$$d_r = d - p = d - \frac{1}{N_t} \tag{8-1.2}$$

In ISO threads, the mean (pitch) diameter d_m of the square and trapezoidal profiles is

$$d_m = d - 0.5p \tag{8-1.3}$$

and the minor diameter is

$$d_r = d - p \tag{8-1.4}$$

where d is the major (nominal standard) diameter of the thread.

Power screws and mating nuts may have a single start or multiple starts that usually do not exceed four, similar to worms. Power screws are characterized by the lead angle λ and lead L. The lead angle λ of a thread can be determined the same way it is determined for worms by using the following formula:

$$\tan\lambda = \frac{L}{\pi d_m} \tag{8-1.5}$$

while

$$L = pN_{st} \tag{8-1.6}$$

where N_{st} is the number of starts as discussed for worm gears in the preceding chapter.

The translational linear velocity of the screw or nut for the UNS type of threads can be determined using the following formula:

$$V = \frac{npN_{st}}{12} \tag{8-1.7}$$

where V is the translational linear velocity in fpm, n is the rotational speed in rpm, N_{st} is the number of starts, and the pitch p is in inches. The formula for the translational linear velocity of the screw or nut for ISO type threads is

$$V = \frac{npN_{st}}{60 \cdot 1000} \tag{8-1.8}$$

where V is the translational linear velocity in $\frac{\text{m}}{\text{s}}$, the rotational speed n is in rpm, and the pitch p is in mm.

The analysis of the mechanics of lifting and lowering of a load by a power screw determines the important parameters of the power screw drive. Figure 8-1.3 presents the schematics of the relationships between the forces acting in a power screw drive during lifting of a load. Figure 8-1.3a shows a simplified front view of one thread having a profile angle 2α and a lead angle λ. In order to simplify the analysis, we apply all acting forces on point A that is indicated in the front view and top view shown in Figure 8-1.3b.

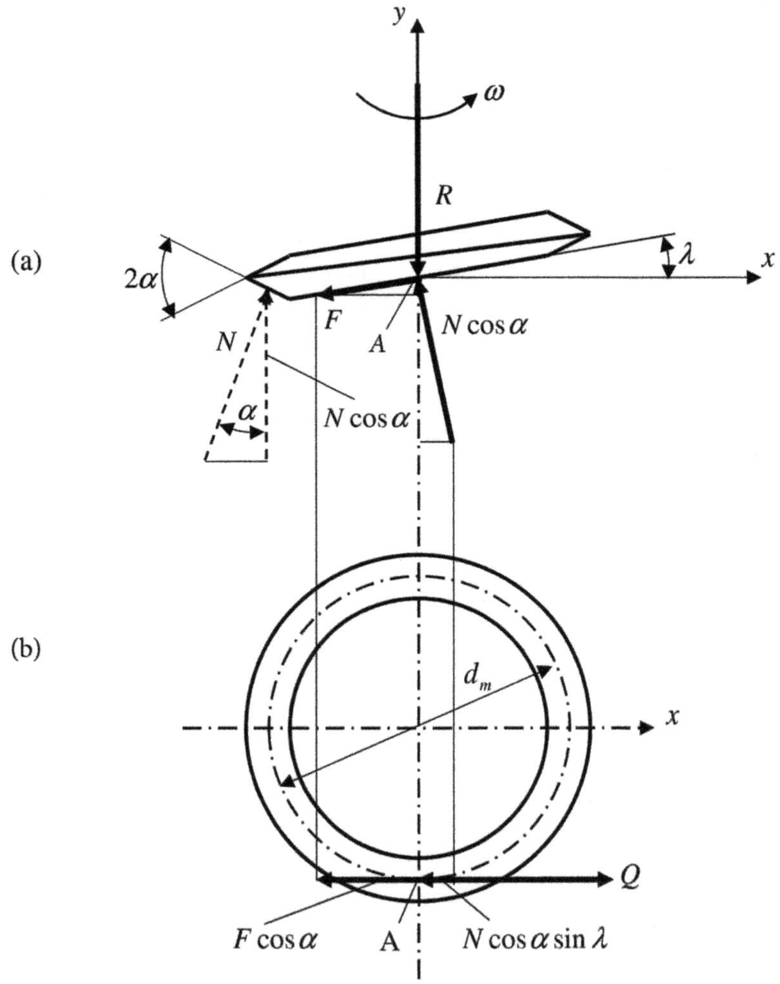

Figure 8-1.3. Forces applied to a power screw.

In order to lift the load R the screw is rotated by the force Q into the direction indicated by the arrow ω. Due to the load R, the nut exerts a normal reaction N and a friction force F that are applied to the screw and to the nut and are distributed over the surfaces of the contacting threads of the screw and nut that are inclined according to the lead angle. The dotted vector N

in Figure 8-1.3b indicates the direction of the normal reaction. And again, for the simplicity of the analysis, the dotted vector $N \cos \alpha$ is transferred to point A and it is shown that this vector is perpendicular to the thread inclined according to the lead angle. Analyzing the equilibrium of the system, we can determine the relationship between the applied forces during lifting of the load. Thus, considering the sum of the projections onto the axis *y*, we may write

$$N \cos\alpha \cos\gamma - F \sin\lambda - R = 0 \tag{8-1.9}$$

Accounting for

$$F = Nf \tag{8-1.10}$$

where *f* is the friction coefficient between the nut and the screw, we can combine the expressions (8-1.9) and (8-1.10) and determine the normal reaction:

$$N = \frac{R}{\cos\alpha \cos\lambda - f \sin\lambda} \tag{8-1.11}$$

Considering the sum of projections of the forces onto the *x* axis, we obtain

$$-F \cos\lambda - N \cos\alpha \sin\lambda + Q = 0 \tag{8-1.12}$$

Combining equations (8-1.10), (8-1.11), and (8-1.12), we have

$$Q = \frac{R(f \cos\lambda + \cos\alpha \sin\lambda)}{\cos\alpha \cos\lambda - f \sin\lambda} \tag{8-1.13}$$

Multiplying expression (8-1.13) by $\frac{d_m}{2}$, we determine the torque T_{lift} that should be applied to the power screw in order to lift the load:

$$T_{lift} = \frac{Rd_m (f \cos\lambda + \cos\alpha \sin\lambda)}{2(\cos\alpha \cos\lambda - f \sin\lambda)} \tag{8-1.14}$$

Factoring in equation (8-1.14) the term $\cos \alpha \cos \lambda$, we may write

$$T_{lift} = \frac{Rd_m \cos\alpha \cos\lambda \left(\frac{f}{\cos\alpha} + \tan\lambda \right)}{2 \cos\alpha \cos\lambda \left(1 - \frac{f}{\cos\alpha} \tan\lambda \right)} = \frac{Rd_m (f_e + \tan\lambda)}{2(1 - f_e \tan\lambda)} \tag{8-1.15}$$

where

$$f_e = \frac{f}{\cos\alpha} \tag{8-1.16}$$

while f_e represents an effective coefficient of friction associated with the profile angle of the thread. In cases of square threads, $\alpha = 0$, $\cos 0 = 1$, and the effective friction coefficient is getting its minimum value that is $f_e = f$. Therefore, the square profile is recommended for power screw drives when the higher efficiency and the lesser frictional wear play an important role.

It is convenient to present the effective friction coefficient using the angle of friction μ; therefore, we may write

$$f_e = \tan\mu \tag{8-1.17}$$

Combining equation (8-1.15) with formula (8-1.17), we obtain
$T_{lift} = \dfrac{Rd_m(\tan\mu + \tan\lambda)}{2(1-\tan\mu\tan\lambda)} = \dfrac{Rd_m}{2}\tan(\lambda+\mu)$, and, therefore,

$$T_{lift} = \frac{Rd_m}{2}\tan(\lambda+\mu) \tag{8-1.18}$$

When lowering the load, the direction of the screw rotation changes to the opposite direction, resulting in the corresponding change in the direction of the friction force and the applied torque. Performing an analysis similar to the analysis we performed for lifting, we obtain

$$T_{lower} = -\frac{Rd_m}{2}\tan(\lambda-\mu) \tag{8-1.19}$$

According to this formula, in cases when $\lambda = \mu$ the load R cannot generate a lowering torque ($T_{lower} = 0$) and the power screw becomes self-locking. This feature is widely used in power screw drives for lifting and lowering loads. It should be emphasized that in actual power screw drives in order to ensure the self-locking condition it is recommended to have $\lambda < \mu$. This may prevent accidental self-lowering when lubricant enters between the sliding threaded surfaces.

Based on formula (8-1.18), we determine the efficiency of the lifting process. By definition, the efficiency coefficient represents the ratio between the output work and input work. Let us analyze this ratio for one revolution of the screw during the lifting process. The ratio of the output and input works reads

$$e = \frac{RL}{2\pi T_{lift}} \tag{8-1.20}$$

where L is lead of the screw thread that according to formula (8-1.15) equals

$$L = \pi d_m \tan \lambda \tag{8-1.21}$$

Combining formulas (8-1.18), (8-1.20), and (8-1.21), we obtain

$$e_{lift} = \frac{\tan \lambda}{\tan(\lambda + \mu)} \tag{8-1.22}$$

The analysis of the efficiency according to formulas (8-1.16), (8-1.17), and (8-1.22) shows that the square thread has the highest efficiency, all other conditions being equal. The reason for this is that the profile angle for the square thread equals zero ($2\alpha = 0$). However, it should be mentioned that square threads are more difficult to fabricate, and square threads have less strength than trapezoidal threads. Based on Figure 8-1.2, it can be seen that the heights of the roots of trapezoidal threads are bigger than the heights of square threads having the same pitches and mean diameters. This results in higher strength for trapezoidal threads with regard to bending and shear stresses.

In some cases, power screw drives include an intermediate collar that causes, so called, collar friction that should be accounted for in the corresponding calculations. Figure 8-1.4 partially shows a power screw 1 having a collar 2 that is loaded by a load P.

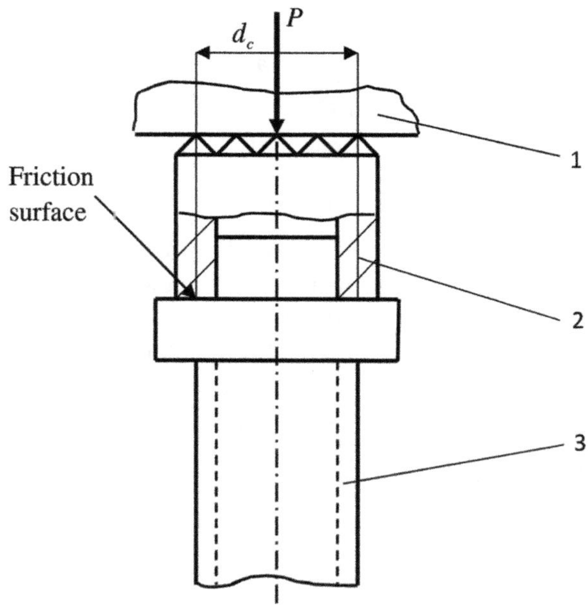

Figure 8-1.4. Collar friction.

Collar friction causes an additional frictional torque that according to some authors can be estimated using the following formula:

$$T_c = \frac{Rf_c d_c}{2} \tag{8-1.23}$$

where f_c is the coefficient of collar friction and d_c is the mean diameter of the collar as it is shown in Figure 8-1.4. Accounting for this formula in the expressions associated with lifting and lowering, we obtain formulas for calculating the total torques required for lifting and lowering the load:

$$T_{tot.\,lift} = \frac{Rd_m}{2}\tan(\lambda + \mu) + T_c \tag{8-1.24}$$

where $T_{tot\ lift}$ is the total torque for lifting the load

and

$$T_{tot.\,lower} = -\frac{Rd_m}{2}\tan(\lambda - \mu) - T_c \tag{8-1.25}$$

where $T_{tot\ lower}$ is the total torque for lowering the load.

With regard to the efficiency of lifting a load, the formula (8-1.22) is modified to the following shape:

$$e_{lift} = \frac{d_m \tan\lambda}{d_m \tan(\lambda + \mu) + d_c f_c} \tag{8-1.26}$$

8-2. Power Screw Drive Calculations

Power screws are manufactured from alloyed steels, while the nuts most often are made of bronze or cast iron. Depending on special requirements, power screws may undergo thermal treatments that cause surface hardening to about 55 HR_c. These screws should be ground after the thermal treatment. Usually, bronze nuts are used for the relatively high translational linear velocity applications, such as lead screws of lathes, milling machines, control mechanisms, etc. For the manual jacks and other low linear velocity applications, cast-iron nuts are recommended.

It should be emphasized that strength calculations of threads are associated with statically undetermined problems. Numerous analytical and experimental studies indicate that the distribution of the load in the threads of a nut and in the corresponding threads of the screw is not uniform. Experiments show that the first thread from the supporting surface of a five-thread nut takes up over 33% of the load, while the fifth thread takes up about less than 8% of the load. But based on the alignment of mating threaded pairs during their operational process, the load distribution in the threads comes close to uniform. This justifies considering in engineering calculations that the load distribution over the mating threads is uniform. This allows for relatively increased heights, specifically for bronze nuts.

The serviceability of a power screw drive depends on the wear resistance of the thread and on the strength of the screw body. This is addressed in the engineering calculations that are presented below.

8-2.1. Bearing Stress Calculations for Power Screw Drives and Determining Power Screw Diameters

The wear resistance of power screws is characterized by the interaction between the sliding surfaces of the threads of the screws and nuts subjected to pressure forces. The stresses that occur on the siding contact surfaces represent bearing stresses (if the contact surfaces are rolling, then the stresses are characterized by contact stresses). Figure 8-2.1 shows a square thread subjected to a concentrated force P.

In reality the load is distributed over the surfaces of the threads of the screw and the mating nut. We assume that the load is distributed uniformly over all mating threads.

Hence, the area of the surface of one ring of the square thread equals

$$A_1 = 0.5\pi d_m p \tag{8-2.1}$$

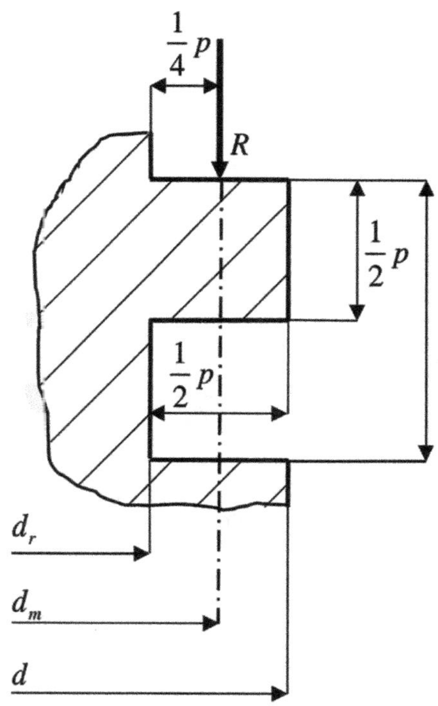

Figure 8-2.1. Schematic of square threads subjected to a load.

The height of the nut is

$$H = jp \tag{8-2.2}$$

where j is the number of threads in the nut. Accounting for formulas (8-2.1) and (8-2.2), we calculate the bearing stress σ_{br} in the threads:

$$\sigma_{br} = \frac{R}{jA_1} \leq S_{br} \tag{8-2.3}$$

where S_{br} is the allowable bearing stress for the materials of the screw-nut pair. Combining equations (8-2.1), (8-2.2) and (8-2.3), we obtain

$$\sigma_{br} = \frac{2R}{\pi d_m H} \leq \frac{S_{br}}{n_{br}} \tag{8-2.4}$$

where n_{br} is the safety factor with regard to bearing stresses.

For preliminary design calculations that allow us to estimate the mean diameter of the screw, we introduce an empiric coefficient ϑ of the nut's height:

$$\vartheta = \frac{H}{d_m} \tag{8-2.5}$$

Substituting this coefficient into expression (8-2.4), we write

$$\sigma_{br} = \frac{2R}{\pi d_m^2 \vartheta} \leq \frac{S_{br}}{n_{br}} \tag{8-2.6}$$

Solving the expression (8-2.6) for d_m, we obtain a formula for calculating the mean diameter of the power screw:

$$d_m = \sqrt{\frac{2Rn_{br}}{\pi \vartheta S_{br}}} \tag{8-2.7}$$

Several authors recommend for a pair of a hardened steel screw and a bronze nut to take the allowable bearing stress S_{br} in the range from 1.6 to 1.9 ksi (from 11 to 13 MPa). In cases of a pair of not hardened steel screw and a bronze nut, the recommended range of bearing stresses is 1.15 to 1.45 ksi (from 8 to 10 MPa); and for a pair of not hardened steel screw with a cast-iron nut, the range is 0.58 to 0.73 ksi (from 4 to 5 MPa).

The coefficient of the nut's height ϑ is recommended by several authors to be from 1.2 to 2.5.

Based on the value of the calculated mean diameter of the screw, we select the standard diameter of the screw, the profile, and the pitch of the thread. For a self-locking power drive, a thread with one start should be selected. In this case, it is necessary to comply with the following condition: $\lambda < \mu$. In cases when the power screw drive will be used to enable the power feed of a lathe or milling machine, or for a similar application, the number of starts should be taken as 2 or 4.

It should be noted that since the nut is made from bronze having significantly less strength than steel, it is required that after determining the pitch diameter of the power screw to verify the strength of the nut with regard to bending and shear stresses accounting for the fatigue phenomenon. The methodology of these verifying calculations is presented in the next subsection.

Example 8-2.1a: Determining the diameter of a power screw.
We need to determine the diameter of a power screw with a square profile for the feeding mechanism of a lathe that develops a force of 10,000 lb. We take a safety factor $n_{br} = 2$.

Solution:
We select a steel screw with no hardening treatment and a bronze nut. This threaded pair has the allowable bearing stress $S_{br} = 1.35$ ksi. Taking the coefficient of the nut's height $\vartheta = 2$ and using formula (8-2.7), we calculate $d_m = \sqrt{\dfrac{2\cdot 10{,}000\cdot 2}{\pi 2\cdot 1{,}350}} = 2.172$ in. For the calculated mean diameter the closest major diameter is $d = 2.5$ in. For this diameter, according to published sources, the number of threads per inch is $N_t = 3$ and, consequently, the pitch $p = 0.333$ in. In this case, the mean diameter $d_m = 2.5 - 0.5\cdot 0.333 = 2.333$ in. The minor diameter of the power screw $d_r = 2.5 - 0.333 = 2.167$ in.

8-2.2. *Bending and Shear Stresses in the Threads of Nuts*

In addition to bearing stresses, threads are subjected to bending and shear stresses. Since the screw is made from steel that significantly exceeds the strength of the bronze nut, we verify the stresses in the nut. Accepting that all mating threads are loaded uniformly, we determine based on Figure 8-2.1 the area of the roots of the threads of the nut subjected to bending and shear stresses:

$$A_2 = 0.5\pi dpj \tag{8-2.8}$$

The moment of inertia I'_{tr} of this area according to formula (1-4.23) is

$$I'_{tr} = \frac{\pi d\left(\frac{p}{2}\right)^3 j}{12} = \frac{\pi dp^3 j}{96} \tag{8-2.9}$$

The bending moment applied to the threads equals

$$M = 0.25Rp \tag{8-2.10}$$

Hence, the bending stress is

$$\sigma_b = \frac{0.25 \cdot 0.25Rp^2}{I'_{tr}} = \frac{6R}{\pi dpj} \tag{8-2.11}$$

The shear (stripping) stress in the tread is calculated based on the following formula:

$$\tau = \frac{R}{A_2} = \frac{2R}{\pi dpj} \tag{8-2.12}$$

Bending and shear stresses act on the same point, creating a plane state of stresses. The equivalent stresses in the threads can be determined using an appropriate theory of failure.

Because power screw drives can be used in a successive way in both directions, the stresses in the threads could be subjected to completely reversed loading cycles and, therefore, the fatigue considerations are applicable to the corresponding stress calculations. Applying the third theory of failure, we may write

$$\sigma_{eq} = \sqrt{\sigma_b^2 + 4\tau^2} \le \frac{S_{fe}}{K_f n_\sigma} \tag{8-2.13}$$

where K_f is the fatigue stress concentration factor and n_σ is the safety factor.

Example 8-2.2a: Verifying calculations of stresses in the threads.
This example is the continuation of the example 8-2.1a that deals with determining the mean diameter of a power screw intended to lift a load R = 10,000 lb. According to this example, the pitch of the thread is p = 0.333, the mean diameter of the power screw d_m = 2.333 in., and the major diameter d = 2.5 in. We need to determine the safety factor n_σ with regard to the strength of the threads.

Solution:
Since the stresses in the threads of the screw and nut are equal, and since the strength of the steel screw significantly exceeds the strength of the bronze nut, it is justifiable to verify the stresses in the nut only. Selecting for the nut the UNS C62300 bronze having S_{ut} = 80 ksi and following the related recommendations presented in Chapter 7, we determine the fatigue strength limit $S_{fe} = 0.16 \cdot 80 = 12.8$ ksi. Using formula (8-2.5), we determine the height of the nut is $H = \vartheta d_m = 2 \cdot 2.333 = 4.666$ in. and the number of threads is $j = \frac{H}{p} = \frac{4.666}{0.333} = 14$. Based on formula (8-2.11), we calculate the bending stress $\sigma_b = \frac{6R}{\pi dpj} = \frac{6 \cdot 10{,}000}{\pi 2.5 \cdot 0.333 \cdot 14} = 1{,}639$ psi.

Using formula (8-2.12), we determine the shear stress $\tau = \frac{2R}{\pi dpj} = \frac{2 \cdot 10{,}000}{\pi 2.5 \cdot 0.333 \cdot 14} = 546.5$ psi.

According to the graph in Figure 5-8.1, we estimate the value of the theoretical stress concentration factor K_t = 3.0. Now taking the notch sensitivity factor q = 0.6 and using formula (1-10.2), we determine the fatigue stress concentration factor K_f = 1 + 0.6(3.0 – 1) =2.2.

Based on formula (8-2.13), we calculate $\sigma_{eq} = \sqrt{1.639^2 + 4 \cdot 0.5465^2} = 1.97$ ksi. Hence, $n_\sigma = \dfrac{S_{fe}}{\sigma_{eq} K_f} = \dfrac{12.8}{1.97 \cdot 2.2} = 2.95$, which is acceptable.

8-2.3. Calculating the Compression and Torsional Stresses and Verifying the Stability of Power Screws

Screws are subjected to axial forces and torques that cause normal and torsion stresses.

The compression stress can be determined on the basis of the following expression:

$$\sigma_{max} = \frac{4R}{\pi d_r^2} \tag{8-2.14}$$

The torsional stress is associated with the lifting torque and can be determined based on expression (1-4.20):

$$\tau_{max} = \frac{T_{tot.lift}\, 0.5 d_r}{J} \tag{8-2.15}$$

where J is the polar moment of inertia and according to formula (1-4.21) we have

$$J = \frac{\pi d_r^4}{32} \tag{8-2.16}$$

Combining formulas (8-1.25), (8-2.15), and (8-2.16), we obtain

$$\tau_{max} = \frac{16 T_{tot.lift}}{\pi d_r^3} \tag{8-2.17}$$

Because normal and torsional stresses are variable stresses that change according to a completely reversed loading cycle, the stress analysis should be based on fatigue considerations.

The combination of normal and torsional stresses causes a plane stress state; therefore, the calculations of the stresses should be performed using an appropriate theory of failure. Applying the third theory of failure and combining formulas (8-2.14), (8-2.16), and (8-2.17), we obtain

$$\sigma_{eq} = \sqrt{\sigma_{max}^2 + 4\tau_{max}^2} \le \frac{S_e}{K_f n_\sigma} \tag{8-2.18}$$

where S_e is the endurance limit, K_f is the fatigue stress concentration factor, and n_σ is the safety factor.

Because power screws are subjected to compression, and in many applications the screws are relatively long, it becomes necessary to verify the stability of the screws by conducting buckling calculations. According to these calculations, the compression stress should not exceed the, so called, critical stress σ_{cr} introduced by L. Euler (1707-1783). Applying the principles of buckling to power screws, we may write

$$\sigma_{cr} = \frac{\pi^2 E d_r^2 \delta_\sigma}{16k^2 l^2} \geq n_e \sigma_{max} \tag{8-2.19}$$

where k is the effective length factor depending on the supports of the screw at its ends, l is the maximum length of the compressed part of the power screw measured between the support and the middle of the nut, n_e is the stability safety factor, and δ_σ is the reduction coefficient of the critical stress depending on the slenderness of the screw. For power screw drives, it is recommended to accept $k = 1$, while the stability safety factor should be $n_e \geq 3$. As seen from expression (8-2.19), the critical stress does not depend on the strength of the power screw material. Experimental investigations indicate that the slenderness of the screw influences the critical stress. Thus, the increase of slenderness from 30 to 150 causes a decrease in the critical stress of about 3.5 times. The slenderness s_r of power screws can be determined using the following formula:

$$s_r = \frac{4kl}{d_r} \tag{8-2.20}$$

The value of the coefficient δ_σ is calculated based on the empiric formula that reads

$$\delta_\sigma = 0.91 - 0.0051\,(s_r - 30) \tag{8-2.21}$$

Example 8-2.3a: Determining the stability safety factor.

We need to calculate the stability safety factor n_e for the power screw from Example 8-2.1a that is loaded by 10,000 lb and has the minor diameter d_r = 2.167 in., assuming that the maximum length of the compressed part of the power screw is 50 in.

Solution:

First, we determine the compression stress in the screw $\sigma_{max} = \frac{4 \cdot 10{,}000}{\pi 2.167^2} = 5{,}105$ psi. Now we calculate the slenderness using formula (8-2.20): $s_r = \frac{4 \cdot 1 \cdot 50}{2.167} = 92.29$. And the critical stress reduction coefficient based on the formula (8-2.21): δ_σ = 0.91 – 0.0051(92.29-30) = 0.592. Applying formula (8-2.19), we determine $\sigma_{cr} = \frac{\pi^2 30 \cdot 10^6 \cdot 2.167^2 \cdot 0.59}{16 \cdot 1^2 \cdot 50^2} = 20{,}613$ psi. Hence, $n_e = \frac{\sigma_{cr}}{\sigma_{max}} = \frac{20{,}613}{5{,}105} = 4.03$, which is acceptable.

8-3. Design Project

We need to design a power screw drive for a milling machine. The required load capacity is 35,000 N. The maximum length of the compressed part of the power screw is 0.8 m.

1. Calculations of the power screw and mating nut.

We select a threaded pair where the screw is made from steel with no hardening treatment and the nut is made from bronze. The allowable bearing stress for this pair of materials is taken S_{bk} = 9.0 MPa. We take the coefficient of the nut's height ϑ = 1.8, the safety factor n_{br} = 2, and applying formula (8-2.7), we calculate $d_m = \sqrt{\dfrac{2 \cdot 35{,}000 \cdot 2}{\pi 1.8 \cdot 9 \cdot 10^6}} = 0.0427$ m (43 mm). We select the standard trapezoidal ISO metric screw having the major diameter d = 48 mm, the pitch p = 8 mm, the mean (pitch) diameter d_m = 44 mm, and the minor diameter d_r = 40 mm. We need to perform verifying stress calculations for the bronze nut. We select for the nut UNS C62400 bronze having S_{ut} = 670 MPa. We calculate the fatigue strength limit S_{fe} for bronze according to formula (7-5.5), hence, S_{fe} = 0.16·670 = 107.2 MPa. Applying formula (8-2.5), we calculate the height of the nut $H = \vartheta d_m = 1.8 \cdot 40 = 72$ mm. The number of threads in the nut is $j = \dfrac{H}{p} = \dfrac{72}{8} = 9$. Since in a milling machine the power screw drive is reversible, the screw and the nut are subjected to a completely reversed loading cycle. According to formula (8-2.11), we calculate the bending stress $\sigma_b = \dfrac{6R}{\pi d_r p j} = \dfrac{6 \cdot 35{,}000}{\pi \cdot 0.040 \cdot 0.08 \cdot 9} = 23{,}222{,}868$ Pa (23.223 MPa). From Figure 5-8.1, we determine that the theoretical stress concentration factor K_t = 3.0. Considering that the notch sensitivity factor q = 0.6 and using formula (1-10.2), we compute the fatigue stress concentration factor: K_f = 1 + 0.6(3.0 – 1) = 2.2. In order to determine the shear stress in the roots of the threads, we use formula (8-2.12), hence, $\tau = \dfrac{35{,}000}{\pi 0.040 \cdot 0.008 \cdot 9} = 3{,}870{,}311$ Pa (3.87 MPa). Applying the third theory of failure according to formula (8-2.13), we calculate

$$\sigma_{eq} = \sqrt{23.223^2 + 4 \cdot 3.87^2} = 24.479 \text{ MPa}.$$

Therefore, $n_\sigma = \dfrac{S_{fe}}{\sigma_{eq} K_f} = \dfrac{107.2}{24.479 \cdot 2.2} = 2.0$, which is acceptable.

2. Schematic design of the power screw drive.

Figure 8-3.1 shows a schematic partial view of the power screw drive, where the right-hand side end of the screw is supported by a pair of back-to-back taper roller bearings 6 that are separated by distance ring 7, and the left-hand end is supported by single roller bearing 3.

The bearings on the right-hand end prevent any axial displacement of the power screw, while the left-hand end of the screw is free to move axially in order to compensate for the thermal changes due to the length of the screw. Nut 2 is installed in the corresponding nest in table 5 and is restricted from rotation and axial displacement.

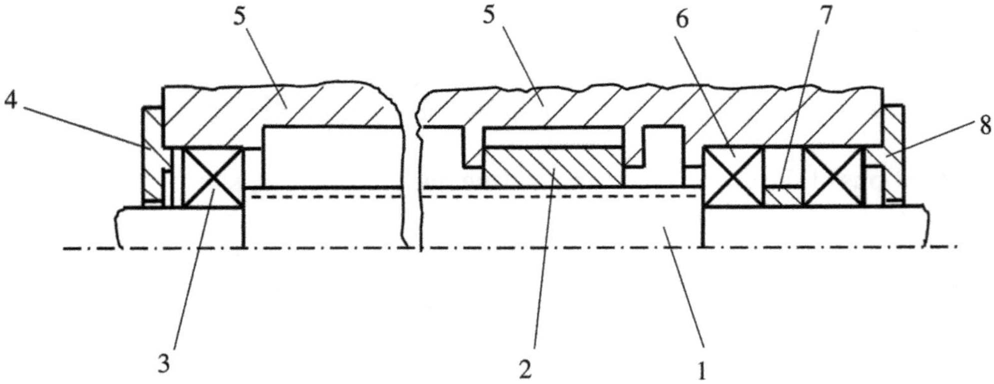

Figure 8-3.1. Schematic view of a power screw drive for a horizontal milling machine.

3. Verifying calculations of the body of the power screw.

The maximum normal stress according to formula (8-12.4) is

$$\sigma_{\max} = \frac{4 \cdot 35{,}000}{\pi 0.040^2} = 27{,}866{,}242 \text{ Pa (27.87 MPa).}$$

In order to calculate the lifting torque and, consequently, the torsion stresses, we need to determine the lead angle λ and the effective friction angle μ. Taking the number of starts $N_{st} = 2$ and using formulas (8-1.5) and (8-1.6), we determine $\tan\lambda = \frac{pN_{st}}{\pi d_m} = \frac{8 \cdot 2}{\pi 44} = 0.116$, or $\lambda = 6.62°$.

Accounting for the fact that the friction coefficient between steel and bronze $f = 0.16$, as well as that half of the trapezoidal thread profile $\alpha = 15°$, and using formulas (8-1.16) and (8-1.17), we have $f_e = \frac{0.16}{\cos 15°} = 0.1656$ and $0.1656 = \tan\mu$, hence, we determine that $\mu = 9.4°$. Based on formula (8-1.18), we calculate the lifting torque $T_{lift} = \frac{35{,}000 \cdot 0.044}{2}\tan(6.62° + 9.4°) = 221.1\,\text{N}\cdot\text{m}$.

Combining formulas (8-2.15) and (8-2.16), we calculate the maximum torsional stress: $\tau_{\max} = \frac{16T_{lift}}{\pi d_r^3} = \frac{16 \cdot 221.1}{\pi 0.04^3} = 17{,}603{,}503$ Pa (17.6 MPa). The fatigue stress concentration factor calculated above ($K_f = 2.2$) is also applicable for the screw. We select for the screw AISI 5150 annealed steel having $S_{ut} = 675$ MPa. Using formulas (1-9.2) and (1-9.3), we determine $k_a = 0.8$ and $k_b = 0.84$, respectively. Taking $k_c = 1$, we have $S_e = 0.8 \cdot 0.84 \cdot 1 \cdot 0.5 \cdot 675 = 226.7$ MPa.

Now applying formula (8-2.18), we calculate $\sigma_{eq} = \sqrt{27.87^2 + 4 \cdot 17.6^2} = 44.9$ MPa. Hence, $n_\sigma = \frac{S_e}{\sigma_{eq}K_f} = \frac{226.7}{44.9 \cdot 2.2} = 2.29$, which is acceptable.

4. Calculations of the stability safety factor of the screw.
The compression stress in the screw σ_{max} = 27.87 MPa. The slenderness of the power screw according to formula (8-2.20) is $s_r = \dfrac{4 \cdot 1 \cdot 0.8}{0.04} = 80$. Using formula (8-2.21), we calculate the critical stress reduction coefficient δ_σ = 0.91 – 0.0051(80 – 30) = 0.655. According to formula (8-2.19), we calculate $\sigma_{cr} = \dfrac{\pi^2 207 \cdot 10^9 \cdot 0.04^2 \cdot 0.655}{16 \cdot 1^2 \cdot 1^2} = 133{,}681{,}387$ Pa (133.68 MPa). Therefore, $n_e = \dfrac{\sigma_{cr}}{\sigma_{max}} = \dfrac{133.68}{27.87} = 4.8$, which is acceptable.

Chapter 8. Problems

8-01. A manual power screw drive has the major diameter of the screw $d = 1.0$ in., the number of starts $N_{st} = 1$, and the number of threads per inch $N_t = 5$. The threaded pair has a trapezoidal profile. Determine the mean and minor diameters of the screw, the lead angle, and the lead.

8-02. A power screw of a lathe has the major diameter of the screw $d = 2.0$ in., the number of threads per inch $N_t = 4$, the number of starts $N_{st} = 2$, and the rotational speed of the screw $n = 60$ rpm. The threaded pair has a square profile. Determine the mean and minor diameters of the screw, the lead angle, the lead, and the translational linear velocity of the nut.

8-03. A power screw of a lathe has the major diameter of the screw $d = 50$ mm, the pitch $p = 6$ mm, and the rotational speed of the screw $n = 60$ rpm. The threaded pair has a square profile. The number of starts $N_{st} = 2$. Determine the mean and minor diameters of the screw, the lead angle, the lead, and the translational linear velocity of the nut.

8-04. A power screw drive has the major diameter of the screw $d = 1.5$ in., the number of threads per inch $N_t = 4$ and the number of starts $N_{st} = 1$. The drive lifts 3,000 lb. The nut is made of bronze and the thread profile angle is $2\alpha = 29°$. Determine the lifting torque and the efficiency of the power screw drive.

8-05. A power screw drive has the major diameter of the screw $d = 36$ mm, the pitch $p = 6$ mm, and the number of starts $N_{st} = 1$. The drive lifts 15,000 N. The nut is made of bronze and the thread profile angle $2\alpha = 30°$. Determine the lifting torque and the efficiency of the power screw drive.

8-06. A power screw drive has the major diameter of the screw $d = 44$ mm, the pitch $p = 7$ mm and the number of starts $N_{st} = 1$. The drive lifts 22,000 N. The collar has a diameter $d_c = 50$ mm and the collar friction coefficient $f_c = 0.1$. The nut is made of bronze and the thread profile angle $2\alpha = 30°$. Determine the total lifting torque and the efficiency of the power screw drive.

8-07. Determine the geometric parameters of a power screw that lifts a load of 35,000 N. Take the allowable bearing stress for the threaded pair screw-nut $S_{br} = 9.3$ MPa, the nut's height coefficient $\vartheta = 2$, and the safety factor $n_{br} = 2$. Also determine the efficiency of the drive if the nut is made from bronze and the number of starts $N_{st} = 1$, while the profile angle is $2\alpha = 30°$.

8-08. Determine the geometric parameters of a power screw that lifts a load of 6,000 lb. Take the allowable bearing stress for the threaded pair screw-nut $S_{br} = 1.35$ ksi, the nut's height coefficient $\vartheta = 2$, and the safety factor $n_{br} = 2$. Determine the equivalent stress acting in the threads. Also determine the efficiency of the drive if the nut is made from bronze and the number of starts $N_{st} = 1$, while the profile angle is $2\alpha = 29°$.

8-09. Determine the geometric parameters of a power screw that lifts a load of 9,000 lb. Take the allowable bearing stress for the threaded pair screw-nut $S_{br} = 1.35$ ksi, the nut's height coefficient $\vartheta = 2$, and the safety factor $n_{br} = 2$. Determine the stability safety factor if the maximum length of the compressed part of the power screw is 50 in.

8-10. Determine the geometric parameters of a power screw that lifts a load of 65,000 N. Take the allowable bearing stress for the threaded pair screw-nut $S_{br} = 9.3$ MPa, the nut's height coefficient $\vartheta = 2$, and the safety factor $n_{br} = 2$. Also determine the stability safety factor if the maximum length of the compressed part of the power screw is 2 m.

CHAPTER 9

Threaded Joints

In most cases, assembling machine elements into units and units into machines involves the use of threaded joints that are nonpermanent connections. Certain components have threaded structures that represent threaded joints; however, the vast majority of nonpermanent joints are created by threaded fasteners, such as bolts, screws, studs, and nuts. In the previous chapter, we discussed the use of threaded elements in power drives, and almost all of the considerations related to the structure of threads and to the analytical aspects of their functionality are applicable to the study of threaded fasteners.

The tremendous need of threaded fasteners in equipment and machinery resulted in the creation of entire industries that manufacture and supply numerous types of fasteners. It has become more cost effective to purchase the fasteners than manufacture them in house. National and international standards determine the nomenclature and the parameters of threaded fasteners. The design process for threaded fasteners determines the diameter, the length, and the material of the bolt, the nut, and the material of associated parts, such as washers.

A plurality of threaded structures that differ by their shapes and geometric parameters is in wide use in many different industries. These structures include external and internal, right-hand and left-hand, single-start, double-start, and multiple-start threads, as well as square, trapezoidal, buttress, triangular, round, tapered, and other thread profiles. Square, trapezoidal, buttress, and triangular threads are used for power screw drives. The pipe industry utilizes cylindrical and tapered treads that produce hermetic connections. Tapered threads are also used for fittings and quick connectors for hydraulic and pneumatic systems.

According to the standards, fastener threads have a triangular profile with a profile angle of 60°. The triangular thread profile is stronger than square or trapezoidal threads due to the larger height of their roots. In addition, the triangular thread has a relatively high effective friction coefficient that enhances the self-locking effect of fasteners. Manufacturing triangular threads is a relatively simple production process. These are the basic reasons why the triangular profile is accepted for joint threads.

Nuts and the heads of screws come in a number of standard shapes that can be handled with the appropriate tools. Most nuts have a hexagonal shape.

Threads are often manufactured using single-point cutting tools on lathes. The mass production of fasteners is based on highly sophisticated equipment and methods, including hobbing, milling, rolling, and others. Threads are also cut using taps and dies.

According to the Unified Thread Standard (UTS) and other standards, there are several grades of accuracy and strength of the fasteners. Comprehensive information regarding threaded fasteners is available in numerous published sources.

9-1. Thread Profiles

The 60° triangular profile thread has become the international standard for fasteners for detachable joints. Figure 9-1.1 presents the structure of this thread and its geometric parameters.

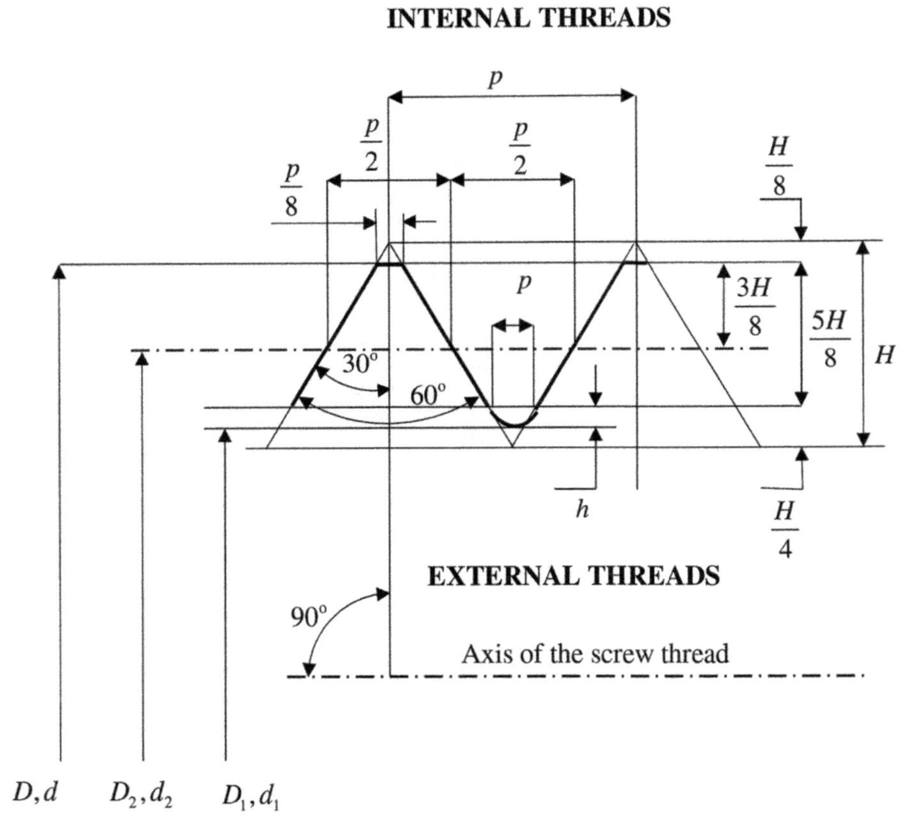

Figure 9-1.1. Screw thread profile.

According to this figure, p is the pitch of the thread, D and d are the major diameters of the nut and the screw, respectively, D_2 and d_2 are the main (pitch) diameters of the nut and the screw (the subscripts "m" and "p" are also used for these diameters), D_1 and d_1 are the minor diameters of the nut and screw (the subscript "r" is also used for these diameters), and H is the height of the profile base triangle. As discussed in the preceding chapter, the pitch (the thread pitch, the axial pitch) of the thread is the distance between identical points of two adjacent thread forms measured parallel to the axis of the threads. According to the American National Standard, the number of threads per inch N_t is the reciprocal of the pitch, hence, $P=\frac{1}{N_t}$ and, consequently, $p=\frac{1}{N_t}$

(in inches). Figure 9-1.1 shows that $\frac{0.5p}{H} = \tan 30° = 0.577350$, hence, $H = 0.866025p$. For practical calculations, we take $H = 0.866\,p$. Therefore, according to Figure 9-1.1 we determine the pitch diameter (mean diameter) and the minor diameter of the bolt (screw), respectively:

$$d_2 = d - \frac{2 \cdot 3H}{8} = d - 2 \cdot 0.375 \cdot 0.866\,p = d - 0.6495\,p \tag{9-1.1}$$

$$d_1 = d - \frac{2 \cdot 5H}{8} = d - 2 \cdot 0.625 \cdot 0.866p - 2h = d - 1.0825\,p - 2h \tag{9-1.2}$$

where for the Standard Unified Inch thread $h = \frac{H}{8} = \frac{0.866p}{8} = 0.10825\,p$ and for the ISO metric thread $h = \frac{H}{12} = \frac{0.866p}{12} = 0.07216\,p$.

According to the UTS, there are three types of triangular threads: the UNC (Unified National Coarse), the UNF (Unified National Fine), and the UNEF (Unified National Extra Fine). The standard designation of threads consists of the nominal (major) diameter and the pitch in threads per inch followed by one of these three identifications—UNC, UNF, or UNEF. For example, 0.75 – 10 UNC (the major diameter d = 0.75 in., the number of threads per inch N_t = 10, and the coarse thread identification UNC. The designation of the thread may also contain the indication of the tolerance class; for example, 0.25 – 28 UNF 2B. Here 2B is the tolerance grade. If the major diameter is smaller than 0.25 in., then the standard designation contains a corresponding integer number from 1 to 12; for example, 6 – 32 UNC (the numeral 6 designates the major diameter of 0.1380 in., N_t = 32, and the thread is coarse).

According to the ISO metric standard, there are also coarse and fine threads. Hence, if the designation of a thread is $M12$, then M stands for metric and it is a coarse thread with the major diameter d = 12 mm. When the designation of the thread is $M12 \cdot 1.25$, it is a fine metric thread with the major diameter d = 12 mm and the pitch p = 1.25 mm.

Here is an example how to use formulas (9-1.1) and (9-1.2). For the UTS tread designated 0.25 – 20 UNC, we have $d_2 = 0.25 - 0.6495 \cdot \frac{1}{20} = 0.2175$ in., and

$d_1 = 0.25 - 1.0825 \cdot \frac{1}{20} - 2 \cdot 0.10825 \cdot \frac{1}{20} = 0.18505$ in.

For ISO metric thread designated $M10 \cdot 1.25$, we have $d_2 = 10 - 0.6495 \cdot 1.25 = 9.188$ mm and $d_1 = 10 - 1.0825 \cdot 1.25 - 2 \cdot 0.07216 \cdot 1.25 = 8.4665$ mm. Because the pitch is not indicated in the designation of coarse metric treads, it is necessary to determine the pitch from corresponding sources.

The lead angle λ and the lead L, and the number of starts N_{st} were introduced in the preceding chapter, and they are the same for the threads of fasteners. Hence, the formula for calculating the lead angle reads $\tan \lambda = \frac{L}{\pi d_m}$, while $L = pN_{st}$.

9-1.1 Bolts, Screws, Studs, Nuts

Fasteners include a variety of mostly standardized machine elements comprising bolts, screws, nuts, washers, and associated components. Figure 9-1.2 shows examples of bolted joints, while Figure 9-1.2a presents a joint, including bolt 1, connected members 2 and 3, and nut 4.

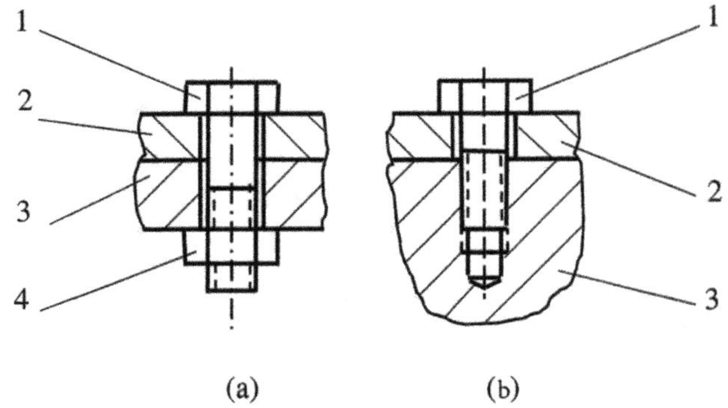

Figure 9-1.2. Bolted joints.

It should be noted that in the vast majority of cases the bolt is placed into the holes of the members with a clearance fit that actually prevents the body of the bolt from being subjected to shear stresses. But in some special cases, the bolt is made to fit into the holes providing alignment for the connected members and, therefore, the bolt is subjected to shear stresses. In these cases, the diameters of the holes in the members as well as the diameter of the body of the bolt should be in compliance with corresponding tolerances for an appropriate fit.

Figure 9-1.2b shows a screw securing the members 2 and 3, while the member 3 has a threaded hole. There are no structural differences between a bolt and a screw. It is just acceptable to call the threaded element "bolt" when it is secured with a nut. Bolts always go with nuts, and when specifying in the documentation the needed number of bolts, this translates into the need for the same number of corresponding nuts. The majority of bolts and screws have hexagon-shaped heads. The majority of nuts, as mentioned above, also have a hexagonal shape.

Figure 9-1.3a shows bolt 1 having a square head. As in Figure 9-1.2, this bolt is joining members 2 and 3 with the help of the nut 4.

Figure 9-1.3b presents a stud 1 that is securely screwed into the internal thread of member 4. This stud and nut 2 are connecting members 3 and 4. The stud is a round bar threaded on both ends.

The height of the hexagonal and square heads is usually equal to $0.7d$, where d is the major (nominal) diameter of the thread. The depth of the thread for a screw or a stud in steel members ranges from $0.8d$ to $1.0d$.

Self-loosening of treaded joints is a well-known issue that may cause undesirable consequences. In order to prevent self-loosening, it is acceptable to use all kind of spring washers, cotter pins, and other means. The problem of self-loosening is very often solved by applying to the threaded surface anaerobic adhesive sealant that solidifies in the joint connection and creates

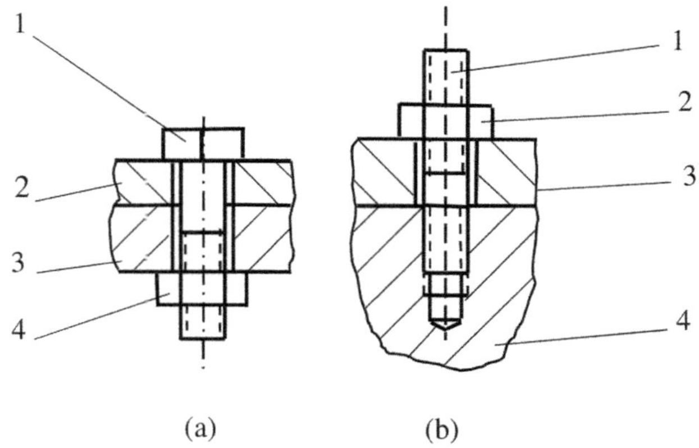

Figure 9-1.3. A joint with a square-head bolt and a joint with a stud and a nut.

a very strong bond between the bolt and nut or between the screw and member. These adhesive sealants come in different grades for corresponding applications. In order to disassemble some of the joints that have solidified adhesive sealants, it is necessary to apply significant heat to the joint. Comprehensive information related to preventing self-loosening washers, cotter pins, and adhesive sealants is available in numerous published sources.

There is a wide variety of manual and mechanical tools intended for tightening a nut or screw. The length of the handles of manual tools depends on the diameter of the thread. It should be kept in mind that relatively small screws that have a diameter of 0.25 in. or less can be easily destroyed by using conventional wrenches, specifically by using universal ratchets having long handles. It is a good practice to use adjustable calibrated wrenches.

9-2. Characteristics of Fasteners and Threaded Joints

The design of a threaded joint is associated with the strength calculations intended to determine the major (nominal) diameter of the fastener (the bolt, the screw, or the stud). These calculations do not include the stress analysis of the threads. The reason behind this is that according to the standards each particular diameter of a bolt has strictly designated pitches that are based on corresponding strength calculations and experimental testing. The standards also specify the height of the nut. According to these standards, the strength of the threads involved in the engagement of the bolt and nut or of the screw (stud) and plate is equivalent or exceeds the strength of the body of the bolt (stud). The cross section of the body of the bolt is determined by the major (nominal) diameter of the bolt.

Usually, upon determining the required dimension of the major diameter and length of the bolt (screw), defining the shape of its head based on design consideration is required. Bolts, nuts, and other fasteners represent off-the-shelf products.

Depending on the characteristics of the loading applied to the bolt, there are several appropriate types of calculations for the diameter of the thread. These types of calculations are associated with the necessity of selecting the appropriate allowable stresses. Some of these calculations are associated with the deformation of the bolts and the members and require determining the stiffness of the bolts and members. These and other related aspects of threaded joint calculations are presented below.

9-2.1. Allowable Stresses for Threaded Fasteners

Bolts and screws are manufactured in compliance with the national standard specifications that indicate their mechanical properties. According to these mechanical properties, bolts and screws are divided into 12 specific grades that are identified in alphabetical order by capital letters from A to L. All technical information related to these grades is presented in the table in Appendix A-6-1.

This standard introduces the definition of the minimum proof strength S_p, representing the maximum tensile stress that a fastener can withstand without any residual (plastic) deformation. The table mentioned above includes the information related to the grade, the range of the diameters, the minimum values of the proof strength, the tensile (ultimate), and the yield strength. The proof strength is the allowable stress for calculation of the fasteners, and it may be mentioned that the relationship between the proof strength and the yield strength can be expressed as follows: $S_p = 0.85S_{yt}$.

It is important to emphasize that according to the standard the corresponding grades include just a certain range of diameter sizes of the fasteners. Therefore, the designer selects the grade based on the calculated diameter of the bolt in compliance with the described above table.

According to the standard, the fasteners are made from low and medium carbon steels and alloyed steels. Some fasteners undergo thermal treatment and quenching.

In cases when the bolts and screw are subjected to shear stress, it is recommended to assign the allowable shear stress based on the following relationship: $S_{sp} = 0.6S_p$.

The proof strength is applicable as an allowable stress for fasteners experiencing static loading. In cases of variable loading, the stress calculations of fasteners should include all relevant fatigue considerations.

It should be mentioned that appropriate information related to standard American National threads and ISO metric threads is presented in Appendixes A-6-2 and A-6-3.

In addition it may be outlined that hexagon and square nuts are divided into three grades, namely 2, 5, and 8, covering diameters from 0.25 to 1.5 in. The mechanical properties of these grades of nuts almost coincide with the same grades of bolts and screws.

9-2.2. Stiffness Coefficients of Joint Elements

In the analysis of certain threaded joints, it is necessary to determine the stiffness coefficients of the involved elements. Figure 9-2.1 shows a schematic drawing of a threaded joint comprising a bolt 1, members 2 and 3, and a nut 4. The stiffness coefficient of the bolt K_b is calculated using the known formula from the course of Strength of Materials (Mechanics of Materials):

$$K_b = \frac{\pi d^2 E_b}{4l_b} \tag{9-2.1}$$

where d is the major (nominal) diameter of the bolt, E_b is the modulus of elasticity of the bolt material, and l_b is the effective length of the bolt as it is shown in Figure 9-2.1.

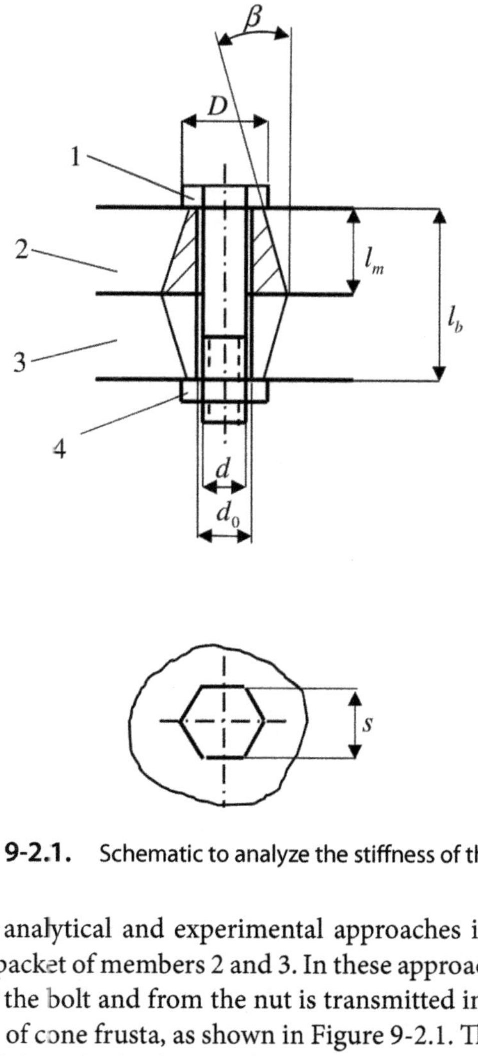

Figure 9-2.1. Schematic to analyze the stiffness of the joint.

Several authors use analytical and experimental approaches in order to determine the stiffness coefficient of the packet of members 2 and 3. In these approaches, it is assumed that the pressure from the head of the bolt and from the nut is transmitted into the depth of the members through two volumes of cone frusta, as shown in Figure 9-2.1. The volume of the frusta do not include the volume of the cylinder having the diameter d_0. Note that if the thickness and the material of the two members are identical then their frusta are mirror images, as shown in Figure 9-2.1. Different authors suggest different values for the angle β (Figure 9-2.1) that varies from 22° to 45°. However, some of these sources indicate that β should be taken at about 30°. For calculating the combined stiffness K_m of the jointed members, it is accepted in this book to use the analytical formula derived by Birger that reads

$$K_m = \frac{0.683 d_0 E_m \tan\beta}{\log \dfrac{(s+d_0)(s-d_0+2l_m \tan\beta)}{(s-d_0)(s+d_0+2l_m \tan\beta)}} \tag{9-2.2}$$

where d_0 is the diameter of the hole for the bolt, s is the width across the flats of the bolt or nut, E_m is the modulus elasticity of the material of the members (the members are made from the same material), while the rest of the parameters are shown in Figure 9-2.1.

Taking $\tan \beta = 0.5$ $(\beta = 26.57°)$, replacing d_o with d, and accounting that

$s = D - 0{,}268d = 2d - 0.268d = 1.732d$,

we simplify formula (9-2.2) to the following shape:

$$K_m \approx \frac{0.34dE_m}{\log \dfrac{3.73(0.732d + l_m)}{2.732d + l_m}} \tag{9-2.3}$$

It should be emphasized that installing in formula (9-2.3) the thickness l_m of one member results in the stiffness coefficient K_m for the packet of members 2 and 3 having a combined thickness of $2l_m$.

In addition, it is recommended for a packet of two relatively thin similar members having the total thickness $2l_m \leq 0.5d$ to calculate the stiffness coefficient of this packet of members using the following formula:

$$K_m = \frac{\pi E_m[(s + 2l_m \tan \beta)^2 - d_0^2]}{4 \cdot 2l_m} \tag{9-2.4}$$

And similarly, as we transformed formula (9-2.2), taking $\tan \beta = 0.5$, replacing d_0 with d, and accounting that $s = 1.732\ d$, we obtain

$$K_m = \frac{0.39E_m[(1.732d + l_m)^2 - d^2]}{l_m} \tag{9-2.5}$$

Figure 9-2-2 shows a threaded joint consisting of stud 1, nut 2, plate 3, gasket, 4 and body 5. It is difficult to determine with high accuracy the effective length of the stud or of the members in order to calculate their stiffness.

Published sources offer some approaches to perform these calculations in an approximate way. One of these ways is presented below. It should be mentioned that in many different approaches it is assumed that the nut is not deformable and, therefore, is not included in the computing of the effective length of the bolt or the thickness of the members (Figure 9-2.1). Some authors suggest that the concept of a non-deformable element having an internal thread could be applied to body 5 in the studded joint or a joint with a screw. Thus, assuming that the body is not deformable, it may be suggested that the effective length of the stud is as shown in Figure 9-2.2: $l_b = l_m + l_0$. With regard to the effective thickness of the member (plate), it is assumed that during the compression of a single plate a mirrored frustum having the thickness of one-half of the plate's thickness appears on each side of it. It represents an imaginary split of the plate into two identical plates with the mirrored frusta, resembling the phenomena described above for two plates (it is shown by the dotted lines in Figure 9-2.2). Therefore, it is suggested to substitute into formula (9-2.3) one-half of the actual thickness of the plate, namely 05 l_m.

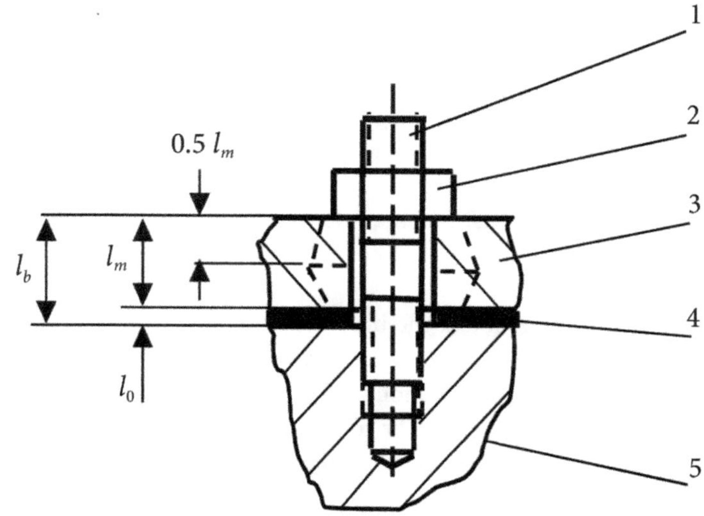

Figure 9-2.2. Effective length and thickness of the elements in cases of a studded joint.

In addition, it should be noted that in cases of connecting a packet containing two or more members when the thickness of each of the members exceeds one-half of the diameter of the fastener, it is suggested in this book to apply the following approach:

1) Using formula (9-2.1), we accept that the effective length of the fastener l_b equals the sum of the plates thicknesses.
2) Using formula (9-2.3), we accept that the effective thickness of the members l_m equals one-half of the sum of all members.

9-3. Calculations of Typical Threaded Joints

Bolts (screws, studs, or similar machine elements) could be subjected to one of the following three related types of stresses:

1) Tensile stress
2) A combination of tensile and torsional stresses
3) Pure shear stress

The threaded joint subjected to tensile stress is a bolt or other element that has an external thread, while a nut or other element having an internal thread is screwed onto the bolt without tightening. In this case, the load applied to the bolt or to the nut develops a tensile stresses in the bolt.

The threaded joint subjected to the combination of tensile and torsion stresses may have the following structures that include:

1) A bolt, two or more members in a packet, and a nut
2) A stud with a nut and a packet of members of which the last one has an internal thread
3) A screw connecting at least two members of which the last one has an internal thread

When these joints are assembled, the bolt (stud, screw) should be subjected to preliminary tightening by applying a certain torque. The threaded joint that experiences shear stresses is an assembly of a packet of members mounted on the bolt, screw, or stud according to an interference fit, while the load applied to the member or members is directed perpendicularly to the axis of the fastener. In this case, the normal stress in the fastener due to the preliminary tightening is negligible. In all three of the above cases, the stresses could be static or variable, and the calculations of them are presented below.

9-3.1. Fasteners Subjected to Tensile Stress

Figure 9-3.1 shows a threaded element 1 that resembles a screw having a major (nominal) diameter d. The threaded element is screwed into a non-movable support 2. A hole in the threaded element allows for a suspended load F.

The tensile stress in the threaded element is

$$\sigma = \frac{F}{A_s} \le \frac{S_P}{n_\sigma} \tag{9-3.1}$$

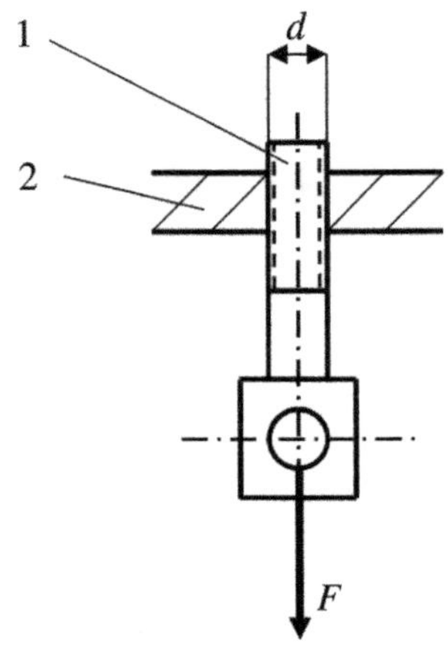

Figure 9-3.1. A threaded element subjected to tensile stress.

where A_s is the tensile stress area, that for Unified inch and ISO metric threads can be calculated based on the following formula:

$$A_s = \frac{\pi(d_1 + d_2)^2}{16} \tag{9-3.2}$$

The values of A_s for Unified inch threads are presented in Appendix A-6-2. With regard to the ISO metric threads, it is recommended to refer to the Appendix A-6-3, which contains the appropriate information including the dimensions of the minor diameters of the threads. Hence, upon determining the mean diameter, it is possible to use formula (9-3.2) in order to compute the tensile stress area A_s that is needed for the calculations.

In addition, it should be emphasized that the fasteners are very often subjected the action of static loading.

Example 9-3.1a: Calculating screw diameter.
We need to determine the diameter of a screw that will lift a load of 7,500 lb. We take the safety factor $n_\sigma = 2$.

Solution:
From the table in Appendix A-6-1, we select the proof strength of the SAE Grade 1 thread that is applicable for the range of diameters from 0.25 to 1.5 inches, $S_P = 33$ ksi. Based on formula (9-3.2), we calculate the tensile stress area: $A_s = \frac{Fn_\sigma}{S_P} = \frac{7{,}500 \cdot 2}{33{,}000} = 0.455 \text{ in}^2$. From Appendix A-6-2, we find that the closest larger tensile stress area is 0.462 in^2. for a screw having a major diameter $d = 0.875$ in., and the $N_t = 9$ threads per inch. Therefore, we select this diameter and pitch for our screw.

9-3.2. Calculating the Required Preload Torque

Preloaded threaded joints are used in a variety of assemblies subjected to static or variable loading. These assemblies comprise a packet of members (one of which could be a gasket) that are clamped together by a group of fasteners, such as bolts with nuts, screws, or studs with nuts. During the assembly of these joints, they are subjected to a predetermined tightening by applying a certain torque to the fasteners, developing preloaded stresses in all elements of the joint. Because of the preload, the bolts and the members become deformed—the bolts are elongated and the members become compressed. The tightening force that is developed by applying the torque is called the preload force, or simply preload, and is denoted by F_i. The analysis of the mechanics of threaded joints determines the value of the required torque in order to achieve the necessary preload. This analysis is presented in the preceding chapter, and formula (8-1.24) is applicable for calculating the required torque. However, this formula should by slightly modified by replacing $T_{tot.lift}$ and R with T and F_i, respectively. In addition, we have to replace in formula (8-1.24) the collar friction diameter d_c with the width across the flats s. As a result we obtain

$$T = \frac{F_i d_2}{2}\tan(\lambda + \mu) + \frac{F_i f_c s}{2} \tag{9-3.3}$$

Considering that the width across the flats equals $s = 1.732d$, we modify formula (9-3.3) to the following:

$$T = \frac{F_i d_2 d}{2d}\tan(\lambda+\mu) + \frac{1.732 F_i d f_c}{2} = F_i d\left[\frac{d_2}{2d}\tan(\lambda+\mu) + \frac{1.732 f_c}{2}\right] \qquad (9\text{-}3.4)$$

Introducing a torque coefficient K_j, we may write

$$K_j = \frac{d_2}{2d}\tan(\lambda+\mu) + 0.866\, f_c \qquad (9\text{-}3.5)$$

Therefore, the formula for calculating the torque in order to obtain the required preload reads

$$T = K_j F_i d \qquad (9\text{-}3.6)$$

where K_j is the torque coefficient that depends on the frictional characteristics of the materials of the joint elements and on the size of the bolt. For low carbon steel fasteners in the range of diameters from 0.25 to 1.0 inch (6 to 24 mm) and for zinc-plated bolts, it is recommended to take $K_j = 0.2$. For lubricated bolts, it is recommended to take $K_j = 0.18$ and for cadmium-plated bolts $K_j = 0.13$. For bolts with low surface finish and without plating, it is recommended to take $K_j = 0.3$.

9-3.3. Calculations of Preloaded Joints Subjected to External Forces

Consider a threaded joint of a hydraulic cylinder whose flange is clamped by a group of bolts and nuts with a cover and a gasket, while the cylinder is connected to a source of compressed fluid. During the assembly of the cover and the gasket to the flange, the joint is subjected to a certain preload, and because of it the bolts become elongated and the packet, which includes the cover, gasket, and flange, becomes compressed. Hence, the bolts become subjected to the tension force that equals the preload force. At the same time, the packet mentioned above is subjected to the compression force that is also equal to the absolute value of the preload force. Hence, when the compressed fluid is filling up the cylinder, the pressure force representing an external force is applied to the cover and causes an additional elongation of the bolts. Thus, when the cover is subjected to the action of the external force, the tension force acting in the bolts becomes increased and it exceeds the preload force previously applied to the bolts. At the same time, due to the additional elongation of the bolts, the packet becomes somewhat released from the compression and the compression force applied to the packet decreases and becomes smaller than the previously applied preload force. Therefore, due to applied external force, the thickness of the compressed packet increases by the same value as the additional elongation of the bolts. These types of threaded joints are preloaded joints subjected to external forces.

A similar joint is used in cases when a group of screws securely connects covers and shims to bearing housings that contain angular-contact or tapered rolling bearings. During the adjustment of the bearings, the screws are subjected to a predetermined preload, while during the operation of the assembly the axial forces play the role of the external forces applied to the joint.

One more example of a preloaded joint subjected to external forces is the assembly of cylinder head, the cylinder block, and the gasket to the block of an internal combustion engine where a group of studs and nuts holds the assembly togeher. During the operation of the engine, the external pressure forces of the ignited gases cyclically load the assembly. These threaded joints are subjected to preload forces applied to the fasteners and to the members during the assembly and to additional external forces that appear during the operation of these assemblies.

These external forces could be static as in cases of pressurized cylinders and axial forces in the bearings, or they could be variable as in internal combustion engines.

A threaded joint that is subjected to the external force is partially shown in Figure 9-3.2, where it is assumed that all bolts in the assembly are loaded uniformly.

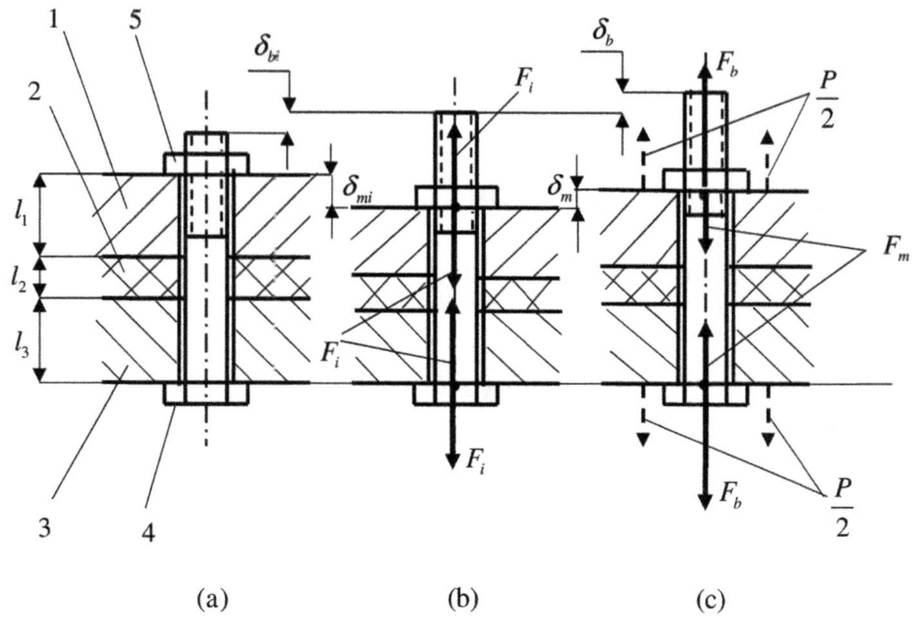

Figure 9-3.2. Schematic of a preloaded joint subjected to an external force.

The analysis of this joint is carried out for one bolt with associated parts of the members related to this bolt. This joint, as shown in Figure 9-3.2a, comprises members 1, 2, and 3 that have before the preload force is applied the thicknesses l_1, l_2, and l_3, of which member 2 represents a gasket. Figure 9-3.2b schematically shows the forces applied to the members and to bolt 4 and indicates the deformations of the bolt and members due to the tightening of nut 5. The bolt is under tension and the members are under the compression of the preload forces F_i. The bolt became longer on the amount of deformation δ_{bi} and the thickness of the packet of members became decreased according to the deformation δ_{mi}. It should be emphasized that the combined stiffness coefficient for the packet of members is determined by using the formula that is introduced in the course of Strength of Materials:

$$\frac{1}{K_m}=\frac{1}{K_1}+\frac{1}{K_2}+\ldots+\frac{1}{K_n} \tag{9-3.7}$$

where K_m is the combined stiffness coefficient of the members, and K_1, K_2, and K_n are the stiffness coefficients according to the ordinal numbers of the members, while n is the number of members. Figure 9-3.2c shows the forces acting in the joint and indicates the deformations of the bolt and members after an external force P is applied. In Figure 9-3.2c the force P is shown in dotted lines because this force is actually already accounted for in the resulting forces F_b and F_m applied to the bolt and to the members, respectively. A certain part of the external force P caused the additional deformation of the bolt δ_b, while the other part of this force decreased the compression force applied to the members, resulting in the increase of their total thickness by δ_m. It is obvious that δ_b must be equal to δ_m. Hence,

$$\delta_b=\delta_m \tag{9-3.8}$$

This condition allows us to determine the values of the forces F_b and F_m that depend on the stiffness characteristics of the joint. Assume that ΔP is a part of the force P that caused the additional elongation deformation δ_b of the bolt. Therefore, the force $P - \Delta P$ causes the increase in thickness of the packet according to the deformation δ_m. This allows us to write

$$\delta_b=\frac{\Delta P}{K_b} \tag{9-3.9}$$

where K_b is the stiffness coefficient of the bolt and

$$\delta_m=\frac{P-\Delta P}{K_m} \tag{9-3.10}$$

Based on equation (9-3.8), we may equate the right-hand sides of formulas (9-3.9) and (9-3.10), and as a result we obtain

$$\frac{\Delta P}{K_b}=\frac{P-\Delta P}{K_m} \tag{9-3.11}$$

Solving expression (9-3.11) for ΔP we have

$$\Delta P=\frac{PK_b}{K_b+K_m} \tag{9-3.12}$$

By adding the force ΔP to the preload force F_i, we determine the maximum tensile force F_b applied to the bolt:

$$F_b = F_i + \frac{PK_b}{K_b + K_m} \tag{9-3.13}$$

Using formula (9-3.12), we determine the part $P - \Delta P$ of the external load that unloads the members:

$$P - \Delta P = P - \frac{PK_b}{K_b + K_m} = \frac{PK_m}{K_b + K_m} \tag{9-3.14}$$

Therefore, accounting for expression (9-3.14), we determine the resulting compressive force applied to the members:

$$F_m = F_i - \frac{PK_m}{K_b + K_m} \tag{9-3.15}$$

Formulas (9-3.13) and (9-3.15) allow us to carry out the engineering calculations of preloaded threaded joints subjected to the action of external forces.

According to formulas (9-3.14) and (9-1.15), it can be seen that if the stiffness of the bolt equals the combined stiffness of the members, then $F_b = F_i + 0.5P$ and $F_m = F_i - 0.5P$. If the stiffness of the bolt is less than the stiffness of the packet, for instance, if $K_b = 0.5K_m$ then $F_b = F_i + 0.333P$ and $F_m = F_i - 0.666\ P$. Therefore, by controlling the stiffness of the elements, we can achieve the desire structural features of the threaded joint. The role of the stiffness of the involved elements and of the forces applied to these elements could be clarified by analyzing the graphs in Figure 9-3.3 that reflect the interaction between the elements of the threaded joint.

Stiffness of these elements is a linear function of the force applied to the element, while the deformation δ is the abscissa and the force F is the ordinate.

The diagrams in Figure 9-3.3 reflect the stiffness of the bolt and the packet of members as functions of applied forces (the ordinates) and the deformations (abscissas). According to this figure, the coefficients of stiffness of the bolt and the packet of members could be determined, respectively, by the following expressions: $K_b = \tan \beta_b$ and $K_m = \tan \beta_m$. Therefore, the straight line OA characterizes the stiffness of the bolt, while the line OB reflects the stiffness of the members. Upon applying preload F_i, which is positive for the bolt, the fastener develops a positive deformation δ_{bi}. For the members the preload force F_i is negative and results in a negative deformation of members $-\delta_{mi}$. After applying the external force P, a part of it ΔP causes an additional positive deformation δ_b of the bolt, while the rest of the external force $P - \Delta P$, which is also positive and applied to the members, also results in a positive deformation of the members δ_m that is equal to the deformation δ_b of the bolt. The diagram shows that the force $P - \Delta P$ has a positive direction and it decreases the preload force F_i that was originally applied to the

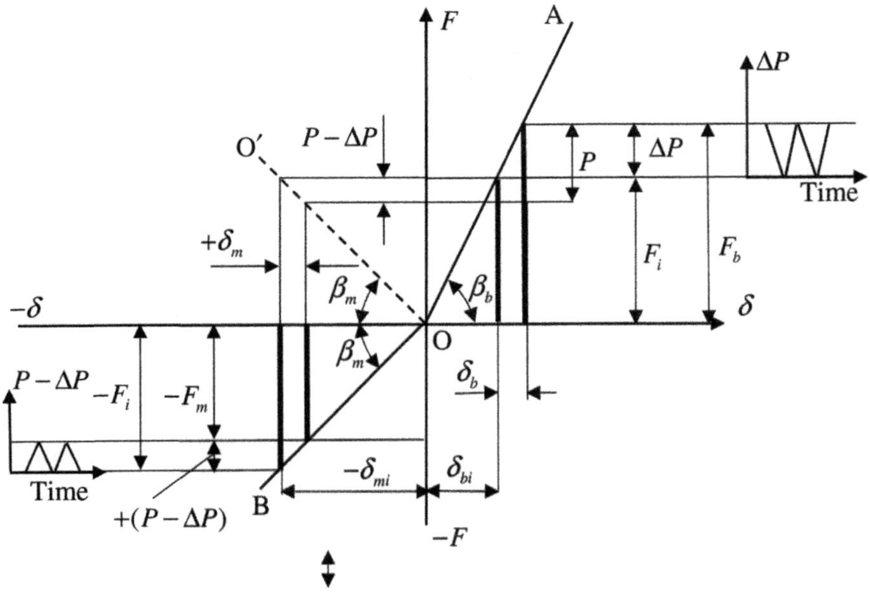

Figure 9-3.3. Diagram of loading and deformations.

members. The diagram shows after applying the external force P, the bolt becomes subjected to the action of the force $F_b = F_i + \Delta P$, while the force $-F_m = -F_i + P - \Delta P$ compresses the members. Actually, in the formula (9-3.15) the force F_m is negative; however, it is acceptable to present it as positive, keeping in mind that it is a compresson force. The dotted line OO′ allows the graphical transfer of the force $P - \Delta P$ in order to be aligned parallel with the force ΔP. This allows adding the force $P - \Delta P$ to the force ΔP and obtain the external force P.

It should be mentioned that in cases when in the threaded joints the external force P alternates between its minimum and maximum values, then the bolts and the members are subjected to variable stresses and, consequently, the related strength calculations should include fatigue considerations. Figure 9-3.3 also presents graphs that characterize the maximum and minimum forces applied to the bolt and members in cases when the external force is alternating from zero to its maximum value. In these graphs, the abscissas represent time, and the ordinates are the forces, while for the bolt the external force fluctuates from zero to ΔP, and for the members the external force fluctuates from zero to $P - \Delta P$. These graphs show that the preload force F_i represents the minimum force for the bolt and the maximum force for the members, while the force F_b is the maximum force for the bolt and the force $-F_m$ is the minimum force applied to the members.

9-3.4. Characteristics of Preload Forces

The preload force plays a significant role in the operation of threaded joints, and it is critical to designate the appropriate preload to particular joints. Preloading of the joint causes it to behave as a solid body. The tightening is achieved by using calibrated torque wrenches. However, it should be kept in mind that the toque determined based on calculations according to formula (9-3.6) reflects with limited accuracy the actual friction conditions between the threads and other involved elements. These conditions depend on many factors that are difficult to take into consideration. Because of this, calibrated torque wrenches are not completely accurate; however, it is still a good practice to use these wrenches. The manufacturers of fasteners and equipment provide the information of the required torques for tightening the corresponding bolts most of the time. This information is based on appropriate experimental data and closely reflects the actual conditions of the preload. In cases when this information is not available, it is recommended to apply some general guidelines for determining the preload of the fasteners. Thus, for reusable joints subjected to static or variable loading, apply the preload force using the following formula:

$$F_i = 0.75 A_s S_P \tag{9-3.16}$$

and for permanent joints (one-time tightening) the preload force is taken according to the formula that reads

$$F_i = 0.9\, A_s S_P \tag{9-3.17}$$

For materials that are not commonly used for standard fasteners, the proof strength can be, as it is mentioned above, calculated from the following expression: $S_P = 0.85 S_{yt}$. It should be emphasized that soft materials should not be used for threaded fasteners.

Threaded joints subjected to shear loading should prevent the sliding displacement of the members by developing sufficient friction forces between the connected members. As a result of this, the applied shear forces are balanced by the corresponding friction forces preventing the occurrence of shear stresses in the bolts. Actually, the bolt preload forces represent the normal forces acting between the members, creating the mentioned friction forces. Therefore, the preload forces should be determined accounting for the frictional interaction between the members.

9-4. Verifying Calculations of Preloaded Threaded Joints Subjected to Static External Loading

In this type of joint the bolts are subjected to the combination of normal (tensile) stresses and torsional stresses, while the members are loaded by the preload and external forces that influence the tightness of the joint. All these considerations are addressed below.

9-4.1. Joint Tightness Factor, Preload Torque, and the Equivalent Stress

The maximum tensile stress acting in the bolt after applying to the joint the external force is obtained from formula (9-3.13); the diagram in Figure 9-3.3 shows the sum of the stress due to preload and the stress due to the part of the external force applied to the bolt. Therefore, recalling formula (9-3.2), we may write

$$\sigma_{\max} = \frac{F_b}{A_s} = \frac{1}{A_s}\left(F_i + \frac{PK_b}{K_b + K_m}\right) \tag{9-4.1}$$

Obviously, the maximum tensile stress should be less than the proof strength of the bolt; therefore, $\sigma_{max} \le \frac{S_P}{n_\sigma}$, where n_σ is the safety factor. It should be clear that the force ΔP according to formula (9-3.13) is the only force that has the potential to increase the preload above the allowable value. The force $P - \Delta P$ according to formula (9-3.14) is potentially capable of decreasing the compressing force F_m to the level that the joint becomes completely separated. In this case, the extreme level of decrease of the compressing force is when $F_m = 0$.

Taking in formula (9-3.15) that $F_m = 0$, we have

$$0 = F_{i\min} - \frac{PK_m}{K_b + K_m} \tag{9-4.2}$$

where $F_{i\,\min}$ is the minimum preload force that represents the remainder of what is left from the initial preload force F_i after applying the external force P. Introducing the coefficient of the external load χ, we write

$$\chi = \frac{K_b}{K_b + K_m} \tag{9-4.3}$$

It is reported that for threaded joints with no gaskets the value of the coefficient of the external load χ is in the range from 0.15 to 0.4. In this case, we obtain stiff members and compliant bolts. For hermetic joints with gaskets it is recommended to take the value the coefficient χ in the range from 0.5 to 0.8, resulting in compliant members and stiff bolts.

Based on formula (9-4.3), we may write that $\frac{K_m}{K_b + K_m} = 1 - \chi$. Combining this formula with expression (9-4.2), we have

$$F_{i\min} = P(1 - \chi) \tag{9-4.4}$$

This formula allows for introducing the criterion for evaluating the tightness of the preloaded joint, indicating that the joint separation occurs when $F_{i\,\min} = 0$. This criterion represents the joint tightness factor n_j that is calculated from the following expression:

$$n_j = \frac{F_m}{P(1 - \chi)} > 1 \tag{9-4.5}$$

In case of static loading, it is advisable to use a joint tightness factor n_j in the range from 1.25 to 2.0. However, in cases of variable loading, the joint tightness factor should be from 2.5 to 4.

Torsional stress is imposed into the bolt during the tightening procedure of preloading. The torsional stress should be determined based on the torque according to formula (8-1.18) that does not account for collar friction resistance. Therefore, replacing in this formula T_{lift} with the maximum torque T_{max} and R with F_i we obtain

$$T_{max} = \frac{F_i d_m}{2} \tan(\lambda + \mu) \tag{9-4.6}$$

The average value of the lead angle λ for threaded fasteners is $\lambda = 2.5°$ and the average value of the friction coefficient between a bolt and nut made of steel is $f = 0.15$. Using formulas (8-1.16) and (8-1.17), we find that $\mu = 9.8°$. Hence, $T_{max} = \frac{F_i d_m}{2}\tan(2.5°+9.8°) = 0.107F_i d_m$. Considering that for these calculations it is justifiable to take that $d_m \approx 1.1d_1$, we may write

$$T_{max} = 0.118F_i d_1 \tag{9-4.7}$$

Based on formula (9-4.7), we calculate the maximum torsion stress:

$$\tau_{max} = \frac{T_{max} d_1}{2J} = \frac{0.118F_i d_1^2}{2J} \tag{9-4.8}$$

Using formula (1-4.21), we may write $J = \frac{\pi d_1^4}{32}$. Therefore, the maximum torsion stress is

$$\tau_{max} = \frac{16 \cdot 0.118F_i}{\pi d_1^2} = \frac{0.6F_i}{d_1^2} \tag{9-4.9}$$

Applying the fourth theory of failure, we have

$$\sigma_{eq} = \sqrt{\left(\frac{P_b}{A_s}\right)^2 + 3\left(\frac{0.6F_i}{d_1^2}\right)^2} \le \frac{S_P}{n_\sigma} \tag{9-4.10}$$

Example 9-4.1a: Verifying calculations of a preloaded threaded joint subjected to static loading.

We need to determine the tightness factor of the joint and the safety factor of stresses in the bolts of a preloaded threaded joint subjected to a static external force. Figure 9-4.1 shows one of the bolts of a threaded joint that consists of a group of uniformly loaded bolts.

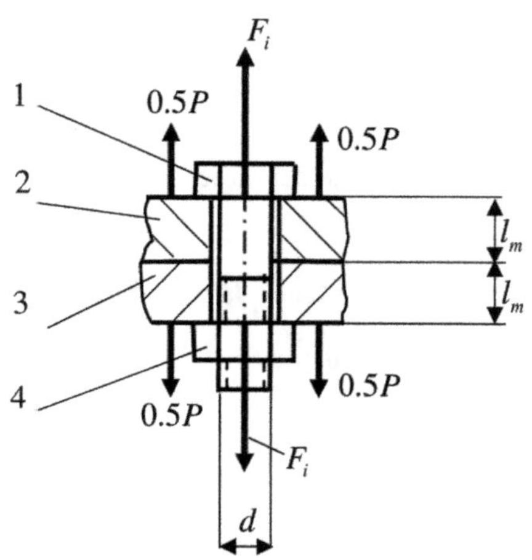

Figure 9-4.1. Preloaded threaded joint under static loading.

The data presented below relates to one bolt. Fastener 1 is a SAE Grade 4 bolt having a major diameter d = 0.375 in. and the pitch p = 0.0625 in. The steel members 2 and 3 have the thickness l_m = 0.5 in. of each. By tightening nut 4 the joint obtains the preload F_i = 1,700 lb. The external load P = 800 lb.

Solution:

From the Appendixes A-6-1 and A-6-2, we determine that the proof strength of the bolt material S_p = 65 ksi and the tensile stress area of the bolt A_s = 0.0775 in². From formula (9-2.1), we determine the stiffness coefficient of the bolt: $K_b = \frac{\pi 0.375^2 \cdot 30 \cdot 10^6}{4 \cdot 2 \cdot 0.5} = 3.3 \cdot 10^6 \frac{\text{lb}}{\text{in.}}$. Using formula (9-2.3), we calculate the stiffness of the members: $K_m = \frac{0.34 \cdot 0.375 \cdot 30 \cdot 10^6}{\log \frac{3.73(0.732 \cdot 0.375 + 0.5)}{2.732 \cdot 0.375 + 0.5}} = 13{,}8 \cdot 10^6 \frac{\text{lb}}{\text{in.}}$. Applying formula (9-4.3), we calculate $\chi = \frac{3.3 \cdot 10^6}{3.3 \cdot 10^6 + 13.8 \cdot 10^6} = 0.193$. Based on formula (9-3.13) and accounting for χ, we determine the force applied to the bolt:

$F_b = F_i + \chi P$ = 1,700 + 0.193·800 = 1,954.4 lb. Therefore, $\sigma_{\max} = \frac{F_b}{A_s} = \frac{1.954.4}{0.0775} = 25{,}218$ psi.

Applying formula (9-3.15), we calculate F_m = 1700 – 800 (1 – 0.193) = 1,054.4 lb. Using formula (9-4.5), we evaluate the tightness of the joint: $n_j = \frac{1{,}054.4}{800(1-0.193)} = 1.633$, which is acceptable.

Now we calculate the minor diameter of the bolt using formula (9-1.2):

$d_1 = 0.375 - 1.0825 \frac{1}{16} - 2 \cdot 0.10825 \frac{1}{16} = 0.294$ in.

Applying expression (9-4.9), we calculate the torsion stress: $\tau_{\max} = \frac{0.6F_i}{d_1^2} = \frac{0.6 \cdot 1{,}700}{0.294^2} = 11{,}825.7$ psi.

Now based on expression (9-4.10), we calculate $\sigma_{eq} = \sqrt{\sigma_{\max}^2 + 3\tau_{\max}^2} = \sqrt{25.22^2 + 3 \cdot 11.8^2} = 32.46$ ksi. Hence, $n_\sigma = \frac{S_p}{\sigma_{eq}} = \frac{65}{32.46} = 2$, which is acceptable.

9-4.2. Design Calculations of Preloaded Joints Subjected to Static Loading

Replacing in formula (9-4.10) the preload force F_i with the maximum force F_b applied to the bolt leads to an increase of the margin of safety of the bolt. According to the fourth theory of failure, the equivalent stress in the bolt has the following shape:

$$\sigma_{eq} = \sqrt{\left(\frac{4F_b}{\pi d_1^2}\right)^2 + 3\left(\frac{0.6F_b}{d_1^2}\right)^2} \leq \frac{S_P}{n_\sigma} \qquad (9\text{-}4.11)$$

Applying some conventional transformations to formula (9-4.11), we write

$\sigma_{eq}=\sqrt{\left(\frac{4F_b}{\pi d_1^2}\right)^2+3\left(\frac{0.6\pi 4F_b}{4\pi d_1^2}\right)^2}=\frac{4F_b}{\pi d_1^2}\sqrt{1+3\left(\frac{0.6\pi}{4}\right)^2}\le\frac{S_P}{n_\sigma}$. This expression allows us to write

$$\frac{1.3\cdot 4F_b}{\pi d_1^2}=\frac{S_P}{n_\sigma} \tag{9-4.12}$$

Solving expression (9-4.12) for d_1, we obtain the formula for determining the minor diameter of the bolt:

$$d_1=1.3\sqrt{\frac{F_b n_\sigma}{S_P}} \tag{9-4.13}$$

It is clear that accepting for formula (9-4.13) the value of the proof strength S_P we should select a certain grade of bolt on a preliminary basis. For preliminary calculations we may take $n_\sigma = 2$.

In cases when the proof strength of the steel is not available, the formula (4-9.13) can be modified to the following shape:

$$d_1=1.3\sqrt{\frac{F_b n_\sigma}{0.85S_{yt}}}=1.41d_1=1.41\sqrt{\frac{F_b n_\sigma}{S_{yt}}} \tag{9-4.14}$$

It should be mentioned that for U.S. standard threads whose diameters range from 0.25 to 1.0 inches the average ratio between the major and minor diameters is $\frac{d}{d_1}=1.275$, while for diameters of ISO metric threads in the range of 6 mm to 24 mm the average ratio is $\frac{d}{d_1}=1.11$.

It should be emphasized that these design calculations are based on the conservative approach that tends to favor the increased margin of safety with respect to the strength of machine elements.

Example 9-4.2a: Determining the diameter of the bolt for a preloaded joint subjected to static loading.

We need to calculate the diameter of the bolt for a preloaded threaded joint subjected to a static load. The input data for this example is taken from the preceding Example 9-4.1a. This will allow us to evaluate the level of accuracy of the presented methodology of the design calculations. Therefore, the maximum axial load applied to the bolt is F_b = 1954.4 lb, the safety factor is n_σ = 2, and the proof strength S_P = 65 ksi for the selected Grade 4 of the bolt.

Solution:

Applying formula (9-4.13), we calculate $d_1=1.3\sqrt{\frac{1{,}954.4\cdot 2}{65{,}000}}=0.318$ in. Hence, multiplying the minor diameter by the above mentioned multiplier 1.275 for U.S. standard threads, we determine the major diameter of the thread: $d = 0.318 \cdot 1.275 = 0.405$ in. The major diameter in the preceding example is $d = 0.375$ in. Comparing these two diameters, we obtain $\frac{0.405}{0.375} = 1.08$. Hence, the calculated diameter exceeds the actual diameter by 8%, which is acceptable. Hence, according to the corresponding standard, we take the closest higher diameter $d = 0.4375$ in.

9-5. Preloaded Threaded Joints Subjected to Variable External Loading

The principles of calculating the basic parameters of preloaded joints subjected to variable external loading are the same as used for static loading. The stiffness coefficients of the bolt and the members are calculated using formulas (9-2.1) and (9-2.3), respectively. The maximum value of the force applied to the bolt is determined based on formula (9-3.13), and the minimum preload force compressing the members is determined using formula (9-3.15). All stress calculations applicable to a joint subjected to static loading should be performed for cases dealing with variable loading.

Now we emphasize the additional aspects that should be accounted for in the calculations of the joint when the external force is cyclically changing its value. The external force P fluctuates between its minimum and maximum values, imposing a random loading cycle into the bolt and the members. Obviously, the members will not fail due to this cycle. The danger is that the bolt may not be strong enough and fatigue failure may occur. The main concern is associated with the strength of the body of the bolt. We do not verify the strength of the threads with regard to fatigue failure.

The worse case with regard to fatigue failure of the bolt is when the external force fluctuates between zero and its maximum value. Referring to formulas (9-3.13) and (9-4.3), we determine the maximum and minimum forces applied during one cycle:

$$F_{max} = F_b = F_i + \chi P_{max} \tag{9-5.1}$$

where P_{max} is the maximum value of the external force, and

$$F_{min} = F_i \tag{9-5.2}$$

Based on formulas (9-5.1) and (9-5.2), we determine the maximum and minimum stresses acting in the bolt during the cycle:

$$\sigma_{max} = \frac{F_i + \chi P_{max}}{A_s} \tag{9-5.3}$$

and

$$\sigma_{min} = \frac{F_i}{A_s} \tag{9-5.4}$$

These considerations allow us to approach the discussion related to the verifying and the design calculations of preloaded joints subjected to variable external loading.

9-5.1. Verifying Calculations of Preloaded Joints Subjected to Variable External Loading

The goal of these verifying calculations is to determine the margin of safety of a bolt experiencing variable stresses of a random loading cycle.

Formulas (9-5.3) and (9-5.4) allow us to calculate the mean stress and the stress amplitude that characterize the cycle:

$$\sigma_m = \frac{F_b + F_i}{2A_s} = \frac{2F_i + \chi P_{max}}{2A_s} \tag{9-5.5}$$

and

$$\sigma_a = \frac{F_b - F_i}{2A_s} = \frac{\chi P_{max}}{2A_s} \tag{9-5.6}$$

Now it should be verified that $\sigma_m + \sigma_a < S_P$. Actually, $\sigma_m + \sigma_a = \sigma_{max}$. All further steps are discussed in Chapter 1 in sections related to fatigue calculations. It should be emphasized that the conventional published resources related to the grades of bolts contain the information of the proof strength S_P as well as of the ultimate strength S_{ut}. Using formulas (1-9.2), (1-9.3), and (1-9.4), we determine the corresponding factors affecting the endurance limit and calculate the endurance limit. While determining $\tan\alpha$ according to formula (1-12.4), the S_{yt} should be replaced with S_P. Using formulas (1-12.7) and (1-12.8), we calculate the main stress limit S_m and the mean stress amplitude limit S_a, respectively.

The stress concentration factor plays a significant role in increasing the tensile stress. The value of the fatigue stress concentration factor K_f varies from 2 to 3.8, depending on the method of thread manufacturing and the grade.

The tightness of the joint subjected to variable loading is verified in the same way as for static loading by using formula (9-4.5). The fatigue strength of the bolt is verified based on the compliance with the condition that is formulated in the following expression:

$$\frac{S_a}{\sigma_a} \geq n_\sigma \tag{9-5.7}$$

It may be noted that some authors report that the fluctuating of tensile stresses in the preloaded fasteners causes reduction of the torsional stresses imposed into the fasteners during the preloading procedure. This may explain why the torsional stresses are not accounted for in the fatigue strength calculations of the preloaded screws.

The example presented below demonstrates the methodology of the verifying calculations.

Example 9-5.1a: Verifying calculations of a preloaded joint subjected to variable external forces.

We need to perform verifying calculations of a preloaded gasketed joint subjected to a variable external load that changes its value from zero to maximum. All screws of the joint are uniformly loaded and the verifying calculations should be performed for screw 1 that is shown in Figure 9-5.1 representing a schematic view of the joint comprising steel members 2 and 4 and a soft gasket 3 having a thickness l_0 = 0.063 in. The gasket is made from polycarbonate gasket material having $E_0 = 0.552 \cdot 10^6$ psi.

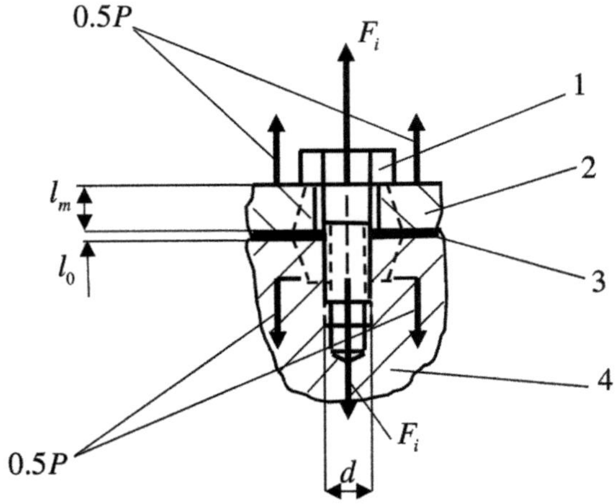

Figure 9-5.1. Preloaded joint subjected to a variable external force.

Member 2 has a thickness l_m = 0.75 in. The screw has the SAE Grade 8 classification, while the major diameter d = 0.625 in., and the number of threads per inch N_t = 11. The screw is made of steel having S_P = 120 ksi, S_{yt} = 130 ksi, and S_{ut} = 150 ksi. The tensile stress area $A_s = 0.226\ \text{in}^2$. The preload F_i = 2,500 lb and the maximum external load P = 1,500 lb.

Solution:

We start with determining the stiffness coefficients of the elements of the joint. It should be mentioned that in cases when the fastener is screwed into the body of a member (member 4) the volume of the compressed material consists of two frusta shown in Figure 9-5.1 by the dotted lines. In this case, the effective length of the screw is $l_b = 2l_m + l_0$. Applying formulas (9-2.1), (9-2.3), and (9-2.5), we calculate $K_b = \dfrac{\pi 0.625^2 \cdot 30 \cdot 10^6}{4(2 \cdot 0.75 + 0.063)} = 5.89 \cdot 10^6 \dfrac{\text{lb}}{\text{in.}}$,

$$k_m = \frac{0.34 \cdot 0.625 \cdot 30 \cdot 10^6}{\log \dfrac{3.73(0.732 \cdot 0.625 + 0.75)}{2.732 \cdot 0.625 + 0.75}} = 24.23 \cdot 10^6 \frac{\text{lb}}{\text{in.}},$$

$$K_0 = \frac{0.39 \cdot 0.552 \cdot 10^6[(1.732 \cdot 0.625 + 0.063)^2 - 0.625^2]}{0.063} = 3.15 \cdot 10^6 \frac{\text{lb}}{\text{in.}}.$$

Applying formula (9-3.7), we have

$$\frac{1}{K'_m} = \frac{1}{K_m} + \frac{1}{K_0} = \frac{1}{24.23 \cdot 10^6} + \frac{1}{3.5 \cdot 10^6} = \frac{3.15 \cdot 10^6 + 24.23 \cdot 10^6}{24.23 \cdot 3.15 \cdot 10^{12}} = \frac{0.35}{10^6} \frac{\text{in.}}{\text{lb}}$$

According to these calculations, we obtain $K'_m = 2.79 \cdot 10^6 \dfrac{\text{lb}}{\text{in.}}$. Using formula (9-4.3), we calculate

$\chi = \dfrac{3.77}{3.77+1.89} = 0.67$. Now we determine the maximum force applied to the screw. Using formula (9-3.13) and accounting the value of χ, we calculate

$F_{max} = F_b = F_i + \chi P = 2{,}500 + 0.67 \cdot 1{,}500 = 3{,}505$ lb. The maximum tensile stress applied to screw is $\sigma_{max} = \dfrac{F_b}{A_s} = \dfrac{3{,}505}{0.226} = 15.508$ psi.

The minimum tensile stress acting in the screw is $\sigma_{min} = \dfrac{F_i}{A_s} = \dfrac{2{,}500}{0.226} = 11.062$ psi. We continue to verify the joint with regard to static loading. Based on formula (9-3.15), we calculate the minimum clamping force applied to the members $F_m = F_i - (1-\chi)P = 2{,}500 - (1-0.67)1{,}500 = 2{,}005$ lb. Applying formula (9-4.5), we evaluate the tightness of the joint: $n_j = \dfrac{2{,}005}{1{,}500(1-0.67)} = 4.05$, which is acceptable.

In order to determine the torsional stress applied to the screw, we need to determine the minor diameter of the screw using formula (9-1.2):

$$d_1 = 0.625 - 1.0825 \cdot \frac{1}{11} - 2 \cdot 0.10825 \cdot \frac{1}{11} = 0.507 \text{ in.}$$

Based on formula (9-4.9), we calculate the torsion stress:

$$\tau_{max} = \frac{0.6F_i}{d_1^2} = \frac{0.6 \cdot 2{,}500}{0.507^2} = 5.838 \text{ psi.}$$

Using the fourth theory of failure according to expression (9-4.10), we write

$$\sigma_{eq} = \sqrt{\sigma_{max}^2 + 3\tau_{max}^2} = \sqrt{15.508^2 + 3 \cdot 5.838^2} = 15.509 \text{ ksi.}$$

Hence, $n_\sigma = \dfrac{S_p}{\sigma_{eq}} = \dfrac{120}{15.509} = 7.74$, which is acceptable.

Now we address the fatigue considerations. Using formulas (1-9.2), (1-9.3), and (1-9.4), we calculate $k_a = 0.73$, $k_b = 0.963$, and $k_c = 0.8$. Therefore, we have $S_e = 0.73 \cdot 0.963 \cdot 0.8 \cdot 0.5 \cdot 150 = 42.18$ ksi. Applying formulas (9-5.5) and (9-5.6), we calculate $\sigma_m = 13{,}285$ psi and $\sigma_a = 2{,}223$ psi. Based on formula (1-12.5) we determine $\tan\beta = \dfrac{\sigma_a}{\sigma_m} = 0.167$. From formula (1-12.4) we obtain $\tan\alpha = \dfrac{42.18(150-120)}{150(120-42.18)} = 0.108$. Since $\tan\beta > \tan\alpha$ we may continue the calculations accounting fatigue considerations. Based on formula (1-12.8), we determine

$$S_a = \frac{S_e S_{ut} \sigma_a}{S_e \sigma_m + S_{ut} \sigma_a} = \frac{42.18 \cdot 150 \cdot 3.54}{42.18 \cdot 21.16 + 150 \cdot 3.54} = 15.73 \text{ ksi.}$$

Taking the fatigue stress concentration factor $K_f = 3$, we calculate $n_\sigma = \dfrac{S_a}{\sigma_a} = \dfrac{15.73}{2.223 \cdot 3} = 2.35$, that is acceptable.

9-5.2. Design Calculations of Preloaded Joints Subjected to Variable Loading

The bolts in the preloaded joints subjected to variable external loading should be designed with elevated factors of safety. The design calculations for these types of joints is based on the approach that is presented above for statically loaded joints; however, for preliminary calculations it is recommended to apply the safety factor $n_\sigma \geq 4$.

To evaluate the accuracy of the design calculations, let us determine the diameter of the screw for the example 9-5.1a presented above. We take from this example the maximum force F_b = 3,505 lb applied to the screw and the proof strength S_P = 120 ksi. Taking $n_\sigma = 4$ and using formula (9-4.13), we calculate $d_1 = 1.3\sqrt{\frac{3{,}505 \cdot 4}{120 \cdot 10^3}} = 0.444$ in. For approximate and preliminary calculations, the major diameter could be determined as the product of multiplying the minor diameter by the coefficient 1.275. Therefore, we may write $d = 0.444 \cdot 1.275 = 0.567$ in. In the example mentioned above, the major diameter of the thread is $d = 0.625$ in., which exceeds by about 10% the calculated diameter. However, selecting the closest larger standard major diameter for the calculated diameter 0.567 in. leads to the diameter of 0.625 in. Therefore, considering that we used the smallest recommended safety factor $n_\sigma = 4$, we may consider the result obtained from the design calculations as satisfactory.

To finalize the decision the designer should carry out the corresponding verifying calculations.

9-6 Threaded Joints Subjected to Shear Forces

There are several types of threaded joints where the members are loaded by forces that are directed perpendicularly to the axes of the bolts, causing shear stresses in these bolts. The most typical cases are considered below.

9-6.1. Bolts Installed in the Holes with an Interference Fit

Figure 9-6.1 shows a threaded joint of flat members 3 and 4 connected to each other by bolt 1 and nut 4 without applying an essential preload.

The members having a thickness *t* are subjected to shear forces F_τ. These forces develop pure stresses in the body of the bolt having a diameter *d*. The verifying calculations of the strength of the bolts are performed using formula (1-4.5):

$$\tau = \frac{4F_\tau}{\pi d^2} \le \frac{0.6S_P}{n_\tau} \tag{9-6.1}$$

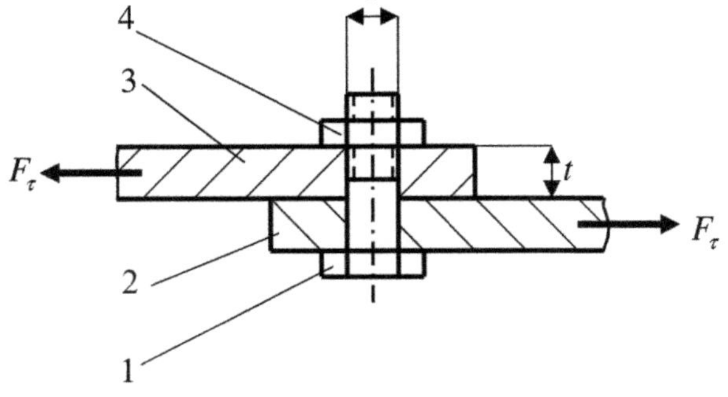

Figure 9-6.1. Bolted joint subjected to shear stresses.

In these cases, it is also necessary to verify the bearing stress in the bolt and in the member that has the smaller thickness. Based on the formula (1-4.31), we write the expression for verifying stresses in the bolt:

$$\sigma = \frac{F_\tau}{dt} \leq \frac{S_P}{n_\sigma} \qquad (9\text{-}6.2)$$

and in the member

$$\sigma = \frac{F_\tau}{dt} \leq \frac{S_{yt}}{n_\sigma} \qquad (9\text{-}6.3)$$

Based on expressions (9-6.1) and (9-6.2), we derive formulas for design calculations that allow determining the major diameter of the fastener with respect to shear stresses:

$$d = 2\sqrt{\frac{F_\tau n_\tau}{0.6\pi S_P}} \qquad (9\text{-}6.4)$$

and with respect to bearing stresses

$$d = \frac{F_\tau n_\sigma}{t S_p} \qquad (9\text{-}6.5)$$

We select the larger of the two bolt diameters calculated according to formulas (9-6.4) and (9-6.5).

9-6.2. Preloaded Bolts Installed with a Clearance Fit

Figure 9-6.2 shows a joint comprising members 2 and 3 that are clamped by bolt 1 and nut 4.

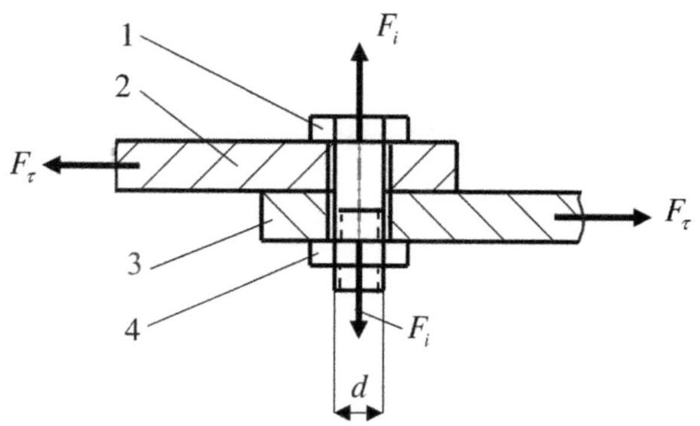

Figure 9-6.2. Preloaded bolt installed with a clearance fit.

The bolt has a diameter d and is preloaded by a force F_i that develops a friction force between the members, preventing them from moving relative to each other under the action of shear forces F_τ. The bolt is placed with a clearance fit. The preload force can be determined from the following formula:

$$F_i = \frac{F_\tau}{f} \tag{9-6.6}$$

where f is the friction coefficient between the members. The verifying calculation should account for the normal and torsion stresses in the bolt. The normal stresses are

$$\sigma_{\max} = \frac{F_\tau}{fA_s} \tag{9-6.7}$$

Based on formula (9-4.9), we determine

$$\tau_{\max} = \frac{0.6F_\tau}{fd_1^2} \tag{9-6.8}$$

Considering formulas (9-6.7) and (9-6.8) and applying the fourth theory of failure, we obtain the expression for verifying calculations:

$$\sigma_{eq} = \sqrt{\left(\frac{F_\tau}{fA_s}\right)^2 + 3\left(\frac{0.6F_\tau}{fd_1^2}\right)^2} \le \frac{S_P}{n_\sigma} \tag{9-6.9}$$

In order to derive an expression for preliminary design calculations of the bolt diameter, we replace in formula (9-6.7) the tensile stress area A_s with the cross-sectional area according to the minor diameter of the bolt $A_1 = \frac{\pi d_1^2}{4}$; therefore, we may write

$$\sigma'_{\max} = \frac{4F_\tau}{\pi fd_1^2} \tag{9-6.10}$$

It should be noted that $\sigma'_{\max}$ slightly exceeds $\sigma_{\max}$ that increases the safety margin of the results of the design calculations. Replacing in expression (9-6.9) $\sigma_{\max}$ with $\sigma'_{\max}$, we may write $\sqrt{\left(\frac{4F_\tau}{\pi fd_1^2}\right)^2 + 3\left(\frac{0.6F_\tau}{fd_1^2}\right)^2} = \frac{S_P}{n_\sigma}$. Solving this equation for d_1, we obtain an expression that is similar to formula (9-4.13), that reads

$$d_1 = 1.3\sqrt{\frac{F_\tau n_\sigma}{fS_P}} \tag{9-6.11}$$

Multiplying the value of the minor diameter by the coefficient mentioned above, we determine the major diameter of the required bolt.

Example 9-6.1a: Calculating a threaded joint subjected to shear stresses.
We must determine the preload force and the diameter of the bolt for the joint presented in Figure 9-6.3 in order to keep the members together. The joint comprises a bolt that is installed with a clearance fit, members 2 and 3 loaded by shear forces $F_\tau = 150$ lb, and a nut 4 that develops the required preload force as a result of tightening. The friction coefficient between the members is $f = 0.15$. We apply a safety factor $n_\sigma = 2$.

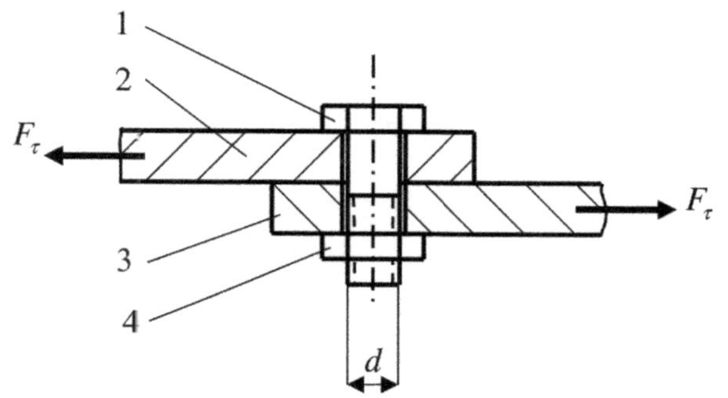

Figure 9-6.3. A threaded joint subjected to shear forces.

Solution:
We start with the calculations of the normal force that is required for creating between the plates the necessary friction forces that could balance the applied shear forces. This normal force actually represents the preload force that is determined the following way: $F_i = \frac{F_\tau}{f} = \frac{1{,}500}{0.15} = 1{,}000$ lb.

Now we determine the diameter of the bolt. From Appendix A-6-1, we select the ASE Grade 4 bolt having S_P = 65 ksi. Using formula (9-6.11), we calculate the minor diameter of the bolt: $d_1 = 1.3\sqrt{\frac{150 \cdot 2}{0.15 \cdot 65 \cdot 10^3}} = 0.228$ in. Multiplying the value of the calculated minor diameter of the bolt by the coefficient 1.275, we determine the major diameter of the required bolt: $d = 0.228 \cdot 1.275 = 0.291$ in. According to Appendix A-6-2, we select the closest larger major diameter $d = 0.3125$ in.

9-7. Design Project

An air compressor having a piston type structure supplies compressed air at the nominal air pressure of 150 psi. We need to design a detachable threaded joint of the block of cylinders with the head and gasket if the diameter of the piston is 6.0 in., and four screws secure the joint for each cylinder. The block of cylinders and the head are made from cast iron having the modulus of elasticity $E_m = 18.6 \cdot 10^6$ psi. The thickness of the member representing the head of the block is $l_m = 1.25$ in., while the thickness of the gasket $l_0 = 0.063$ in., and the modulus elasticity of the material of the gasket $E_0 = 0.552 \cdot 10^6$ psi. The safety factor for determining the diameter of the screw $n_\sigma = 6$.

The partial view of the threaded joint is shown in Figure 9-7.1 where the block of cylinders 1, the head 3, and gasket 4 are connected by a group of screws 2 (just one screw is shown).

It is obvious that this threaded joint is subjected to variable external loading representing a random cycle.

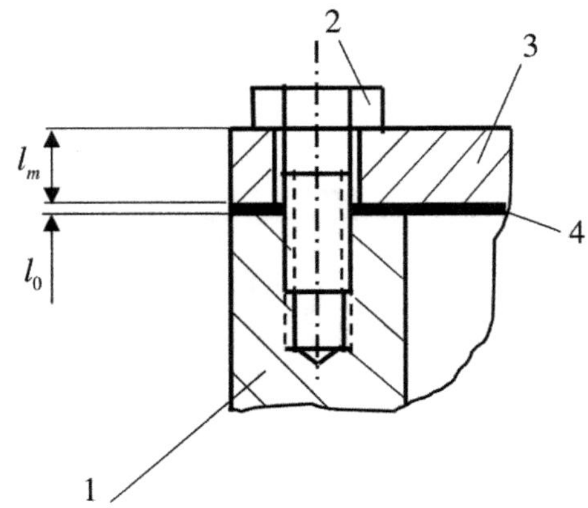

Figure 9-7.1. Partial view of a threaded joint of a cylinder block, including the head and gasket.

1. Determining the diameter of the screw.

Since the screws in our case are reusable, we may estimate the preload force of the screw based on the following recommendation mentioned above: $F_i = 0.75A_sS_p$. Therefore, the additional force as a part of the external force applied to the screw should not exceed $\Delta P = 0.25\,A_sS_p$. This design version of the joint is characterized by the recommended maximum value of the preload force and the corresponding maximum value of the additional part of the external force applied to the screw. The external compressed air force cyclically applied to the joint including the four screw is $P' = 150 \cdot \frac{\pi 6^2}{4} = 4{,}590$ lb, where 150 psi is the air pressure and 6 in. is the diameter of the piston. Taking for a hermetic gasketed joint the average recommended value of the external force coefficient $\chi = 0.65$, we determine the additional external force ΔP applied to each screw, hence, $\Delta P = \frac{\chi P}{4} = \frac{0.65 \cdot 4{,}590}{4} = 746$ lb. Therefore, $P = \frac{\Delta P}{\chi} = \frac{746}{0.65} = 1{,}148$ lb. According to formula (9-5.1), the maximum force applied to the screw is $F_{\max} = F_b = F_i + \Delta P = 0.75A_sS_p + \Delta P$.

Since $F_{\max} = A_sS_p$, we may write $0.25A_sS_p = 746$ lb. Selecting the SAE Grade 7 screw, we find the strength of the material of the screw $S_p = 105{,}000$ psi and the ultimate strength $S_{ut} = 133{,}000$ psi. Therefore, we obtain $0.25 \cdot 105{,}000A_s = 746$ lb. Accounting for the required safety factor $n_\sigma = 6$, we determine the tensile stress area of the screw: $A_s = \frac{746 \cdot 6}{0.25 \cdot 105{,}000} = 0.171\ \text{in.}^2$ From the Appendix A-6-2, we determine the closest larger tensile stress area is 0.226 in.^{2} that corresponds to the major diameter of the screw $d = 0.625$ in.

2. Determining the parameters of the joint.

Because the external force is predetermined and cannot be changed, in order to keep the accepted ratio between the preload force F_i and the external force ΔP we have to determine the preload force based on the calculated tensile stress area: $A_s = 0.171$ in.^{2} Hence, we calculate the preload force: $F_i = \frac{0.75A_sS_p}{n_\sigma} = \frac{0.75 \cdot 0.171 \cdot 105{,}000}{6} = 2{,}244$ lb. Now we calculate the stiffness coefficients of the elements of the joint. In our case, the effective length of the screw is $l_b = l_m + l_0$. Applying formula (9-2.1), we compute $K_b = \frac{\pi 0.625^2 \cdot 30 \cdot 10^6}{4(1.25+0.063)} = 7 \cdot 10^6 \frac{\text{lb}}{\text{in.}}$. Accounting for the above suggestions, we determine the effective thickness of the plate (the head of the block of cylinders) $l'_m = 0.5l_m = 0.5 \cdot 1.25 = 0.625$ in. Hence, we substitute l'_m into formula (9-2.3) and determine $k_m = \frac{0.34 \cdot 0.625 \cdot 18.6 \cdot 10^6}{\log \frac{3.73(0.732 \cdot 0.625 + 0.625)}{2.732 \cdot 0.625 + 0.625}} = 16.585 \cdot 10^6 \frac{\text{lb}}{\text{in.}}$ Applying formula (9-2.5), we calculate $K_0 = \frac{0.39 \cdot 0.552 \cdot 10^6[(1.732 \cdot 0.625 + 0.063)^2 - 0.625^2]}{0.063} = 3.149 \cdot 10^6 \frac{\text{lb}}{\text{in.}}$. Using formula (9-3.7), we determine

$$\frac{1}{K'_m} = \frac{1}{K_m} + \frac{1}{K_0} = \frac{1}{16.585 \cdot 10^6} + \frac{1}{3.149 \cdot 10^6} = \frac{3.149 \cdot 10^6 + 16.585 \cdot 10^6}{16.585 \cdot 3.149 \cdot 10^{12}} = \frac{0.378}{10^6}\frac{\text{in.}}{\text{lb}}.$$

According to these calculations, we obtain $K'_m = 2.65 \cdot 10^6 \frac{\text{lb}}{\text{in.}}$. Based on formula (9-4.3), we calculate $\chi = \frac{7}{7+2.65} = 0.725$. Using formula (9-3.13), we determine the maximum force applied to the screw: $F_{max} = F_b = F_i + \chi P = 2{,}244 + 0.725 \cdot 1{,}148 = 3{,}076$ lb. Now according to formula (9-3.15), we compute the minimum compression force applied to members: $F_m = 2{,}244 - (1 - 0.725)1{,}148 = 1{,}928$ lb. Using formula (9-4.5), we calculate the joint tightness factor: $n_j = \frac{1{,}978}{1{,}148(1-0.725)} = 6.265$, which is acceptable.

3. Calculating the stresses in the screw.

The minimum stress is $\sigma_{min} = \frac{F_i}{A_s} = \frac{2{,}244}{0.226} = 9{,}929$ psi, while the maximum stress is $\sigma_{max} = \frac{F_{max}}{A_s} = \frac{3{,}076}{0.226} = 13{,}611$ psi. Now we determine the minor diameter of the screw in order to determine the torsion stress due to the preload: $d_1 = 0.625 - 1.0825 \cdot \frac{1}{11} - 2 \cdot 0.10825 \cdot \frac{1}{11} = 0.507$ in.

Using formula (9-4.9), we calculate the torsion stress:

$$\tau_{max} = \frac{0.6F_i}{d_1^2} = \frac{0.6 \cdot 2{,}244}{0.507^2} = 5{,}238 \text{ psi.}$$

Applying the fourth theory of failure according to expression (9-4.10), we have

$$\sigma_{eq} = \sqrt{\sigma_{\max}^2 + 3\tau_{\max}^2} = \sqrt{13{,}611^2 + 3 \cdot 5{,}238^2} = 16{,}358 \text{ ksi.}$$

Now we verify the strength of the screw with regard to static loading:

$$n_\sigma = \frac{0.75 S_p}{\sigma_{eq}} = \frac{0.75 \cdot 105}{16{,}358} = 4.814, \text{ which is acceptable.}$$

4. Verifying calculations of variable stresses for a random cycle.

Based on formulas (1-9.2), (1-9.3), and (1-9.4), we compute, respectively, $k_a = 0.919$, $k_b = 0.963$, and $k_c = 0.8$. Therefore, we have $S_e = 0.793 \cdot 0.963 \cdot 0.8 \cdot 0.5 \cdot 105 = 32.07$ ksi. Now using formulas (9-5.5) and (9-5.6), we compute $\sigma_m = \frac{13{,}611 + 9{,}929}{2} = 11{,}770$ psi and $\sigma_a = \frac{13{,}611 - 9{,}929}{2} = 1{,}841$ psi. Applying formula (1-12.4), we calculate $\tan\alpha = \frac{32.07(133-105)}{133(105-32.07)} = 0.093$. Using formula (1-12.5), we determine $\tan\beta = \frac{\sigma_a}{\sigma_m} = \frac{1{,}841}{13{,}611} = 0.135$. Since $\tan\beta > \tan\alpha$, therefore, we may continue the verifying calculations for the random cycle. Hence, applying formula (1-12.8), we determine the stress amplitude limit: $S_a = \frac{S_e S_{ut} \sigma_a}{S_{ut}\sigma_a + S_e \sigma_m} = \frac{32.07 \cdot 133 \cdot 1.841}{133 \cdot 1.841 + 32.07 \cdot 13.611} = 11.525$ ksi. We take the fatigue stress concentration factor $K_f = 3$, and we calculate $n_\sigma = \frac{S_a}{\sigma_a K_f} = \frac{11.525}{1.841 \cdot 3} = 2.09$, which is acceptable.

Chapter 9. Problems

9-01. Determine the stiffness coefficient of the national UTS bolts having the following major diameters and lengths:

d	l_b
5/16	0.8
3/8	0.9
9/16	1.2
7/8	1.35
1 1/8	1.5
1 3/4	1.75

9-02. Determine the stiffness coefficient of a packet of two steel members that are joined by the ISO metric bolts having the following major diameters, while the thicknesses of the members l_m are:

d	l_m
6	5
10	8
16	12
22	15
27	16
42	22

9-03. For a detachable joint using SAE Grade 1 bolts having the major diameter $d = 0.375$ in. take the torque coefficient $K_j = 0.3$ and determine the preload torque.

9-04. A detachable preloaded threaded joint subjected to static external loading comprises two steel members having the thickness $l_m = 0.6$ in. each and a group of uniformly loaded SAE Grade 7 bolts having the major diameter $d = 0.625$ in. and the pitch $p = 0.09$ in. Determine the joint tightness factor and the safety factor of the bolts with regard to the equivalent normal stress if the preload and external forces applied to the joint per one bolt are $F_i = 5{,}500$ lb and $P = 2{,}500$ lb, respectively.

9-05. A detachable preloaded threaded joint subjected to static external loading comprises two steel members having the thickness $l_m = 0.1$ in. each and a group of uniformly loaded SAE Grade 7 bolts having the major diameter $d = 0.500$ in. and the pitch $p = 0.0679$ in. Determine the tightness factor of the joint and the safety factor with respect to the equivalent stress of the bolts if the preload and external forces applied to the joint per one bolt are $F_i = 3{,}200$ lb and $P = 1{,}200$ lb, respectively. Pay attention that $2l_m < 0.5\,d$.

9-06. Based on the specifications according to the SAE Grade 7 bolts, calculate the major diameter of the bolt for a detachable threaded joint subjected to an external static load $P = 3{,}500$ lb and having the relationship between the external and preload forces as follows: $P = 0.3F_i$. Take the external load coefficient $\chi = 0.25$ and the safety factor $n_\sigma = 2$.

9-07. Calculate the major diameter of the bolt for a detachable threaded joint subjected to external static load $P = 16{,}000$ N and having the relationship between the external and preload forces as follows: $P = 0.28\ F_i$. Take the external load coefficient $\chi = 0.8$, the safety factor $n_\sigma = 2$, and the yield strength $S_{yt} = 725$ MPa.

9-08. A preloaded threaded joint with a gasket is subjected to a variable external loading and includes a group of uniformly loaded SAE Grade 8 bolts having the major diameter $d = 0.5625$ in. The joint includes two steel members having the thickness $l_m = 0.7$ in. each and a soft gasket having the thickness $l_0 = 0.063$ in. and the modulus of elasticity $E_0 = 0.552 \cdot 10^6$ psi. Determine the joint tightness factor and the fatigue related normal stress safety factor of the bolts if each bolt is subjected to the preload force 2,400 lb and to the external force that fluctuates between zero and 1,500 lb.

9-09. A preloaded threaded joint subjected to variable external loading consists of a stud, steel plate, gasket, nut, and a rigid body as shown in Figure 9p-09. The thickness of the plate is $l_m = 32$ mm and the thickness of the gasket is $l_0 = 1.6$ mm. The modulus of elasticity of the gasket material is 3 GPa. The joint comprises four bolts that are uniformly loaded and have the major diameter $d = 16$ mm, the pitch $p = 2$ mm, the yield strength $S_{yt} = 897$ MPa, and the ultimate strength $S_{ut} = 1{,}035$ MPa. Determine the joint tightness factor and the fatigue related normal stress safety factor of the bolts if each bolt is subjected to the preload force $F_i = 11{,}000$ N and the external force fluctuates between zero and $P_{max} = 8{,}000$ N.

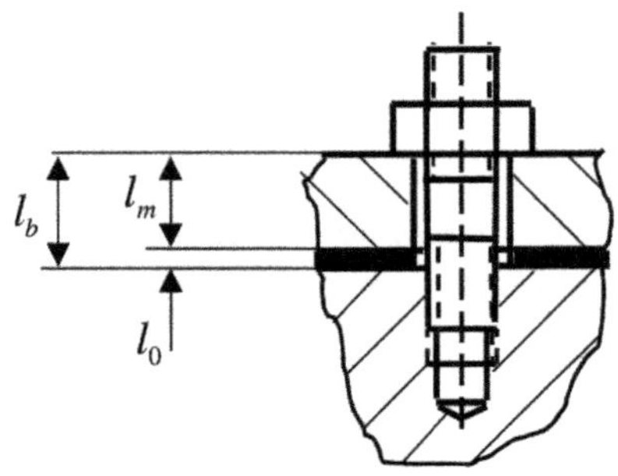

Figure 9p-09

9-10. A preloaded threaded joint with a gasket subjected to variable external loading comprises a group of uniformly loaded bolts having the major diameter $d = 12$ mm, the yield strength $S_{yt} = 897$ MPa, and the ultimate strength $S_{ut} = 1{,}035$ MPa. The joint includes two steel members having the thickness $l_m = 18$ mm each and a soft gasket having the thickness $l_0 = 1.8$ mm and the modulus of elasticity $E_0 = 2 \cdot 10^9$ GPa. Determine the joint tightness factor and the fatigue related normal stress safety factor of the bolts if each bolt is subjected to the preload force $F_i = 6{,}000$ N and the external force fluctuates between zero and 4,000 N.

CHAPTER 10

Riveted Joints

Permanent joints play a significant role in industry, and riveted joints represent one of the types of these joints. The main application of riveted joints is with connecting together sheet metal components. Welding successfully took over many industrial areas where permanent joints are employed; however, in certain industries such as aircraft building, shipbuilding, bridge building, and others riveted joints still dominate among permanent joint applications. Riveted joints in comparison with welded joints possess an essential advantage: their high durability against vibrating loading. This feature is extremely critical for the industries mentioned above. There are other advantages of riveting that justify their applications. For example, a weld may have a normal appearance, but the parent metal could be destroyed during the welding process, and it is complicated to examine the quality of the weld without special, costly instrumentation. The welding process is associated with localized melting of the parent material that leads to reduction of the content of carbon near the weld, resulting in lowering of the mechanical properties of the parent material in the area of the joint. Welding causes warping of the jointed components that very often requires applying special fabrication techniques in order to straighten the joint. In addition, it should be mentioned that it is complicated to weld color metals.

Riveted joints are free of these disadvantages, and in addition certain types of rivets create a permanent joint in cases when it is possible to access just one surface of the connecting elements.

10-1. Types of Rivets, Riveted Joints, and Riveting Processes

The types of rivets and their geometric parameters are standardized by the American National Standards Institute (ANSI), by ISO, and by the national standards of many other countries. There are two basic types of rivets: solid and tubular. The heads of rivets come in a variety of different shapes. The commonly used types of heads are shown in Figure 10-1.1.

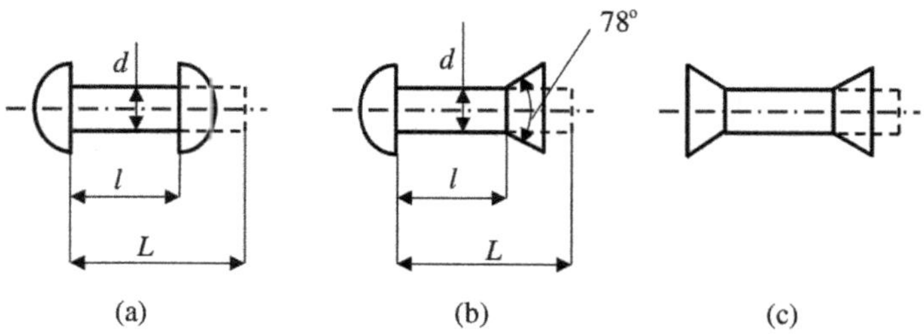

Figure 10-1.1. Types of rivets.

Parameters of rivets are presented in standards mentioned in Appendixes A-7-1 and A-7-2. These rivets come in shank diameters from 0.094 to 1.5 inches and in shank lengths from 0.5 to 5.875 inches. Figure 10-1.1a shows a round head rivet having an initial shank length L that forms a closing of a similar round head. The shank of the rivet has a diameter d, while the grip l equals the combined thickness of the jointed members. Figure 10-1.1b shows a round head rivet having the shank length L that forms a countersunk closing head that has the cone angle of 78°. The diameter of the shank of this rivet is d and the grip is l. The rivet shown in Figure 10-1.1c has countersunk heads on both ends. The rest of the parameters of this rivet are similar to the corresponding parameters of the rivets discussed above.

The head forming process for relatively small rivets, usually up to 0.3125 in. (8 mm), represents cold riveting. During cold riveting, the diameter of the shank expands and takes up all of the clearance between the shank and the hole. Larger diameter rivets are driven by the hot riveting process that requires preheating the rivet to about 1,800°F (1,000°C). During the hot forming process, the shank also expands; however, after cooling the shank shrinks in diameter and creates a clearance in the joint. It should be mentioned that during cooling the rivet imposes into the jointed members essential compression that develops friction forces preventing movement of the jointed members.

The general classification of riveted joints is associated with applications, including machine element joints, structural joints, and pressure vessel joints. In this book, we are focused on machine element and structural riveted joints.

There are two types of riveted joints: lap joints and butt joints. Figure 10-1.2a presents a lap joint comprising elements 1 and 6 that are connected by one row of rivets 2, 3, and 4 (the rivets 2 and 4 are presented by their center lines). The elements (members) are loaded by shear forces F. This joint is called single-row joint.

Lap joints may have several rows of rivets. The distance between two centers of the adjacent rivets as shown in Figure 10-1.2b is the pitch p of the joint.

In Figure 10-1.3 is shown a single-row butt joint comprising elements (members) 1 and 5, rivets 2 and 3, and cover plates (butt straps) 4 and 6.

The shear forces F are applied to the elements. In a butt joint, the shear of the rivet occurs simultaneously in two cross-sectional areas of the rivet; therefore, butt joints represent double shear joints. These joints may also comprise several rows.

Usually, rivets are made from low carbon soft steel, aluminum, and copper alloys.

10-2. Calculations of Riveted Joints

The verifying and design calculations of riveted joints are associated with determining the stresses and the strength of the rivets as well as the jointed members (plates). Very often in engineering practice, the design problems consist in determining the diameter of the rivet for the given thickness of the plates. For preliminary calculations, an empiric formula for determining the diameter of the rivet depending on the thickness of the plate may be recommended. Hence, for American Standard Rivets (in inches) the relationship between the diameter d and the thickness is $d = \alpha\sqrt{t}$,

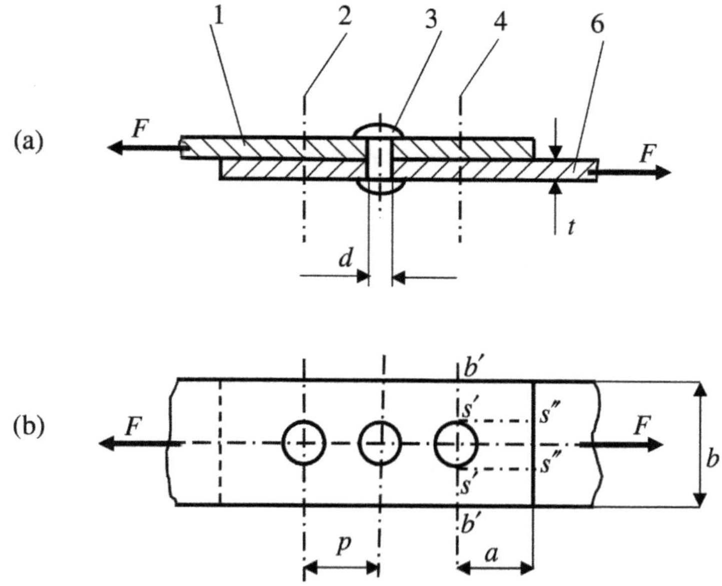

Figure 10-1.2. A lap joint.

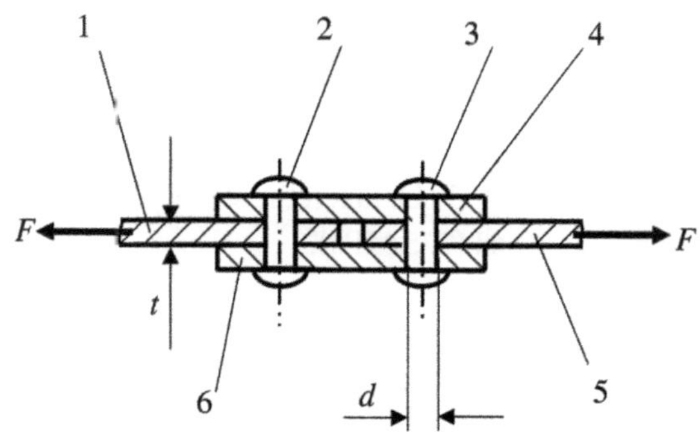

Figure 10-1.3. A butt joint.

where α is in the range from 1.2 to 1.4, while for ISO standard rivets (in millimeters) α is in the range from 6 to 7.

In actual joints, the load applied to the plates is not evenly distributed among the rivets in the joints. For instance, in a single-row joint the rivets on both ends of the row are more loaded than the rivets in the middle. However, it is accepted in stress calculations to consider that the load applied to the rivets in the joint is evenly distributed among the rivets.

The stress analysis of a joint includes calculations of shear and bearing stresses on the rivets, while for the plates the calculations are on tensile, shear, and bearing stresses.

10-2.1. Verifying Stress Calculations of Rivets

It is acceptable to consider that riveted joints are subjected to static loading. For a lap joint, the shear stress is calculated from the formula

$$\tau = \frac{F}{A_\tau N} \leq \frac{S_{sy}}{n_\tau} \tag{10-2.1}$$

where $A_\tau = \frac{\pi d^2}{4}$ is the cross-sectional area of the rivet, N is the number of rivets in the joint, S_{sy} is the shear yield strength, and n_τ is the safety factor with respect to shear stress. Considering that in a butt joint the failure of a rivet occurs when the rivet is sheared simultaneously in two cross sections, the formula for shear stress calculations in this case reads

$$\tau = \frac{F}{A_\tau N} \leq \frac{S_{sy}}{n_\tau} \tag{10-2.2}$$

The bearing stress for both types of joints is calculated from the following formula:

$$\sigma = \frac{F}{dtN} \leq \frac{S_{yt}}{n_\sigma} \tag{10-2.3}$$

where t is the thickness of the plate, and n_σ is the safety factor. It should be kept in mind that for lap joints if one of the plates is thinner, the thickness of the thinner plate should be taken for expression (10-2.3).

It is recommended for low carbon steel rivets to apply reduced values of allowable stresses. Based on these recommendations, it should be taken the shear yield strength S_{sy} = 15 ksi (105 MPa) and the tensile yield strength S_{yt} = 40 ksi (275 MPa).

10-2.2. Design Calculations of Rivets

Rearranging expression (10-2.1), we obtain the formula for determining the diameter of the rivet for a lap joint with respect to shear stresses in the rivet:

$$d = 2\sqrt{\frac{Fn_\tau}{\pi N S_{sy}}} \tag{10-2.4}$$

Similarly, based on expression (10-2.2), we derive the formula for design calculations for a butt joint rivet subjected to shear stresses:

$$d = 1.41\sqrt{\frac{Fn_\tau}{\pi N S_{sy}}} \tag{10-2.5}$$

From expression (10-2.3), we obtain the formula for determining the diameter of the rivet for lap and butt joints accounting for the bearing stresses:

$$d = \frac{Fn_\sigma}{tNS_{yt}} \tag{10-2.6}$$

Comparing the values of the diameters according to formulas (10-2.4) and (10-2.6) for the lap joint, we select the larger diameter for the rivet. Similarly, we compare the values according to formulas (10-2.5) and (10-2.6) and select the larger diameter of the rivet for the butt joint.

10-2.3 Calculating the Stresses in the Plates

The plates should be calculated for bearing, shear, and tensile stresses. The calculations of bearing stresses could be performed similarly to the rivets applying the following expression:

$$\sigma_{bp} = \frac{F}{dtN} \le \frac{S_{yt}}{n_\sigma} \tag{10-2.7}$$

where σ_{bp} is the bearing stress in plate.

For lap joints, the thickness of the thinner plate should be accounted for in this expression.

With lap joints and butt joints, the plates could be torn apart across the lines *s′b′* that are shown in Figure 10-1.2b. In this case, the tensile stress area $A_{ts} = t(b - Nd)$, where N is the number of rivets in the section *b′b′* (in Figure 10-1.2 $N = 1$). Therefore, the stress in the plate should be determined according to the following expression:

$$\sigma_{pl} = \frac{F}{t(b - Nd)} \le \frac{S_{yt}}{n_\sigma} \tag{10-2.8 }$$

where σ_{pl} is the tensile stress in the plate. For the expressions (10-2.7) and (10-2.8), the yield strength S_{yt} should be taken for the material of the plate.

The plate in the lap joint as well as in the butt joint could fail due to shear stresses along the lines *s′s″* that are shown in Figure 10-1.2b. In this case, the shear stress area $A_{ss} = 2t(a - 0.5d)$. Therefore, the shear stress in the plate τ_{pl} should not exceed the allowable stress of the material of the plate, and we can write

$$\tau_{pl} = \frac{F}{2(a - 0.5d)tN} \le \frac{S_{sy}}{n_\tau} \tag{10-2.9}$$

It should be emphasized that in this expression S_{sy} should be taken for the actual material of the plate, usually $S_{sy} = 0.577S_{yt}$.

10-3. Calculations of a Group of Rivets in a Joint Subjected to Eccentric Loading

In cases when the force applied to a joint comprising a group of rivets does not pass through the centroid of this group of rivets, the joint becomes subjected to the combined action of this force and a moment (torque) that is equal to the product of multiplying this force by the distance between the force and the centroid. This distance characterizes the measure of eccentricity of the applied load that represents the eccentric loading.

Figure 10-3.1 shows a lap joint comprising plates 1 and 2 that are connected by four round head rivets *A*, *B*, *C*, and *D*. Plate 1 is rigidly attached to a non-movable support. Force *F* is applied to the plate 2 on the distance *L* from the centroid *O*. For simplicity, the rivets are located in the corners of a square having the length of its side equal to $2a$.

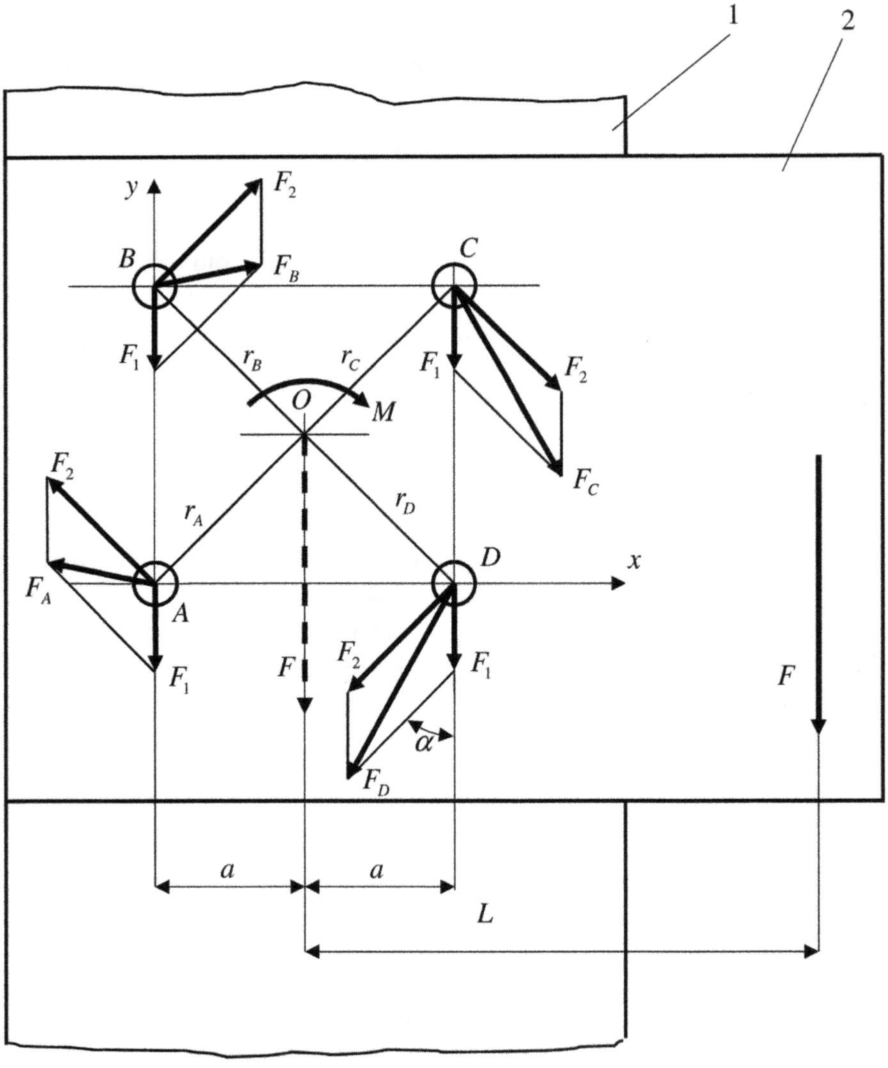

Figure 10-3.1. A riveted joint subjected to eccentric loading.

The distances between the rivets and the centroid are equal to each other, so we may write $r_A = r_3 = r_C = r_D = r$. It is permissible to transfer the force F to the point O (the dotted vector), while adding a corresponding moment M, as shown in Figure 10-3.1. The moment equals

$$M = FL \tag{10-3.1}$$

The force F is evenly distributed among these four rivets; therefore, each rivet is subjected to the same shear force F_1, as shown in Figure 10-3.1, and we may write

$$F_1 = \frac{F}{N} \tag{10-3.2}$$

where N is the number of rivets (in our case $N = 4$). The shear stress caused by the force F_1 can be calculated based on expression (10-2.1):

$$\tau_1 = \frac{F_1}{A_\tau} \tag{10-3.3}$$

However, each rivet is subjected to an additional shear force due to the applied moment. These forces are denoted in Figure 10-3.1 by F_2, and they are perpendicular to the corresponding radii representing the distances to the centroid and are directed in the same direction as the moment M. In our case, these forces are equal to each other, while in a general case these forces may differ from each other. The sum of moments of these forces about the centroid located in point O is equal to the applied moment M. Therefore, for a general case we may write

$$\sum_N F_i r_i = M \tag{10-3.4}$$

where F_i is the additional force (related to the moment) applied to the ith rivet having the ordinal number i, while r_i is the distance between this rivet and the centroid. In our case, we may write $F_2 r_A + F_2 r_B + F_2 r_C + F_2 r_D = FL$. Obviously, in a general case these forces and their distances may not be equal to each other, however, this does not change the methodology of determining the additional force applied to the rivet. The value of the additional force that the rivet receives from the moment is proportional to its distance to the centroid. Therefore, for a general case we may write

$$\frac{F_i}{r_i} = \frac{F_{i+1}}{r_{i+1}} = \ldots = \frac{F_N}{r_N} \tag{10-3.5}$$

For our case, we have $\frac{F_2}{r_A} = \frac{F_2}{r_B} = \frac{F_2}{r_C} = \frac{F_2}{r_D} = \frac{F_2}{r}$. Combining equations (10-3.4) and (10-3.5), we obtain

$$F_i = \frac{M r_i}{\sum_N r_i^2} \tag{10-3.6}$$

In this case, we may write $F_2 = \frac{Mr}{4r^2} = \frac{M}{4r}$. But for most cases, we must determine the additional (moment related) force for each rivet. The resultant shear force applied to each rivet represents a vector sum of the two forces—namely, the force related to the applied force (F_1 in our case) and the force related to the moment (F_2 in our case). As shown in Figure 10-3.1, the most loaded rivets in our case are the rivets indicated by C and D. But in general, the decision of which rivet is most loaded should be based on the calculations related to each rivet. (In some cases, it may not be obvious which rivet is loaded more than the others.) In addition, it should be emphasized that the coordinates of the centroid for most general cases should be determined using the following formulas:

$$X_O = \frac{\sum_N A_i x_i}{\sum_N A_i} \tag{10-3.7}$$

where A_i and x_i are respectively the cross-sectional area and the abscissa of the centroid of the ith rivet, and N is the number of rivets; and

$$Y_O = \frac{\sum_N A_i y_i}{\sum_N A_i} \tag{10-3.8}$$

where y_i is the ordinate of the ith rivet.

In the majority of actual particular riveted joints, the diameters in the group of rivets are the same. This allows us to simplify the formulas (10-3.7) and (10-3.8) to the following shape

$$X_O = \frac{\sum_N x_i}{N} \tag{10-3.9}$$

and

$$Y_O = \frac{\sum_N y_i}{N} \tag{10-3.10}$$

Applying these formulas to our case for the coordinate axes shown in Figure 10-3.1, we have $X_O = \frac{0+0+2a+2a}{4} = a$ and $Y_O = \frac{0+2a+0+2a}{4} = a$.

Now we consider the stresses in the rivets.

Taking for our case F = 4,000 lb, a = 4 in., L = 15 in., and t = 0.5 in., we perform the verifying calculations of the joint. For preliminary calculations, we take $d = 1.2\sqrt{t} = 1.2\sqrt{0.5} = 0.848$ in. The closest larger standard diameter of the rivet is d = 0.875 in.

$F_1 = \frac{F}{4} = \frac{4{,}000}{4} = 1{,}000$ lb. Now we determine the parameters that are needed for the calculations, hence, $r = \sqrt{2a^2} = \sqrt{2 \cdot 4^2} = 5.66$ in., $M = FL = 4{,}000 \cdot 15 = 60{,}000$ lb-in. and $F_2 = \frac{M}{4r} = \frac{60{,}000}{4 \cdot 5.66} = 2{,}650$ lb. According to Figure 10-3.1, we determine the resultant force applied to the rivet D, hence, $F_D = \sqrt{F_1^2 + F_2^2 - 2F_1F_2 \cos(180° - \alpha)}$. As it can be seen from this figure, in our case $\alpha = 45°$; therefore, $F_D = \sqrt{1{,}000^2 + 2{,}650^2 - 2 \cdot 1{,}000 \cdot 2{,}650 \cos(180° - 45°)} = 3{,}431$ lb. Now we determine the shear stress in the most loaded rivet $\tau = \frac{4 \cdot 3{,}431}{\pi 0.875^2} = 5{,}911$ psi. Taking the recommended value of S_{sy} = 15 ksi, we calculate the safety factor $n_\tau = \frac{15}{5.911} = 2.54$, which is acceptable. Now we determine the bearing stress in the rivet $\sigma = \frac{3{,}431}{0.5 \cdot 0.875} = 7{,}842$ psi. The recommended value of yield strength for the rivet is S_{yt} = 40 ksi; therefore, the safety factor with regard to bearing stresses is $n_\sigma = \frac{40}{7.842} = 5.1$, which is acceptable. It should be mentioned that the bearing stress in the plate is the same as in the rivet and there is no need to verify the stress in the plate since the plate usually is made of a stronger material than the rivet.

A general case of a riveted joint comprising a group of randomly located rivets subjected to the action of an eccentric inclined force is presented in the following example.

Example 10-3.1a: A riveted joint subjected to an eccentric inclined force.
A riveted lap joint presented in Figure 10-3.2 comprises plate 1 secured to a non-movable support, plate 2 jointed to plate 1 by the rivets A, B, and C, the coordinates of which are shown in the system of axes x, y.

Plate 2 is loaded in point G by a force F = 20 kN that is inclined by 30° to the y axis. The thickness of the plates is t = 15 mm.

It is required to select the rivets and perform verifying calculations applying the safety factors $n_\sigma = n_\tau = 2$.

Solution:
Using formulas (10-3.9) and (10-3.10), we determine the coordinates of the centroid O:

$$X_O = \frac{\sum_N x_i}{N} = \frac{40 + 110 + 60}{3} = 70 \text{ mm and } Y_O = \frac{\sum_N y_i}{N} = \frac{30 + 130 + 80}{3} = 80 \text{ mm.}$$

We resolve the force F into components
$F_x = F \sin 30° = 20 \sin 30° = 10$ kN and $F_y = F \cos 30° = 20 \cos 30° = 17.32$ kN.
Now we transfer the force F with its components F_x and F_y into the point O (the dotted vectors in the figure) and add the moments $M(F_x)$ and $M(F_y)$ that are caused, respectively, by the forces

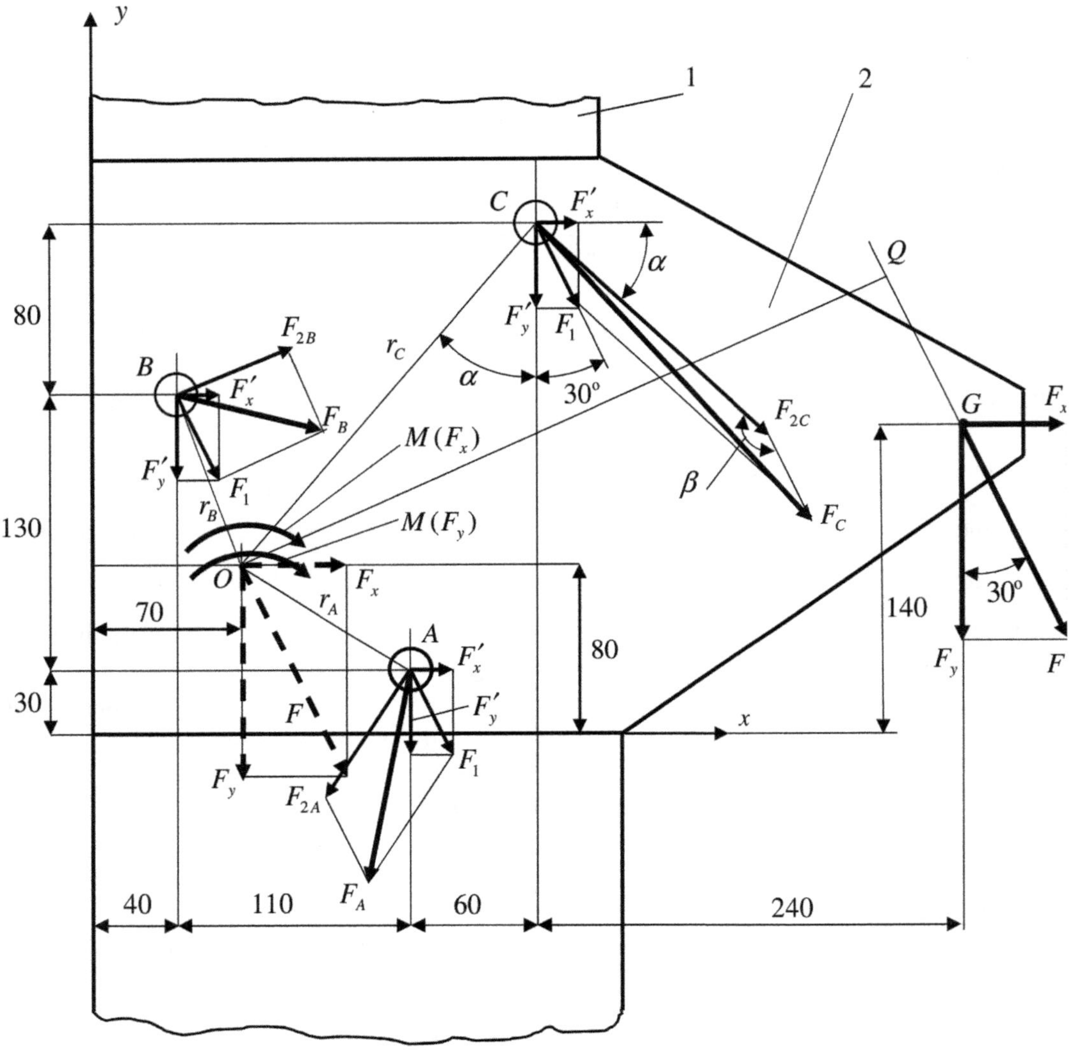

Figure 10-3.2. A riveted joint subjected to an inclined eccentric force.

F_x and F_y. Hence, $M(F_x) = 10 \cdot 10^3 (140 - 80) = 600 \cdot 10^3$ N · mm and
$M(F_y) = F_y(40 + 110 + 60 + 240 - 70) = 17.32 \cdot 10^3 \cdot 380 = 6{,}582 \cdot 10^3$ N · mm.
The total moment applied to the rivets is
$M = M(F_x) + M(F_y) = 600 \cdot 10^3 + 6{,}582 \cdot 10^3 = 7{,}182 \cdot 10^3$ N · mm.

It is obvious that we obtain the same result multiplying the force F by its distance OQ to the centroid; however, it is sometimes not so simple to determine this distance. Actually, for this case, this distance is $OQ = 359.1$ mm, and when we multiply it by $F = 20$ kN, we obtain $M = 7{,}182{,}000$ N · mm (obviously, the same result as received above).

Next, we determine the distances between the rivets and the centroid: $r_A = \sqrt{(150-70)^2 + (80-30)^2} = 94.34$ mm, $r_B = \sqrt{(70-40)^2 + (130+30-80)^2} = 85.44$ mm, and $r_C = \sqrt{(40+110+60-70)^2 + (80+130+80-80)^2} = 252.39$ mm. Using formula (10-3.6), we calculate the additional forces applied to each rivet. Thus, $F_{2A} = \frac{Mr_A}{r_A^2 + r_B^2 + r_C^2}$, hence, $F_{2A} = \frac{7{,}182 \cdot 10^3 r_A}{94.34^2 + 85.44^2 + 252.39^2} = 89.887 r_A = 89.887 \cdot 94.34 = 8{,}480$ N, $F_{2B} = 89.887 \cdot 85.44 = 7{,}680$ N, and $F_{2C} = 89.887 \cdot 252.39 = 22{,}687$ N. Now we analyze the forces applied to each rivet. Actually, the force F is evenly distributed between the rivets and the inclined force F_1 applied to each rivet represents the corresponding part of the force F, therefore, $F_1 = \frac{F}{N} = \frac{20 \cdot 10^3}{3} = 6{,}667$ N. In Figure 10-3.2, the force F_1 is resolved into components F'_x and F'_y that in some cases simplifies the calculations. Comparing the values of the forces applied to each rivet, it becomes obvious that the rivet C is by far the most loaded rivet. In order to calculate the value of F_C, we need to determine the angles α and β that are indicated in Figure 10-3.2 from which we have $\tan\alpha = \frac{40+110+60-70}{30+130+80-80} = \frac{140}{160} = 0.825$, therefore, $\alpha = 39.52°$. It is seen from this figure that $\beta = \frac{360° - 2[90° - (30° + 39.52°)}{2} = 159.52°$. Now we calculate the force applied to the rivet C: $F_C = \sqrt{F_1^2 + F_{2C}^2 - 2F_1F_{2C}\cos\beta}$, hence, $F_C = \sqrt{6{,}667^2 + 22{,}687^2 - 2 \cdot 6{,}667 \cdot 22{,}687 \cos 159.52°} = 29{,}026$ N. Based on the recommendation above, we select the diameter of the rivet $d = 7\sqrt{t} = 7\sqrt{15} = 27.11$ mm. We take $d = 28$ mm. Now we verify the stresses in the rivet $\tau = \frac{4 \cdot 29{,}026}{\pi 28^2} = 47.16\,\text{MPa}$. The recommended $S_{sy} = 105$ MPa; therefore, $n_\tau = \frac{105}{47.16} = 2.23$, which exceeds the required safety factor. Calculating the bearing stresses we have $\sigma = \frac{29{,}026}{28 \cdot 15} = 69.11$ MPa. The above recommended $S_{yt} = 275$ MPa; therefore, $n_\sigma = \frac{275}{69.11} = 3.98$, which also exceeds the requirements. The plates are made from stronger material and there is no need to verify their bearing stresses.

Chapter 10. Problems

10-01. Figure 10-1.2 shows a lap joint having three rivets in row ($N = 3$), while the diameter of the rivet $d = 0.625$ in. The thickness of the steel plates $t = 0.3125$ in., their width $b = 1.8$ in., and the distance from the axis of the last rivet to the end of the plate $a = 0.85$ in. The plates are loaded by a force $F = 7{,}000$ lb. Determine the stresses in the rivet and in the plate.

10-02. Figure 10-1.2 shows a lap joint having three rivets in row ($N = 3$), while the diameter of the rivet $d = 14$ mm. The thickness of the steel plates $t = 7.4$ mm, their width $b = 45$ mm, and the distance from the axis of the last rivet to the end of the plate $a = 22$ mm. The plates are loaded by a force $F = 28{,}000$ N. Determine the stresses in the rivet and in the plate.

10-03. Figure 10-1.2 shows a lap joint having three rivets in row ($N = 3$). The plates have the thickness $t = 0.28$ in. and are loaded by a force $F = 7{,}000$ lb. Taking the safety factor $n_t = n_\sigma = 2$, determine the diameter of the rivet.

10-04. Figure 10-1.2 shows a lap joint having three rivets in row $N = 3$ ($N = 3$). The plates have the thickness $t = 7.14$ mm and are loaded by a force $F = 36{,}000$ N. Taking the safety factor $n = 2$, determine the diameter of the rivet.

10-05. Figure 10-1.3 shows a butt joint having the diameter of rivets $d = 0.438$ in. All the plates have the thickness $t = 0.28$ in. The plates are loaded by a force $F = 2{,}000$ lb. Determine the stresses in the rivet.

10-06. Figure 10-1.3 shows a butt joint having the diameter of rivets $d = 10$ mm. All the plates have the thickness $t = 7.14$ mm. The plates are loaded by a force $F = 7{,}500$ lb. Determine the stresses in the rivet.

10-07. Figure 10-1.3 shows a butt joint having the plates of the thickness $t = 0.26$ in. The plates are loaded by a force $F = 4{,}500$ lb. Taking the safety factor $n_\tau = n_\sigma = 2$, determine the diameter of the rivet.

10-08. Figure 10-1.3 shows a butt joint having the plates of the thickness $t = 6.745$ mm. The plates are loaded by a force $F = 19{,}000$ N. Taking the safety factor $n_\tau = n_\sigma = 2$, determine the diameter of the rivet.

10-09. Figure 10p-09 presents a riveted lap joint comprising two plates 1 and 2 and three rivets located in the points A, B, and C. The joint is loaded by an eccentric force $F = 5{,}000$ lb applied on the distance $l = 3a$ from the point C. The distances between the rivets are shown in the figure, while $a = 3.0$ in. The diameter of the rivet $d = 0.875$ in. The thickness of the plates $t = 0.412$ in. Determine the stresses in the most loaded rivet.

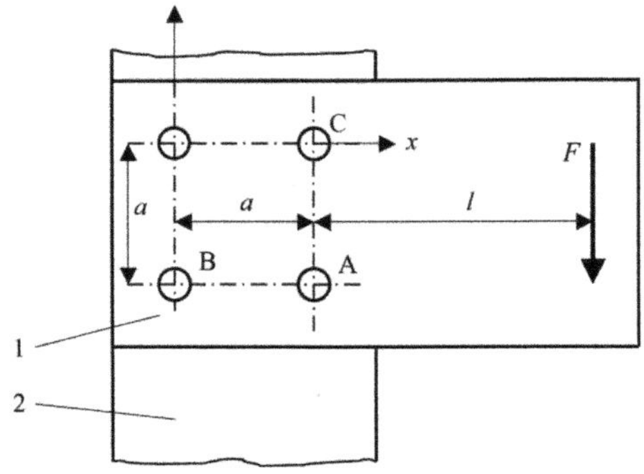

Figure 10p-09. Lap joint subjected to an eccentric load.

10-10. Figure 10p-10 presents a riveted lap joint comprising two plates 1 and 2 and three rivets located in the points A, B, and C. The joint is loaded by an eccentric force $F = 10{,}000$ N that is applied on the distance $l = 2.5a$ from the point C. The distances between the rivets are shown in the figure, while $a = 60$ mm. The diameter of the rivet $d = 20$ mm. The thickness of the plates $t = 9$ mm. Determine the stresses in the most loaded rivet.

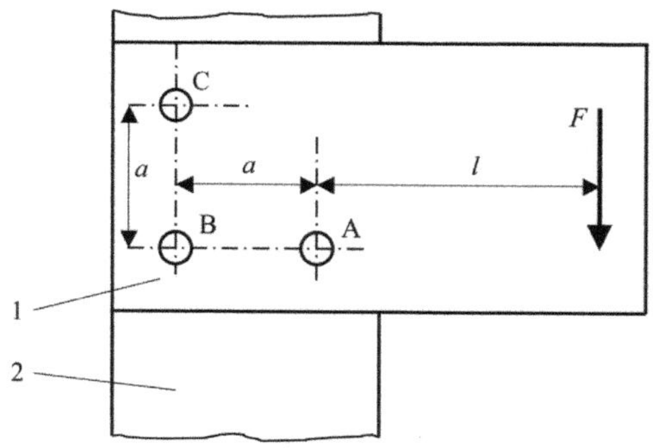

Figure 10p-10. Lap joint subjected to an eccentric load.

CHAPTER 11

Welded Joints

Along with riveted joints, welded joints belong to the category of permanent joints. Welds connect two or more plates or other elements, creating a joint. Actually, a weld by itself does not exist separately from the machine elements or structural components that it is holding together; however, the weld possesses parameters of a machine element that has real geometric characteristics and mechanical properties. There are several types of welds and all of them consist of metallic materials; they have dimensions, and they have different levels of strength. It is important to emphasize that welds connect together different elements in many different ways; they have a large number of applications in all kind of engineering systems.

It should be underlined that welding technology is not just another alternative to obtaining a permanent joint. Welding revolutionized the design and manufacturing processes by opening up endless opportunities to avoid costly machining operations and preventing the use of expensive metal cutting and metallurgical tooling. Welding allows the overlaying of worn out surfaces of components that become usable after inexpensive machining. Repairing cracks by welding costly components provides a second life to machinery and equipment.

The main disadvantages of welding consist in the difficulty to inspect the quality of the weld, the warping of the connected members, and the recrystallization of the structure of the parent material in the proximity of the weld that may result in reduced strength. However, it should be mentioned that in the majority of welded joints low carbon steels are used that are not highly sensitive to the reheating processes.

Welding represents an advanced field of manufacturing including numerous types of processes and equipment. Welding is an exciting and important engineering profession by itself.

The basic principle of welding consists in heating the connecting parts to the molten state, providing the conditions so that they can fuse together. The majority of welding processes utilizes a consumable electrode (filler) that builds up the weld. Thus, the molten pool consists of joining metal and an electrode. The source of heat could be a gas flame, an electric arc, or a laser beam. In some cases, the heating is achieved by a combination of high pressure and friction of the joining elements. The interaction of the atmosphere with the molten metal results in oxidation of the weld and absorbing some chemical elements that, as it is mentioned above, reduces to a certain degree the mechanical properties of the weld and of the parent material just near the weld.

The welding processes include all kind of shielding techniques to protect the weld from contact with the atmosphere. Different shielding techniques are used for different types of welding processes. In the industry, there are close to 100 different welding manufacturing processes. All welding processes can be subdivided in three categories: manual, semi-automatic, and automatic welding. Welded joints that could be considered joints for general applications are predominantly based on manual and semi-automatic welding processes that are characterized by relatively short welds, which usually do not exceed several feet, while in rare cases several yards. Automated welding processes produce long welds.

Shielded arc welding processes are used in the majority of industrial applications. The shielded metal arc welding (SMAW) process is based on applying a consumable coated metal electrode as the filler. During the welding process, the electrode coating vaporizes, producing a shielding gas that protects the molten metal. During cooling, a layer of slag forms on the weld, which is easily removed. Arc welding with the use of an electrode is often called stick welding.

The gas metal arc welding (GMAW) process is based on a manually controlled device (gun) that includes a nozzle that feeds an electrode that is a non-coated wire that comes from a wire coil. This nozzle is connected to a bottle with pressurized inert gas that flows to the wire for shielding the weld. This method is also called MIG (Metal Inert Gas) welding and it has a widespread application.

Gas tungsten arc welding (GTAW) process, also known as TIG (Tungsten Inert GAS), welding is based on applying a non-consumable tungsten electrode for managing the arc. The electrode is held in the torch that also supplies the inert gas to the weld. Usually, the operator manually feeds a wire electrode to the area of the weld.

It should be mentioned that there is one more type of welding: contact welding that is widely used for connecting sheet metal. Contact welding is performed on special welding machines, and the welding process is based on holding the connecting sheets between two non-consumable electrodes that instantaneously compress these two sheets together, while transmitting the electrical power that creates a melted spot in the sheets.

11-1. Types of Welded Joints

There are two types of welded joints: butt and fillet joints. These types of joints differ in their shapes and the corresponding weldments have different structures. It is important to mention that these types of joints are characterized by different methodologies of engineering calculations.

In Figure 11-1.1 are presented schematic drawings of some typical butt joints. The weldment consists of two metal plates 1 and 2 having the same thickness *t*. The plate edges that are facing each other could undergo machining to become beveled or could remain square (without any preliminary machining). Depending on the thickness of the plates, the edges should be preliminarily machined or kept square. In Figure 11-1.1a is shown a single V-joint that is usually applied to plates that have thickness up to 0.75 in. (20 mm).

The angle of the groove is usually from 60 to 70°. A double V-joint for the plates having a thickness of over 0.75 in. (20 mm) is shown in Figure 11-1.1b. The convex part of the weld causes an interruption of continuity of the shape of the weldment that results in essential stress

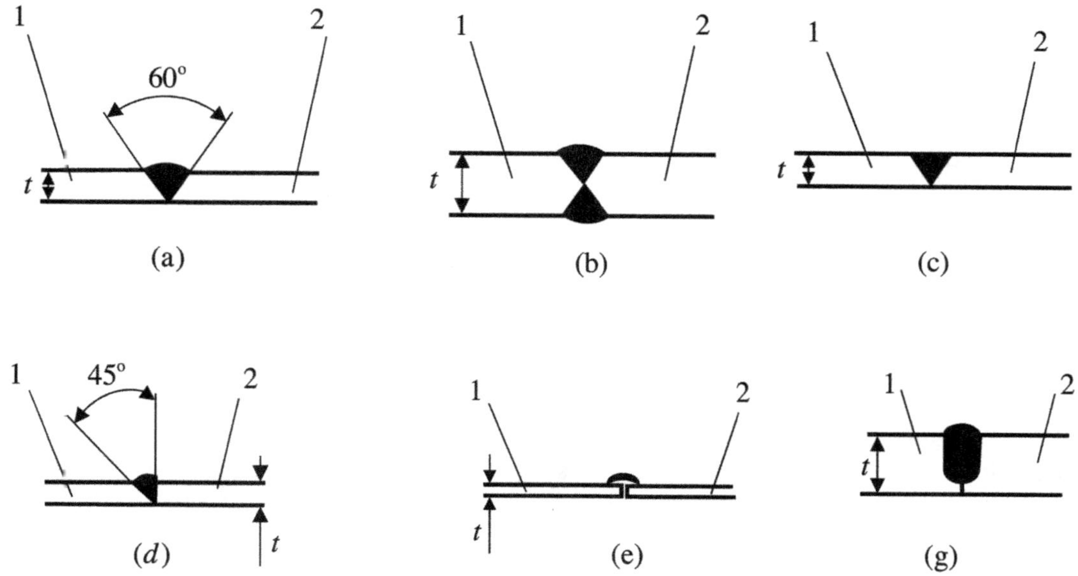

Figure 11-1.1. Butt joints.

concentration. In order to reduce the stress concentration, it is recommended to machine the convex part of the weld as shown in Figure 11-1-1c. A single-bevel joint having a groove angle of about 45° is presented in Figure 11-1.1d. These types of welds are usually recommended for plates having a thickness from 0.185 to 0.375 in. (from 4.5 to 10 mm). Square joints shown in Figure 11-1.1e are applied to plates having a thickness up to 0.25 in. (6 mm). Figure 11-1.1g shows a single U-joint that is applicable up to 0.75 in. (20 mm) of plate thickness. The double U-joint (not shown in the figure) is used for plates having a thickness of over 0.75 in. (20 mm). Single J-joints and double J-joints (not shown in the figure) are applicable for thicknesses from 0.5 to 0.75 in. (from 12.5 to 20 mm) and over 0.75 in. (20 mm), respectively.

Figure 11-1.2 shows schematic drawings of basic fillet joints.

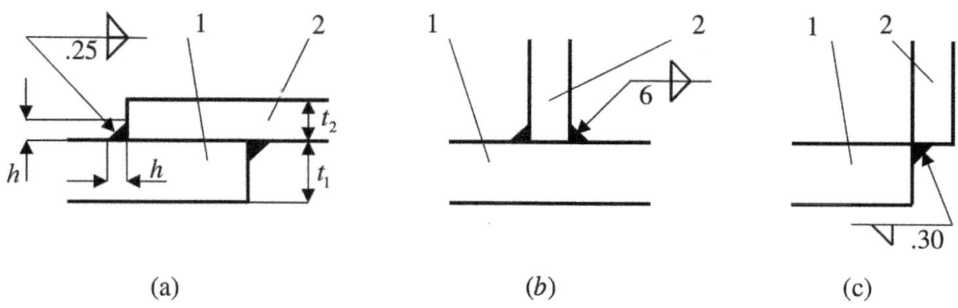

Figure 11-1.2. Fillet joints.

A fillet lap joint is presented in Figure 11-1.2a, where plates 1 and 2 of different thicknesses t_1 and t_2 are welded by two welds. A very important parameter of the fillet joints is the so-called length of the leg h of the weld. The leg of the weld is usually smaller than the thickness of the thinner plate. In general, the legs of different welds performed on different sides of the weldment may have different lengths. However, in the majority of actual cases, the legs of welds in the same weldment have equal lengths. It should be clear that the length of the weld's leg cannot be longer than the thickness of the thinner plate.

It is convenient to use the standard drawing symbols of welds to clarify the specifications of the welds. The symbols indicate the location of the weld, specify the length of the leg, and provide other required information. The two right triangles on the reference line of the arrow symbol in Figure 11-1.2a indicate that there is a weld on the side of the joint where the arrow is pointing, as well as on the other side of the joint, and the length of the legs is h = 0.25 in. A fillet T-joint is shown in Figure 11-1.2b, where the plates 1 and 2 of different thickness are welded together by two welds having legs h = 6 mm, as it is specified in the arrow symbol. In Figure 11-1.2c is shown a fillet corner joint connecting plates 1 and 2 of different thickness by one weld that according to the arrow symbol has the leg h = 0.30 in. Needless to say that the thickness of the plates may vary regardless of the length of the legs of the weld and the structure of the fillet weld.

11-2. Strength Calculations of Welded Joints Subjected to Static Loading

The methodologies of strength calculations of welded joints are different for butt joints and fillet joints. The main difference is that the strength calculations of butt joints are based on the mechanical properties of the welded (parent) materials, while the strength calculations of fillet joints are based on the mechanical properties of the electrodes that are used for the joint.

11-2.1. Strength Calculations of Butt Joints

Usually, significant stress concentrations appear in the vicinity of a welded joint that play a destructive role, specifically in cases of variable loading. In order to eliminate the stress concentration, it is required for the butt joint weldment to remove the convex part of the butt joint and face it with the surface of the plates that prevents the shape discontinuity of the plate. In this case, the two plates welded together by a butt joint are considered as a homogenous plate, and its strength is determined by the geometric parameters of the welded plates and the mechanical properties of the material that the plates are made from. It should be mentioned that this is justifiable because the strength of the electrodes essentially exceeds the strength of the welded (parent) metals. Therefore, the stress calculations of a butt joint are based on the principles that are applicable to any other component having the same shape.

Figure 11-2.1 shows a welded butt joint 1 loaded by a axial force F and a bending moment M.

It does not matter what kind of a welding process was used to manufacture the joint. The dimensions of the weldment are shown in the drawing. The stresses acting in this plate consist of

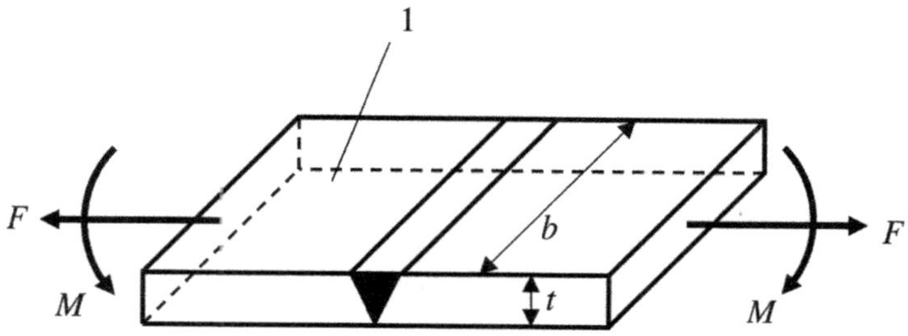

Figure 11-2.1. Butt joint subjected to a force and a moment.

the normal stress σ_F caused by the force F and the bending stress σ_M imposed by the moment M. These two stresses act in the same direction; therefore, the resultant stress σ is a superposition of these stresses. Hence, we write

$$\sigma = \sigma_F + \sigma_M \tag{11-2.1}$$

The normal stress σ_F based on formula (1-4.1) is determined as follows:

$$\sigma_F = \frac{F}{A} = \frac{F}{bt} \tag{11-2.2}$$

where A is the cross-sectional area of the plate.

Accounting for formula (1-4.22), we calculate the bending stress:

$$\sigma_M = \frac{Mt}{2I} = \frac{12Mt}{2bt^3} = \frac{6M}{bt^2} \tag{11-2.3}$$

where I is the moment of inertia of the cross-sectional area of the plate, and it is determined using formula (1-4.23). Combining formulas (11-2.1), (11-2.2), and (11-2.3), we obtain an expression for determining the stress in the welded joint that allows us to verify and evaluate the magnitude of the strength of the joint. Hence, verifying calculations for a butt joint are performed using the following expression:

$$\sigma = \frac{1}{bt}(F + \frac{6M}{t}) \leq \frac{S_{yt}}{n_\sigma} \tag{11-2.4}$$

where S_{yt} is the yield strength of the material of the plate and n_σ is the safety factor. It should be mentioned that applying to this expression appropriate transformations we derive formulas for corresponding design calculations of butt joints.

11-2.2. Strength Calculations of Fillet Joints

In the vast majority of welded fillet joints, the plates or other parts of the weldment are by far stronger than the corresponding welds. In general, a fillet joint could be destroyed by the failure of the weld. Therefore, unlike butt joints, the stress calculations of fillet joints are based on determining the stresses in the welds that consist of the material of the electrode. Obviously, these stresses should not exceed the allowable stresses for the material of the weld, or in other words, the electrode should possess the corresponding strength. It is extremely important to emphasize that while butt joints are typically subjected to normal stresses, fillet joints are typically subjected to shear stresses. Numerous investigations associated with the failure of fillet joints strongly suggest that the failure of the weld occurs in the plane of the bisect of the weld that is indicated in Figure 11-2.2 by the line *BB*.

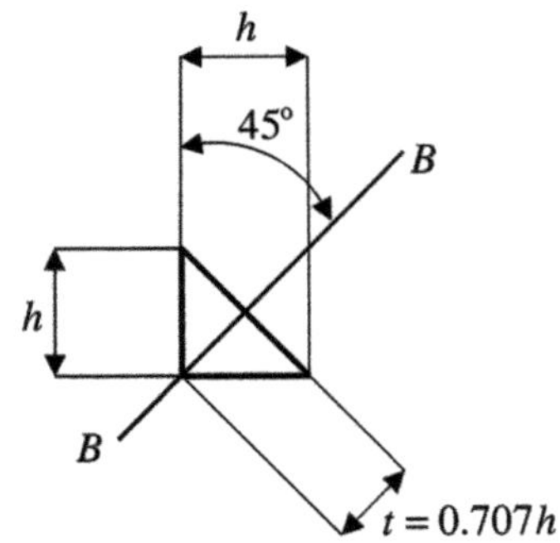

Figure 11-2.2. The weld throat.

This figure shows the cross-sectional area of a fillet weld that has two equal legs of length h and the weld throat $t = 0.707h$ located on the bisect and is the shortest distance from the vertex of the triangle to its basis. The throat area equals the product bt, where b is the length of the weld. The throat area is the smallest area of the weld related to the resistance of the weld to the applied loading.

To illustrate the concept of the weld throat with respect to the stress calculations of fillet joints, we consider the following example. In Figure 11-2.3 is shown a drawing of a fillet lap joint comprising plates 1 and 2 that are welded together by two welds of length b and having the legs of

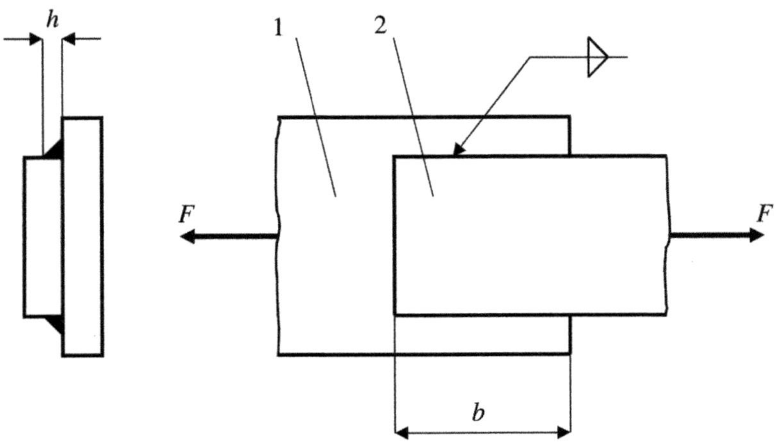

Figure 11-2.3. Fillet lap joint.

the welds equal to h. These plates are loaded by shear forces F. We need to determine the stresses in the welds.

Based on formula (1-4.5), we write $\tau = \frac{F}{2b0.707h} = 0.707\frac{F}{bh}$, therefore,

$$\tau = 0.707\frac{F}{bh} \le \frac{S_{sy}}{n_\tau} \tag{11-2.5}$$

where S_{sy} is the yield strength of the material of the electrode, b is the length of the joint on one side of the joint shown in Figure 11-2.3, and n_τ is the safety factor.

Applying to this expression some regular algebraic procedures, we derive appropriate equations for design calculations of similar fillet joints.

It is justifiable to consider a fillet joint subjected to the combined action of a shear force and bending moment as it is shown in Figure 11-2.4.

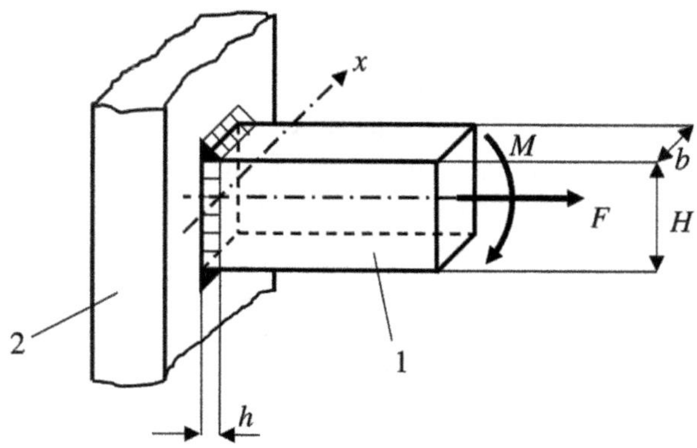

Figure 11-2.4. Fillet joint loaded by a shear force and a bending moment.

In this figure, plate 1 is welded to the non-movable plate 2 by a round fillet joint. Plate 1 is loaded by a shear force F and moment M. The shear stress in the joint represents a superposition of the shear stresses resulted from the force τ_F and from the moment τ_M, thus we may write

$$\tau = \tau_F + \tau_M \le \frac{S_{sy}}{n_\tau} \tag{11-2.6}$$

The shear stress caused by the force is calculated according to formula (1-4.5), while accounting for the parameters 0.707h, b, and H from Figure 11-2.4. Hence, we have

$$\tau_F = \frac{F}{2 \cdot 0.707\,h(b+H)} \tag{11-2.7}$$

In order to determine the shear stress τ_M, we need first to calculate the moment of inertia I of the welds about axis x. The following expression contains the corresponding sum of the moments of inertia of the welds $I = \dfrac{2 \cdot 0.707\, hH^3}{12} + \dfrac{2b(0.707\, h)^3}{12} + \dfrac{2 \cdot 0.707\, hbH^2}{4}$.

Analyzing this equation, it can be shown that for realistic values of the involved parameters, the second term in comparison with the first and third terms is ignorable. Therefore, it is justifiable to accept

$$I = 0.1178\, hH^3 + 0.3535\, hbH^2 \tag{11-2.8}$$

Therefore, the shear stresses caused by moment can be determined from the following formula:

$$\tau_M = \frac{MH}{2I} = \frac{M}{2(0.1178hH^2 + 0.3535\, hbH)} \tag{11-2.9}$$

Combining expressions (11-2.6), (11-2.7), and (11-2.9), we determine the shear stresses in the fillet joint subjected to the action of the shear force and the bending moment.

With regard to allowable stresses in the fillet joints, the American Welding Society (AWS) developed numerous standards related to the parameters of electrodes and other aspects of welding. According to the AWS numbering system, Class E electrodes for arc welding have a prefix E followed by four or five digits. The first two or three digits indicate the minimum ultimate tensile strength of the electrode. For instance, the electrodes E6010 and E10018 have S_{ut} = 60 ksi (415 MPa) and S_{ut} = 100 ksi (690 MPa), respectively. However, the electrode E6012 has S_{ut} in the range from 68 to 78 ksi. For calculations associated with static loading, it is necessary to know the yield strength of the electrodes. Based on the AWS specifications, for the electrode E60XX the yield strength S_{yt} is 50 ksi, while for the electrode E70XX that has S_{ut} = 70 ksi, the yield strength S_{yt} is 57 ksi. Appendix A-8-1 presents the data regarding a group of Class E electrodes. Comprehensive data related to mechanical properties of electrodes are available in numerous published sources.

Example 11-2.1a: Calculating the width of a welded butt joint subjected to static loading.

The butt joint presented in Figure 11-2.1 is loaded by the force F = 10,000 lb and the moment M = 3,000 lb-in. The thickness of the plates t = 0.5 in. The plates are made of AISI 1030 annealed steel having S_{ut} = 80 ksi and S_{yt} = 50 ksi. Taking the safety factor n_σ = 2, we determine the width b of the plates.

Solution:

Applying to expression (11-2.4) the corresponding algebraic procedures, we obtain the formula for calculating the width of a butt joint:

$$b = \frac{n_\sigma}{tS_{yt}}\left(F + \frac{6M}{t}\right) \tag{11-2.10}$$

By the way, solving formula (11-2.10) for the safety factor, we obtain

$$n_\sigma = \frac{bt^2 S_{yt}}{Ft + 6M} \tag{11-2.11}$$

Using formula (11-1-6), we obtain $b = \frac{2}{0.5 \cdot 50 \cdot 10^3}\left(10{,}000 + \frac{6 \cdot 3{,}000}{0.5}\right) = 3.68$ in.

Example 11-2.1b: Calculating the thickness of a welded butt joint subjected to static loading.

Now for the same butt joint shown in Figure 11-2.1 with the same loading factors mentioned above, we take the width of the plates $b = 4.0$ in. It is required to determine the thickness t of the plate.

Solution:

Performing corresponding transformations with expression (11-2.4), we have $\frac{F}{bt} + \frac{6M}{bt^2} = \frac{S_{yt}}{n_\sigma}$.

Simplifying this equation, we write $Ftn_\sigma + 6Mn_\sigma = bt^2 S_{yt}$ or $t^2 bs_{yt} - tFn_\sigma - 6Mn_\sigma = 0$. Solving this equation for t, we obtain $t = \frac{Fn_\sigma}{2bS_{yt}} \pm \sqrt{\left(\frac{Fn_\sigma}{2bS_{yt}}\right)^2 + \frac{6Mn_\sigma}{bS_{yt}}}$. Since $\frac{Fn_\sigma}{2bS_{yt}} < \sqrt{\left(\frac{Fn_\sigma}{2bS_{yt}}\right)^2 + \frac{6Mn_\sigma}{bS_{yt}}}$, the formula for determining t reads

$$t = \frac{Fn_\sigma}{2bS_{yt}} + \sqrt{\left(\frac{Fn_\sigma}{2bS_{yt}}\right)^2 + \frac{6Mn_\sigma}{bS_{yt}}} \tag{11-2.12}$$

Substituting the related data into formula (11-2.12), we calculate

$$t = \frac{10{,}000 \cdot 2}{2 \cdot 4 \cdot 50 \cdot 10^3} + \sqrt{\left(\frac{10{,}000 \cdot 2}{2 \cdot 4 \cdot 50 \cdot 10^3}\right)^2 + \frac{6 \cdot 3{,}000 \cdot 2}{4 \cdot 50 \cdot 10^3}} = 0.477 \text{ in.}$$

Example 11-2.1c: Calculating the width of the weld of a fillet lap joint subjected to static shear forces.

In the joint presented in Figure 11-1.5, the welds are built up by using the electrode E60XX having the $S_{yt} = 345$ MPa. The leg of the weld $h = 6$ mm, and the joint is loaded by a shear force $F = 40$ kN. Taking $n_\tau = 2.5$, we determine the width of the weld b.

Solution:

Based on expression (11-2.5), we derive the formula for calculating the length of the weld:

$$b = \frac{0.707\, Fn_\tau}{hS_{sy}} \tag{11-2.13}$$

Substituting into formula (11-2.13) that $S_{sy} = 0.577 S_{yt}$, we calculate

$$b = \frac{0.707 \cdot 40 \cdot 10^3 \cdot 2.5}{0.577 \cdot 6 \cdot 415} = 59.1 \text{ mm.}$$

11-2.3. Fillet Joint of a Round Bar Subjected to a Combined Loading

Figure 11-2.5 shows a welded fillet joint of a round bar 1 and a non-movable plate 2. The bar has a diameter d, while the length of the weld leg is h. The bar is subjected to the action of an axial force F_1, shear force F_2, moment M, and torque T.

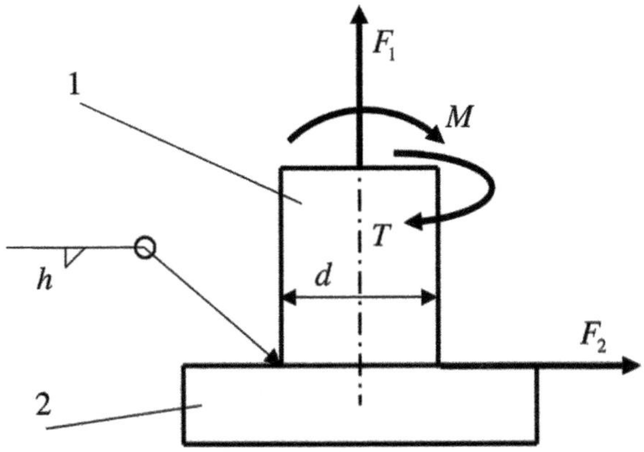

Figure 11-2.5. A fillet joint subjected to a combined loading.

We need to calculate the stresses acting in the weld. In order to compute the shear stresses due to the forces, we need to determine the cross-sectional area related to these forces. Accounting for the weld throat, we determine the cross-sectional area in the following way: $A = \pi\,(d + 0.707\,h)0.707h = \pi 0.707\,dh\left(1 + \frac{0.707\,h}{d}\right)$.

Since in the majority of real situations $\frac{0.707\,h}{d} = 1$, we may write that

$$A = \pi 0.707\,dh \tag{11-2.14}$$

Therefore, the shear stresses due to the applied forces are, respectively,

$$\tau_{F_1} = \frac{F_1}{\pi 0.707\,dh} \tag{11-2.15}$$

and

$$\tau_{F_2} = \frac{F_2}{\pi 0.707\,dh} \tag{11-2.16}$$

The shear stress resulted from the axial force F_1 is directed perpendicularly to the shear stress caused by the shear force F_2 and their resultant stress can be determined based on vector summation.

In order to calculate the torsion stress we use the formula presented in the course of Strength of Materials for thin walled hollow shafts. Replacing in this formula the thickness of the wall by the weld throat, we have $J = 0.25\pi(d + 0.707h)0.707h(d^2 + 0.707^2h^2)$. Applying to this expression similar procedures as for the equation of the cross-sectional area, we obtain

$$J = 0.177\pi d^3 h \tag{11-2.17}$$

The moment of inertia for calculating the shear stresses due to the moment M equals $I = 2J$, hence,

$$I = 0.354\pi d^3 h \tag{11-2.18}$$

The shear stresses caused by the torque and moment are

$$\tau_T = \frac{T0.5\,d}{0.177\pi d^3 h} = \frac{2.825\,T}{\pi d^2 h} \tag{11-2.19}$$

and

$$\tau_M = \frac{5.65\,M}{\pi d^2 h} \tag{11-2.20}$$

The directions of the stresses caused by the axial force and the moment are the same, and, therefore, these stresses can be added to each other. Hence, according to formulas (11-2.15) and (11-2.20), we have

$$\tau_{F_1} + \tau_M = \frac{F_1}{\pi 0.707 dh} + \frac{5.65M}{\pi d^2 h} \tag{11-2.21}$$

The stresses caused by the shear force and the torque are acting in the same direction and should be added according to the formulas (11-2.16) and (11-2.19):

$$\tau_{F_2} + \tau_T = \frac{F_2}{\pi 0.707 dh} + \frac{2.825T}{\pi d^2 h} \tag{11-2.22}$$

The resulting stress acting in the fillet weld represents a vector sum of the stresses according to formulas (11-2.21) and (11-2.22), hence, we write

$$\tau = \sqrt{\left(\frac{F_1}{\pi 0.707 dh} + \frac{5.65M}{\pi d^2 h}\right)^2 + \left(\frac{F_2}{\pi 0.707 dh} + \frac{2.825T}{\pi d^2 h}\right)^2} \tag{11-2.23}$$

11-3. Fatigue Calculations of Welded Joints

All principles related to fatigue calculations of machine elements discussed in Chapter 1 are applicable for calculations of welded joints subjected to variable loading. The fatigue stress concentration factor K_f depends on the type of the welded joint and varies from 1.0 to 4.0. For butt joints where the concave part of the joint is machined off, the fatigue stress concentration factor is $K_f = 1$. In general, for butt joints this coefficient may be taken in the range between 1.2 and 1.4. For fillet lap joints, the fatigue stress concentration factor varies from 1.5 to 3.0, while the higher values should be taken for welds that are parallel to the applied load (parallel welds).

It is important to realize that fatigue calculations of fillet joints subjected to the action of repeated and random cycles of tangential (shear and torsional) stresses are not considered in this book. The calculations of variable stresses acting in fillet joints are presented only for the completely reversed cycle because it is the most destructive loading cycle. Therefore, the variable stresses in the welded fillet joints subjected to the action of completely reversed shear stress cycles should be based on the fatigue shear stress endurance limit $S_{se} = 0.577S_e$ and on the maximum value of the loading factor. It is recommended to determine the safety factor for each particular case.

Example 11-3.1a: Welded butt joint subjected to variable loading.
In this case, contrary to Example 11-2.1a that deals with the calculations of a welded butt joint under static loading, we apply to the same joint shown in Figure 11-2.1 a completely reversed loading cycle. Our goal is to determine the length of the welds accounting for fatigue considerations and compare the results to those obtained for static loading.

The joint is loaded by the axial force having the absolute value $F = |10{,}000|$ lb and by the moment having the absolute value $M = |3{,}000|$ lb-in. The thickness of the plates $t = 0.5$ in. The plates are made from AISI 1030 annealed steel having $S_{ut} = 80$ ksi and $S_{yt} = 50$ ksi. The safety factor $n_\sigma = 2$.

Solution:
We start by determining the endurance limit of the plate. Using formulas (1-9.2) and (1-9.3), we calculate $k_a = 0.87$ and $k_b = 0.963$. For the axial normal stresses according to expression (1-9.4), the load factor $k_c = 0.8$, while for the normal stresses at bending this factor, according to the same expression, is $k_c = 1$.

The endurance limit associated with the axial force $S_{eF} = 0.87 \cdot 0.963 \cdot 0.8 \cdot 0.5 \cdot 80 = 26.8$ ksi, while with the bending moment $S_{eM} = 0.87 \cdot 0.963 \cdot 0.5 \cdot 80 = 33.51$ ksi. Since the convex part of the joint is removed, we take $K_f = 1$. Transforming the expression (11-2.10) by taking instead of yield strengths S_{yt} the corresponding endurance limits S_{eF} and S_{eM}, we obtain $b = \frac{n_\sigma}{t}\left(\frac{F}{S_{eF}} + \frac{6M}{tS_{eM}}\right) = \frac{2}{0.5}\left(\frac{10{,}000}{26{,}800} + \frac{6 \cdot 3{,}000}{0.5 \cdot 33{,}510}\right) = 5.788$ in. In case of static loading, the length of the welds is 3.68 in. Hence, in the case of a completely reversed loading cycle the length of the weld is 1.573 times longer as shown from $\frac{5.788}{3.68} = 1.573$.

Example 11-3.1b: Determining the width of a fillet lap joint subjected to a completely reversed loading cycle.
We take the same joint in Figure 11-2.3 as in example 11-2.1c that is associated with static loading and assume that the joint is subjected to a completely reversed loading cycle. The shear force

$F = |40|$ kN is cyclically changing its value from minimum to maximum. As in the previous example, the goal is to compare the width b of the welds for the cases of static and variable loadings. The rest of the input data is the same as in example 11-2.1c.

The joint uses the electrode E60XX having the $S_{ut} = 415$ MPa. The length of the leg of the weld $h = 6$ mm and the safety factor $n_\tau = 2.5$. We need to determine the width of the weld b.

Solution:

Based on expression (11-2.5), we calculate the maximum shear stress in the welds: $\tau_{max} = 0.707\frac{F}{bh} = 0.707\frac{40\cdot10^3}{6b} = \frac{4{,}713}{b}$ and $\tau_{min} = -\frac{4{,}713}{b}$.

It is important to note that the factors k_a and k_b are related to the weld, not to the plates (we do not calculate the plates). Applying formulas (1-9.2), (1-9.3) and (1-9.4), we determine $k_a = 0.93$, $k_b = 1$, and $k_c = 0.577$, respectively. Therefore, $S_e = 0.93\cdot1\cdot0.577\cdot0.5\cdot415 = 111.35$ MPa. Taking the fatigue stress concentration factor $K_f = 3$ and accounting for the value of τ_{max}, we may write $\frac{\tau_{max}n_\tau K_f}{b} \le S_e$. Hence, $\frac{4{,}713\cdot2.5\cdot3}{b} \le 111.35$ MPa and $b = \frac{4{,}713\cdot2.5\cdot3}{111.35} = 317.44$ mm. Comparing this result with its static counterpart, that is 49.2 mm, we obtain that for a completely reversed cycle the width of the weld should be 6.45 times longer than for the case of static loading: $\left(\frac{317.44}{49.2} = 6.45\right)$.

It should be emphasized that the fatigue calculations of welded joints presented above are based on the assumption that the stresses in the welds are distributed uniformly. However, as many authors report, the stresses on both ends of the welds parallel to the applied force are significantly higher than in the middle of the weld. This, of course, takes place in cases of static as well as in case of variable loading. Note: as we see in the example presented above, in cases of variable stresses the width of the weld is about 6.45 times longer than in cases of static stresses. From the point of view of uniformity of stresses along the weld, the difference puts the fatigue case at a disadvantage. Some sources recommend that the width of the parallel weld should be limited to $50h$. For our case, the maximum width of the weld would be $50\cdot6 = 300$ mm, which is slightly less than the calculated width (317.44 mm); however, it should be kept in mind that, where it is possible, during the design of welded fillet joints parallel welds should be reasonably short. For the considered case, it would be reasonable to apply the E120XX electrode having the $S_{ut} = 828$ MPa, that would reduce twice the width of the weld. In addition, it is advisable to apply higher safety factors for welded joints subjected to variable stresses.

11-4. Fillet Lap Joints Subjected to an Eccentric Force

In cases when a fillet lap joint is subjected to the action of an eccentric force, the welds become exposed to shear and torsional stresses. These stresses are components of the resultant tangential stress that is determined from the vector sum of the components. Therefore, the calculations of this kind of a weld comprise the procedures related to determining the values of these components and their resultant, and revealing the most loaded point or points of the welds. Figure 11-4.1 shows a fillet lap joint comprising plates 1 and 2, while plate 2 is loaded in the point G by an eccentric force F = 7,500 N. The weld is performed using an electrode E70XX having S_{yt} = 380 MPa. All geometric parameters related to the joint are presented in Figure 11-4.1.

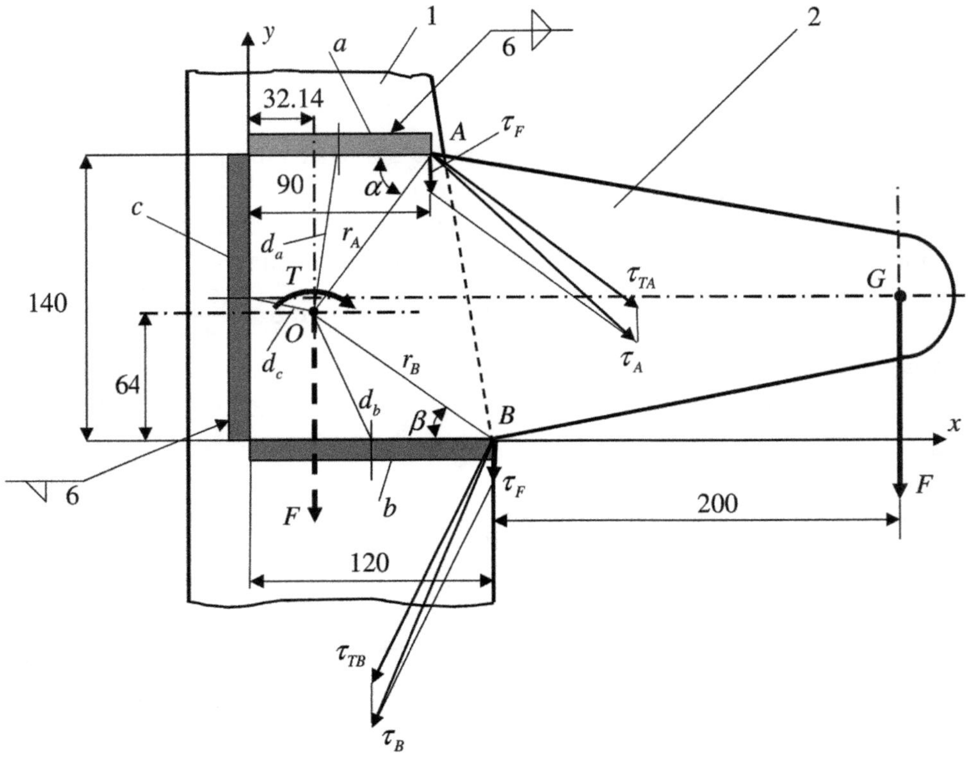

Figure 11-4.1. A fillet lap joint subjected to an eccentric force.

It is required to determine the safety factor n_τ.

First we determine the centroid of the welds by applying formulas (10-3.7) and (10-3.8) to the welds:

$$X_O = \frac{\sum_N A_i x_i}{\sum_N A_i} \tag{11-4.1}$$

where A_i and x_i are the throat area and the abscissa of the centroid of the ith weld, respectively, and N is the number of welds, and

$$Y_O = \frac{\sum_N A_i y_i}{\sum_N A_i} \tag{11-4.2}$$

where y_i is the ordinate of the centroid of the ith weld. In cases when all welds have the same throat, as in this case, these formulas can be simplified as follows:

$$X_O = \frac{\sum_N b_i x_i}{\sum_N b_i} \tag{11-4.3}$$

where b_i is the length of ith weld, and

$$Y_O = \frac{\sum_N b_i y_i}{\sum_N b_i} \tag{11-4.4}$$

Hence, using formulas (11-4.3) and (11-4.4), we calculate $X_0 = \frac{90 \cdot 45 + 120 \cdot 60}{90 + 140 + 120} = 32.14$ mm and $Y_0 = \frac{90 \cdot 140 + 140 \cdot 70}{90 + 140 + 120} = 64$ mm. Point O represents the centroid of the welds. We transfer the force F into the centroid O (the doted vector in the figure), while adding the torque $T = 17{,}500\ (200 + 120 - 32.14) = 5{,}037{,}550$ N·mm. Now we determine the shear stress caused by the force $\tau_F = \frac{F}{A} = \frac{17{,}500}{0.707 \cdot 6(90 + 140 + 120)} = 11.78$ MPa. The shear stress τ_F is acting in each point of the welds. The drawing in Figure 11-4.1 is made close to the scale that allows estimating the farthest distances of the points of the welds from the centroid. In these points, we obtain the highest torsional stresses. In our case, the farthest points are A and B where we attached the stresses τ_F for further analysis. In order to determine the torsional stresses, we have first to determine the total polar moment of inertia of all welds about the centroid. To avoid possible confusion in performing this stage of calculations, we suggest the following step-by-step sequence of calculations. We start by remembering that the polar moment of inertia J of an area around its centroid equals the sum of the axial moments (or moments) of inertia $J = I_x + I_y$. In this formula, I_x is the moment of inertia of this area about axis x that passes through the centroid of this area, and I_y is the moment of inertia of this area about axis y that passes through the centroid of this area. The next rule is that the polar moment of inertia J_O of an area in the plane of this area about a point O (that is in this case the centroid of the welds) in this plane equals $J_O = J + Ad_2$, where A is the value of the area and d is the distance between the centroid of the area and the point O. Now we proceed with the corresponding calculations of our case. Denoting the welds by a, b, and c, we begin the step-by-step calculations.

1) Determining the polar moments of inertia of each weld around its centroid. $J_a = I_{xa} + I_{ya} = \frac{b_a(0.707h)^3}{12} + \frac{0.707hb_a^3}{12} = \frac{0.707hb_a[b_a^2 + (0.7h)^2]}{12}$, where $b_a = 90$ mm and $h = 6$ mm. Hence, $J_a = 0.3535b_a\ (b_a^2 + 18) = 0.3535 \cdot 90\ (90^2 + 18) = 258{,}274$ mm^4. Similarly, $J_b = 0.3535 \cdot 120(120^2 + 18) = 611{,}612$ mm^4. In addition, finally $J_c = 0.3535 \cdot 140(140^2 + 18) = 970{,}895$ mm^4.

2) Determining the distances between the centroid O and the centroids of each weld. We ignore the thickness of the weld and take the distance from the centroid O to the middle of the length of the weld. This error is ignorable. Therefore, the distance d_a to the weld a is $d_a = \sqrt{(45-32.14)^2 + (140-64)^2} = 77.1$ mm; the distance d_b to the weld b is $d_b = \sqrt{(60-32.14)^2 + 64^2} = 69.8$ mm; and the distance d_c to the weld c is $d_c = \sqrt{(70-64)^2 + 32.14^2} = 32.7$ mm.
3) Determining the polar moments of inertia of each weld about the centroid O. For the weld d we have $J_{Oa} = J_a + A_a d^2{}_a = 258{,}274 + 90 \cdot 0.707 \cdot 6 \cdot 77.1^2 = 2{,}527{,}731$ mm^4; similarly for the weld b we obtain $J_{Ob} = 611{,}612 + 120 \cdot 0.707 \cdot 6 \cdot 69.8^2 = 3{,}091{,}675$ mm^4; and for the weld c we have $J_{Oc} = 970{,}895 + 140 \cdot 0.707 \cdot 6 \cdot 32.7^2 = 1{,}605{,}925$ mm^4.
4) Calculating the total polar moment of inertia of the weld group about its centroid O.

The total polar moment of inertia equals
$J_O = J_{Oa} + J_{Ob} + J_{Oc} = 2{,}527{,}731 + 3{,}091{,}675 + 1{,}605{,}925 = 7{,}225{,}331$ mm^4.

Now we continue the solution of the problem. It can be seen from the Figure 11-4.1 that the points A and B are the most loaded points because they are most distant from the centroid. In addition, the angles between the directions of the pair of stresses τ_F and τ_{TA} as well as of the pair of stresses τ_F and τ_{TB} are acute angles. In order to calculate the components of the torsional stresses in the points A and B, we need to determine the values of the distances r_A and r_B. Hence, $r_A = \sqrt{(90-32.14)^2 + (140-64)^2} = 95.52$ mm, and $r_B = \sqrt{(120-32.14)^2 + 64^2} = 108.7$ mm.

The torsional stress in point A is $\tau_{TA} = \dfrac{Tr_a}{J_O} = \dfrac{5{,}037{,}550 \cdot 95.52}{7{,}225{,}331} = 66.6$ MPa and in point B the torsional stress is $\tau_{TB} = \dfrac{Tr_b}{J_O} = \dfrac{5{,}037{,}550 \cdot 108.7}{7{,}225{,}331} = 75.8$ MPa. It can be seen from Figure 11-4.1 that $\tan\alpha = \dfrac{140-64}{90-32.14} = 1.31$; therefore, α = 52.64°. This figure shows that the angle in point A between the stresses τ_F and τ_{TA} is also α. Therefore, applying the law of cosines we calculate the resultant shear stress τ_A in point A: $\tau_A = \sqrt{\tau_{TA}^2 + \tau_F^2 - 2\tau_{TA}\,\tau_F \cos(180° - \alpha)}$. Hence, $\tau_A = \sqrt{66.6^2 + 11.78^2 - 2 \cdot 66.6 \cdot 11.78 \cos(180° - 52.64°)} = 74.34$ MPa. From Figure 11-4.1, we determine that $\tan\beta = \dfrac{64}{120-32.14} = 0.728$; therefore, β = 36.07°. This figure also shows that the angle in point B between the stresses τ_{TB} and τ_F also equals β. Hence, applying the law of cosines we calculate the resultant shear stress τ_B acting in point B: $\tau_B = \sqrt{\tau_{TB}^2 + \tau_F^2 - 2\tau_{TB}\tau_F \cos(180° - \beta)}$. Therefore, substituting the corresponding data we obtain $\tau_B = \sqrt{75.8^2 + 11.78^2 - 2 \cdot 75.8 \cdot 11.78 \cos(180° - 36.07°)} = 85.6$ MPa. Since $\tau_B > \tau_A$ (as expected), the point B is the most loaded. Taking $S_{sy} = 0.577\ S_{yt}$, we calculate $S_{sy} = 0.577 \cdot 380 = 219.26$ MPa. Therefore, $n_\tau = \dfrac{S_{sy}}{\tau_B}$, hence $n_\tau = \dfrac{219.26}{85.6} = 2.56$, which is acceptable.

Chapter 11. Problems

11-01. Determine the stresses in the welded joint shown in Figure 11p-01 if $F_1 = 2F$, $F_2 = 4F$, $M = 6Fl$, $t = 0.05l$, and the width of the weld $b = 0.4l$ (b is not shown in the figure).

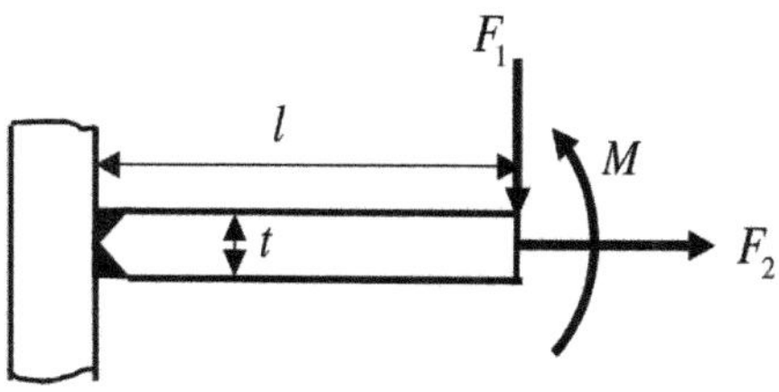

Figure 11p-01

11-02 Taking the safety factor $n_\sigma = 2$, determine the width b of the welded joint shown in Figure 11p-02 (b is not shown in the figure). The joint is subjected to a completely reversed loading cycle. The maximum absolute values of the applied forces are $F_1 = F$, $F_2 = 2F$, $F_3 = 3F$, and $F_4 = 4F$. The thickness of the plate $t = 0.25a$. The plates are made from AISI 1015 annealed steel having $S_{yt} = 45{,}500$ psi and $S_{ut} = 61{,}000$ psi. Take $F = 200$ lb and $a = 4$ in. Note: You will need to modify formula (11-2.10) as it is presented in example 11-3.1a.

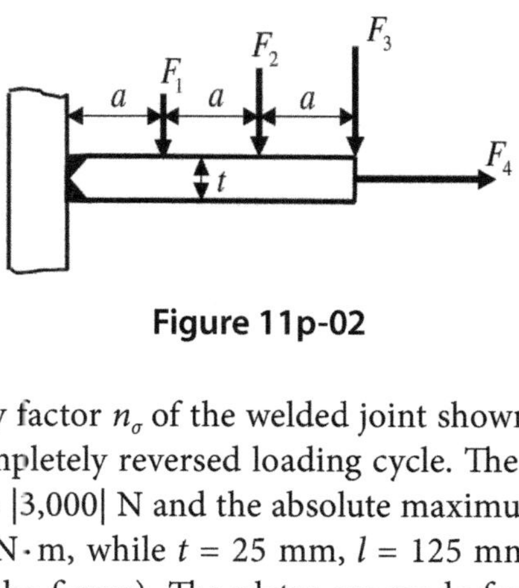

Figure 11p-02

11-03. Determine the safety factor n_σ of the welded joint shown in Figure 11p-03. The joint is subjected to a completely reversed loading cycle. The absolute maximum value of the applied force $F = |3{,}000|$ N and the absolute maximum value of the of the applied moment $M = |400|$ N·m, while $t = 25$ mm, $l = 125$ mm, and the width $b = 90$ mm (b is not shown in the figure). The plates are made from AISI 1015 annealed steel having $S_{yt} = 314$ MPa and $S_{ut} = 421$ MPa.

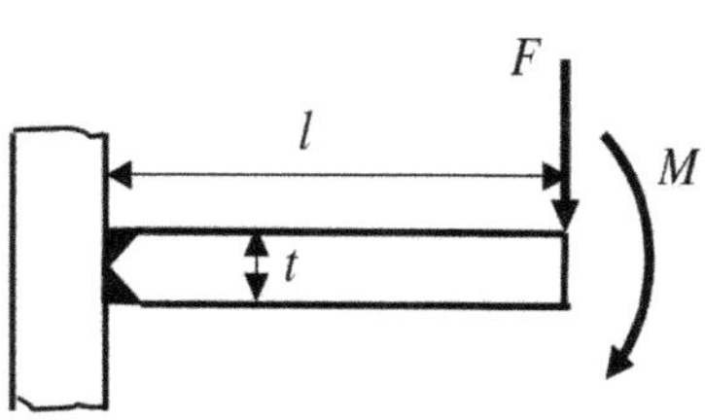

Figure 11p-03

11-04. Determine the stresses in the joint shown in Figure 11p-04. Plate 2 is welded to the non-movable plate 1.

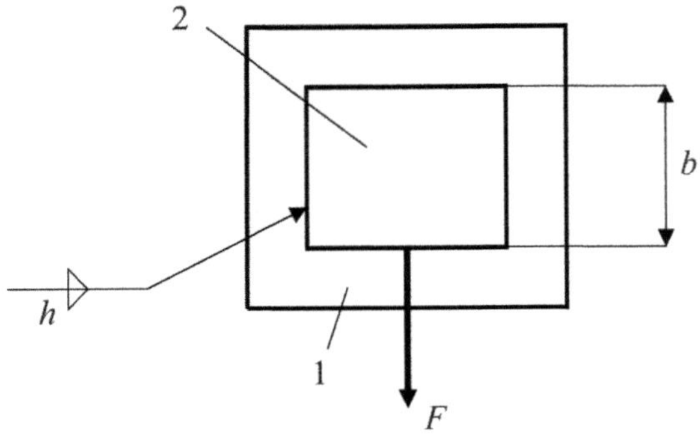

Figure 11p-04

11-05. Figure 11p-05 shows a pipe welded to a non-movable plate. Determine the stresses in the joint.

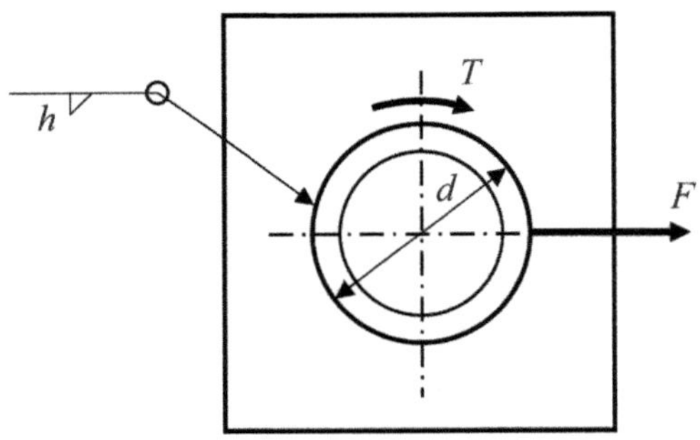

Figure 11p-05

11-06. Figure 11p-06 shows a round bar having the diameter $d = 2.0$ in. The bar is welded to the non-movable plate using the electrode E70XX having the yield strength $S_{yt} = 55$ ksi. The applied forces $F_1 = 8{,}200$ lb and $F_2 = 10{,}500$ lb. The length of the leg of the weld $h = 0.3$ in. Determine the safety factor n_τ.

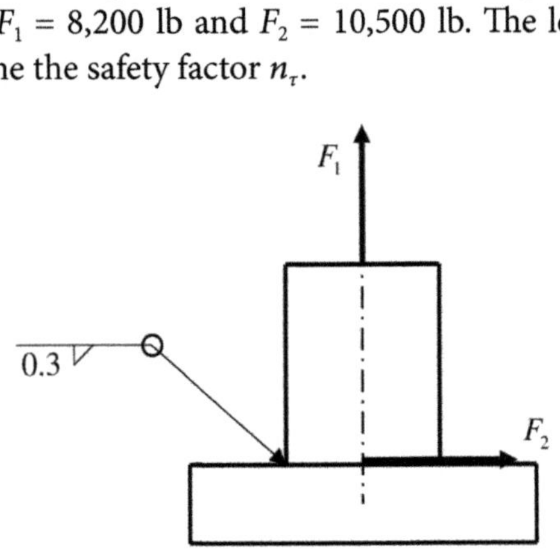

Figure 11p-06

11-07. Figure 11p-07 shows a round bar having the diameter d = 50 mm. The bar is welded to the non-movable plate using the electrode E70XX. The joint is subjected to a completely reversed cycle that is caused by a torque having its maximum absolute value $T_{max} = |225|$ N·m. The length of the leg of the weld h = 6 mm. Take K_f = 3.5 and determine the safety factor n_t.

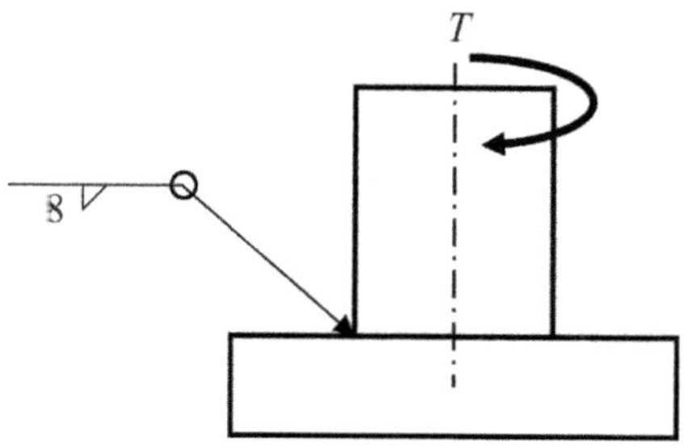

Figure 11p-07

11-08. Determine the safety factor n_τ of the welded joint shown in Figure 11p-08. The joint is subjected to a completely reversed cycle caused by the alternating moment the maximum absolute value of which $M = |2{,}000|$ lb-in., while b = 3.0 in., H = 5.0 in., and h = 0.3 in. The welding is performed using the electrode E60XX having S_{ut} = 60 ksi. Take K_f = 4.

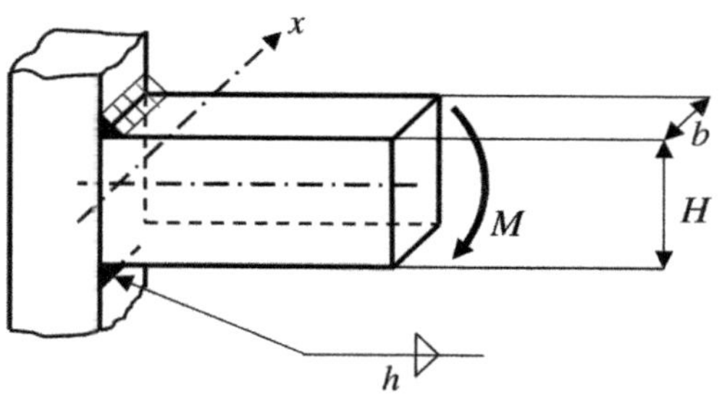

Figure 11p-08

11-09. Determine the safety factor n_τ of the welded joint shown in Figure 11p-09. The plates 1 and 2 are welded using electrode E60XX. The force is causing a completely reversed cycle having it maximum absolute value $F = |3{,}000|$ lb, while b = 4.0 in., h = 0,25 in., and K_f = 3.

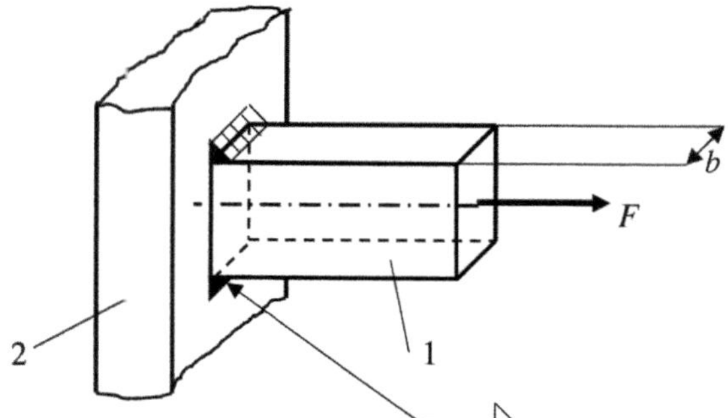

Figure 11p-09

11-10. Figure 11p-10 shows a weldment consisting of plates 1 and 2 joined by using an electrode E80XX. The joint is subjected to a completely reversed cycle due to the action of an alternating force the maximum absolute value of which $F = |2{,}000|$ N and an alternating moment the maximum absolute value of which $M = |85|$ N·m. Determine the safety factor n_τ if $H = 50$ mm, $b = 25$ mm, $h = 5$ mm, and $K_f = 3$.

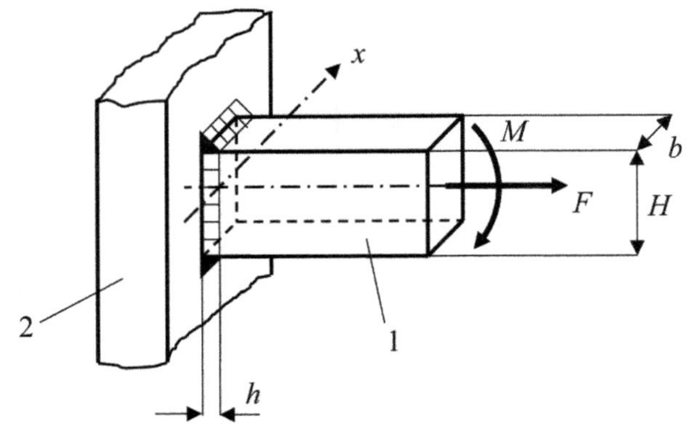

Figure 11p-10

CHAPTER 12

Springs

The functionality of numerous mechanisms and machines is often associated with the action of elastic links, or springs. Springs are used in a tremendous variety of industrial applications. Elastic links, or springs, enable the reciprocating movement of some machine elements, such as latches, valves, pistons, and similar components. Springs are broadly used as shock absorbers and vibration suppressors. In many industrial applications, elastic links store potential energy. Springs are widely used in the watch industry for balance wheels and winding mechanisms. Spring applications are almost unlimited and nearly any device, mechanism, and machine includes springs.

There are numerous industries in many countries that specialize in manufacturing a wide variety of types of springs.

12-1. Types of Springs

Springs can be divided into different types. One of the most widespread type is the helical compression coil springs that may have one of a variety of end structures. Figure 12-1.1 shows different end structures for helical compression coil springs.

These springs may have different types of ends: closed (squared) but not ground, closed (squared) and ground, open (plain) but not ground, and open (plain) and ground (not shown in Figure 12-1.1).

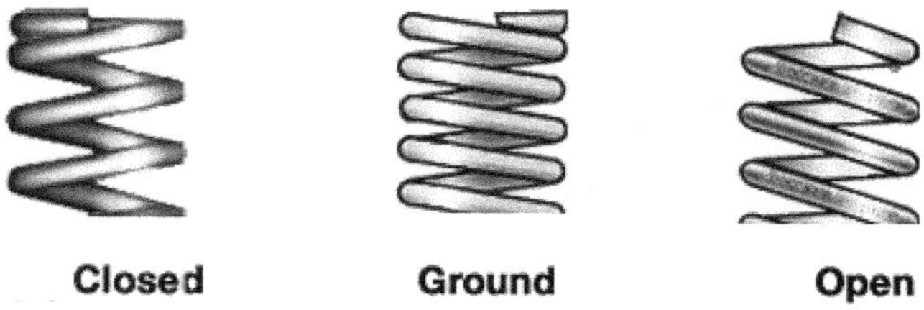

Figure 12-1.1. Typical end structures for helical compression coil springs. (Courtesy of Acxess Spring Company)

Figure 12-1.2 shows another type of coil spring: the helical extension coil spring.

Comparing compression and extension springs shows that in the unloaded state the coils in a compression spring do not touch each other, while the coils in an extension spring are in contact with one another. This is a result of the appropriate tension that is applied to the spring

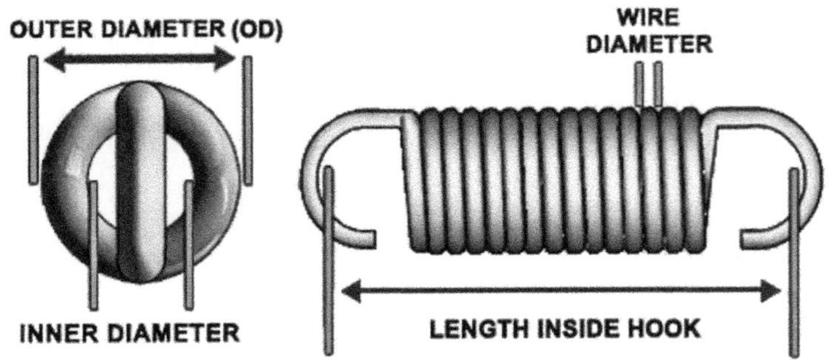

Figure 12-1.2. Extension helical coil spring. (Courtesy of Axcess Spring Company)

wire during the manufacturing process. This type of extension spring is called a closed extension spring. Figure 12-1.3 shows the different types of hooks of extension springs. The differences are self-explanatory.

Figure 12-1.4 shows a compression conical coil spring and its parameters.

Figure 12-1.5 shows torsion springs and their parameters.

The types of springs shown in the figures above represent the springs most used as machine elements for general applications. There are other types of springs that are used for specific applications, such as leaf springs, Belleville springs, and others.

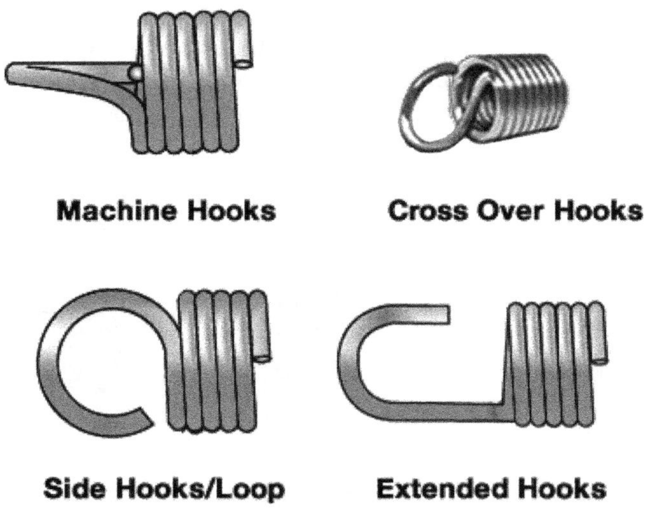

Figure 12-1.3. Types of hooks of extension coil springs. (Courtesy of Acxess Spring Company)

12-2. Parameters and Basic Calculations of Helical Springs

The basic parameters of helical springs and corresponding engineering calculations of springs are discussed below.

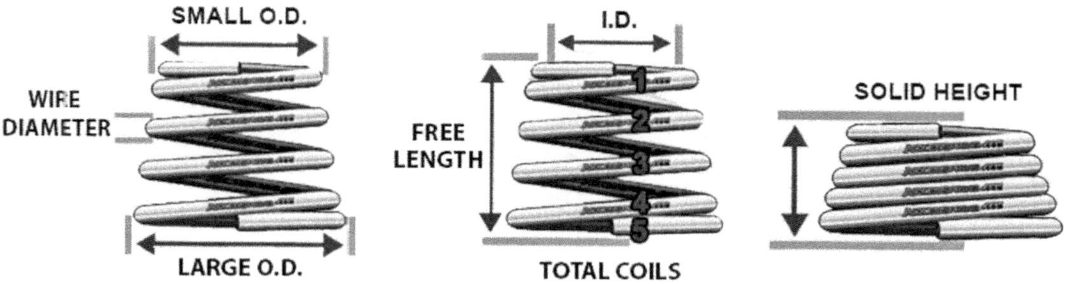

Figure 12-1.4. Conical coil compression spring. (Courtesy of Acxess Spring Company)

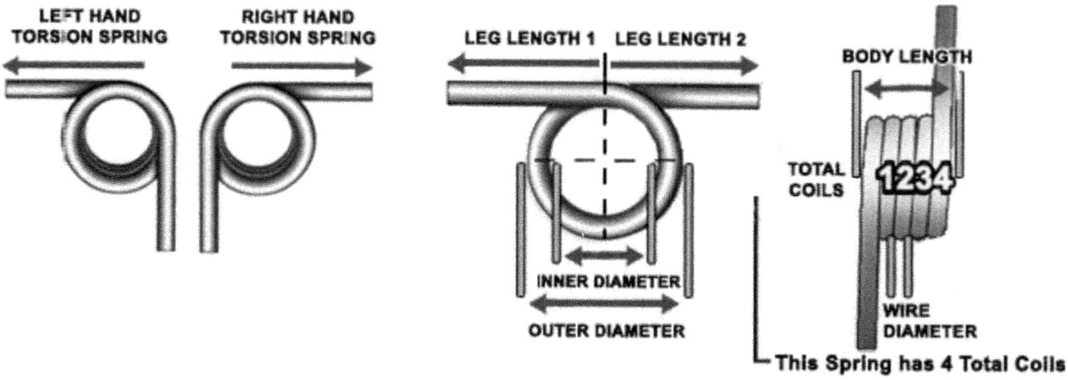

Figure 12-1.5. Torsion springs. (Coutesy of Acxess Spring Company)

12-2.1. Parameters of Helical Springs

The structure of compression helical springs is associated with the configuration of the spring ends. These configurations define the number of coils in the springs. There are two basic types of end configurations for compression helical springs: open and closed (squared). The open and closed ends could be ground or not ground. These features define the number of coils in the spring. With regard to functinality, the coils could be active or inactive. The active coils, unlike inactive coils, are deformed during the operation of the spring. Actually, the inactive coils form the ends of the springs. The total number of coils N_{tot} includes the number of active coils N as well as the number of the inactive coils.

Figure 12-2.1 shows a helical comression spring in the unloaded (free) state with open not ground ends.

In this spring all coils are active; therefore, $N = N_{tot}$. The number of active coils can be determined based on the operational requirements of the spring. However, in order to prevent a collision between the coils it is usually acceptable to increase the calculated number by a couple of additional active coils.

Figure 12-2.2 shows a compression helical spring in the unloaded (free) state with closed not ground ends. (As noted, this figure and figure 12-2.1 have similar parameters.)

Type below the caption of Figure 12-2.1 the following:

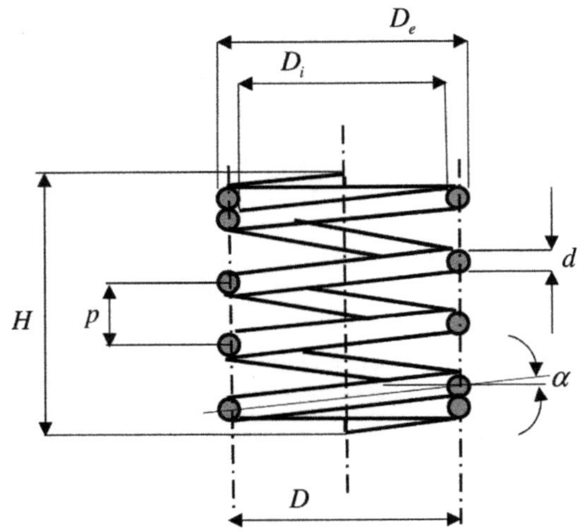

Figure 12-2.1. A compression helical spring with open not ground ends.

For the spring shown in Figure 12-2.2 and having closed not ground ends the total number of coils equals $N_{tot} = N + 2$, where N is the calculated number of active coils. Obviously, this number does not include the mentioned above additional number of active coils that is needed to prevent the collision of the coils.

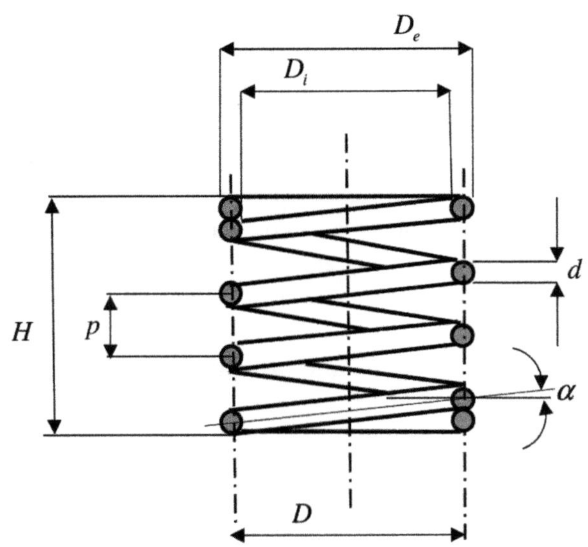

Figure 12-2.2. A compression helical spring with closed not ground ends.

The coil spring is characterized by its mean diameter D, external diameter D_e, internal diameter D_i, diameter of the wire d, coil pitch p, and the helix angle α. During the deformation of the spring, the values of these parameters, shown in the unloaded state of the springs in Figures 12-2.1 and 12-2.2, change with exception of the diameter of the wire d. The relationships between these parameters for the unloaded state of the spring can be determined from these figures. The coil pitch p and the helix angle α are interrelated according to the following expression:

$$p = \pi D \tan\alpha \tag{12-2.1}$$

The free height (free length) H for springs with closed (squared) not ground ends is calculated from the following formula:

$$H = pN + 3d \tag{12-2.2}$$

where N is the number of active coils. The number of total coils N_{tot} for these springs is determined from the formula $N_{tot} = \frac{H-3d}{p} + 2$. The solid height (the spring is completely compressed) H_{sh} for these springs is computed from the formula $H_{sh} = d\,(N_{tot} + 1)$. Because there are several options for the configurations of the ends of compression helical springs, the corresponding parameters of these springs are calculated from formulas that account for the variety of these ends. These formulas are presented in Appendix A-9-1.

In general, accounting for just the active coils (not including the coils forming the ends), the length of the wire of a coil spring is

$$L = \frac{\pi DN}{\cos\alpha} \tag{12-2.3}$$

An important parameter of springs is the spring index C that is the ratio between the mean diameter and the diameter of the wire:

$$C = \frac{D}{d} \tag{12-2.4}$$

The spring index is related to the characteristics of the coil curvature. Usually, the spring index is taken in the following limits $4 \le C \le 12$. Some authors recommend that for springs having the wire diameter $d \le 0.1$ in. ($d \le 2.5$ mm) to take $5 \le C \le 12$; while for the diameters $0.11 \le d \le 0.2$ in. ($3 \le d \le 5$ mm) it is recommended to take $4 \le C \le 10$; and for diameters $0.25 \le d \le 0.5$ in. ($6 \le d \le 12$ mm) to take $4 \le C \le 9$. Applying $C < 3$ imposes high stress concentrations into the coils, which is highly undesirable. It should be mentioned that in some applications the spring index approaches 16. It is important to mention that according to the *Machinery's Handbook* (30th edition) the best structural proportions of springs are achieved at the spring index in the range from 7 to 9.

Some applications require relatively long springs that are prone to buckling. In these cases, in order to prevent buckling the springs are enclosed in sleeves or round bars (long pins) are installed inside the springs.

12-2.2. Stresses in Helical Springs Subjected to Static Loading

The stress analysis of the coils of compression and extension springs is based on the same principles and the same expressions. Figure 12-2.3 shows a compression coil spring loaded by an axial force *F*.

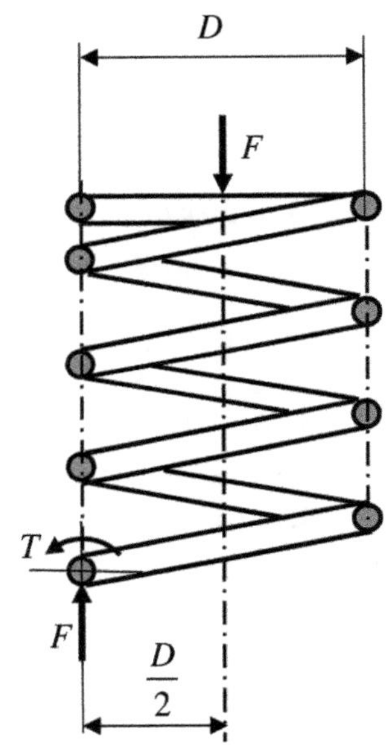

Figure 12-2.3. Schematic for stress calculations in a coil spring.

We apply the method of cuts to reveal the stresses acting in the coil. We cut the spring in two pieces and analyze the equilibrium of the remaining part as illustrated in this figure. In order to simplify the analysis, we assume that the coil is perpendicular to the axis of the spring, in other words, that the helix angle $\alpha = 0$. In actual springs the helix angle is about $\alpha \le 12°$. More rigorous calculations show that accounting for the helix angle does not make much of a difference. The equilibrium of the remaining part of the spring is preserved if we apply to the cross-sectional area of the wire the shear force F and the torque $T = \frac{FD}{2}$. Consequently, the shear stresses in the cross-sectional area of the coil represent a superposition of the direct shear stress with the torsional stress. Based on formulas (1-4.5) and (1-4.20), we may write

$$\tau_{max} = \frac{F}{A} + \frac{T0.5d}{J} \qquad (12\text{-}2.5)$$

Substituting into this formula the corresponding parameters, we have

$$\tau_{\max}=\frac{4F}{\pi d^2}+\frac{8FD}{\pi d^3}=\frac{4F}{\pi d^2}\left(1+\frac{2D}{d}\right) \tag{12-2.6}$$

Combining this expression with formula (12-2.4), we obtain an expression for verifying calculations of the stresses in a spring subjected to static loading:

$$\tau_{\max}=\frac{4F}{\pi d^2}(1+2C)\le\frac{S_{sy}}{n_\tau} \tag{12-2.7}$$

Appropriately transforming this equation, we obtain the formula for calculations of the diameter of the wire of the coil spring subjected to static loading:

$$d=2\sqrt{\frac{Fn_\tau}{\pi S_{sy}}(1+2C)} \tag{12-2.8}$$

12-3. Calculations of Springs Subjected to Variable Stresses

The expression (12-2.7) does not account for the effect of the wire curvature on the stresses in the wire, as mentioned above, with regard to the spring index. The curvature of the wire that is associated with the spring index causes an additional shear stress in the wire. According to some authors, the influence of the spring index on the stresses in the wire could be accounted for by applying a stress concentration factor. This stress concentration plays a significant role in the strength of the spring subjected to variable stresses. Different sources suggest different empiric formulas for the shear stress correction factors related to the spring index. All these formulas are based on the slightly different spring index correction factors; however, the calculations according to these formulas are characterized by close results. In this book, we adopted the following empiric formula for the spring index correction factor:

$$K_\tau=\frac{C+0.25}{C-1} \tag{12-3.1}$$

Combining expressions (12-2.7) and (12-3.1), we obtain an expression for verifying calculations of the stresses acting in the springs subjected to a completely reversed loading cycle:

$$\tau_{\max}=\frac{4F_{\max}}{\pi d^2}(1+2C)\left(\frac{C+0.25}{C-1}\right)\le\frac{S_{se}}{n_\tau} \tag{12-3.2}$$

From expression (12-3.2), we derive the formula for calculating the diameter of the wire for a spring subjected to the action of a completely reversed loading cycle:

$$d=2\sqrt{\frac{F_{\max}n_\tau}{\pi S_{se}}(1+2C)\left(\frac{C+0.25}{C-1}\right)} \tag{12-3.3}$$

12-4. Stiffness Coefficient (Spring Rate) of Springs

One of the most important characteristics of a spring is its coefficient of stiffness or spring rate coefficient (spring rate) that is the ratio between the applied force and the deflection (deformation) of the spring:

$$K = \frac{F}{l} \tag{12-4.1}$$

where K is the spring rate and l is the deflection (elongation or shrinkage) of the spring. It should be emphasized that since the deformations of springs are elastic deformations, the spring rate K is a linear function of the deformation l. The graph of the spring rate is shown in Figure 12-4.1, where the straight line OB reflects the spring rate according to formula (12-4.1).

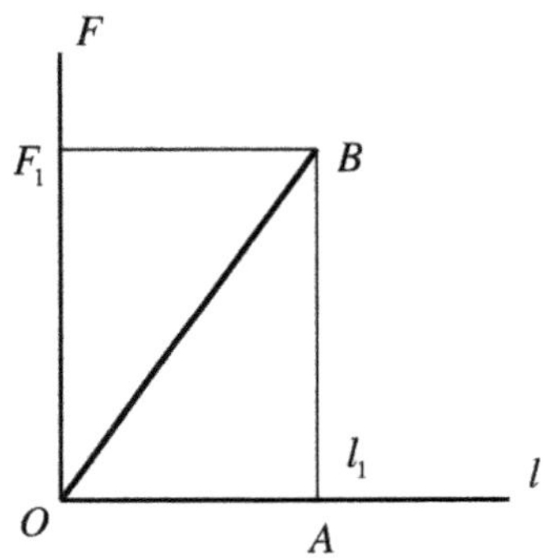

Figure 12-4.1. Graph of the spring stiffness coefficient.

The area of the triangle OAB equals the energy of deformation that the spring accumulates as a result of the action of the static force F_1 during the deformation l_1. Therefore, the potential energy is $E_1 = 0.5F_1l_1$, or in general we may write

$$E = 0.5\,Fl \tag{12-4.2}$$

The potential energy of a deformed coil spring caused by a static force F can be determined based on the corresponding torque $T = 0.5FD$ without accounting for the action of the shear force that creates an ignorable amount of deformation energy. Based on the principles discussed in the course Strength of Materials (Mechanics of Materials), we write

$$U = \frac{T^2 L}{2GJ} \tag{12-4.3}$$

where L is the total length of spring wire—calculated from formula (12-2.3)—G is the modulus of rigidity, and J is the polar moment of inertia of the wire according to formula (1-4.21). Since the helix angle of the coil $\alpha \le 12°$ and $\cos 12° \approx 0.98$, it is justifiable to simplify formula (12-2.3) to the following shape:

$$L \approx \pi DN \tag{12-4.4}$$

Therefore, accounting for these considerations, we present formula (12-4.3) in the following way:

$$U = \frac{4F^2D^3N}{Gd^4} \tag{12-4.5}$$

Equating the left-hand side of the formula (12-4.2) to the right-hand side of the formula (12-4.5), we have $0.5Fl = \frac{4F^2D^3N}{Gd^4}$. Solving this equation for the deflection (deformation) l, we write

$$l = \frac{8FD^3N}{Gd^4} \tag{12-4.6}$$

Combining formulas (12-4.1) and (12-4.6), we obtain the formula for calculating the spring rate or stiffness coefficient of the spring:

$$K = \frac{Gd^4}{8D^3N} \tag{12-4.7}$$

It is important to emphasize that normally the operation of compression and extension helical springs is associated with the initial preload that is imposed into the spring during the assembly process. Actually, the spring obtains an initial deflection because of the preload and, consequently, has an initial stress. This initial or preload force is the minimum force $F_{\min}$ applied to the spring. Therefore, the external force that is the maximum force $F_{\max}$ applied to the spring causes additional deflection. The relationship between the minimum and maximum forces is expressed by the following formula:

$$F_{\min} = F_{\max}\delta_{os} \tag{12-4.8}$$

where δ_{os} is the preload coefficient that is recommended to be taken in the range from 0.1 to 0.3. Accounting for the above presented considerations, we transform formula (12-4.6) and obtain the following expression for calculating the operational displacement l_{os}:

$$l_{os} = \frac{8F_{\max}(1-\delta_{os})D^3N}{Gd^4} \tag{12-4.9}$$

Based on this expression, we derive the formula for calculating the number of active coils:

$$N = \frac{l_{os}Gd^4}{8F_{\max}(1-\delta_{os})D^3} \tag{12.4.10}$$

For compression springs, the calculated value of the number of active coils, as mentioned above, should be increased by about two coils and rounded off to one-half or a whole coil. This is required in order to prevent the coils touching during the compression of the spring.

12-5. Spring Materials and Allowable Stresses

The majority of springs are manufactured from spring-dedicated materials, such as high carbon spring steels and alloy spring steels. Springs working in corrosive environments are made from stainless spring steels and cooper-base spring alloys. These materials come in wire and flat strip shapes. In this book, we consider wire shape spring materials that represent high carbon spring steels.

The most widely used wire material is a high carbon spring steel called music wire that according to ASTM A228 contains from 0.80 to 0.95% carbon. The diameters of music wires are available in the range from 0.005 to 0.125 in. (from 0.125 to 3.175 mm). Music wire can be also found in larger diameters up to 0.1875 in. (4.75 mm). The operating temperature for music wire does not exceed 250°*F* (120°*C*).

For larger diameters than those available in music wires, the high carbon spring steel that is Oil-Tampered MB Grade ASTM 229 steel containing from 0.60 to 0.70% carbon is used. In addition, for these diameters Oil-Tampered HB Grade SAE 1080 steel containing from 0.75 to 0.85% carbon is also used. These steels are available in diameters from 0.125 to 0.5 in. (3.175 to 12.5 mm) and applicable for working temperatures up to 350°*F* (177°*C*).

The Hard-Drawn MB Grade ASTM A227 steel contains from 0.60 to 0.70% carbon and is available in diameters from 0.031 to 0.5 in. (from 0.8 to 12.5 mm) and is applicable at the temperature up to 250°*F* (120°*C*).

It is very important to emphasize that the mechanical properties, specifically the ultimate and yield strength, of wire spring steels, unlike regular steels, significantly depend on the diameter of the wire. The specific values of these parameters decrease with the increase of the diameter of the wire. Therefore, for each particular diameter of wire it is necessary to determine its allowable stresses. This makes the spring related strength calculations more complex. However, numerous published sources contain the needed information related to allowable stresses of spring wire metals. Appendix A-9-2 includes graphs of the allowable torsional stresses of spring wire metals for compression coil springs, depending on the diameter of the wire and operational conditions of the spring. It should be underlined that for extension springs these allowable stresses should be reduced by about 10 to 15% .

The operational conditions play an extremely important role in determining the allowable torsional strength of the springs. The graphs in Appendix A-9-2 are given for three different operational conditions that correspond to the following three types of services: light, average, and severe.

The light service is limited to 10^4 cycles of deformation, or in other words, the allowable stresses according to this type of service are applicable for calculations of springs subjected to static loading. The light service type of springs are used in safety devices that are activated in case of overloads, etc.

The average service allows 10^6 cycles of deformations that is the finite life cycle for springs subjected to variable stresses. The average service group of springs is characterized by numerous industrial applications of springs in all kinds of machine tools and electrical devices, etc.

The severe type of service allows 10^7 cycles of deformations that corresponds to the infinite life of springs subjected to variable stresses. Thus, the severe service group of springs is associated with the working processes of spring-loaded valves in internal combustion engines, compressors, and similar machines. Impact loading applied to the springs in pneumatic hammers and similar machines represents severe service as well. Therefore, the allowable stress according to the severe type of service represents the endurance limit of the spring wire metal.

In may be noted that for springs subjected to variable stresses it is justifiable to use spring wires that went through the shot-peening treatment that increases the strength of the wire up to 30%.

12-6. Design of Helical Coil Springs Subjected to Static Loading

The designs of compression and extension springs are similar, but with some particular differences that are indicated below.

The design process begins with the formulation of the requirements of the operational conditions and the basic characteristics of the spring. The operational conditions are associated with the type of loading that usually determines the type of service conditions of the spring—light, average, or severe. Light service is associated with static loading, while average and severe with dynamic loading.

The basic input characteristics comprise the initial (minimum) load F_{min} and the final (maximum) load F_{max} applied to the spring. For compression and extension springs, the initial load is the preload that is applied to the spring during the assembly of the unit that employs the spring. However, for closed extension springs, the preload could be the residual compression force imposed on the spring during fabrication. Another input parameter is the operational (working) deflection l_{os}, during which the applied force increases from its minimum to its maximum. Obviously, the maximum loads applied to the spring are interrelated with the operational deflection through the stiffness coefficient of the spring.

Normally, the safety factor n_τ is assigned at the beginning of the design process.

Usually, the dimensions of the space allocated for the spring are also given at the start of the design process. In some cases, the value of the maximum force is unknown at the beginning of the design process, and it should be determined according to the given operational deflection.

It is helpful to follow the design procedure by considering the steps of calculations of an actual compression spring. During the design calculations, we need to determine the following basic parameters of the spring: d, D, F_{max}, N, α, p, H, L, K, and τ_{max}. We should also determine the configuration of the spring ends.

To demonstrate the steps of the design process, we consider a spring having the operational displacement $l_{os} = 0.25$ in., for which the safety factor $n_\tau = 2$, and the approximate dimensions of the space where the spring is supposed to operate are given. The spring is projected to work in light service operational conditions.

In some cases, it is desirable that because of the allocated space for the spring the mean diameter of the spring should not exceed a certain size. Assume that in this case the mean diameter should not exceed 0.5 in.

Following the above recommendations, we select the spring index $C = 8$. For this spring index it is recommended to use the diameters of wires below 0.1 in. to below 0.5 in. Hence, we select from Appendix A-9-2 (Figure 3) the hard-drawn steel wire QQ-W-428 having $d = 0.04$ in. and $S_{sy} = 145$ ksi. Hence, $d = 0.04$ in. and $D = 8 \cdot 0.04 = 0.32$ in. (since 0.32<0.5, the desired condition is achieved).

Rearranging formula (12-2.8), we calculate the force that should be applied to the spring:

$$F_{max} = \frac{\pi d^2 S_{sy}}{4 n_\tau (1 + 2C)} \qquad (12\text{-}6.1)$$

Taking $n_t = 2$, we have

$F_{max} = \dfrac{\pi 0.04^2 \cdot 145 \cdot 10^3}{4 \cdot 2 (1 + 2 \cdot 8)} = 5.36$ lb. Transforming formula (12-4.10), we may write

$$N = \frac{l_{os} d G}{8 F_{max} (1 - \delta_{os}) C^3} \qquad (12\text{-}6.2)$$

Taking $\delta_{os} = 0.2$ and considering that $l_{os} = 0.25$ in. and $G = 12 \cdot 10^6$ psi, we apply formula (12-6.2) to calculate the number of active coils: $N = \dfrac{0.25 \cdot 0.04 \cdot 12 \cdot 10^6}{8 \cdot 5.36 (1 - 0.2) 8^3} = 6.83$. Following the above recommendations, we take the numbers of coils $N = 9.0$. The helix angle of springs, as mentioned above, is usually less than 12°. Taking for our case the helix angle $\alpha = 10°$, we calculate using formula (12.2.1) the pitch of the spring $p = \pi D \tan\alpha = \pi 0.32 \tan 10° = 0.177$ in. Assuming that the spring has closed not ground ends, the total number of coils increases to 11.

Now we calculate the free height of the spring using formula (12-2.2):
$H = 0.177 \cdot 9.0 + 3 \cdot 0.04 = 1.713$ in.
Considering that the total number of coils is 11, we calculate from formula (12-2.3) the length of the wire: $L = \dfrac{\pi 0.32 \cdot 11}{\cos 10°} = 11.22$ in.

According to formula (12-4.1), we determine the spring rate coefficient: $K = \dfrac{5.36}{0.25} = 21.44 \dfrac{\text{lb}}{\text{in.}}$.

Applying formula (12-2.7), we calculate the stress in the wire: $\tau_{max} = \dfrac{4 \cdot 5.36}{\pi 0.04^2}(1 + 2 \cdot 8) = 72{,}548$ psi.

Thus, we determined all of the basic parameters of the spring.

12-7. Design of Helical Coil Springs Subjected to Variable Loading—Fatigue Calculations

The springs used in the majority of applications are subjected to variable stresses that are characterized by low life, finite life, and infinite live cycles. These cycles are equivalent to light, average, and severe service operational conditions described above. Average and severe service conditions are associated with fatigue strength. The graphs reflecting the allowable torsional stresses for spring wires for corresponding service conditions are presented in Appendix A-9-2. These graphs of allowable torsional stresses for the average and severe service operational conditions illustrate the dependence of the fatigue strength on the diameters of the spring metal wires. For severe service, these graphs indicate the fatigue strength of the wires subjected to maximum torsional stresses when the number of cycles is 10^7. The characteristics of these allowable stresses coincide with the definition of the endurance limit of steel subjected to the completely reversed loading cycle. In this book, the allowable torsional stresses related to severe service conditions are considered as the endurance limits for the metal spring wires. Hence, the springs are subjected to all kinds of variable stresses; however, the most severe variable stresses are associated with the completely reversed loading cycle, and this cycle is used in this book for the fatigue stress calculations of springs.

In general, the calculations of helical coil springs subjected to static and dynamic loading are similar. The differences are the calculations for maximum shear stresses, the maximum applied force, and determining the characteristics of the allowable stresses. For dynamic loading of the springs, the calculations of the maximum stress and force include the effect of the curvature associated with the spring index, while the allowable stresses are determined by the average or severe service operational conditions.

For the finite life cycle, we take the endurance limit based on the allowable torsional stress of the average service, while for the infinite life cycle we take the endurance limit based on the allowable stresses for severe service operational conditions. In this book, we assume that the variable stresses acting in the springs are characterized by the completely reversed cycle, which is the most destructive loading cycle.

In order to demonstrate the sequence of steps for fatigue calculations of a helical spring, we consider an example associated with the design of a helical compression spring. In the previous section, we presented the design calculations of a spring for light service subjected to static loading. Now we design the same spring for severe service that is equivalent to the operational conditions characterized by an infinite cycle of variable stresses. Thus, we design a spring for the following input data: the required deflection $l_{os} = 0.25$ in., the preload coefficient $\delta_{os} = 0.2$, the safety factor $n_\tau = 2$, and spring index $C = 8$. We select the same hard-drawn steel wire QQ-W-428 having $d = 0.04$ in. However, for this case, we take from the graph for severe service that is presented in Appendix A-9-2 (Figure 3) the allowable torsional stress $S_{se} = 110$ ksi.

In order to calculate the maximum force F_{max} applied to the wire, we rearrange the expression (12-3.2) that accounts for the stress concentration coefficient and obtain the following formula:

$$F_{max} = \frac{\pi d^2 S_{se}(C-1)}{4n_\tau(1+2C)(C+0.25)} \tag{12-7.1}$$

Using formula (12-7.1), we calculate $F_{max} = \frac{\pi 0.04^2 \cdot 110 \cdot 10^3 (8-1)}{4 \cdot 2(1+2 \cdot 8)(8+0.25)} = 3.448\ \text{lb}$. As we did for the static calculations, we take $\delta_{os} = 0.2$ and applying formula (12-6.2), we determine the number of active coils: $N = \frac{0.25 \cdot 0.04 \cdot 12 \cdot 10^6}{8 \cdot 3.448(1-0.2)8^3} = 10.62$. Rounding off the number of active coils to 11 and adding two more coils , the total number of active coils is 13. Taking the helix angle $\alpha = 10°$ and considering that $D = 0.32$ in., we calculate from formula (12.2.1) the pitch of the spring $p = \pi D \tan\alpha = \pi 0.32 \tan 10° = 0.177$ in. In order to make the closed not ground ends of the spring, it is required to add two coils, so we increase the number of total coils to 15. Now we calculate the free height of the spring, substituting into formula (12-2.2) the number of active coils $N = 13$. Hence, $H = 0.177 \cdot 13 + 3 \cdot 0.04 = 2.421$ in. The length of the wire including the 15 coils is calculated from formula (12-2.3): $L = \dfrac{\pi 0.32 \cdot 15}{\cos 10°} = 15.305$ in. According to formula (12-4.1), we determine the spring rate coefficient: $K = \dfrac{3.44}{0.25} = 13.76 \dfrac{\text{lb}}{\text{in.}}$. Applying expression (12-3.10), we determine $\tau_{max} = \dfrac{4 \cdot 3.448}{\pi 0.04^2}(1 + 2 \cdot 8)\left(\dfrac{8 + 0.25}{8 - 1}\right) = 55{,}003$ psi.

It is important to compare the basic parameters of these two springs calculated for static and variable loading. Hence, the number of active coils and the maximum load for the spring with static loading are 6.83 and 5.36 lb, respectively; while for the spring subjected to variable stresses, these parameters are 10.62 and 3.448 lb, respectively.

This concludes the fatigue calculations of the spring.

12-8. Impact Loading of Helical Compression Springs

Compression springs experience impact loading when used in shock absorbers and similar mechanisms. One of the main concerns with the design of these springs is to prevent the coils colliding during the deflection of the spring. Colliding of the coils could be prevented by increasing the number of active coils that is determined by the maximum deflection of the spring. Therefore, the maximum deflection of the spring due to impact loading should be determined. As it is known, an impact is the product of multiplying a force by the time of the duration of the impact. During the impact, the force tends to infinity and the corresponding time is infinitesimally close to zero, while neither the force nor the time can be measured. However, the same impact is equal to the momentum that is the product of multiplying the involved mass by its instantaneous initial velocity obtained by the mass as a result of the impact. Fortunately, these two parameters, the mass and the initial velocity, can be determined. This allows us to compose the differential equation of motion of the mass attached to the spring. Usually, the mass of the spring is ignorable. Figure 12-8.1 presents a schematic that simulates the action of a shock absorber comprising a spring having a spring rate coefficient K.

The spring is rigidly attached to a non-movable support and to the mass m that obtains an initial velocity v_0 as a result of the impact. The weight of gravity W_g of this mass causes an initial or static deflection s_0 of the spring.

For convenience, the axis y is directed downward. The differential equation describing the motion of the mass after the impact reads

$$m \frac{d^2 y}{dt^2} + Ky = W_g \tag{12-8.1}$$

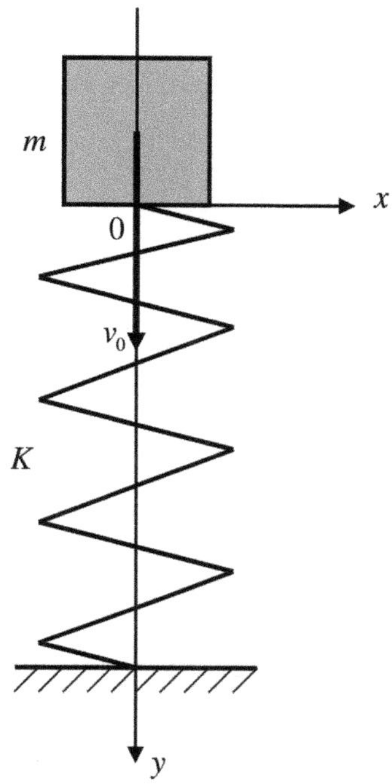

Figure 12-8.1. Schematic of a spring with an attached mass having an initial velocity.

where y is the mass displacement that is equivalent to the deflection of the spring, and t is the running time. The initial conditions of motion are

$$\text{for } t=0 \quad y=s_0, \frac{dy}{dt}=v_0 \tag{12-8.2}$$

Dividing the equation (12-8.1) by m, we have

$$\frac{d^2y}{dt^2}+\omega^2=g \tag{12-8.3}$$

where ω is the natural frequency of the vibrational system representing the mass attached to the spring and g is the acceleration of gravity, while

$$\omega^2=\frac{K}{m} \tag{12-8.4}$$

and

$$g=\frac{W_g}{m} \tag{12-8.5}$$

The solution of the differential equation of motion (12-8.3) at initial conditions of motion (12-8.2) reads

$$y=\frac{1}{\omega^2}\left[g+\sin(\omega t-\beta)\sqrt{(g-s_0\omega^2)^2+v_0^2\omega^2}\right] \tag{12-8.6}$$

while

$$\sin\beta=-\frac{(g-s_0\omega^2)^2}{(g-s_0\omega^2)^2+v_0^2\omega^2} \tag{12-8.7}$$

The maximum displacement of the mass m occurs when its velocity becomes equal to zero. Taking the first derivative from equation (12-8.6) with respect to time t, we obtain

$$\frac{dy}{dt}=\frac{1}{\omega}\cos(\omega t-\beta)\sqrt{(g-s_0\omega^2)^2+v_0^2\omega^2} \tag{12-8.8}$$

where

$$\cos\beta=\frac{v_0^2\omega^2}{(g-s_0\omega^2)^2+v_0^2\omega^2} \tag{12-8.9}$$

The velocity of the mass becomes equal to zero when $\cos(\omega t-\beta)=0$. In this case,

$$\sin(\omega t-\beta)=\pm 1 \tag{12-8.10}$$

Combining equations (12-8.6) and (12-8.10) and taking for our case $\sin(\omega t-\beta)=1$, we calculate the maximum displacement s_{max} of the mass m:

$$s_{max}=\frac{1}{\omega^2}\left[g+\sqrt{(g-s_0\omega^2)^2+v_0^2\omega^2}\right] \tag{12-8.11}$$

Actually, $l_{os}=s_{max}$, therefore, using formula (12-8.11) we can determine the maximum deflection of the spring as a result of an impact and avoid colliding of the coils.

Example 12-8.1a: Design of a helical compression spring subjected to impact loading
The operational conditions of a spring subjected to impact loading correspond to the conditions of severe service. In order to design this spring, the input data should contain the value of the initial velocity that the mass attached to the spring obtains as the result of the impact.

The task is to design a helical compression spring subjected to impact loading. The schematic of this spring with the attached mass is presented in Figure 12-8.1. This mass obtains an initial velocity $v_0=2\,\frac{\text{m}}{\text{s}}$ as a result of the impact.

Solution:

Following the recommendations presented above, we take $C = 8$. We select the music wire QQ-Q-470, ASTM A228 from Appendix A-9-2 (Figure 2). The diameter of the wire is $d = 5$ mm and the allowable torsional stress is $S_{se} = 625$ MPa. The selection of the diameter of the wire is based on the previous design experience. Obviously, this decision may not be final, but it allows starting the design process. We may expect to perform several iterations before we finalize the design. However, the design steps remain the same for all iterations. We take the safety factor $n_\tau = 2$.

Based on formula (12-7.1), we calculate the maximum force that could be applied to this spring: $F_{max} = \dfrac{\pi 5^2 \cdot 625 \cdot (8-1)}{4 \cdot 2(1+2 \cdot 8)(8+0.25)} = 306.1$ N. Taking $\delta_{os} = 0.1$, we calculate the minimum force applied to the spring: $F_{min} = \delta_{os}\, F_{max} = 0.1 \cdot 306.15 = 30.61$ N. Based on formula (12-8.5), we determine the mass attached to the spring using the following expression: $m = \dfrac{F_{min}}{g}$; therefore, the attached mass is $m = \dfrac{30{,}61}{9.81} = 3.12$ kg. Analyzing equation (12-8.11), we notice that the initial static deflection equals $s_0 = \dfrac{mg}{K}$. Therefore, we may rearrange the equation (12-8.11) to the following shape:

$$s_{max} = \frac{1}{\omega^2}\left[g + \sqrt{\left(g - \frac{mgK}{Km}\right)^2 + v_0^2\omega^2}\right] = \frac{g + v_0\omega}{\omega^2} = \frac{m(g + v_0\omega)}{K} \tag{12-8.12}$$

It should be emphasized that the maximum operational displacement l_{os} cannot exceed s_{max}, hence $l_{os} \le s_{max}$. Accounting for formula (12-6.12), we rearrange formula (12-4.1) and obtain

$$K = \frac{F_{max}}{s_{max}} \tag{12-8.13}$$

Combining formulas (12-8.4), (12-8.12), and (12-8.13), we have

$$K = \frac{(F_{max} - F_{min})^2}{m v_0^2} \tag{12-8.14}$$

It should be noted that when we use formula (12-8.14), we substitute for the U.S. customary units the force in pounds (*lb*), the mass in slugs $\left(\dfrac{\text{lb-s}^2}{\text{ft}}\right)$, and the velocity in feet per second $\left(\dfrac{\text{ft}}{\text{s}}\right)$, while for the SI units we substitute the force in newtons (N), the mass in kilograms (kg), and the velocity in meters per second $\left(\dfrac{\text{m}}{\text{s}}\right)$. Keeping this in mind and using formula (12-8.14), we have $K = \dfrac{(306.15 - 30.615)^2}{3.12 \cdot 2^2} = 6{,}083\,\dfrac{\text{N}}{\text{m}}$. It is more convenient for our calculations to take $K = 6.083\,\dfrac{\text{N}}{\text{mm}}$. Therefore, the maximum deflection is $s_{max} = \dfrac{F_{max}}{K} = \dfrac{306.15}{6.083} = 50.33$ mm. Using formula (12-6.2) and accounting that $G=80\cdot 10^3\,\frac{\text{N}}{\text{mm}^2}$, we determine the number of active coils: $N=\frac{50.33\cdot 5\cdot 80\cdot 10^3}{8\cdot 306.15\,(1-0.1)8^3}=17.83$. Rounding off to 18 coils and taking two more coils, according to the above recommendations, we have 20 active coils. Considering that $D = dC = 5\cdot 8 = 40$ mm and taking the helix angle $\alpha = 5°$, we apply formula (12-2.1) and calculate the pitch of the spring: $p = \pi 40 \tan 5° = 10.99$ mm; we take $p = 11$ mm. Adding two coils that are needed for arranging closed not ground ends, we calculate the free length of the spring according to formula (12-2.2), hence, we have $H = 11\cdot 20 + 3\cdot 5 = 235$ mm. Using formula (12-2.3), we calculate the length of the wire: $L=\frac{\pi 40\cdot 22}{\cos 5°}=2{,}774$ mm. Based on expression (12-3.2) we determine the stresses in the spring: $\tau_{\max}=\frac{4\cdot 306.15}{\pi d^2}(1+2\cdot 8)\left(\frac{8+0.25}{8-1}\right)=312$ MPa. Recalling that $S_{se} = 625$ MPa, we have $n_\tau=\frac{625}{312}=2.003$ This result should be expected since the calculations are performed for the accepted safety factor $n_\tau = 2$.

All required parameters of the spring are determined, and there is no need for iterations.

However, if there are some particular requirements that are not met by the current calculations, then some corrections or iterations should be performed based on the formulas presented above.

Chapter 12. Problems

12-01. A helical compression spring has the wire diameter $d = 0.2$ in., the helix angle $\alpha = 10°$, the number of active coils $N = 11$, and the spring index $C = 8$. Determine the total number of coils of the spring, its solid height, and its free height if the spring has open not ground ends.

12-02. A helical compression spring has the wire diameter $d = 0.3$ in., the helix angle $\alpha = 9°$, the number of active coils $N = 14$, and the spring index $C = 7$. Determine the total number of coils of the spring, its solid height, and its free height if the spring has closed not ground ends.

12-03. A helical compression spring has the diameter of the wire $d = 0.2$ in., the spring index $C = 8$, and it is loaded by a static force $F = 80$ lb. Determine the safety factor n_τ if the spring is made from music wire QQ-Q-470.

12-04. A helical compression spring has the diameter of the wire $d = 4$ mm, the spring index $C = 7$, and it is loaded by a static force $F = 200$ N. Determine the safety factor n_τ if the spring is made from a hard drawn steel wire QQ-W-428.

12-05. A helical compression spring has the spring index $C = 7$, and it is loaded by a static force $F = 65$ lb. Taking the safety factor $n_\tau = 2$, determine the diameter d of the wire if the material of the wire has $S_{sy} = 115$ ksi.

12-06. Taking the safety factor $n_\tau = 2$, determine the basic parameters of a helical compression spring for light service operational conditions. Take the spring index $C = 8$ and helix angle $\alpha = 10°$. The spring wire has the diameter $d = 0.06$ in., and the torsional allowable stress $S_{sy} = 140$ ksi. The operational displacement of the spring $l_{os} = 0.3$ in. and the preload coefficient is $\delta_{os} = 0.15$. The spring has open not ground ends.

12-07. Taking the safety factor $n_\tau = 2$ determine the basic parameters of a helical compression spring for average service operational conditions. Take the spring index $C = 8$ and helix angle $\alpha = 10°$. The spring wire has the diameter $d = 0.08$ in., and the torsional allowable stress $S_{se} = 120$ ksi. The operational displacement of the spring $l_{os} = 0.35$ in., and the preload coefficient is $\delta_{os} = 0.15$. The spring has open not ground ends.

12-08. Taking the safety factor $n_\tau = 2$ determine the basic parameters of a helical compression spring for severe service operational conditions. Take the spring index $C = 8$ and helix angle $\alpha = 10°$. The spring wire has the diameter $d = 0.1$ in., and the torsional allowable stress $S_{se} = 95$ ksi. The operational displacement of the spring $l_{os} = 0.4$ in. and the preload coefficient is $\delta_{os} = 0.15$. The spring has open not ground ends.

12-09. A helical compression spring shown in Figure 12-8.1 is subjected to impact loading and obtains, as a result of the impact, an initial velocity $v_0 = 3.5\ \frac{\text{ft}}{\text{s}}$. The spring index $C = 10$, the helix angle $\alpha = 8°$, the diameter of the wire $d = 0.2$ in., and the torsional allowable stress $S_{se} = 90$ ksi. Taking the safety factor $n_\tau = 2$ and the preload coefficient $\delta_{os} = 0.15$, determine the basic parameters of the spring. The spring has open not ground ends.

12-10. A helical compression spring shown in Figure 12-8.1 is subjected to impact loading and obtains, as a result of the impact, an initial velocity $v_0 = 1.5\ \frac{\text{m}}{\text{s}}$. The spring has the spring index $C = 9$, the helix angle $\alpha = 8°$, the diameter of the wire $d = 5$ mm, and the torsional allowable stress $S_{se} = 620$ MPa. Taking the safety factor $n_\tau = 2$ and the preload coefficient $\delta_{os} = 0.15$, determine the basic parameters of the spring. The spring has open not ground ends.

Appendixes

Appendix A-1

Appendix A-1-1

Quantities and Their Units

Quantities	Units	
	U.S. Customary System	International System (SI)
Energy	ft-lb	N·m
Force	lb	N (newton)
Length	ft	m
Mass	$\frac{\text{lb-s}^2}{\text{ft}}$	kg
Moment, Torque	lb-ft	N·m
Power	$\frac{\text{ft-lb}}{\text{s}}$	$\frac{\text{N·m}}{\text{s}}$
Stress	$\frac{\text{lb}}{\text{in.}}$ (psi)	$\frac{\text{N}}{\text{m}^2}$ (pascal)
Angular velocity	$\frac{\text{rad}}{\text{s}}$	$\frac{\text{rad}}{\text{s}}$
Linear velocity	$\frac{\text{ft}}{s}$	$\frac{\text{m}}{\text{s}}$

Appendix A-1-2

Conversion from U.S. Customary System to International System (SI)

U.S. Customary units	International (SI) units
Energy, ft-lb	1.36 N·m
Force, lb	4.45 N
Mass, $\frac{\text{lb-s}^2}{\text{ft}}$	14.6 kg
Power, horsepower hp, $\left(550\frac{\text{ft-lb}}{\text{s}}\right)$	746 W
Stress, ksi	6.9 MPa
Velocity, $\frac{\text{ft}}{\text{s}}$	$0.305\frac{\text{m}}{\text{s}}$

Appendix A-1-3

Modulus of Elasticity and Modulus of Rigidity of Some of Materials

Material	Modulus of elasticity, E		Modulus of rigidity, G	
	psi	Pa	psi	Pa
Aluminum alloys	$10.5\cdot10^6$	$73\cdot10^9$	$4\cdot10^6$	$28\cdot10^9$
Bronze	$15\cdot10^6$	$110\cdot10^9$	$5.6\cdot10^6$	$40\cdot10^9$
Cast iron	$18\cdot10^6$	$125\cdot10^9$	$7\cdot10^6$	$45\cdot10^9$
Polyethylene	$0.15\cdot10^6$	$2.7\cdot10^9$		
Steel	$30\cdot10^6$	$207\cdot10^9$	$12\cdot10^6$	$80\cdot10^9$

Appendix A-1-4

Mechanical Properties of Steels

AISI No.	Treatment	Strength Tensile lb/in.²	Strength Yield lb/in.²	Elongation, Percent	Reduction in Area, Percent	Hardness, Bhn	Impact Strength (Izod), ft-lb
1015	As-rolled	61,000	45,500	39.0	61.0	126	81.5
	Normalized (1700 F)	61,500	47,000	37.0	69.6	121	85.2
	Annealed (1600 F)	56,000	41,250	37.0	69.7	111	84.8
1020	As-rolled	65,000	48,000	36.0	59.0	143	64.0
	Normalized (1600 F)	64,000	50,250	35.8	67.9	131	86.8
	Annealed (1600 F)	57,250	42,750	36.5	66.0	111	91.0
1022	As-rolled	73,000	52,000	35.0	67.0	149	60.0
	Normalized (1700 F)	70,000	52,000	34.0	67.5	143	86.5
	Annealed (1600 F)	65,250	46,000	35.0	63.6	137	89.0
1030	As-rolled	80,000	50,000	32.0	57.0	179	55.0
	Normalized (1700 F)	75,000	50,000	32.0	60.8	149	69.0
	Annealed (1550 F)	67,250	49,500	31.2	57.9	126	51.2
1040	As-rolled	90,000	60,000	25.0	50.0	201	36.0
	Normalized (1650 F)	85,500	54,250	28.0	54.9	170	48.0
	Annealed (1450 F)	75,250	51,250	30.2	57.2	149	32.7
1050	As-rolled	105,000	60,000	20.0	40.0	229	23.0
	Normalized (1650 F)	108,500	62,000	20.0	39.4	217	20.0
	Annealed (1450 F)	92,250	53,000	23.7	39.9	187	12.5
1060	As-rolled	118,000	70,000	17.0	34.0	241	13.0
	Normalized (1650 F)	112,500	61,000	18.0	37.2	229	9.7
	Annealed (1450 F)	90,750	54,000	22.5	38.2	179	8.3
1080	As-rolled	140,000	85,000	12.0	17.0	293	5.0
	Normalized (1650 F)	146,500	76,000	11.0	20.6	293	5.0
	Annealed (1450 F)	89,250	54,500	24.7	45.0	174	4.5
1095	As-rolled	140,000	83,000	9.0	18.0	293	3.0
	Normalized (1650 F)	147,000	72,500	9.5	13.5	293	4.0
	Annealed (1450 F)	95,250	55,000	13.0	20.6	192	2.0
1117	As-rolled	70,600	44,300	33.0	63.0	143	60.0
	Normalized (1650 F)	67,750	44,000	33.5	63.8	137	62.8
	Annealed (1575 F)	62,250	40,500	32.8	58.0	121	69.0
1118	As-rolled	75,600	45,900	32.0	70.0	149	80.0
	Normalized (1700 F)	69,250	46,250	33.5	65.9	143	76.3
	Annealed (1450 F)	65,250	41,250	34.5	66.8	131	78.5
1137	As-rolled	91,000	55,00	28.0	61.0	192	61.0
	Normalized (1650 F)	97,000	57,500	22.5	48.5	197	47.0
	Annealed (1450 F)	84,750	50,000	26.8	53.9	174	36.8
1141	As-rolled	98,000	52,000	22.0	38.0	192	8.2
	Normalized (1650 F)	102,500	58,750	22.7	55.5	201	38.8
	Annealed (1500 F)	86,800	51,200	25.5	49.3	163	25.3
1144	As-rolled	102,000	61,000	21.0	41.0	212	39.0
	Normalized (1650 F)	96,750	58,000	21.0	40.4	197	32.0
	Annealed (1450 F)	84,750	50,250	24.8	41.3	167	48.0

Mechanical Properties of Steels *(continued)*

AISI No.	Treatment	Strength: Tensile (lb/in.^{2})	Strength: Yield (lb/in.^{2})	Elongation, Percent	Reduction in Area, Percent	Hardness, Bhn	Impact Strength (Izod), ft-lb
1340	Normalized (1600 F)	121,250	81,000	22.0	62.9	248	68.2
	Annealed (1475 F)	102,000	63,250	25.5	57.3	207	52.0
3140	Normalized (1600 F)	129,250	87,000	19.7	57.3	262	39.5
	Annealed (1500 F)	100,000	61,250	24.5	50.8	197	34.2
4130	Normalized (1600 F)	97,000	63,250	25.5	59.5	197	63.7
	Annealed (1585 F)	81,250	52,250	28.2	55.6	156	45.5
4140	Normalized (1600 F)	148,000	95,000	17.7	46.8	302	16.7
	Annealed (1500 F)	95,000	60,500	25.7	56.9	197	40.2
4150	Normalized (1600 F)	167,500	106,500	11.7	30.8	321	8.5
	Annealed (1500 F)	105,750	55,000	20.2	40.2	197	18.2
4320	Normalized (1640 F)	115,000	67,250	20.8	50.7	235	53.8
	Annealed (1560 F)	84,000	61,625	29.0	58.4	163	81.0
4340	Normalized (1600 F)	185,500	125,000	12.2	36.3	363	11.7
	Annealed (1490 F)	108,000	68,500	22.0	49.9	217	37.7
4620	Normalized (1650 F)	83,250	53,125	29.0	66.7	174	98.0
	Annealed (1575 F)	74,250	54,000	31.3	60.3	149	69.0
4820	Normalized (1580 F)	109,500	70,250	24.0	59.2	229	81.0
	Annealed (1500 F)	98,750	67,250	22.3	58.8	197	68.5
5140	Normalized (1600 F)	115,000	68,500	22.7	59.2	229	28.0
	Annealed (1525 F)	83,000	42,500	28.6	57.3	167	30.0
5150	Normalized (1600 F)	126,250	76,750	20.7	58.7	255	23.2
	Annealed (1520 F)	98,000	51,750	22.0	43.7	197	18.5
5160	Normalized (1575 F)	138,750	77,000	17.5	44.8	269	8.0
	Annealed (1495 F)	104,750	40,000	17.2	30.6	197	7.4
6150	Normalized (1600 F)	136,250	89,250	21.8	61.0	269	26.2
	Annealed (1500 F)	96,750	59,750	23.0	48.4	197	20.2
8620	Normalized (1675 F)	91,750	51,750	26.3	59.7	183	73.5
	Annealed (1600 F)	77,750	55,875	31.3	62.1	149	82.8
8630	Normalized (1600 F)	94,250	62,250	23.5	53.5	187	69.8
	Annealed (1550 F)	81,750	54,000	29.0	58.9	156	70.2
8650	Normalized (1600 F)	148,500	99,750	14.0	40.4	302	10.0
	Annealed (1465 F)	103,750	56,000	22.5	46.4	212	21.7
8740	Normalized (1600 F)	134,750	88,000	16.0	47.9	269	13.0
	Annealed (1500 F)	100,750	60,250	22.2	46.4	201	29.5
9255	Normalized (1650 F)	135,250	84,000	19.7	43.4	269	10.0
	Annealed (1550 F)	112,250	70,500	21.7	41.1	229	6.5
9310	Normalized (1630 F)	131,500	82,750	18.8	58.1	269	88.0
	Annealed (1550 F)	119,000	63,750	17.3	42.1	241	58.0

Source: Bethlehem Steel Corp. and Republic Steel Corp. as published in 1974 DATABOOK issue of the American Society for Metals' *METAL PROGRESS* magazine and used with permission.

Appendix A-1-5

Source: R. E. Peterson, "Design Factors for Stress Concentration," *Machine Design*, vol. 23, 1951. For other stress concentration charts, see Lipson and Juvinall, *The Handbook of Stress and Strength*, The Macmillan Co., 1963.

Theoretical Stress Concentration Factors K_t

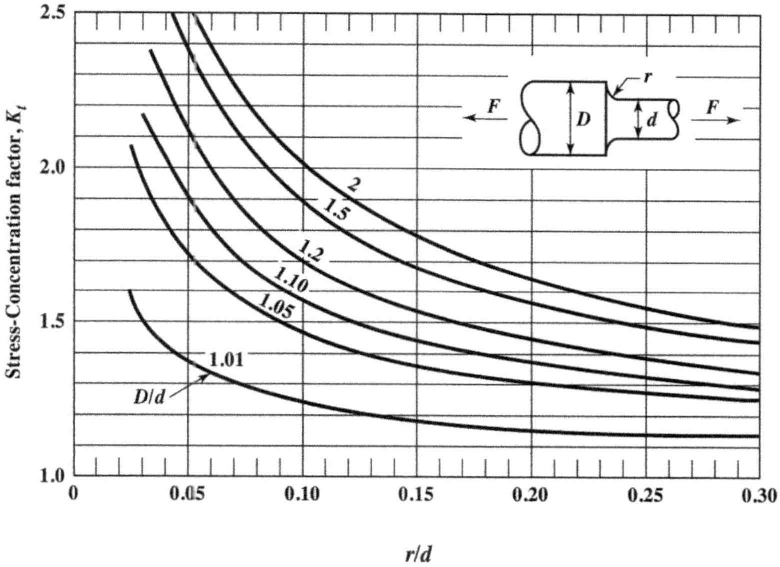

Figure 1. Stress-concentration factor K_t for a filleted shaft in tension.

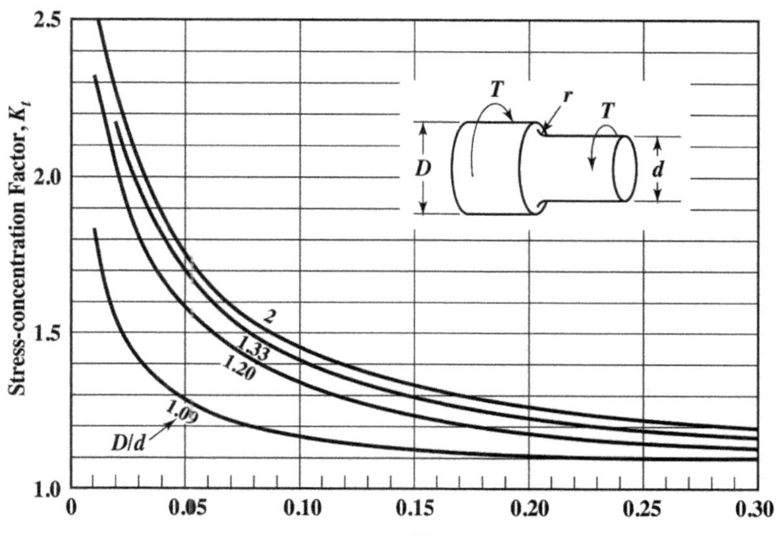

Figure 2. Stress-concentration factor K_t for a filleted shaft in torsion.

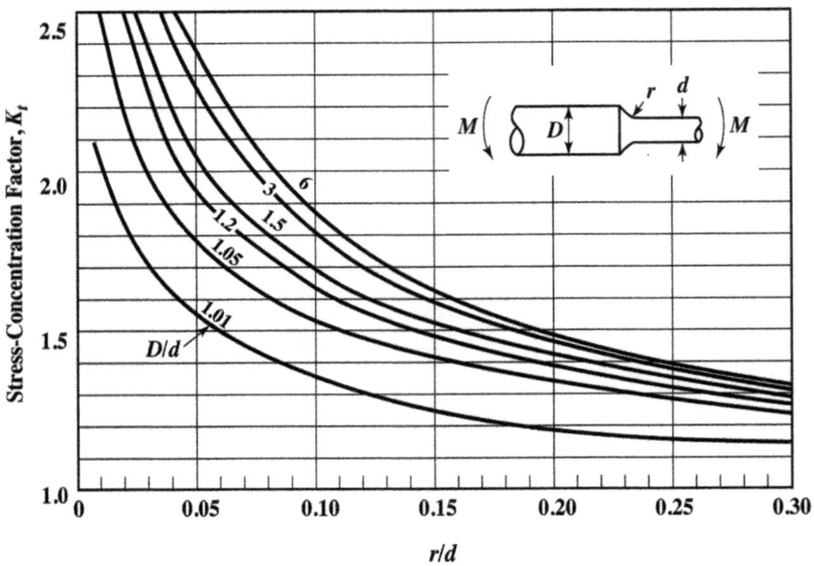

Figure 3. Stress-concentration factor K_t for a shaft with shoulder fillet in bending.

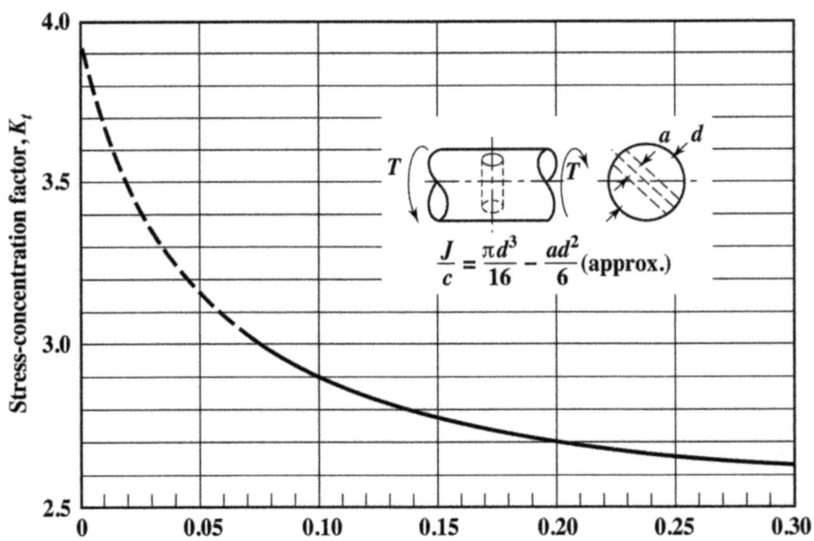

Figure 4. Stress-concentration factor K_t for a shaft with a transverse hole in torsion.

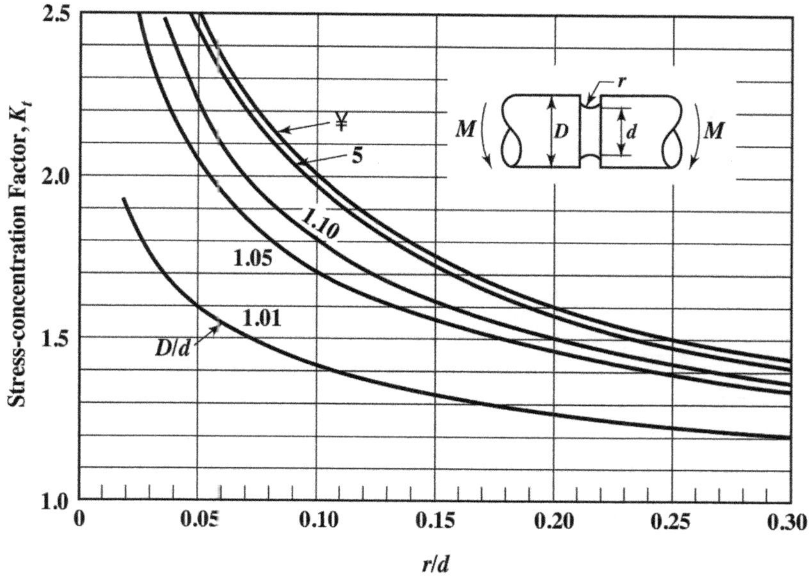

Figure 5. Stress-concentration factor K_t for a grooved shaft in bending.

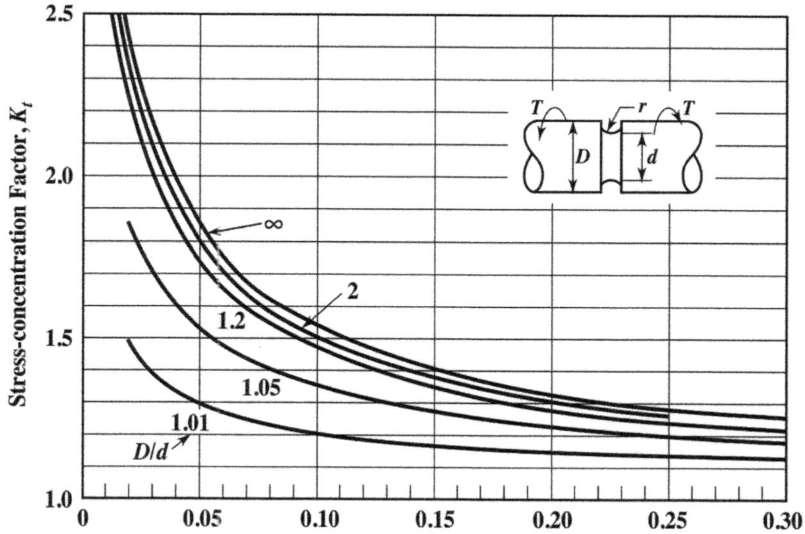

Figure 6. Stress-concentration factor K_t for a grooved shaft in torsion.

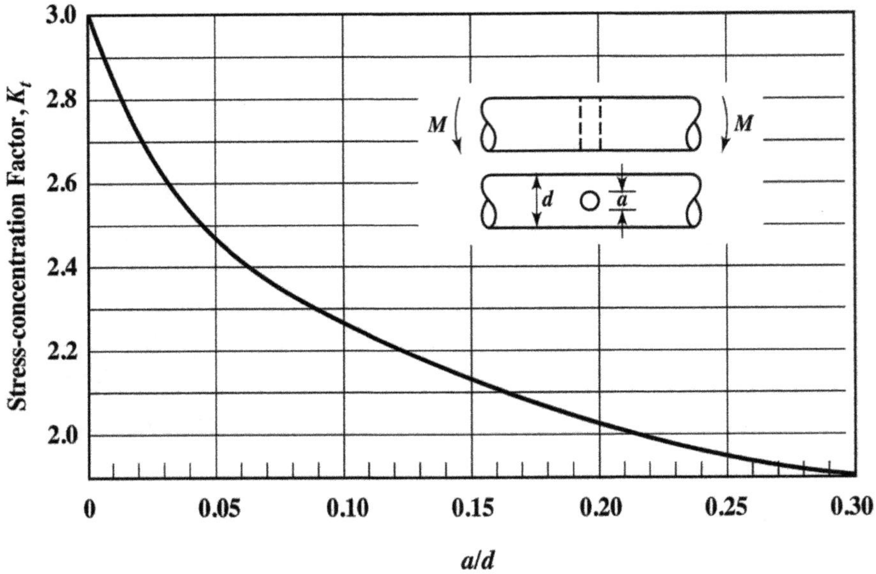

Figure 7. Stress-concentration factor K_t for a shaft with a transverse hole in bending.

Appendix A-2

Source: Machinery's Handbook, 30th Edition, E. Oberg et al., Norwalk, CT: Industrial Press, 2016, *industrialpress.com*

Appendix A-2-1

PV Factor

Bearing Material	Pressure, *P* (Maximum)		Linear Velocity (Maximum)		*PV* (Maximum)	
	psi	MPa	fpm	$\frac{m}{s}$	psi·fpm	MPa·$\frac{m}{s}$
Cast Bronze	4,000	26.7	750	46	75,000	31.5
Nylon	1,000	6.9	1,000	61	150,000	63.0
Phenolics	6,000	41.4	2,500	153	15,000	6.3
Porous Bronze	4,500	31.0	1,500	92	50,000	21.0
Porous Iron	8,000	55.2	800	49	50,000	21.0
Porous Metals	8,000	55.2	1,500	92	50,000	21.0
Reinforced Teflon	2,500	17.2	1,000	61	15,000	6.3

Appendix A-2-2

Values of X and Y Coefficients for Radial Ball Bearings and Angular-Contact Ball Bearings having the contact angle of $\alpha = 5°$ when $\frac{F_a}{F_r} > e = \frac{F_a}{C_0}$.

$e = \frac{F_a}{C_0}$	X	Y
0.19	0.56	2.30
0.22	0.56	1.99
0.26	0.56	1.71
0.28	0.56	1.56
0.30	0.56	1.45
0.34	0.56	1.31
0.38	0.56	1.15
0.42	0.56	1.04
0.44	0.56	1.00

Appendix A-2-3

Values of X and Y Coefficients for Angular-Contact Ball Bearings having the contact angles in the range from $\alpha = 10°$ to $\alpha = 40°$ when $\frac{F_a}{F_r} > e = \frac{F_a}{C_0}$.

Contact Angle, α, deg.	$e = \frac{F_a}{C_0}$	X	Y
10	0.29	0.46	1.88
10	0.32	0.46	1.71
10	0.36	0.46	1.52
10	0.38	0.46	1.41
10	0.40	0.46	1.34
10	0.44	0.46	1.23
10	0.45	0.46	1.10
10	0.54	046	1.01
15	0.38	0.44	1.47
15	0.40	0.44	1.40
15	0.43	0.44	1.30
15	0.46	0.44	1.23
15	0.47	0.44	1.19
15	0.50	0.44	1.12
15	0.55	0.44	1.02
15	0.56	0.44	1.00
20	0.57	0.43	1.00
25	0.68	0.41	0.87
30	0.80	0.39	0.76
40	1.14	0.35	0.57

Appendix A-2-4

Typical Bearing Life for Various Design Applications

Uses	Design life in hours	Uses	Design life in hours
Agricultural equipment	3000–6000	Gearing units	
Aircraft equipment	500–2000	Automotive	600–5000
Automotive		Multipurpose	8000–15000
Race car	500–800	Machine tools	20000
Light motorcycle	600–1200	Rail Vehicles	15000–25000
Heavy motorcycle	1000–2000	Heavy rolling mill	> 50000
Light cars	1000–2000	Machines	
Heavy cars	1500–2500	Beater mills	20000–30000
Light trucks	1500–2500	Briquette presses	20000–30000
Heavy trucks	2000–2500	Grinding spindles	1000–2000
Buses	2000–5000	Machine tools	10000–30000
Electrical		Mining machinery	4000–15000
Household appliances	1000–2000	Paper machines	50000–80000
Motors ≤ ½ hp	1000–2000	Rolling mills	
Motors ≤ 3 hp	8000–10000	Small cold mills	5000–6000
Motors, medium	10000–15000	Large multipurpose mills	8000–10000
Motors, large	20000–30000	Rail vehicle axle	
Elevator cables sheaves	40000–60000	Mining cars	5000
Mine ventilation fans	40000–50000	Motor rail cars	16000–20000
Propeller thrust bearings	15000–25000	Open-pit mining cars	20000–25000
Propeller shaft bearings	> 80000	Streetcars	20000–25000
Gear drives		Passenger cars	26000
Boat gearing units	3000–5000	Freight cars	35000
Gear drives	> 50000	Locomotive outer bearings	20000–25000
Ship gear drives	20000–30000	Locomotive inner bearings	30000–40000
Machinery for 8 hour service which are not always fully utilized	14000–20000	Machinery for short or intermittent operation where service interruption is of minor importance	4000–8000
Machinery for 8 hour service which are fully utilized	20000–30000	Machinery for intermittent service where reliable operation is of great importance	8000–14000
Machinery for continuous 24 hour service	50000–60000	Instruments and apparatus in frequent use	0–500

Appendix A-3

Appendix A-3-1

Key size versus shaft diameter in inches can be found in the following standard: ASME B17.1 - 1967 (R2013)

Appendix A-3-2

Specifications for metric keys and keyways can be found in the following standard: BS - 1: 1972

Appendix A-4

Appendix A-4-1

Classical Profiles of V-Belts

Classical Profiles of V-Belts	US System (inches)			ISO System (mm)		
	Width	Height	Angle	Width	Height	Angle
A	½	5/16	38°	13	8	40°
B	21/32	13,32	38°	17	11	40°
C	⅞	17/32	38°	22	14	40°
D	1-¼	¾	38°	32	19	40°
E	1-½	29/32	38°	38	25	40°

Source: Machinery's Handbook, 30th Edition, E. Oberg et al., Norwalk, CT: Industrial Press, 2016, *industrialpress.com*

Appendix A-4-2

Classical V-Belt Standard Datum Length *ANSI/RMA IP-20, 1988* (All dimensions in inches)

Standard Length Designation[a]	Standard Datum Lengths				Permissible Deviations from Std. Datum Length	Matching Limits for One Set
	Cross Section					
	A, AX	B, BX	C, CX	D		
26	27.3	...	...	...	+0.6, −0.6	0.15
31	32.3	...	...	...	+0.6, −0.6	0.15
35	36.3	36.8	...	...	+0.6, −0.6	0.15
38	39.3	39.8	...	...	+0.7, −0.7	0.15
42	43.3	43.8	...	...	+0.7, −0.7	0.15
46	47.3	47.8	...	...	+0.7, −0.7	0.15
51	52.3	52.8	53.9	...	+0.7, −0.7	0.15
55	56.3	56.8	...	...	+0.7, −0.7	0.15
60	61.3	61.8	62.9	...	+0.7, −0.7	0.15
68	69.3	69.8	70.9	...	+0.7, −0.7	0.30
75	75.3	76.8	77.9	...	+0.7, −0.7	0.30
80	81.3	...	...	...	+0.7, −0.7	0.30
81	...	82.8	83.9	...	+0.7, −0.7	0.30
85	86.3	86.8	87.9	...	+0.7, −0.7	0.30
90	91.3	91.8	92.9	...	+0.8, −0.8	0.30
96	97.3	...	98.9	...	+0.8, −0.8	0.30
97	...	98.8	...	...	+0.8, −0.8	0.30
105	106.3	106.8	107.9	...	+0.8, −0.8	0.30
112	113.3	113.8	114.9	...	+0.8, −0.8	0.30
120	121.3	121.8	122.9	123.3	+0.8, −0.8	0.30
128	129.3	129.8	130.9	131.3	+0.8, −0.8	0.30
144	...	145.8	146.9	147.3	+0.8, −0.8	0.30
158	...	159.8	160.9	161.3	+1.0, −1.0	0.45
173	...	174.8	175.9	176.3	+1.0, −1.0	0.45
180	...	181.8	182.9	183.3	+1.0, −1.0	0.45
195	...	196.8	197.9	198.3	+1.1, −1.1	0.45
210	...	211.8	212.9	213.3	+1.1, −1.1	0.45
240	...	240.3	240.9	240.8	+1.3, −1.3	0.45
270	...	270.3	270.9	270.8	+1.6, −1.6	0.60
300	...	300.3	300.0	300.8	+1.6, −1.6	0.60
330	...	...	330.9	330.8	+2.0, −2.0	0.60
360	...	...	380.9	360.8	+2.0, −2.0	0.60
540	...	...	...	540.8	+3.3, −3.3	0.90
390	...	...	390.9	390.8	+2.0, −2.0	0.75
420	...	...	420.9	420.8	+3.3, −3.3	0.75
480	...	...	...	480.8	+3.3, −3.3	0.75
600	...	...	...	600.8	+3.3, −3.3	0.90
660	...	...	...	660.8	+3.3, −3.3	0.90

[a] To specify belt size use the Standard Length Designation prefixed by the letter indicating the cross section, e.g., B90.

Source: Machinery's Handbook, 30th Edition, E. Oberg, et al., Norwalk, CT: Industrial Press, 2016, industrialpress.com.

Appendix A-4-3

Minimum datum diameters of sheaves for classical V-belts

Cross Section Letter Designation	Minimum Datum Diameter (inches)
A	3.0
B	5.4
C	9.0
D	13.0
E	21.0

Source: Machinery's Handbook, 30th Edition, E. Oberg et al., Norwalk, CT: Industrial Press, 2016, *industrialpress.com*

Appendix A-4-4

Speed (velocity) ratio correction factor $i = \frac{\omega_1}{\omega_2} = \frac{d_2}{d_1}$

Speed (velocity) Ratio Range	Speed (velocity) Ratio Correction Factor – K_{SR}
1.00 – 1.01	1.0000
1.02 – 1.04	1.0112
1.05 – 1.07	1.0226
1.08 – 1.10	1.0394
1.11 – 1.14	1.0463
1.15 – 1.20	1.0586
1.21 – 1.27	1.0711
1.28 – 1.39	1.0840
1.40 – 1.64	1.0972
Over 1.64	1.1106

Source: Machinery's Handbook, 30th Edition, E. Oberg et al., Norwalk, CT: Industrial Press, 2016, *industrialpress.com*

Appendix A-4-5

Standard Sheave Outside Diameters *ANSI/RMA IP-22, 1983* (All dimensions in inches)

3V			5V			8V		
Nom	Min	Max	Nom	Min	Max	Nom	Min	Max
2.65	2.638	2.680	7.10	7.087	7.200	12.50	12.402	12.600
2.80	2.795	2.840	7.50	7.480	7.600	13.20	13.189	13.400
3.00	2.953	3.000	8.00	7.874	8.000	14.00	13.976	14.200
3.15	3.150	3.200	8.50	8.346	8.480	15.00	14.764	15.000
3.35	3.346	3.400	9.00	8.819	8.960	16.00	15.748	16.000
3.55	3.543	3.600	9.25	9.291	9.440	17.00	16.732	17.000
3.65	3.642	3.700	9.75	9.567	9.720	18.00	17.717	18.000
4.00	3.937	4.000	10.00	9.843	10.000	19.00	18.701	19.000
4.12	4.055	4.120	10.30	10.157	10.320	20.00	19.685	20.000
4.50	4.409	4.480	10.60	10.433	10.600	21.20	20.866	21.200
4.75	4.646	4.720	10.90	10.709	10.880	22.40	22.047	22.400
5.00	4.921	5.000	11.20	11.024	11.200	23.60	23.622	24.000
5.30	5.197	5.280	11.80	11.811	12.000	24.80	24.803	25.200
5.60	5.512	5.600	12.50	12.402	12.600	30.00	29.528	30.000
6.00	5.906	6.000	13.20	13.189	13.400	31.50	31.496	32.000
6.30	6.299	6.400	14.00	13.976	14.200	35.50	35.433	36.000
6.50	6.496	6.600	15.00	14.764	15.000	40.00	39.370	40.000
6.90	6.890	7.000	16.00	15.748	16.000	44.50	44.094	44.800
8.00	7.874	8.000	18.70	18.701	19.000	50.00	49.213	50.000
10.00	9.843	10.000	20.00	19.685	20.000	52.00	51.969	52.800
10.60	10.433	10.600	21.20	20.866	21.200	63.00	62.992	64.000
12.50	12.402	12.600	23.60	23.622	24.000	71.00	70.866	72.000
14.00	13.976	14.200	25.00	24.803	25.200	79.00	78.740	80.000
16.00	15.748	16.000	28.00	27.953	28.400	99.00	98.425	100.000
19.00	18.701	19.000	31.50	31.496	32.000	...	...	...
20.00	19.685	20.000	37.50	37.402	38.000	...	...	...
25.00	24.803	25.200	40.00	39.370	40.000	...	...	...
31.50	31.496	32.000	44.50	44.094	44.800	...	...	...
33.50	33.465	34.000	50.00	49.213	50.000	...	...	...
...	...	...	63.00	62.992	64.000	...	...	...
...	...	...	71.00	70.866	72.000	...	...	...

Source: Machinery's Handbook, 30th Edition, E. Oberg, et al., Norwalk, CT: Industrial Press, 2016, industrialpress.com.

Appendix A-4-6

Length Correction Factors

Std. Length Designation	Cross Section			
	A, AX	B, BX	C, CX	D
26	0.78	...	...	...
31	0.82	...	...	...
35	0.85	0.80	...	...
38	0.87	0.82	...	...
42	0.89	0.84	...	...
46	0.91	0.86	...	...
51	0.93	0.88	0.80	...
55	0.95	0.89	...	...
60	0.97	0.91	0.83	...
68	1.00	0.94	0.85	...
75	1.02	0.96	0.87	...
80	1.04	...	...	...
81	...	0.98	0.89	...
85	1.05	0.99	0.90	...
90	1.07	1.00	0.91	...
96	1.08	...	0.92	...
97	...	1.02	...	...
105	1.10	1.03	0.94	...
112	1.12	1.05	0.95	...
120	1.13	1.06	0.96	0.88
128	1.15	1.08	0.98	0.89
144	...	1.10	1.00	0.91
158	...	1.12	1.02	0.93
173	...	1.14	1.04	0.94
180	...	1.15	1.05	0.95
195	...	1.17	1.08	0.96
210	...	1.18	1.07	0.98
240	...	1.22	1.10	1.00
270	...	1.24	1.13	1.02
300	...	1.27	1.15	1.04
330	...	...	1.17	1.06
360	...	...	1.18	1.07
390	...	...	1.20	1.09
420	...	...	1.21	1.10
480	...	...	...	1.13
540	...	...	...	1.15
600	...	...	...	1.17
660	...	...	...	1.18

Source: Machinery's Handbook, 30th Edition, E. Oberg, et al., Norwalk, CT: Industrial Press, 2016, industrialpress.com.

Appendix A-4-7

Arc of Contact Correction Factor

Arc of contact, φ_1, small Sheave (deg.)	Correction Factor, K_φ
180	1.00
174	0.99
169	0.97
163	0.96
157	0.94
151	0.93
145	0.91
135	0.89
133	0.87
127	0.85
120	0.82
113	0.80
106	0.77
99	0.73
91	0.70
83	0.65

Source: Machinery's Handbook, 30th Edition, E. Oberg et al., Norwalk, CT: Industrial Press, 2016, *industrialpress.com*

Appendix A-4-8

Power Rating Formulas for B, C, and D Size V-belt Drives

$$P_B = d_1 n_1 10^{-3}\left[1.769 - \frac{4.372}{d_1} - 3.081\cdot 10^{-4}(d_1 n_1 10^{-3})^2 - 0.3658\cdot \log(d_1 n_1 10^{-3})\right] + 4.372 n_1 10^{-3}\left(1 - \frac{1}{K_{SR}}\right)$$

$$P_C = d_1 n_1 10^{-3}\left[3.325 - \frac{12.07}{d_1} - 5.828\cdot 10^{-4}(d_1 n_1 10^{-3})^2 - 0.6886\cdot \log(d_1 n_1 10^{-3})\right] + 12.07 n_1 10^{-3}\left(1 - \frac{1}{K_{SR}}\right)$$

$$P_D = d_1 n_1 10^{-3}\left[7.160 - \frac{43.21}{d_1} - 1.384\cdot 10^{-4}(d_1 n_1 10^{-3})^2 - 1.4544\cdot \log(d_1 n_1 10^{-3})\right] + 43.21 n_1 10^{-3}\left(1 - \frac{1}{K_{SR}}\right)$$

Source: Machinery's Handbook, 30th Edition, E. Oberg et al., Norwalk, CT: Industrial Press, 2016, *industrialpress.com*

Appendix A-4-9

Standard Thickness and Width of Common Rubber, Leather, and Cotton Flat Belts

Thickness		Width	
in.	**mm**	**in.**	**mm**
0.197	5.0	From 1.375 to 2.480	From 35 to 63
0.256	6.5	From 3.540 to 5.512	From 90 to 140
0.315	8.0	From 3.543 to 8.819	From 90 to 224

Appendix A-4-10

Allowable stresses of flat belts depending on the ratio $r_b = \frac{d_1}{t}$, where d_1 is the diameter of the smaller (driving) pulley and t is the thickness of the belt.

	$r_b = 30$		$r_b = 50$		$r_b = 100$	
Material	**ksi**	**MPa**	**ksi**	**MPa**	**ksi**	**MPa**
Rubber	0.32	2.2	0.33	2.3	0.35	2.4
Leather	0.28	1.9	0.33	2.3	0.38	2.6
Cotton	0.23	1.6	0.26	1.8	0.28	1.9

Appendix A-5

Appendix A-5-1

American Standard specifies the following coarse and fine pitches for gears:

a) coarse pitches: 1; 1.5; 2; 2.25; 2.5; 3; 3.5; 4; 4.5; 5; 5.5; 6; 7; 8; 9; 10; 11; 12; 14; 16; 18;
b) fine pitches: 20; 24; 32; 40; 48; 64; 80; 96; 120; 150; 200.

Appendix A-5-2

ISO Standard specifies the following modules for gears:

1) from 0.3 mm to 0.8 mm through 0.1 mm;
2) from 1.0 mm to 3.0 mm through 0.25 mm;
3) from 3.0 mm to 7.0 mm through 0.5 mm;
4) from 8.0 mm to 16.0 mm through 1.0 mm;
5) from 18 mm to 30 mm through 2.0 mm,
6) from 33.0 mm to 45.0 mm through 3.0 mm;
7) from 50 mm through 100 mm through 5 mm.

Appendix A-6

Appendix A-6-1

Grade Identification Marks and Mechanical Properties of Bolts and Screws

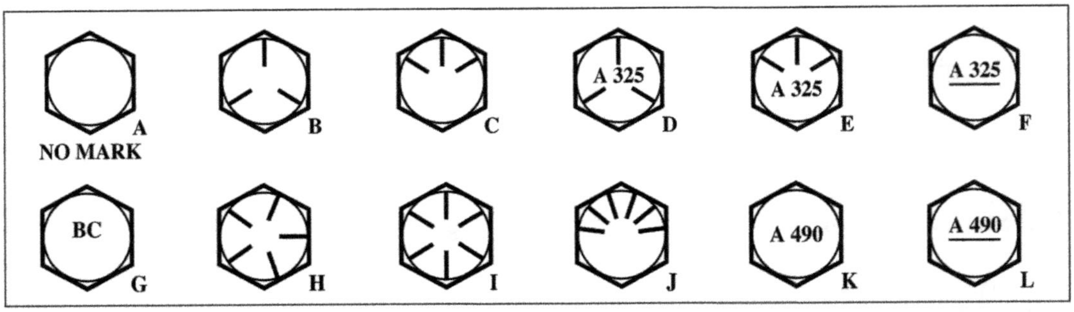

Identifier	Grade	Size (in.)	Min. Strength (10^3 psi)			Material & Treatment
			Proof	Tensile	Yield	
A	SAE Grade 1	¼ to 1½	33	60	36	1
	ASTM A307	¼ to 1½	33	60	36	3
	SAE Grade 2	¼ to ¾	55	74	57	1
		⅞ to 1½	33	60	36	
	SAE Grade 4	¼ to 1½	65	115	100	2, a
B	SAE Grade 5, ASTM A449	¼ to 1	85	120	92	2,b
	ASTM A449	1⅛ to 1½	74	105	81	
	ASTM A449	1¾ to 3	55	90	58	
C	SAE Grade 5.2	¼ to 1	85	120	92	4, b
D	ASTM A325, Type 1	½ to 1	85	120	92	2,b
		1⅛ to 1½	74	105	81	
E	ASTM A325, Type 2	½ to 1	85	120	92	4, b
		1⅛ to 1½	74	105	81	
F	ASTM A325, Type 3	½ to 1	85	120	92	5, b
		1⅛ to 1½	74	105	81	
G	ASTM A354, Grade BC	¼ to 2½	105	125	109	5, b
		2⅝ to 4	95	115	99	
H	SAE Grade 7	¼ to 1½	105	133	115	7, b
I	SAE Grade 8	¼ to 1½	120	150	130	7, b
	ASTM A354, Grade BD	¼ to 2½	120	150	130	6, b
		2⅝ to 4	105	140	115	
J	SAE Grade 8.2	¼ to 1	120	150	130	4, b
K	ASTM A490, Type 1	½ to 1½	120	150	130	6, b
L	ASTM A490, Type 3					5, b

Material Steel: 1—low or medium carbon; 2—medium carbon; 3—low carbon; 4—low-carbon martensite; 5—weathering steel; 6—alloy steel; 7—medium-carbon alloy. Treatment: a—cold drawn; b—quench and temper.

Source: Machinery's Handbook, 30th Edition, E. Oberg, et al., Norwalk, CT: Industrial Press, 2016, industrialpress.com.

Appendix A-6-2

Coarse-Thread Series, UNC

Sizes	Major Dia., d	Threads Per Inch N_t	Tensile Stress Area, A_s
Numbers or inches	Inches	Number	Square inches
1	0.0730	64	0.00263
2	0.0860	56	0.00370
3	0.0990	48	0.00487
4	0.1120	40	0.00604
5	0.1250	40	0.00796
6	0.1380	32	0.00909
8	0.1640	32	0.0140
10	0.1900	24	0.0175
12	0.2160	24	0.0242
¼	0.2500	20	0.0318
5/16	0.3125	18	0.0524
⅜	0.3750	16	0.0755
7/16	0.4375	14	0.1063
½	0.5000	13	0.1419
9/16	0.5625	12	0.182
⅝	0.6250	11	0.226
¾	0.7500	10	0.334
⅞	0.8750	9	0.462
1	1.0000	8	0.606
1⅛	1.1250	7	0.763
1¼	1.2500	7	0.969
1⅜	1.3750	6	1.155
1½	1.5000	6	1.405
1¾	1.7500	5	1.90
2	2.0000	4½	2.50
2¼	2.250	4½	3.25
2½	2.5000	4	4.00
2¾	2.7500	4	4.93
3	3.0000	4	5.97
3¼	3.2500	4	7.10
3½	3.500	4	8.33
3¾	3.7500	4	9.66
4	4.0000	4	11.08

Appendix A-6-3

Metric Coarse Threads

Designation	Major Diameter mm	Pitch mm	Minor Diameter mm
M1.0	1.00	0.25	0.693
M1.1	1.10	0.25	0.793
M1.2	1.20	0.25	0.893
M1.4	1.40	0.30	1.032
M1.6	1.60	0.35	1.171
M1.8	1.80	0.35	1.371
M2	2.00	0.40	1.509
M2.2	2.20	0.45	1.648
M2.5	2.50	0.45	1.948
M3	3.00	0.50	2.387
M3.5	3.50	0.60	2.764
M4	4.00	0.70	3.141
M4.5	4.50	0.75	3.580
M5	5.00	0.80	4.019
M6	6.00	1.00	4.775
M7	7.00	1.00	5.773
M8	8.00	1.25	6.466
M9	9.00	1.25	7.466
M10	10.00	1.50	8.160
M11	11.00	1.50	9.160
M12	12.00	1.75	9.853
M14	14.00	2.00	11.549
M16	16.00	2.00	13.546
M18	18.00	2.50	14.933
M20	20.00	2.50	16.933
M22	22.00	2.5	18.933
M24	24.00	3.00	20.319
M27	27.00	3.00	23.319
M30	30.00	3.50	25.706
M33	33.00	3.50	28.706
M36	36.00	4.00	31.093
M39	39.00	4.00	34.093
M42	42.00	4.50	36.479
M45	45.00	4.50	39.479
M48	48.00	5.00	41.866
M52	52.00	5.00	45.866
M56	56.00	5.50	49.252
M60	60.00	5.50	53.252
M64	64.00	6.00	56.639
M66	66.00	6.00	60.639

Appendix A-7

Appendix A-7-1

Rivet Lengths for Forming Round and Countersunk Heads

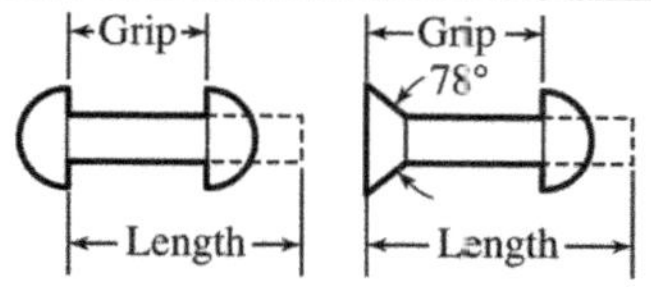

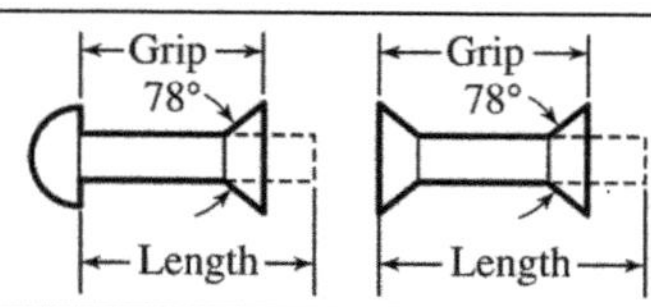

Grip in Inches	To Form Round Head — Diameter of Rivet in Inches — Length of Rivet in Inches						
	½	⅝	¾	⅞	1	1⅛	1¼
½	1⅝	1⅞	1⅞	2	2⅛	. . .	. . .
⅝	1¾	2	2	2⅛	2¼	. . .	. . .
¾	1⅞	2⅛	2⅛	2¼	2⅜	. . .	. . .
⅞	2	2¼	2¼	2⅜	2½	. . .	. . .
1	2¼	2⅜	2⅜	2½	2⅝	2¾	2⅞
1⅛	2⅜	2½	2½	2⅝	2¾	2⅞	3
1¼	2½	2⅝	2⅝	2¾	2⅞	3	3⅛
1⅜	2⅝	2¾	2¾	2⅞	3	3⅛	3¼
1½	2⅞	3	3	3⅛	3¼	3⅜	3½
1⅝	3	3⅛	3⅛	3¼	3⅜	3½	3½
1¾	3⅛	3¼	3¼	3½	3⅝	3¾	3¾
1⅞	3¼	3⅜	3⅜	3⅝	3¾	3⅞	3⅞
2	3½	3½	3⅝	3¾	3⅞	4	4
2⅛	3⅝	3⅝	3¾	3⅞	4	4⅛	4⅛
2¼	3¾	3⅞	3⅞	4	4⅛	4¼	4¼
2⅜	4	4	4	4⅛	4¼	4⅜	4⅜
2½	4⅛	4⅛	4⅛	4¼	4⅜	4½	4½
2⅝	4¼	4¼	4¼	4⅜	4½	4⅝	4⅝
2¾	4⅜	4⅜	4⅜	4½	4⅝	4¾	4¾
2⅞	4⅝	4⅝	4⅝	4⅝	4¾	4⅞	5
3	. . .	4¾	4¾	4⅞	5	5	5⅛
3⅛	. . .	4⅞	4⅞	5	5⅛	5¼	5¼
3¼	. . .	5	5	5⅛	5¼	5⅜	5⅜
3⅜	. . .	5⅛	5⅛	5¼	5⅜	5½	5½
3½	. . .	5⅜	5⅜	5⅜	5½	5⅝	5⅝
3⅝	. . .	5½	5½	5½	5⅝	5¾	5¾
3¾	. . .	5⅝	5⅝	5⅝	5¾	5⅞	5⅞
3⅞	. . .	5¾	5¾	5¾	5⅞	6	6
4	. . .	. . .	5⅞	6	6	6⅛	6¼
4⅛	. . .	. . .	6	6⅛	6¼	6⅜	6⅜
4¼	. . .	. . .	6⅛	6¼	6½	6½	6½
4⅜	. . .	. . .	6⅜	6½	6½	6⅝	6⅝
4½	. . .	. . .	6½	6⅝	6⅝	6¾	6¾
4⅝	. . .	. . .	6⅝	6¾	6¾	6¾	6⅞
4¾	. . .	. . .	6¾	6⅞	6⅞	7	7
4⅞	. . .	. . .	6⅞	7	7	7⅛	7⅛
5	. . .	. . .	. . .	7⅛	7⅛	7¼	7¼
5⅛	. . .	. . .	. . .	7¼	7¼	7⅜	7⅜
5¼	. . .	. . .	. . .	7⅜	7⅜	7½	7½
5⅜	. . .	. . .	. . .	7⅝	7⅝	7¾	7⅝
5½	. . .	. . .	. . .	7¾	7¾	7⅞	7⅞
5⅝	. . .	. . .	. . .	7⅞	7⅞	8	8
5¾	. . .	. . .	. . .	8	8	8⅛	8⅛
5⅞	. . .	. . .	. . .	8⅛	8⅛	8¼	8¼

Grip in Inches	To Form Countersunk Head — Diameter of Rivet in Inches — Length of Rivet in Inches						
	½	⅝	¾	⅞	1	1⅛	1¼
½	1	1	1⅛	1¼	1¼	. . .	. . .
⅝	1⅛	1¼	1¼	1⅜	1⅜	. . .	. . .
¾	1⅜	1⅜	1⅜	1½	1½	. . .	. . .
⅞	1½	1½	1½	1⅝	1⅝	. . .	. . .
1	1⅝	1⅝	1⅝	1¾	1¾	1⅞	1⅞
1⅛	1¾	1¾	1⅞	1⅞	1⅞	2	2
1¼	2	2	2	2	2	2⅛	2⅛
1⅜	2⅛	2⅛	2⅛	2¼	2¼	2⅜	2⅜
1½	2¼	2¼	2¼	2⅜	2⅜	2½	2½
1⅝	2⅜	2⅜	2⅜	2½	2½	2⅝	2⅝
1¾	2⅝	2⅝	2⅝	2⅝	2⅝	2¾	2¾
1⅞	2¾	2¾	2¾	2¾	2¾	2⅞	2⅞
2	2⅞	2⅞	2⅞	2⅞	2⅞	3	3
2⅛	3⅛	3	3	3	3	3⅛	3⅛
2¼	3¼	3⅛	3⅛	3⅛	3¼	3¼	3¼
2⅜	3⅜	3⅜	3⅜	3⅜	3⅜	3⅜	3⅜
2½	3½	3½	3½	3½	3½	3⅝	3⅝
2⅝	3¾	3⅝	3⅝	3⅝	3⅝	3¾	3¾
2¾	3⅞	3¾	3¾	3¾	3¾	3⅞	3⅞
2⅞	4	3⅞	3⅞	3⅞	3⅞	4	4
3	. . .	4⅛	4⅛	4⅛	4⅛	4⅛	4⅛
3⅛	. . .	4¼	4¼	4¼	4¼	4¼	4¼
3¼	. . .	4⅜	4⅜	4⅜	4⅜	4⅜	4⅜
3⅜	. . .	4½	4½	4½	4½	4½	4½
3½	. . .	4⅝	4⅝	4⅝	4⅝	4⅝	4⅝
3⅝	. . .	4¾	4¾	4¾	4¾	4⅞	4⅞
3¾	. . .	5	5	5	5	5	5
3⅞	. . .	5⅛	5⅛	5⅛	5⅛	5⅛	5⅛
4	. . .	. . .	5¼	5¼	5¼	5¼	5¼
4⅛	. . .	. . .	5⅜	5⅜	5⅜	5⅜	5⅜
4¼	. . .	. . .	5½	5½	5½	5½	5½
4⅜	. . .	. . .	5⅝	5⅝	5⅝	5⅝	5⅝
4½	. . .	. . .	5¾	5¾	5¾	5¾	5¾
4⅝	. . .	. . .	6	6	6	6	6
4¾	. . .	. . .	6⅛	6⅛	6⅛	6⅛	6⅛
4⅞	. . .	. . .	6¼	6¼	6¼	6¼	6¼
5	. . .	. . .	. . .	6⅜	6⅜	6⅜	6⅜
5⅛	. . .	. . .	. . .	6½	6½	6½	6½
5¼	. . .	. . .	. . .	6⅝	6⅝	6⅝	6⅝
5⅜	. . .	. . .	. . .	6¾	6¾	6¾	6¾
5½	. . .	. . .	. . .	6⅞	6⅞	6⅞	6⅞
5⅝	. . .	. . .	. . .	7	7	7	7
5¾	. . .	. . .	. . .	7¼	7¼	7¼	7¼
5⅞	. . .	. . .	. . .	7⅜	7⅜	7⅜	7⅜

Source: Machinery's Handbook, 30th Edition, E. Oberg, et al., Norwalk, CT: Industrial Press, 2016, industrialpress.com.

Appendix A-7-2

American National Standard specifies the following small and large diameters of rivets:

a) small diameters: 1/16; 3/32; 1/8; 5/32; 3/16; 7/32; 1/4; 9/32; 5/16; 11/32; 3/8; 13/32; 7/10;
b) large diameters: 1/2; 5/8; 3/4; 7/8; 1.0; 1 1/8; 1 1/4; 1 3/8; 1 1/2.

Appendix A-8

Appendix A-8-1

Mechanical properties of Class Eelectrodes

Class	Ultimate Strength		Yield Strength	
	ksi	MPa	ksi	MPa
E60XX	60	415	50	345
E70XX	70	483	55	380
E80XX	80	552	67	462
E90XX	90	621	77	531
E100XX	100	690	87	600
E110XX	110	759	95	656
E120XX	120	828	107	738

Appendix A-9

Source: Machinery's Handbook, 30th Edition, E. Oberg et al., Norwalk, CT: Industrial Press, 2016, *industrialpress.com*

Appendix A-9-1

Formulas related to the types of ends for compression springs

Features Or Parameters	Open or Plain not ground	Open or Plain ground	Closed or Squared not ground	Closed or Squared ground
Pitch (p)	$p=\frac{H-d}{N}$	$p=\frac{H}{N_{tot}}$	$p=\frac{H-3d}{N}$	$p=\frac{H-2d}{N}$
Solid Height (H_{sh})	$H_{sh}=d(N_{tot}+1)$	$H_{sh}=dN_{tot}$	$H_{sh}=d(N_{tot}+1)$	$H_{sh}=dN_{tot}$
Number of Active Coils (N)	$N=N_{tot}$ $N=\frac{H-d}{p}$	$N=N_{tot}-1$ $N=\frac{H}{p}-1$	$N=N_{tot}-2$ $N=\frac{H-3d}{p}$	$N=N_{tot}-2$ $N=\frac{H-2d}{p}$
Number of Total Coils (N_{tot})	$N_{tot}=\frac{H-d}{p}$	$N_{tot}=\frac{H}{p}$	$N_{tot}=\frac{H-3d}{p}+2$	$N_{tot}=\frac{H-2d}{p}+2$
Free Height (H)	$H=pN_{tot}+d$	$H=pN_{tot}$	$H=pN+3d$	$H=pN+2d$

Appendix A-9-2

Allowable torsional stresses for some metal spring wires.

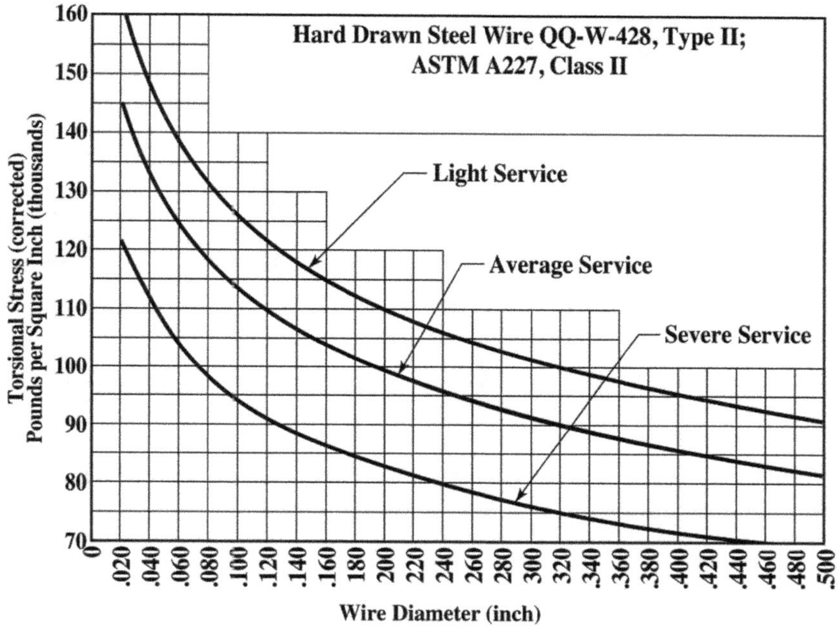

Figure 1. Allowable Working Stresses for Compression Springs—Hard Drawn Steel Wire.

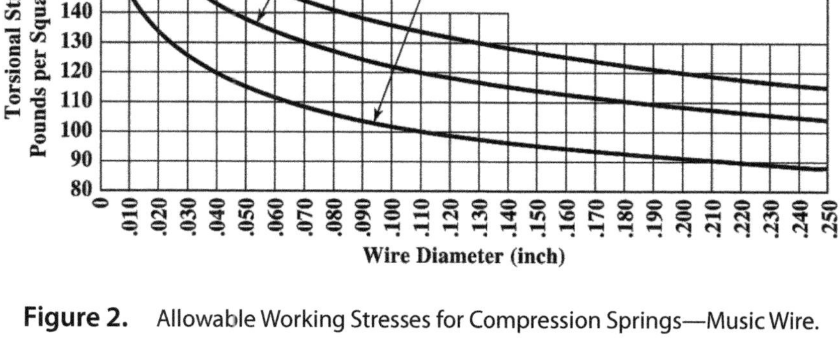

Figure 2. Allowable Working Stresses for Compression Springs—Music Wire.

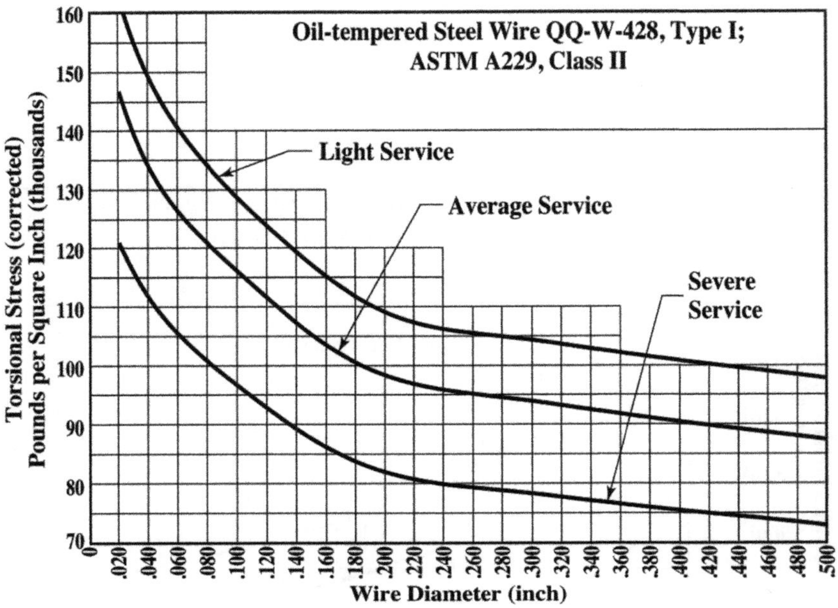

Figure 3. Allowable Working Stresses for Compression Springs—Oil-Tempered.

Bibliography

Artobolevskii, I. I. *Theory of Machines and Mechanisms.* Russian ed. Nauka, 1975.

Budynas, R. G. and K. J. Nisbett. *Shigley's Mechanical Engineering Design.* 8th ed. McGraw-Hill, 2008.

Burr, A. G. and J. B. Cheatham. *Mechanical Analysis and Design.* 2nd ed. Prentice Hall, 1995.

Creamer, R. H. *Machine Design.* 2nd ed. Addison-Wesley Publishing Company, 1976.

Darkov, A. V. and G. S. Shpiro. *Strength of Materials.* Russian ed. Vysshaya Shkola, 1969.

Dmitriev, V. A. *Machine Elements.* Russian ed. Sudostroenie, 1970.

Dobrovolsky, V. K., Zablonsky, S. Mak, A. Radchik, and L. Erlich, *Machine Elements.* Mir Publishers, 1974.

Erdman, A. G. and G. N. Sandor. *Mechanism Design.* 2nd ed. Prentice Hall, Inc., 1991.

Esposito, A. and J. R. Thrower, *Machine Design*, 2nd ed. Delmar Publishers, Inc., 1991.

Faires, V. M. *Design of Machine Elements.* The Macmillan Publishing Company, 1955.

Gere, G. M. *Mechanics of Materials.* 5th ed. Brooks/ Cole, 2001.

Hamrock, B. J., B. Jacobson, and S. R. Schmid. *Fundamentals of Machine Elements.* WCB/ McGraw-Hill, 1999.

Hindhede, U., J. R. Zimmerman, R. B. Hopkins, R. J. Erisman, W. C. Hull, and J. D. Lang. *Machine Design Fundamentals.* John Wiley & Sons, 1983.

Iosilevich, G. B. *Machine Elements.* Russian ed. Mashinostroenie, 1988.

Ivanov, M. N. *Machine Elements.* Russian 3rd ed. Vysshaya Shkola, 1976.

Juvinall, R. C. and K. M. Marshek. *Fundamentals of Machine Component Design.* 2nd ed. John Wiley & Sons, 1983.

Kolstee, H. M. *Machine Design for Mechanical Technology.* Holt, Rinehart and Winston, 1984.

Levinson, I. J. *Machine Design.* Reston Publishing Company, Inc., 1978.

Machinery's Handbook, 30th ed. Industrial Press, 2016.

Mott, R. L. *Machine Elements in Mechanical Design.* 2nd ed. Macmillan Publishing Company, 1992.

Myszka, D. N. *Machines and Mechanisms.* Prentice-Hall, Inc., 1999.

Norton, R. L. *Machine Design.* Prentice-Hall, Inc., 1996.

Orthwein, W. C. *Machine Component Design.* West Publishing Company, 1990.

Reshetov, D. H. *Machine Elements.* Russian 3rd ed. Mashinostroenie, 1975.

Shigley, J. E. and C. R. Mischke. *Mechanical Engineering Design.* McGraw-Hill, Inc., 1989.

Shigley, J. E. *Mechanical Engineering Design.* McGraw-Hill, Inc., 1972.

Shigley, J. E., C. R. Mischke, and R. J. Budynas. *Mechanical Engineering Design*. McGraw-Hill, Inc., 2004.

Spektor, M. B. *Applied Dynamics in Engineering*. Industrial Press, 2016.

Spektor, M. B. *Solving Engineering Problems in Dynamics*. Industrial Press, 2014.

Spotts, M. F. , T. E. Shoup, and L. E. Hornberger. Design of Machine Elements, 8th ed. Pearson, 2003.

Walton, J. W. *Engineering Design*. West Publishing Company, 1991.

Answers

A general note

It is advisable to refer to the corresponding chapters of the book in order to figure out the appropriate recommendations related to the values of certain parameters (factors, coefficients, numbers, characteristics, etc.) required to solve some of the problems. Very often, the parameters fall within certain ranges resulting in adequate answers to the problems. Therefore, some of the answers presented below are given only for comparison.

Chapter 1

1-01. $V_{max} = |3F|; M_{max} = |10Fa|$
1-02. $V_{max} = 1.67F; M_{max} = 1.56Fa$
1-03. $V_{max} = 1.49F; M_{max} = 0.745Fa$
1-04. $V_{max} = 3.28F; M_{max} = 3.0Fa$
1-05. $V_{max} = 3.46F; M_{max} = 7.57Fa$
1-06. $T_3 = -15T_0; T_{max} = |9T_0|$
1-07. $d = 0.937$ in.
1-08. $S_{yt} = 379.5$ MPa, AISI 1095 steel, annealed
1-09. $n_\sigma = 2.62$
1-10. For torsion $S_f = 28.91$ ksi
For bending $S_f = 46.24$ ksi.

Chapter 2

2-01. Porous Bronze, $PV = 1.75\ \text{MPa}\cdot\dfrac{\text{m}}{\text{s}}$
2-02. Cast Bronze, $PV = 2.63\ \text{MPa}\cdot\dfrac{\text{m}}{\text{s}}; V_{max} = 1.64\dfrac{\text{m}}{\text{s}}$
2-03. Porous Bronze, $PV = 50{,}000$ psi · fpm; $V_{max} = 93.75$ fpm
2-04. Porous Bronze, $PV = 1.75\ \text{MPa}\cdot\dfrac{\text{m}}{\text{s}}; F_{max} = 2.2$ kN
2-05. SKF No. 6307; $L_{10h} = 11{,}512$ hours
2-06. $L_{10h} = 11{,}790$ hours

2-07. L_{10h} = 9,260 hours
2-08. SKF No. NU 205 ECP, L_{10h} = 14,700 hours
2-09. L_{10h} = 9,248 hours
2-10. SKF No. 32318 J2, L_{10h} = 14,000 hours

Chapter 3

3-01. σ = 275.5 MPa
3-02. σ_{eq} = 34.4 MPa
3-03. σ_{eq} = 23.4 ksi
3-04. σ_{eq} = 106.5 MPa
3-05. d = 22 mm for n = 3,600 rpm; d = 27.3 mm for n = 1,800 rpm;
d = 31.2 mm for n = 1,200 rpm
3-06. d = 1.232 in.
3-07. n_σ = 2.47
3-08. n_σ = 2.37 for the completely reversed cycle; n_σ = 3.56 for the repeated cycle
3-09. n_σ = 2.29 for the completely reversed cycle; n_σ = 3.47 for the repeated cycle
3-10. h = 0.027 mm; θ = 0.00065 rad; n_{cr} = 5,760 rpm

Chapter 4

4-01. φ_1 = 146.52°; L = 2,317 mm
4-02. Flat belt: φ_1 = 90°, F_1 = 160 lb
V-belt: φ_1 = 180°, F_1 = 1,797 lb
4-03. σ_b = 25 MPa
4-04. R = 2,260 N
4-05. σ_c = 8.96 psi; σ_b = 366.7 psi
4-06. R = 41.22 lb
4-07. R = 741.2 N
4-08. P_B = 7 hp
4-09. P_C = 20 hp
4-10. P_D = 51 hp

Chapter 5

5-01. n_4 = 200 rpm; i_{14} = 9
5-02. Pinion σ = 93.6 MPa; Gear σ = 69.2 MPa
5-03. Pinion σ = 26,307 psi; Gear σ = 21,668 psi; n = 2.4
5-04. m = 2.5 mm; Gear S_{ut} = 775.5 MPa
5-05. σ_H = 43,507 psi; Gear H_B = 241; Pinion H_B = 302
5-06. C = 117 mm; Gear H_B = 197; Pinion H_B = 255
5-07. Pinion σ = 11,692 psi; Gear σ = 9,971 psi

5-08. Pinion $\sigma = 15{,}704$ psi; Gear $\sigma = 13{,}474$ psi; $n = 2.656$
5-09. $m_n = 1.5$ mm; Gear $S_{ut} = 849$ MPa
5-10. $\sigma_H = 43{,}746$ psi; Gear $H_B = 241$; Pinion $H_B = 302$

Chapter 6

6-01. Pinion $\sigma = 106$ MPa; Gear $\sigma = 85.5$ MPa
6-02. Pinion $n_b = 3.66$; Gear $n_b = 5.93$
6-03. Pinion $n_b = 2.67$; Gear $n_b = 3.425$
6-04. Pinion $n_b = 2.38$; Gear $n_b = 2.9$
6-05. $P = 10 \frac{1}{\text{in.}}$
6-06. $P = 11 \frac{1}{\text{in.}}$
6-07. $P = 12 \frac{1}{\text{in.}}$
6-08. $m = 2.75$ mm
6-09. $\sigma_H = 317.3$ MPa
6-10. $m_n = 3.5$ mm; $d_1 = 80.5$ mm; $b = 41.3$ mm; Gear $H_B = 235$; Pinion $H_B = 302$

Chapter 7

7-01. $q = 0.784 < 1$; no need for a cooling device
7-02. $n_b = 2.4$
7-03. $P = 6 \frac{1}{\text{in.}}$
7-04. $m = 6$ mm
7-05. $\sigma_H = 22{,}552$ psi
7-06. $\sigma_H = 65.2$ MPa
7-07. $n_{Hw} = 3.35$
7-08. $n_{Hw} = 3.12$
7-09. $d_2 = 16.667$ in.
7-10. $d_2 = 432$ mm

Chapter 8

8-01. $d_m = 0.9$ in.; $d_r = 0.8$ in.; $\lambda = 17.66°$; $L = 0.2$ in.
8-02. $d_m = 1.875$ in.; $d_r = 1.75$ in.; $\lambda = 4.86°$; $L = 0.5$ in.; $V = 0.5$ fpm
8-03. $d_m = 47$ mm; $d_r = 44$ mm; $\lambda = 4.65°$; $L = 12$ mm; $V = 0.012 \frac{\text{m}}{\text{s}}$
8-04. $T_{lift} = 454.2$ lb-in.; $e = 0.257$
8-05. $T_{lift} = 55.84\ \text{N} \cdot \text{m}$; $e = 0.256$

8-06. $T_{tot.lift} = 154.2\ \text{N}\cdot\text{m}$; $e = 0.159$
8-07. $d = 55$ mm; $p = 9$ mm; $d_m = 50.5$ mm; $d_r = 46$ mm; $e = 0.253$
8-08. $\sigma_{eq} = 3{,}340$ psi; $e = 0.258$
8-09. $d = 2.25$ in.; $p = 0.333$ in.; $d_m = 2.0833$ in.; $d_r = 1.917$ in.; $n_e = 4.6$
8-10. $d = 70$ mm; $p = 4$ mm; $d_m = 68$ mm; $d_r = 66$ mm; $n_e = 3.57$

Chapter 9

9-01. For $d = \frac{5}{16}$ in. $K_b = 2.87\cdot 10^6\ \frac{\text{lb}}{\text{in.}}$; for $d = 1.75$ in. $K_b = 41.2\cdot 10^6\ \frac{\text{lb}}{\text{in.}}$
9-02. For $d = 6$ mm $K_b = 90.3\cdot 10^3\ \frac{\text{N}}{\text{mm}}$; for $d = 42$ mm $K_b = 468\cdot 10^3\ \frac{\text{N}}{\text{mm}}$
9-03. $T = 210$ lb-in.
9-04. $n_j = 1.8$; $n_\sigma = 2.52$
9-05. $n_j = 1.98$; $n_\sigma = 2.58$
9-06. $d = 0.875$ in.
9-07. $d = M27$
9-08. $n_j = 4.16$; $n_\sigma = 2.1$
9-09. $n_j = 4.5$; $n_\sigma = 2.22$
9-10. $n_j = 4.66$; $n_\sigma = 2.4$

Chapter 10

10-01. $\tau = 7{,}609$ psi; $\sigma_{br} = 11{,}946$ psi; $\sigma_{pl} = 9{,}531$ psi; $\tau_{pl} = 6{,}946$ psi; $\sigma_{bp} = 11{,}946$ psi
10-02. $\tau = 60{,}7$ MPa; $\sigma = 90$ MPa; $\sigma_{pl} = 61$ MPa; $\tau_{pl} = 42$ MPa; $\sigma_{bp} = 90$ MPa
10-03. $d = 0.63$ in.
10-04. $d = 17.1$ mm
10-05. $\tau = 6{,}640$ psi; $\sigma = 16{,}308$ psi
10-06. $\tau = 47.8$ MPa; $\sigma = 105$ MPa
10-07. $d = 0.865$ in.
10-08. $d = 20$ mm
10-09. Rivet A; $\tau = 6.15$ ksi; $\sigma_{br} = 11.7$ ksi
10-10. Rivet C; $\tau = 44.2$ MPa; $\sigma = 77.2$ MPa

Chapter 11

11-01. $\sigma = 24{,}200\dfrac{F}{l^2}$
11-02. $b = 5.51$ in.
11-03. $n_\sigma = 2.04$
11-04. $\tau = \dfrac{0.707F}{hb}$

11-05. $\tau = \frac{0.45}{dh}\left(F + \frac{2T}{d}\right)$
11-06. $n_\tau = 3.17$
11-07. $n_\tau = 2.5$
11-08. $n_\tau = 2.25$
11-09. $n_\tau = 2.3$
11-10. $n_\tau = 2.2$

Chapter 12

12-01. $H = 10$ in.; $N_{tot} = 11$; $H_{sh} = 2.4$ in.
12-02. $H = 14.9$ in.; $N_{tot} = 16$; $H_{sh} = 5.1$ in.
12-03. $n_\tau = 2.53$
12-04. $n_\tau = 3.34$
12-05. $d = 0.15$ in.
12-06. $F_{max} = 11.64$ lb; $K = 38.8\ \frac{\text{lb}}{\text{in.}}$; $\tau_{max} = 70$ ksi
12-07. $F_{max} = 15$ lb; $K = 43\ \frac{\text{lb}}{\text{in.}}$; $\tau_{max} = 59{,}8$ ksi
12-08. $F_{max} = 18.6$ lb; $K = 46.5\ \frac{\text{lb}}{\text{in.}}$; $\tau_{max} = 47.5$ ksi
12-09. $F_{max} = 59.1$ lb; $m = 0.275\ \frac{\text{lb}\cdot\text{s}^2}{\text{ft}}$; $S_{max} = 0.778$ in.
12-10. $F_{max} = 277$ N; $m = 4{,}235$ kg; $S_{max} = 47.6$ mm

Index

References to figures are in italics.

A

ABMA. *See* American Bearing Manufacturing Association (ABMA)
addendum, 191, 192, 193, 229
AGMA. *See* American Gear Manufacturers Association (AGMA)
American Bearing Manufacturing Association (ABMA), 88
American Gear Manufacturers Association (AGMA), 206, 209, 291
American Society for Testing Materials (ASTM), 25, 41
American Welding Society (AWS), 382
angle of intersection, 251
angle of rotation, 122–128, 245–247, 304–305
angular-contact bearings, 82, 84–85, 86, 97
arc of contact correction factor, 428
assemblies, 2–3
  prototypes, 39–40
ASTM. *See* American Society for Testing Materials (ASTM)
AWS. *See* American Welding Society (AWS)
axial pitch, 280
axles, 101, 102
  stress calculations for, 110
  verifying calculations for static stresses, 116

B

Babbitts, 74
base diameters, 188
bearing stresses, 24
  calculations for power screw drives, 317–319
bearings
  angular-contact bearings, 82, 84, 86, 97
  design considerations, 96–98
  dynamic load ratings, 90–96
  journal bearings, 75–78
  radial roller bearings, 81–82
  role of the bore diameter and service life in selecting, 88–96
  rolling bearings, 80–88
  sliding bearings, 73–80
  static load ratings, 92
  taper rolling bearings, 82, 84, 86, 97
  typical bearing life for various design applications, 421
belt transmissions
  basic parameters, 150–153
  characteristics of, 153
  design project, 172–180
  forces acting on, 158–165
  overview, 149–150
  stresses in, 165–168
  tensioners, 158
  types of belts, 153–158
  See also specific belt types
bending moments
  caused by a moment, 12–16
  caused by an external force, 9–12
  See also bending stresses
bending stresses, 17–20
  alternating bending and alternating torsion, 120–122

bending stresses (*Cont.*)
  alternating bending and steady torsion, 118–120
  calculations for bevel gears, 255–259
  calculations for worm gears, 290–294
  design calculations for, 216–221
  in helical gears, 231–232
  of a rectangular bar, 21–22
  in spur gears subjected to static loading, 209–212
  in the threads of nuts, 319–321
  verifying calculations for in spur gears subjected to variable loading, 212–216
  verifying calculations of for completely reversed cycle, 215–216
  verifying calculations of for helical gears, 232–235
  verifying calculations of for repeated cycle, 213–215
bending testing, 42
  See also fatigue testing
bevel gears
  bending stress calculations for, 255–259
  design calculations of for bending stresses, 259–262
  design project, 265–272
  geometry of and acting forces, 253–255
  overview, 251–252
  surface durability of, 262–265
bolts, 330–331
  grade identification marks and mechanical properties, 434
  installed in the holes with an interference fit, 352–353
  preloaded bolts installed with a clearance fit, 353–355
bore diameter, 88–89
broaching, 195
butt joints
  riveted, 362, 364–365
  welded, 376–379, 382–383, 385–386

C

calculations, 5–7
camshafts, 101
cast iron bearings, 75
centrifugal forces, 128–129, 163–164
circular pitch, 189, 191, 253
class electrodes, mechanical properties of, 438
coarse threads, metric, 436
coarse-thread series, UNC, 435
cold riveting, 362
collar friction, 315–316
combined loading, fillet joint of a round bar subjected to, 384–385
completely reversed cycle, 54, 206
  calculating fatigue strength for a finite life cycle, 67–68
  design calculations based on bending stresses for, 237–238
  design calculations for bending stresses for, 220–221
  design calculations of bevel gears for a completely reversed cycle of bending stresses, 261–262
  verifying calculations of bending stresses for, 215–216
  verifying calculations of bending stresses for helical gears subjected to, 234–235
  verifying calculations of bending stresses in bevel gears, 258–259
compound gear trains, 200–201
compression conical coil springs, 396, *397*
compression springs, formulas related to the types of ends for, 438
compression stress, calculating, 321–322
conjugate profiles, 185, 186
Conrad-type bearings, 82
contact angles, values of X and Y coefficients, 419–420
contact lines factor, 231
contact ratio, 193–195

contact stresses, 22–24, 206, 221–222
design calculations of, 296–298
verifying calculations of, 222–223, 296
contact welding, 376
copper-lead bearings, 75
couplings, 101, 133, 135–136
critical speed
calculation of the critical rotational speed, 247
of shafts, 128–132
crossed helical gears, 184
cylindrical gears, 183–184

D
dedendum, 191, 192, 193
deep-groove ball bearings, 82
deflection, 122–128, 246–247, 304–305
design basics, overview, 2–5
design calculations, 102
for bending stresses, 216–221
for bending stresses of helical gears, 236–238
of bevel gears based on surface durability, 263–265
of bevel gears for bending stresses, 259–262
of the parameters of an enclosed helical gear set, 241–242
of preloaded joints subjected to variable loading, 351–352
of a shaft subjected to bending and torsion, 112–116
of the shafts and bearings, 242–244
for surface durability, 224–227
for unidirectional stress state, 111–112
of worm gear sets based on bending stresses, 292–294
of worm gears that experience contact stresses, 296–298
diametral pitch, 191, 228, 253

DIN. *See* German Institute for Standardization (DIN)
distortion energy, theory of maximum distortion energy, 30–31
driven pulleys, 150
driving pulleys, 150, 160–161
dynamic load ratings, 90–96

E
eccentric force, fillet lap joints subjected to, 387–390
elasticity, 5–6
enclosed transmissions, 207
endurance limits, 44–45
calculating, 49
effect of the mean stress on, 56–65
factors influencing, 45–50
engineering calculations, 110–111, 204–212
epicyclic gear train, 201–204
Euler, L., 162, 187
external helical cylindrical gears, 184
external spur gears, 183

F
failure theories
modified Mohr theory of failure, 31–39
overview, 28–29
theory of maximum distortion energy, 30–31
theory of maximum normal stresses, 29
theory of maximum shear stresses, 29–30
theory of maximum strain, 29
fasteners
characteristics of, 331–335
subjected to tensile stress, 336–337
fatigue, 40–41
endurance limit, 44–50
fatigue calculations
design of helical coil springs subjected to variable loading, 406–408
of welded joints, 385–387

fatigue failure, 41, 89, 348
criteria of, 57–63
fatigue testing, 41–44
fillet joints
welded, 376, 377–378, 379–382, 383–385
See also fillet lap joints; fillet T-joints
fillet lap joints, 378, 380, 383–385, 386–390
fillet T-joints, 378
finite life cycle, 65–68, 407
flat belts, 153–155
allowable stresses of, 429
design of, 168–171
standard thickness and width of, 429
stresses in, 165–166
fluctuating cycle, 55

G
gas metal arc welding (GMAW), 376
gas tungsten arc welding (GTAW), 376
Gates® Corporation, 169–170
gear, trains, 198–204
gears
basics of teeth engagement, 185–187
design project, 241–248
enclosed transmissions, 207
engineering calculations of, 204–212
manufacturing methods, 195–198
materials, 204
open transmissions, 207
structure and parameters of meshing gears, 187–189
types of, 183–184
See also specific gear types
generating line, 189
Gerber line, 57, 59–60
German Institute for Standardization (DIN), 206
gib head keys, 134
See also keys
Goodman line, 57, 58–60
grade identification marks, 434

H
heat calculations, 299–300
helical compression coil springs, 395
impact loading of, 408–412
helical extension coil springs, 395, *396*
helical gears, 184, 227–231
helical springs
design of helical coil springs subjected to static loading, 405–406
design of helical coil springs subjected to variable loading, 406–408
parameters of, 397–400
stresses in helical springs subjected to static loading, 400–401
herringbone gears, 184
herringbone helical gears, 227–228
Hertz, H., 22
Hertz theory, 222
Hertzian stresses, 22
See also contact stresses
Hertzman stresses, 221–222
hobbing, 197–198
Hooke's Law, 17
hot riveting, 362
hydrodynamic lubrication, 78
hydrostatic lubrication, 77–78

I
impact loading, 408–412
infinite life cycle, 65
See also endurance limits
instantaneous angular velocity, 185
instantaneous linear velocity, 185
interference, 193–195
internal loading, procedures for determining factors, 8–16
internal spur gears, 184
International Standardization Organization (ISO), 88, 206
International System (SI), conversion from U.S. Customary system to, 416

involute engagement, 187–189
ISO. *See* International Standardization Organization (ISO)

J

Japanese Gear Manufacturers Association (JGMA), 206
JGMA. *See* Japanese Gear Manufacturers Association (JGMA)
joints, permanent. *See* riveted joints; welded joints
joints, riveted. *See* riveted joints
joints, welded. *See* welded joints
journal bearings, 75–78

K

key size, vs. shaft diameter, 422
keys
  calculations of, 134–135
  overview, 133–134
keyseats, 135, *136*
keyway size, 423

L

lap joints
  riveted, 362, 364–366
  See also fillet lap joints
lead, 280
legs, 378
length correction factors, 427
Lewis, W., 209
line of action, 189
load factor (k_c), 48
loading factors, 8–9

M

Machine Design Elements and Assemblies course, 4, 6–7
Machinery's Handbook, 172, 399
Marin's equation, 46
mean diameter, 309
mean stress, 55–56
  effect of on the endurance limit, 56–65
mechanical engineering
  design basics, 2–7
  overview, 1–2
mechanical engineers, practice, 1–2
mechanical properties of bolts and screws, 434
mechanical properties of class electrodes, 438
mechanical properties of steels, 417–418
MIG (Metal Inert Gas) welding. *See* gas metal arc welding (GMAW)
minor diameter, 309
miscellaneous-effects factor (k_e), 48–49
Modified Goodman line, 59–65, 213–214
modified Mohr theory of failure, 31–39
modulus of elasticity, 294, 416
modulus of rigidity, 402, 416

N

non-uniform bending, 112
normal planes, 228
normal stresses, 16–17
  theory of maximum normal stresses, 29
nuts, 330–331

O

open transmissions, 207

P

parallel helical gears
  forces in, 230–231
  geometry of, 228–230
  overview, 227–228
  surface durability calculations for, 238–241
Petroff, N., 76
pinions, 187–188
  reactions, bending and torque diagrams for the shaft of the pinion, 244–245
  verifying calculations of the diameter of the shaft for, 245–246

pitch circles, 187–188
pitch diameter, 253, 309
pitting, 89
plain gib keys, 134
See also keys
plane stresses
analysis of the static plane stress state, 103–110
calculating safety factor for a shaft in plane stress state conditions, 50
calculations of variable stresses at a plane stress state, 118–122
reactions in supports, shear forces, bending moments and torque diagrams, 7–16
planetary gear train, 201–204
plastic bearings, 75
Poisson's ratio, 17, 23
polar moment of inertia, 111, 389–390, 402
porous metals, 75
power drives, 183
power rating formulas, for B, C, and D size V-belt drives, 428
power screw drives, 309
calculations, 316–322
design project, 323–325
determining power screw diameters, 319
mechanics, 309–316
preload forces, characteristics of, 342–343
preloaded threaded joints, 337–338
calculations of subject to external forces, 338–342
subjected to variable external loading, 348–352
verifying calculations of subject to static external loading, 343–348
pressure angles, 188, 192
principal stresses, 26–28
prototypes, 39–40
pure bending, 112
PV factor, 78–80, 419

Q

quantities, and their units, 415

R

rack, 189
radial roller bearings, 81–82
random cycle, 54
determining the diameter of a bar for a random cycle of bending loading, 65
determining the diameter of a shaft for a random cycle of axial loading, 64–65
Rayleigh-Ritz methodology, 132
reducers, 183
repeated cycle, 54, 206
calculating fatigue strength for a finite life cycle, 68
design calculations based on bending stresses for, 236–237
design calculations for bending stresses for, 218–219
design calculations of bevel gears for a repeated cycle of bending stresses, 259–260
determining the safety factor for, 63–64
verifying calculations of bending stresses for, 213–215
verifying calculations of bending stresses for helical gears subjected to, 232–234
verifying calculations of bending stresses in bevel gears, 257–258
retaining rings and nuts, 101
reversed loading cycle, determining width of a fillet lap joint subjected to, 386–387
riveted joints, 361–362
American Standard Rivets, 362
calculating stresses in the plates, 365
calculations, 362–363
calculations of a group of rivets in a joint subjected to eccentric loading, 365–371
classification of, 362
design calculations, 364–365

ISO standard rivets, 363
subjected to an eccentric inclined force, 369–371
types of, 362
verifying stress calculations, 364
riveting
cold, 362
hot, 362
rivets
American National Standards, 438
rivet lengths for forming round and countersunk heads, 437
standards, 362
types of, 361
rolling bearings, 97–98
calculating service life for based on dynamic load ratings, 92–96
life ratings, 89–90
overview, 80–81
types of, 81–88
round belts, 153

S

safety factors, 25–26, 49–50
for a repeated cycle, 63–64
screw diameter, calculating, 337
screws, 330–331
grade identification marks and mechanical properties, 434
service life, 89
calculating for rolling bearings based on dynamic load ratings, 92–96
shaft diameter, vs. key size, 422
shafts
angle of rotation, 122–128
calculations for static stresses, 111–116
calculations of variable stresses in a unidirectional stress state, 117
couplings, 133, 135–136
critical speed of, 128–132
deflection, 122–132
design calculations, 102, 111–116
design project, 136–143
engineering calculations of, 110–111
keys, 133–135
overview, 101–102
verifying calculations, 102, 116
shear forces
caused by a moment, 12–16
caused by an external force, 9–12
shear stresses, 26–27, 335–336
at bending, 20
of a rectangular bar, 21–22
theory of maximum shear stresses, 29–30
in the threads of nuts, 319–321
See also tangential stresses
sheaves
minimum datum diameters of sheaves for classical V-belts, 425
standard sheave outside diameters, 426
shielded metal arc welding (SMAW), 376
simple gear train, 198–200
single-tooth milling, 195
size factor (k_b), 47–48
sliding bearings
materials for, 74–75
overview, 73–74
PV factor for, 78–80
sliding friction, 75–78
Soderberg line, 57, 58
speed ratio correction factor, 425
spiral bevel gears, 184
spring rate, 402–404
springs
calculations of springs subjected to variable stresses, 401
design of helical coil springs subjected to static loading, 405–406
materials and allowable stresses, 404–405
overview, 395
parameters of helical springs, 397–400
stiffness coefficient (spring rate) of, 402–404

springs (*Cont.*)
stresses in helical springs subjected to static loading, 400–401
types of, 395–396
spur gears, 183–184
basic geometric parameters of, 189–193
bending stresses in spur gears subjected to static loading, 209–212
forces in spur gear engagement, 207–209
surface durability calculations for, 221–227
verifying calculations for bending stresses in spur gears subjected to variable loading, 212–216
static load ratings, 92
static loading
bending stresses in spur gears subjected to, 209–212
calculating thickness of a welded butt joint subjected to, 383
calculating width of a welded butt joint subjected to, 382–383
design calculations of preloaded joints subjected to, 346–348
design of helical coil springs subjected to, 405–406
verifying calculations of preloaded threaded joints subjected to, 345–346
static shear forces, calculating the width of weld of a fillet lap joint subjected to, 383
steel
materials, 204
mechanical properties of, 417–418
stiffness coefficients, 332–335
of springs, 402–404
straight bevel gears, 184
strain, theory of maximum strain, 29
Strength of Materials course, 5–6
stress amplitude, 55–56
stress calculations, 7–9
loading factors, 8–9
stress concentration, 50–53
theoretical factors K_t, 212, 430–433
stress cycles, types of variable stress cycles, 53–56
stresses
units of, 25
See also specific stress types
studs, 330–331
surface durability calculations
for parallel helical gears, 238–241
for spur gears, 221–227
of worm gear sets, 294–298
surface finish factor (k_a), 46–47
synchronous belts, 154

T

tangential stresses, 16–17
See also shear stresses
taper rolling bearings, 82, 84, 86, 97
tapered keys, 134
See also keys
teeth, minimum number of, 193–195
teeth generating method, 195–198
temperature factor (k_d), 48
tensile stress, 335–336
fasteners subjected to, 336–337
verifying calculations of preloaded threaded joints subjected to static external loading, 343–348
tensioners, 158
testing for fatigue, 41–44
testing steels, 25–26
theory of maximum distortion energy, 30–31
theory of maximum normal stresses, 29
theory of maximum shear stresses, 29–30
theory of maximum strain, 29
threaded fasteners, allowable stresses of, 332
threaded joints
calculations, 335–343
characteristics of, 331–335
design project, 355–358
overview, 327–328
preloaded threaded joints subjected to variable external loading, 348–352
subjected to shear forces, 352–355

thread profiles, 328–331
verifying calculations of preloaded threaded joints subjected to static external loading, 343–348
throat, 380
TIG (Tungsten Inert Gas) welding. *See* gas tungsten arc welding (GTAW)
timing belts, 154
T-joints. *See* fillet T-joints
tooth form factor, 232
toothed power transmissions, 183, 187
contact ratio, interference, and minimum number of teeth, 193–195
torque, calculating the required preload torque, 337–338
torque diagrams, 14–16
torsion bars, 112
torsion shafts, 112
torsion springs, 396, *397*
torsional stresses, 20–21, 335–336, 389–390
allowable for some metal spring wires, 439–440
alternating bending and alternating torsion, 120–122
alternating bending and steady torsion, 118–120
calculating, 321–322
verifying calculations of preloaded threaded joints subjected to static external loading, 343–348
train ratio, 198
train value, 198, 200
transverse circular pitch, 228
transverse diametral pitch, 228, 229
transverse planes, 228

U

UNC, 329
UNEF, 329
UNF, 329
Unified National Coarse. *See* UNC
Unified National Extra Fine. *See* UNEF
Unified National Fine. *See* UNF
U.S. Customary system, conversion from to International System (SI), 416

V

variable loading
design of helical coil springs subjected to, 406–408
verifying calculations for bending stresses in spur gears subjected to, 212–216
welded butt joint subjected to, 386
variable stress cycles
calculations of springs subjected to variable stresses, 401
calculations of variable stresses at a plane stress state, 118–122
calculations of variable stresses in a unidirectional stress state, 117
types of, 53–56
V-belts, 153, 154, 155–158
classical profiles of, 423
classical standard datum length, 424
minimum datum diameters of sheaves for classical V-belts, 425
selecting, 171–172
stresses in, 166
velocity ratio, 253
velocity ratio correction factor, 425
verifying calculations, 102
for bending stresses acting in teeth of the gears, 242
of bending stresses for a completely reversed cycle, 215–216
of bending stresses for helical gears, 232–235
of bending stresses for repeated cycles, 213–215
for bending stresses in spur gears subjected to variable loading, 212–216
of bevel gears based on surface durability, 262–263
for a completely reversed cycle of bending stresses of bevel gears, 258–259
of contact stresses, 239–241

verifying calculations (*Cont.*)
of the diameter of the shaft for the pinion, 245–246
of preloaded joints subjected to variable external loading, 349–351
for a repeated cycle of bending stresses in bevel gears, 257–258
for static stresses, 116
of stresses in the threads, 320–321
of surface durability, 222–223
for worm gear bending stresses, 292
of worm gears that experience contact stresses, 296
von Mises theory, 30–31

W

wedging effect of friction, 156
weld throat, 380
welded joints, 375
fatigue calculations, 385–387
fillet joint of a rounded bar subjected to a combined loading, 384–385
strength calculations of subjected to static loading, 378–385
types of, 376–378
welding, 375–376
white metal alloys, 74
Woodruff keys, 134
See also keys
worm gears, 184, 275–277
bending stress calculations for, 290–294
calculations of worm gear surface durability, 294–298
design project, 298–306
efficiency and heat calculations for worm gear drives, 286–289
forces acting on worm gear sets, 289–290
geometry of, 277–285
materials and manufacturing of, 286
worms, 275–277
four-start, *278*
materials and manufacturing of, 286
one-start and two-start, *279*
single-start, *277*